Advanced Operation and Maintenance in Solar Plants, Wind Farms and Microgrids

Advanced Operation and Maintenance in Solar Plants, Wind Farms and Microgrids

Editors

Luis Hernández-Callejo
Maria del Carmen Alonso García
Sara Gallardo Saavedra

MDPI • Basel • Beijing • Wuhan • Barcelona • Belgrade • Manchester • Tokyo • Cluj • Tianjin

Editors

Luis Hernández-Callejo
Ingeniería Agrícola y Forestal
Universidad de Valladolid
Valladolid
Spain

Maria del Carmen Alonso García
Energy
CIEMAT
Madrid
Spain

Sara Gallardo Saavedra
Ingeniería Agrícola y Forestal
Universidad de Valladolid
Valladolid
Spain

Editorial Office
MDPI
St. Alban-Anlage 66
4052 Basel, Switzerland

This is a reprint of articles from the Special Issue published online in the open access journal *Applied Sciences* (ISSN 2076-3417) (available at: www.mdpi.com/journal/applsci/special_issues/ advanced_renewables).

For citation purposes, cite each article independently as indicated on the article page online and as indicated below:

LastName, A.A.; LastName, B.B.; LastName, C.C. Article Title. *Journal Name* **Year**, *Volume Number*, Page Range.

ISBN 978-3-0365-4062-7 (Hbk)
ISBN 978-3-0365-4061-0 (PDF)

Contents

About the Editors

Luis Hernández-Callejo

Prof. Dr. Luis Hernández-Callejo is Electrical Engineer at Universidad Nacional de Educación a Distancia (UNED, Spain), Computer Engineer at UNED and Ph.D. at Universidad de Valladolid (Spain). Professor and researcher at the Universidad de Valladolid. His areas of interest are: renewable energy, microgrids, photovoltaic energy, wind energy, smart cities, and artificial intelligence. Prof. Dr. Hernández Callejo is an editor in numerous scientific journals, and is a guest editor in many Special Issues. He has directed four Doctoral Theses, and at the moment, he is directing six Doctoral Theses. He is a professor in wind energy, solar energy, microgrids, and he collaborates with many universities in Spain and in the rest of the world.

Maria del Carmen Alonso García

Dr. Carmen Alonso-García is a Staff Scientist in the Photovotlaic Solar Energy Unit of Centro de Investigaciones Energéticas, Medio-ambientales y Tecnológicas (CIEMAT) in Madrid, Spain. She has performed research in photovoltaic solar energy since 1990, focusing her specialization in the characterization and modelling of PV generators, operation and degradation of different technology modules and systems, accelerated testing and circularity of the systems. She has been involved in at least 41 R+D+I projects in the field of photovoltaic solar energy, being the scientist in charge in many of them. With a number of publications in indexed journals, conferences and scientific dissemination media, her contributions cover a broad spectrum within solar photovoltaics, from basic research to applied research and demonstration projects. She has been actively involved in standardization committees and other relevant photovoltaic solar energy international groups.

Sara Gallardo Saavedra

Dr. Sara Gallardo Saavedra is a professor and researcher at the Campus Duques de Soria of the University of Valladolid (Spain). Her research focuses on the detection, characterization and classification of defects in photovoltaic (PV) modules through the use of thermography, electroluminescence, I-V curves and visual analysis. She has participated in numerous national and international R+D+I projects, carrying out an active dissemination of the results, with high regularity in the scientific production, including scientific publications in high impact factor journals, book chapters, and contributing to congresses on advanced maintenance in PV. She has made a predoctoral and a postdoctoral stay in the Unit of Solar PV Energy in the Energy Department of the Energy Research Center, Environment and Technology (CIEMAT) in Madrid and the researcher has collaborated with different institutions as University of Gavle in Sweden, with the Universidad del Valle in Colombia, with the National Polytechnic Institute of Mexico and with the University of Cuenca in Ecuador.

Preface to "Advanced Operation and Maintenance in Solar Plants, Wind Farms and Microgrids"

The development of renewable generation plants and microgrids is a reality. Every day more facilities of this type are springing up, and their advance requires new research studies. Among renewable energy, solar plants (photovoltaic, thermal, and hybrid) and wind plants have had the greatest impact in recent years. With regard to microgrids, these scenarios are integrators of local generation sources (renewable or nonrenewable) and are facilities that promote energy sustainability.

In any case, both renewable generation plants and microgrids require technological development tools, mainly with regard to their operation and maintenance. Therefore, this Special Issue includes reviews, research articles, case studies, and technical notes on "Advanced Operation and Maintenance in Solar Plants, Wind Farms and Microgrids"

For solar plants, wind farms, and microgrids, the articles focus on one of the following topics:
- Artificial intelligence and data mining;
- Simulations related to operation and maintenance;
- Development of software and/or SCADA for operation and maintenance;
- New development of sensors and hardware for application to operation and maintenance;
- New operation and maintenance techniques;
- Hybrid photovoltaic and thermal systems;
- Storage for operation or maintenance.

Luis Hernández-Callejo, Maria del Carmen Alonso García, and Sara Gallardo Saavedra
Editors

Article

Modeling and Control of a Microgrid Connected to the INTEC University Campus

Miguel Aybar-Mejía [1,*], **Lesyani León-Viltre** [2,3], **Félix Santos** [3,4], **Francisco Neves** [5,*], **Víctor Alonso Gómez** [6] **and Deyslen Mariano-Hernández** [1]

1   Engineering Area, Instituto Tecnológico de Santo Domingo, Santo Domingo 10602, Dominican Republic; deyslen.mariano@intec.edu.do
2   Faculty of Electrical Engineering, University Central "Marta Abreu" de Las Villas, Santa Clara 50100, Cuba; lesyani@uclv.edu.cu or Lesyani.Leon@intec.edu.do
3   Basic Sciences Area, Instituto Tecnológico de Santo Domingo, Santo Domingo 10602, Dominican Republic; felix.santos@intec.edu.do
4   Centre for Energy Studies and Environmental Technologies (CEETA), University Central "Marta Abreu" de Las Villas, Santa Clara 50100, Cuba
5   Department of Electrical Engineering and Power Systems, University Federal de Pernambuco, UFPE, Recife 50670-901, Brazil
6   Department of Physics, University of Valladolid, Duques de Soria, 42004 Soria, Spain; victor.alonso.gomez@uva.es
*   Correspondence: miguel.aybar@intec.edu.do (M.A.-M.); francisco.neves@ufpe.br (F.N.); Tel.: +1-809-567-9271 (M.A.-M.); +55-81-2126-8000 (F.N.)

Abstract: A smart microgrid is a bidirectional electricity generation system—a type of system that is becoming more prevalent in energy production at the distribution level. Usually, these systems have intermittent renewable energy sources, e.g., solar and wind energy. These low voltage networks contribute to decongestion through the efficient use of resources within the microgrid. In this investigation, an energy management strategy and a control scheme for DG units are proposed for DC/AC microgrids. The objective is to implement these strategies in an experimental microgrid that will be developed on the INTEC university campus. After presenting the microgrid topology, the modeling and control of each subsystem and their respective converters are described. All possible operation scenarios, such as islanded or interconnected microgrids, different generation-load possibilities, and state-of-charge conditions of the battery, are verified, and a seamless transition between different operation modes is ensured. The simulation results in Matlab Simulink show how the proposed control system allows transitions between the different scenarios without severe transients in the power transfer between the microgrid and the low voltage network elements.

Keywords: microgrid; control sýstem; storage system; wind turbine; primary control

Citation: Aybar-Mejía, M.; León-Viltre, L.; Santos, F.; Neves, F.; Gómez, V.A.; Mariano-Hernández, D. Modeling and Control of a Microgrid Connected to the INTEC University Campus. *Appl. Sci.* **2021**, *11*, 11355. https://doi.org/10.3390/app112311355

Academic Editor: Amjad Anvari-Moghaddam

Received: 4 November 2021
Accepted: 25 November 2021
Published: 30 November 2021

1. Introduction

The constant growth in renewable energy generation and the integration of these systems into large-scale grids represents a challenge for the proper functioning of the electrical system [1]. Grid integration requirements have therefore become a significant concern as renewable energy sources, such as wind and solar photovoltaic (PV) systems, slowly begin to replace conventional plants [2]. The proper control and operation of electric microgrids integrated into the electrical network can improve the electrical system's power quality, stability, and reliability [3].

In [4], a hybrid AC/DC microgrid has been incorporated into a low voltage AC distribution system. A power control scheme is presented to improve the system's stability after load steps when the microgrid operates in island mode. However, the authors of this study do not consider the presence of mini wind power sources in the microgrid. Further, they do not describe the control algorithm for islanded operation, nor the batteries being fully charged—a condition that requires that distributed generators (DG) units abandon maximum power point tracking (MPPT) and share the total load in proportion to their available primary powers.

A power management system for hybrid AC/DC microgrids, designed to optimize a cost function that considers the maximum utilization of renewable resources, minimal usage of fuel-based generators, extending the lifetime of batteries, and limited utilization of the main power converter between the AC and DC micro-grids is presented in [5]. However, the power management algorithm needs several input variables which are usually not available in most microgrids, e.g., PV and wind primary energies, namely, demanded power (sum of load power, power line loss, and power loss due to the circulating current), the state-of-charge (SOC) of the battery banks, and some statistical and dynamical operational limits.

The management and optimization of a hybrid AC/DC microgrid are still open issues, as affirmed in [6]. In this study, the authors propose a model predictive power and voltage control (MPPVC) method for providing microgrid control under different possible scenarios with smooth transients. However, the precise details about the control of the DG units for off-MPPT operation while ensuring proper total load sharing are not given.

In this paper, a power management scheme for a hybrid AC/DC wind–PV–ESS microgrid is proposed. The only input variables necessary for the converters' control are those usually measured, the state of the microgrid (interconnected or not), and the battery SOC. The control schemes of the converters for all operation scenarios are described in detail. The results obtained in this paper serve as a basis for a hybrid AC/DC microgrid comprising a PV system, an energy storage system (ESS) using lithium-ion batteries, DC loads, a wind microturbine with a permanent-magnet synchronous generator system (PMSG), and AC loads that are under development as part of a project financed by the Ministerio de Educación Superior, Ciencia y Tecnología (MESCyT) of the Dominican Republic which will be installed on the campus of the Instituto Tecnológico de Santo Domingo Instituto de Santo Domingo (INTEC). The objective of the project is to implement and verify the behavior of the hybrid microgrid through the intelligent management of the distributed generation (DG) units and loads and the collection of information. While the microgrid is under construction as part of the energy management and control system (EMCS) design, its behavior is studied through a detailed simulation system. In the present study, the sizes of the photovoltaic, wind, and storage systems are the same as those of the equipment available for installation, and energy production is estimated based on forecasting data. However, implementing the control and management of this system is not limited to the specific elements of the microgrid. The results can be extrapolated to other low voltage infrastructures to meet the voltage and electrical frequency conditions. This project aims to promote the applications of hybrid DG systems integrated into electrical microgrids.

In the present work, we evaluate a control strategy that allows power transitions between the microgrid connected to the university campus and all possible operating scenarios, such as islanded or interconnected microgrids, different generation-load possibilities, and state-of-charge conditions of the battery. The converters have a low number of switches but allow adequate control of each element of the microgrid, so that the GD units operate in maximum power point tracking (MPPT) mode or promote the balance between the generated power, loads, and battery charge/discharge powers.

The overall management strategy, together with the proposal of control schemes for the DG units for off-MPPT operation while ensuring proper total load sharing, are the main original contributions of this article.

The rest of the paper is organized as follows. The proposed topology and modes of operation are described in Section 2. Models were constructed to enable the design of the system control loops, the PV, PMSG-wind, and ESS systems, together with the respective interface converters, and these are presented in Section 3. The control strategies for the interface converters are explained in Section 4. The operational scenarios used to verify and validate the proposed control strategies are described in Section 5, and the corresponding simulation results are presented in Section 6. The conclusions of the study are drawn in Section 7.

2. Microgrid Topology and Modes of Operation

Figure 1 presents a schematic diagram of the PV–wind–ESS hybrid AC/DC microgrid. The photovoltaic panels, lithium-ion battery, and PMSG-based wind system are connected to the microgrid DC bus via DC/DC converters. A bidirectional inverter is used to control the energy flow between the DC and AC microgrid buses. AC loads are connected to the microgrid at 220 V, 60 Hz AC bus, through which the microgrid is connected to the central power system. The capacity of the microgrid simulated for a PV system is 500 Wp, wind turbine capacity is 2 kW, and lithium storage capacity is 5 kWh. A simple boost converter has been chosen as the PV system interface, and a low-cost two-switch bidirectional converter topology was used for the proper control of the ESS.

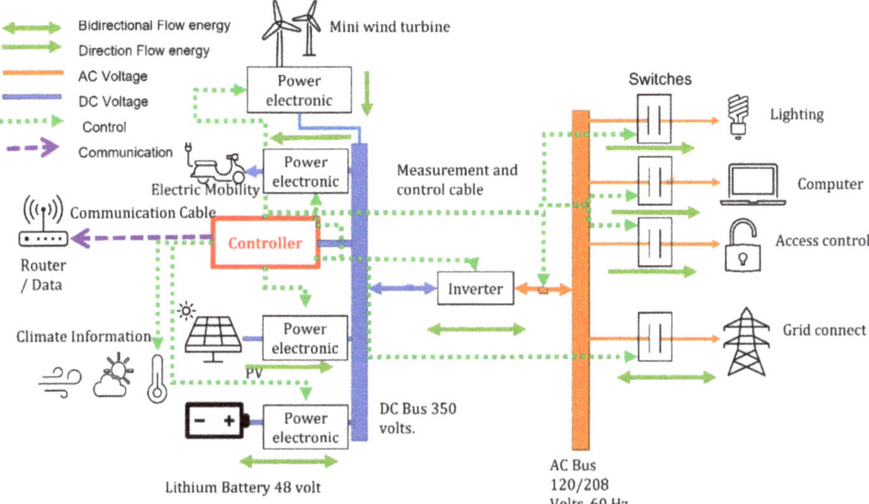

Figure 1. Topology of the PV–wind–ESS hybrid microgrid.

The DG and ESS interface converters must allow different control modes, depending on whether the microgrid is connected to the central power system or not. Furthermore, with respect to island mode, the converters' control strategies depend on the batteries' state of charge (SoC) and the total DG units' instantaneous injected power versus total load demand. Thus, some quantities necessary for the decision about the microgrid mode of operation must be sent to a central control unit, which decides the DG unit's operation mode (under MPPT or with reduced generation) and ESS operation (charging or supplying the instantaneous power deficit).

For long-term operation, climatology information obtained for the project's location in the university was used (18°29′15.9″ N 69°57′48.5″ W).

When the microgrid is connected to the primary grid, the PMSG wind and PV systems must be controlled to maximize the use of available primary energy, i.e., MPPT is adopted. Furthermore, if the battery is not fully charged, its converter controller should determine the instantaneous power to be absorbed. Therefore, for operation in under-connected mode, the inverter between the DC and AC buses must transfer to (or absorb from) the AC microgrid bus the difference between the power generated and consumed by the battery and DC loads.

On the other hand, loads should preferably be fed by the DG sources when the microgrid is islanded. In case of insufficient generation, the batteries should complement the power to meet the load's demand. However, if the generation exceeds the power required by the loads there are two possible situations that might result: (i) the battery is fully charged, in which case the DG systems can no longer operate at the maximum power point MPP and must share the total power consumed by the loads proportionally to their nominal powers; or (ii) the battery can be charged, allowing the DG systems to operate at the MPP, and the excess of generated power is used for battery charging.

It is essential to mention that the topology and control strategies must ensure proper operation under the different conditions described above and provide smooth transitions between operating modes. The energy management strategy described is represented in the flow diagram of Figure 2.

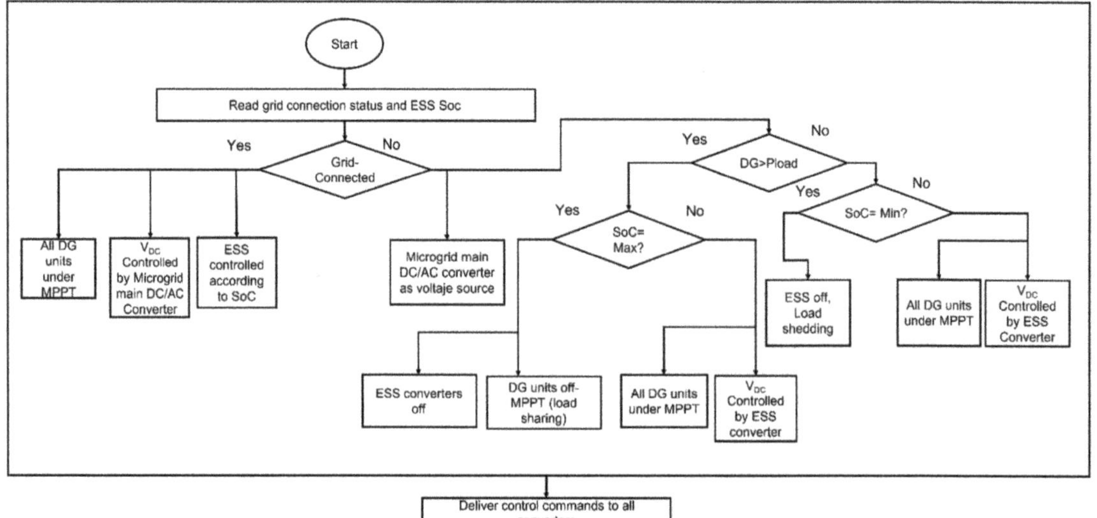

Figure 2. Microgrid energy management strategy.

3. Modeling of the Microgrid

In this section, the models of the main microgrid elements—the PMSG wind system, the PV module, and the battery—are briefly described.

3.1. PV System

The PV module can be characterized by a nonlinear I–V curve that changes according to the local temperature and irradiance conditions. Among the different proposals to accurately represent the PV module, the most used is the single-diode model, shown in Figure 3, due to its good compromise between simplicity and accuracy. Furthermore, it is possible to estimate the model's parameters based on the data provided in the panel manufacturer datasheet [7]. Table 1 describes the characteristics of the one-diode model elements.

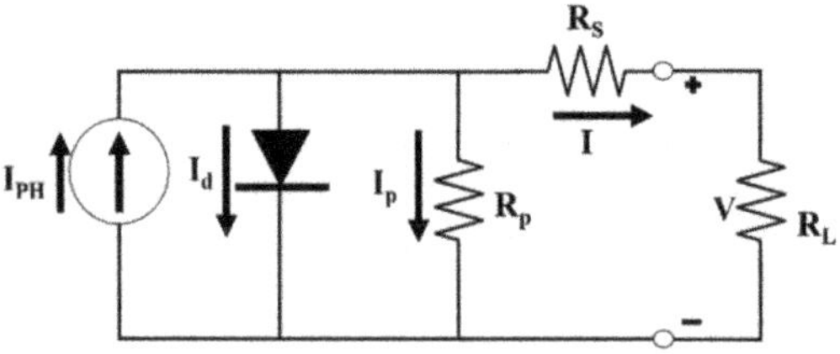

Figure 3. Equivalent circuit of the photovoltaic cell.

Table 1. Model of photovoltaic modules.

Model	Feature
One-diode model	Considers the photogenerated current and the diffusion diode current, which correspond to the electronic conduction phenomena in the neutral zone of the semiconductor.
Series resistance	Represents the losses of the metal contacts of the module.
Parallel resistance	Represents losses from eddy currents circulating in the module.
Diode	Represents the recombination of carriers in the semiconductor charge zone.
I_{PH}	Current generated by the PV cell.

Equations (1)–(4) demonstrate the equation of the PV panel for the light-generated photocurrent as presented in [7]. The output current of the single diode model is a function of the output voltage, and it can be described as:

$$I = I_g - I_{sat}\left[e^{\left(\frac{V+IR_s}{Vt}\right)} - 1\right] - \frac{V + IR_s}{R_p} \tag{1}$$

$$V_t = \frac{N_s A k t}{q} \dots \tag{2}$$

where V and I are the module output voltage and current, I_g is the photogenerated current, I_{sat} is the reverse saturation current of the diode, V_t is the thermal voltage, q is the electron charge, A is the ideality factor of the diode, k is the Boltzmann constant, T is the module temperature, N_s is the number of series-connected cells forming the PV module, and R_s and R_p are the series and the parallel resistances, respectively.

The PV module manufacturers do not directly provide the five parameters of the single-diode model. For this reason, it is difficult to determine those parameters using simple analytical methods. Instead, all datasheets provide the following information for the standard test conditions (STC): open-circuit voltage (Voc), short-circuit current (Isc), MPP voltage (Vmp), MPP current (Imp), a temperature coefficient for Voc (kV), a temperature coefficient for Isc (ki), and maximum power (Pmp). An analysis of the electrical model in three operation points of the I–V curve (Isc, Voc, and Pmp) allows the unknown parameters of the electrical model to be related to the datasheet information.

Several authors propose novel photovoltaic parameter estimation methods. This article uses the method proposed in [7], which involves finding the five unknown parameters that guarantee the absolute minimum error between the P–V curves generated by the electrical model and the P–V curves provided by the manufacturers' datasheets for different external conditions, such as temperature and irradiance. I_g is given by (3), I_{sat} is obtained from (4):

$$I_g = \left[I_{sc,STC} + k_i(T - T_R)\right]\frac{S}{1000}. \tag{3}$$

$$I_{sat} = \frac{I_g - \frac{V_{oc}}{R_p}}{e^{\frac{V_{oc}}{Vt}} - 1} \dots \tag{4}$$

A complete scan of all possible values of A (from 1 to 2, with a step of 0.01) and R_s (from 0 to 2 Ω, with a step of 1 mΩ) is made.

3.2. Wind Turbine and Permanent Magnet Synchronous Generator

The instantaneous power delivered to the PMSG axis by the wind turbine, as in the system described in [8], can be represented by:

$$P = \frac{1}{2}\rho A V^3 C_p(\lambda, \beta), \tag{5}$$

where ρ is the air density, A is the area swept by the turbine rotor blades, V is the instantaneous wind speed and $C_p(\lambda, \beta)$ is the efficiency power conversion factor, which is a nonlinear function of the tip speed ratio $\lambda = \omega_t R / V$ and the pitch blades angle β. Since the manufacturers of wind turbines do not give information about power conversion factor characteristics, some typical empirical curves are often used for representing wind turbines in power system studies, such as the one presented in [8]:

$$C_p(\lambda, \beta) = 0.22\left(\frac{116}{\lambda_i} - 0.4\beta - 5\right)e^{\frac{-12.5}{\lambda_i}}, \tag{6}$$

where:

$$\frac{1}{\lambda_i} = \frac{1}{\lambda + 0.08\beta} - \frac{0.035}{\beta^3 + 1}. \tag{7}$$

The PMSG was represented using the model in the dq reference frame rotating at a rotor electrical angular speed $\omega_r = (P/2)\omega_t$:

$$v_{sd} = R_s i_{sd} + \tfrac{d}{dt}\lambda_{sd} - \omega_r \lambda_{sq}$$

$$v_{sq} = R_s i_{sq} + \tfrac{d}{dt}\lambda_{sq} + \omega_r \lambda_{sd} \ , \tag{8}$$

$$\tfrac{2J}{P}\tfrac{d}{dt}\omega_r = T_e - T_m - \tfrac{2b}{P}\omega_r$$

where v_{sd}, v_{sq}, i_{sd}, i_{sq}, λ_{sd}, and λ_{sq} are the direct and quadrature axes components of the stator voltage, current, and flux space vectors, respectively. T_e and T_m are the electromagnetic and mechanical torque machines, and J and b are rotor inertia and viscous friction coefficients. The flux–current relations and the electromagnetic torque equation complete the model:

$$\lambda_{sd} = L_{sd} i_{sd} + \Lambda$$

$$\lambda_{sq} = L_{sq} i_{sq} \ , \tag{9}$$

$$T_e = \tfrac{3}{2}\tfrac{P}{2}\left(\lambda_{sd} i_{sq} - \lambda_{sq} i_{sd}\right) = \tfrac{3}{2}\tfrac{P}{2}\left[\Lambda i_{sq} + \left(L_{sd} - L_{sq}\right)i_{sd} i_{sq}\right]$$

where Λ is the permanent magnet flux.

3.3. Battery Model

Mathematical modeling and dynamic simulation of battery storage systems can be challenging due to their nonlinear nature. The published literature has presented several thermal models of lithium-ion battery packs [9,10].

In [9], a suitable, convenient dynamic battery model is presented that can be used to model a general battery storage system. The proposed dynamic battery model can analyze the effect of temperature, cyclic charging/discharging, and voltage stabilization effects. Temperature makes a difference to three battery parameters in an equivalent electrical circuit battery model. These are the polarization voltage K, the battery constant E_0, and the exponential coefficients, A and B. The dynamic battery model can be described as [9]:

$$E_{(q,t)} = X_{E0}(T).E_0 - X_K.K\left(\frac{Q}{Q-q}\right) + A\,exp^{-X_B B_q} \ \dots \tag{10}$$

where $E_{(q,t)}$ = no load voltage (V), E_0 = battery constant voltage (V), K = polarization voltage (V), Q = battery capacity (Ah), A and B = exponential constants, and q = charge or extracted capacity (Ah).

The thermal effect on A has been ignored to make the model simpler, since A and B are highly related. For this reason, the mathematical relationships are:

$$X_n = f(T) = A + BT + CT^2..\tag{11}$$

$$n = f(K, E_0, B.C)\tag{12}$$

4. Control Strategies

The converters of the microgrid DG and battery units and the converter between the DC and AC buses must be controlled according to the modes of operation described in Section 2. The control scheme implemented in each converter is presented in this section. Procedures for tuning these controllers are also explained.

4.1. PV Converter Control

As already mentioned in Section 2, the PV generation system should operate in MPPT mode in three possible situations: (i) when the microgrid is in the connected mode; (ii) when the microgrid is islanded, and the total power demanded by the loads is greater than the total power produced by the DG units (the battery supplies the difference); and (iii) when the microgrid is islanded, and the power produced by the DG units is greater than the power demanded by the loads but the battery is not fully charged (so it can absorb the exceeding generated power). However, if the microgrid is islanded, the DG units' produced power is more significant than that demanded by the loads, and the battery is fully charged, then the MPPT cannot be imposed, and the generated power must be reduced so that it equals the total load power. In this last case, the load power is shared among the DG units proportionally to the DG units' rated powers.

4.1.1. PV Converter MPPT Method

The electrical power supplied by PV cells is a non-linear function of voltage, current, temperature, and solar irradiance. This non-linearity makes it challenging to obtain the operating point at which its maximum power is extracted, as this point varies throughout the day due to variations in irradiance and temperature. The objective of any MPPT algorithm is to find the voltage to be applied to the PV string terminals that results in the maximum possible generated power for the momentary conditions of irradiance and temperature.

Several techniques for determining MPP have been proposed over the years. These techniques vary in complexity, speed of convergence, voltage fluctuation around the MPP, and computational cost. The MPPT technique chosen for this project was the very popular perturb and observe (P&O) algorithm due to its simplicity and acceptable performance in most situations [11]. The P&O MPPT technique changes the PV array voltage in one direction and observes the variation in the output power. If the generated power increases, then the voltage perturbation continues in the same direction. However, if the output power reduces, the voltage perturbation occurs in the opposite direction. The voltage disturbance process is repeated periodically, each T_{MPPT}, and the array voltage oscillates around the MPP. This oscillation can be minimized by reducing the size of the disturbance, but minimal disturbances make the technique slow to track the MPP. The parameters of the P&O MPPT method are T_{MPPT} and the magnitude of the voltage perturbation is ΔV. Typical choices of these parameters are $\Delta V = 0.5\%$ of the PV array open-circuit voltage and $T_{MPPT} = 4\tau_{Vdc}$, where τ_{Vdc} is the time constant of the PV array output voltage variation.

After the reference value of the PV array, DC voltage is determined, and the DC/DC boost converter synthesizes it. Thus, the duty cycle of the boost converter is determined from the DC bus voltage of the microgrid and the reference PV array voltage, calculated according to the P&O algorithm for achieving MPPT.

4.1.2. PV Converter Control for Sharing the Microgrid Load Power

As has already been mentioned, if the microgrid is islanded, the battery is fully charged, and the total DG power exceeds the microgrid AC + DC load power, then each DG source must reduce the generated power until the load power demand is shared among the DG units. Ideally, this sharing should be proportional to the maximum power delivered by each DG unit.

If the DG power is greater than the load power demand, considering that no power can be transferred to the grid or the battery, the excess power flows to the DC bus capacitors bank, increasing the DC bus voltage. Thus, based on the information that the microgrid is islanded and that the battery state-of-charge is complete, the DC bus voltage increase indicates that the power generated by each DG unit must be reduced.

A typical voltage–power curve for a PV system is shown in Figure 4. However, it can be observed that, below the MPP voltage, the voltage–power relation can be considered approximately linear.

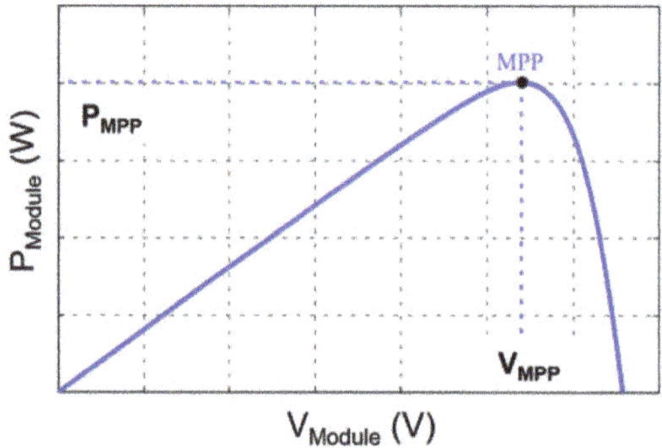

Figure 4. Typical P–V characteristic of a PV string.

The strategy proposed here to make the DG units share the total power required by the microgrid loads is based on the DC bus voltage increase per unit of the maximum allowed voltage increase, expressed as:

$$\Delta V_{DC,pu} = \frac{\Delta V_{DC}}{\Delta V_{DC}^{MAX}} = \frac{V_{DC} - V_{DC}^{rated}}{V_{DC}^{MAX} - V_{DC}^{rated}}. \tag{13}$$

Assuming that on the left side of the P–V characteristic, the PV-generated power is approximately proportional to the PV string output voltage, this voltage can be reduced linearly with the DC bus voltage increase. Since the DC bus voltage becomes constant only when the generated and load powers are equal, the steady-state condition ensures that, for some DC bus acceptable overvoltage, the PV output voltage ensures DG and load powers to match. The PV output voltage command can then be obtained, as shown in Figure 5.

If other PV units were connected to the microgrid, their output voltages would be calculated using the scheme presented in Figure 5. Thus, the generated powers would be approximately proportional to each PV unit's maximum power available.

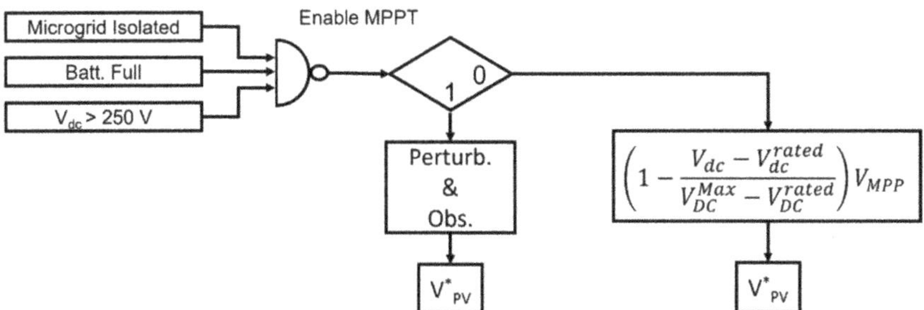

Figure 5. PV output voltage references calculation for MPPT and for reduced generation to track load power demand.

4.2. PMSG-Based Wind Turbine Control

4.2.1. Control for Ensuring MPPT

The PMSG stator terminals are connected to a three-leg AC/DC voltage source converter. If the PMSG rated stator voltage is low enough, the AC/DC converter can directly connect to the microgrid DC bus, otherwise a transformer between the PMSG terminals and the AC/DC converter or a DC/DC converter between the AC/DC converter and the microgrid DC bus would be necessary.

The main objective of the AC/DC converter is to optimize the power extracted from the turbine for any incoming wind speed. The wind microturbine has fixed-pitch blades at angle β. Thus, according to Equations (5)–(7), the turbine efficiency factor $C_p(\lambda, \beta)$ depends only on the tip speed ratio $\lambda = (\omega_t R)/V$. For each wind speed V, there is an angular speed of the turbine rotor ω_t that maximizes $C_p(\lambda, \beta)$, allowing the maximum available power to be extracted. This optimum angular speed is obtained from:

$$\omega_t^{opt} = \frac{\lambda^{opt} V}{R} \tag{14}$$

The PMSG reference rotor speed (in electrical radians per second) can then be obtained from $\omega_r^* = (P/2)\omega_t^{opt}$. A vector control scheme is then applied to impose the reference rotor speed.

The PMSG model is in a rotor-oriented reference frame, i.e., the d-axis is aligned with the permanent magnet flux. The reference d-axis stator current is then set to zero since a negative value would reduce the main flux and a positive value could cause magnetic saturation. Furthermore, if $i_{sd} = 0$, the electromagnetic torque becomes $T_e = \frac{3}{2}\frac{P}{2}\Lambda i_{sq}$, so it can be controlled through i_{sq}. The overall control scheme is depicted in Figure 6.

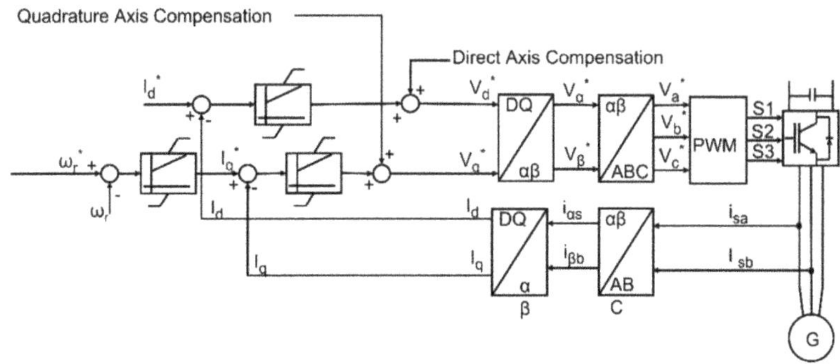

Figure 6. Control of the PMSG.

4.2.2. Control for Proportional Load Sharing

As explained in Section 3, when the microgrid is islanded, if the battery is full, a full charge and the DG units' generated power exceeds the total load power demand, and the excessive produced energy starts to flow to the microgrid DC bus capacitors, making the DC voltage rise. A scheme can again be applied to reduce the PMSG output power proportionally to the DC bus voltage increase per unit of the maximum allowed voltage increment.

Based on the wind generation efficiency characteristic shown in Figure 7, it can be observed that if the tip speed ratio λ is increased from the optimum value for MPP ($\lambda^{opt} = 6.9$), the efficiency factor C_p decreases from the maximum value (0.42), becoming zero when $\lambda \cong 15.5$. Therefore, an adjustment in the value of the tip speed ratio can be made, according to Figure 8, to force a reduction of generated power in proportion (approximately) to the increase in the DC bus voltage increase.

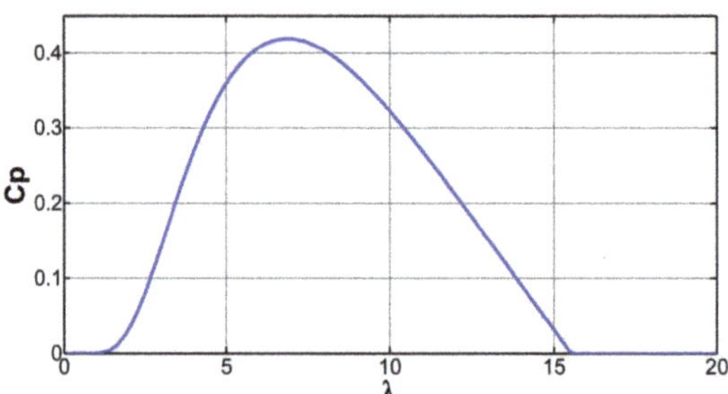

Figure 7. Wind turbine typical efficiency characteristic, considering the fixed blades' pitch angle.

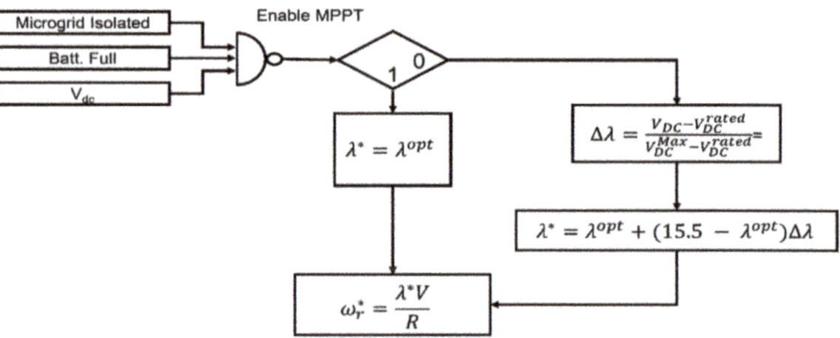

Figure 8. Wind generation schemes for MPPT or adaptation to load power demand.

It is essential to mention that, similarly to the PV power reduction method proposed, wind power is reduced from the level corresponding to the MPP condition in a linear relation with the microgrid DC bus voltage increase per unit of the maximum DC bus voltage variation. Thus, the total load power-sharing among all DG units will be approximately proportional to their respective instantaneous MPPs.

4.3. Control of the Microgrid Inverter

The main objective of the microgrid inverter is to transfer to the microgrid AC bus the excess of power produced by the DG units connected to the microgrid DC bus (or absorb the deficit). Neglecting the system losses, if the net power P_{INV_in} injected into the microgrid DC bus (produced by DG units minus the power consumed by the loads and battery being charged) is higher (lower) than the active power P_{INV_out} delivered to the microgrid AC bus by the inverter, the difference is absorbed (delivered) by the microgrid DC bus capacitors, and the microgrid DC bus voltage rises (falls). Therefore, the control objective of the microgrid inverter can be accomplished by regulating its DC bus voltage. The effect of P_{INV_in} and P_{INV_out} on the microgrid DC bus voltage can be written as:

$$P_{INV_in} - P_{INV_out} = \frac{d}{dt}\left(\frac{1}{2}CV_{DC}^2\right) = \frac{1}{2}C\frac{d}{dt}\left(V_{DC}^2\right) \tag{15}$$

According to (15), P_{INV_out} can be used to control V_{DC}^2 (and therefore to control V_{DC}). Due to the linear relationship between $(P_{INV_in} - P_{INV_out})$ and V_{DC}^2, and since the reference voltage V_{DC}^* is constant in a steady-state condition, then a proportional–integral (PI) controller is adopted, taking the error $\left(V_{DC}^{2*} - V_{DC}^2\right)$ as input and the negative of the inverter output power $\left(-P_{INV_out}^*\right)$ as the output control action. P_{INV_in} is considered a disturbance to be rejected by the controller.

As a second goal, the reactive power delivered by the microgrid AC side should be maintained equal to zero ($Q_{INV_out}^* = 0$) to avoid unnecessary reactive current components flowing through the inverter.

The above shows that it is necessary to regulate the inverter output of active and reactive powers to achieve the objectives. According to the instantaneous power theory [12], the active and reactive powers delivered to the AC bus of the microgrid can be written, in space-vector $\alpha\beta$ Clarke coordinates, in terms of the inverter AC output current vector $\vec{i}$ and microgrid AC bus voltage vector $\vec{v}_{BUS}$ as follows:

$$\vec{s}_{INV} = p_{INV} + jq_{INV} = \vec{v}_{BUS}\,\vec{i}' = (v_{BUS\alpha} + jv_{BUS\beta})(i_\alpha - ji_\beta). \tag{16}$$

From (16), the inverter output current components necessary to impose the reference values of active and reactive powers are

$$\begin{bmatrix} i_\alpha^* \\ i_\beta^* \end{bmatrix} = \frac{1}{\left|\vec{v}_{BUS}\right|^2}\begin{bmatrix} v_{BUS\alpha} & v_{BUS\beta} \\ v_{BUS\beta} & -v_{BUS\alpha} \end{bmatrix}\begin{bmatrix} P_{INV_out}^* \\ 0 \end{bmatrix}, \tag{17}$$

Alternatively, using complex space vector notation:

$$\vec{i}^* = P_{INV_out}^*\frac{\vec{v}_{BUS}}{\left|\vec{v}_{BUS}\right|^2}. \tag{18}$$

Assuming the reference active power $P_{INV_out}^*$ is constant (or slowly varying), if the voltages at the microgrid AC bus are not balanced or contain harmonic components, the currents calculated through (18) would contain the same levels of unbalance and harmonic contamination. For this reason, it is generally preferred to calculate the reference current vector using the positive-sequence fundamental-frequency (FFPS) vector component of $\vec{v}_{BUS}$:

$$\vec{i}^* = P_{INV_out}^*\frac{\vec{v}_{BUS}^{+1}}{\left|\vec{v}_{BUS}^{+1}\right|^2}. \tag{19}$$

Using (19), the active and reactive powers delivered to the microgrid AC bus might contain oscillations, but their mean values correspond to the reference components. Furthermore, the inverter output phase currents will be balanced and sinusoidal. Finally, to ensure the inverter's desired behavior, inner control loops regulate the inverter α and β current components. The inverter is connected to the microgrid AC bus through an inductive filter, modeled as an LR circuit where L and R are the filter inductance and internal resistance. The following vector equation expresses the relationship between the filter voltages and current in the stationary $\alpha\beta$ reference frame:

$$\vec{v}_{INV} - \vec{v}_{BUS} = R\vec{i} + L\frac{d\vec{i}}{dt}. \tag{20}$$

Thus, the inverter output voltage vector $\vec{v}_{INV}$ can be used to regulate the current vector $\vec{i}$. In this case, the microgrid AC voltage vector $\vec{v}_{BUS}$ is a known disturbance (since it is necessarily measured for computing the current reference), allowing a feedforward compensation.

In [13,14], several current controllers used in grid-connected converters are described and compared. Advanced control techniques are necessary whenever the voltage at the grid point of common coupling (PCC) contains harmonic components and/or unbalance once the controller is required to reject these disturbances. In these cases, the most popular control strategies recommend using PI controllers in multiple DQ reference frames [15], second-order PR controllers in parallel [16], PI controllers with resonant controllers in a rotating DQ reference frame [17], stationary-frame controllers using the space-vector Fourier transform [18], or repetitive controllers (in a real domain [19] or a complex domain [20–22]).

This work used a proportional action and some resonant controllers in parallel due to the acceptable performance and relatively simple implementation.

The external loop of the inverter control scheme is shown in Figure 9. If the internal controller is designed to be considerably faster than the external one, then the external controller can be designed neglecting the internal dynamics, i.e., considering $\vec{i}/\vec{i}^{*} = 1$.

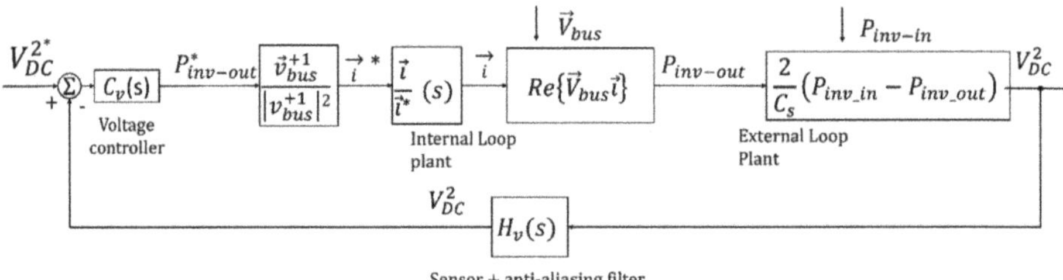

Figure 9. Simplified block diagram of the microgrid inverter's external control loop.

The block diagram of the described internal control loop (α and β current controllers) is presented in Figure 10.

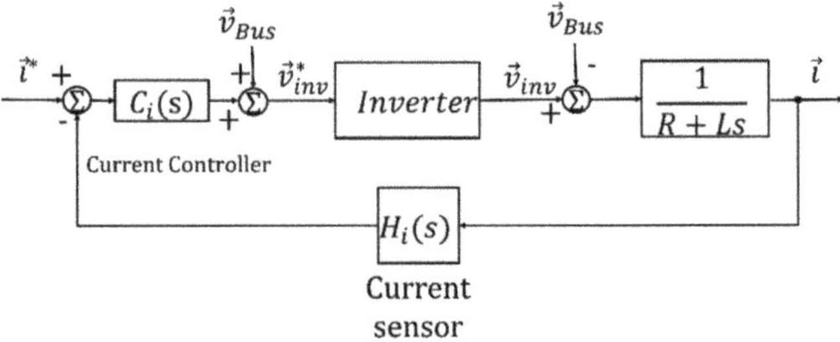

Figure 10. Simplified block diagram of the internal current control loop of the microgrid inverter.

The internal current control loop must have a high bandwidth and high gains at the fundamental frequency and the harmonic frequencies corresponding to typical harmonic disturbances of the grid. These specifications can be met using proportional action and second-order sinusoidal integrators (SSIs) as resonant controllers in parallel. The selection of the appropriate proportional gain enables the imposition of the desired dynamic response to the system, while the resonant terms, implemented in parallel, guarantee a low steady-state error.

A resonant term is included at the fundamental frequency to track the FFPS reference current with a low steady-state error. Resonant terms are also implemented in the low-order harmonic frequencies (3rd, 5th, 7th, and 9th), considering that these are usual components of disturbances in the voltage. In this context, when selecting the maximum harmonic component h = 9, the minimum 0 dB crossover frequency (fPM) required is 540 Hz (an fPM around 1500 Hz was adopted).

In addition to the phase margin (PM) and the gain margin (GM), the sensitivity index η, which defines the minimum distance from the Nyquist diagram to the critical point $(-1, 0)$, is also verified. As a design criterion, we considered GM $\geq$ 3 dB, $\eta \geq 0.3$, and PM $\geq 30°$. The proportional controller gain needed to ensure the desired crossover frequency is $k_p = 0.09899$. Then, the SSI-based resonant controllers are added in parallel and tuned with the harmonic components h = 6k $\pm$ 1 until h = 9. The same resonant gain of $k_i = 1$ was chosen for all resonant units, and it was observed that the desired phase margin was maintained. Concerning the inevitable delay in digital implementation, a computational delay of two samples was considered. The Bode and Nyquist diagrams of the resulting current control loop are presented in Figure 11a,b, respectively, demonstrating that the design criteria were met.

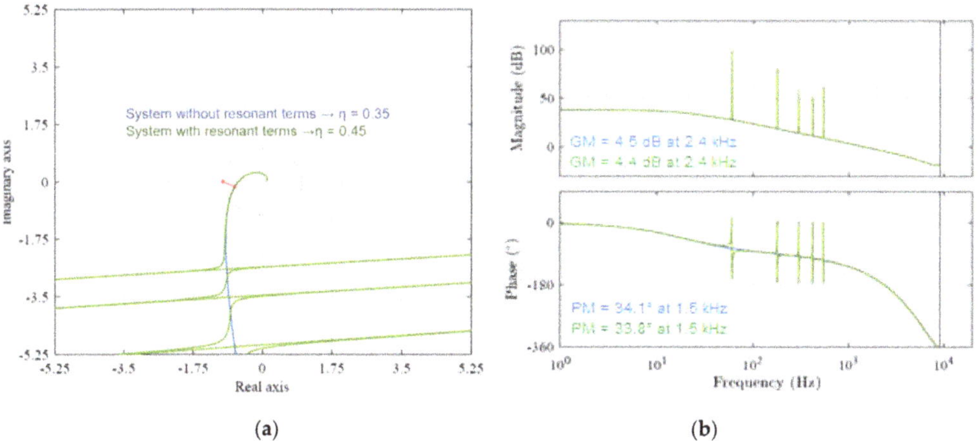

Figure 11. The Bode and Nyquist diagrams of the resulting current control loop. (**a**) Nyquist diagrams of the open-loop system with computational time delay compensation. (**b**) Bode diagrams of the open-loop system with computational time delay compensation. Results were obtained for the proportional controller (blue) and P-SSI in parallel (green).

After defining the internal current control loop gains, the much slower external DC bus voltage loop tuning was considered. The 0 dB crossover frequency of the external loop was chosen so that

$$f_{cv} < \frac{f_{ci}}{10} \tag{21}$$

where f_{cv} and f_{ci} represent the 0 dB crossover frequencies of the external and internal control loops, respectively. This design criterion is selected so that the DC bus voltage regulation does not compromise the dynamics of the current control. Therefore, for $f_{ci} = 1500$ Hz, $f_{cv} < 150$ Hz must be chosen. For practical reasons, in order to avoid the too frequent saturation of the controller output, $f_{cv} = 6$ Hz was adopted. Furthermore, the indices GM $\geq$ 3 dB and PM $\geq 30°$ were considered as design specifications. The frequency response of the open-loop transfer function for $k_i = 1.0562$ and proportional controller gain

$k_p = 0.084$, which is the value that results in the desired crossover frequency $f_{cv} = 6$ Hz, is shown in Figure 12. The compensated system has a gain margin GM = 65 dB and phase margin PM = 71.5°, meeting the design specifications.

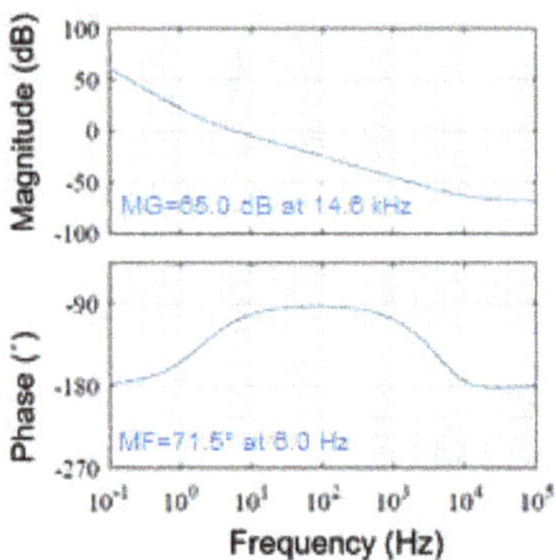

Figure 12. Bode diagrams of the DC bus voltage open-loop control system.

4.4. Synchronization with the Electrical Network

In many practical situations, the distribution network exhibits distorted and unbalanced voltages. In these cases, synchronization with the network strongly influences the performance of the entire control scheme of the power converter connected to the network. Determining the correct value of the fundamental-frequency positive-sequence (FFPS) component of the microgrid bus voltage vector is essential for reasonable control of the inverter. This information is obtained through a phase-locked loop (PLL) algorithm.

Several PLL schemes for three-phase systems have been proposed in recent years, the synchronous reference frame PLL (SRF-PLL) being the most popular [23,24]. However, due to its poor performance when the voltages contain harmonics and/or unbalance, preferable alternatives have been presented [25]. In this paper, the PLL scheme that uses the generalized delayed signal cancelation (GDSC) method as a pre-filter to eliminate the effects of unbalances and harmonic components eventually present in the microgrid AC bus voltage was used [26,27]. The choice was based on the fast and accurate determination of the FFPS component of a three-phase signal, even under severe unbalanced and distorted conditions.

5. Operational Scenarios Used for the Validation of the Proposed System

First, the power flow balance is assumed, i.e., regardless of the state of the distributed generation sources, battery, and load:

$$P_{PV} + P_{WT} \pm P_{ESS} - P_{load} \pm P_{grid} = 0 \tag{22}$$

The power generated photovoltaic panel = P_{PV}; the power generated by the mini-wind generator = P_{WT}, the power injected or absorbed by the storage system = P_{ESS}, the total load power = P_{load}, and the power injected into or absorbed from the electrical network of the university = P_{grid}.

Depending on the microgrid operational condition (interconnected or isolated), the instantaneous generated power, the total load power, and battery state-of-charge, the converters must behave according to the modes of operation described.

For interconnected conditions, it is assumed that the battery does not inject any power into the electrical network; in addition, the PV and WT are maintained at their maximum power points.

- Interconnected mode, DG > P_{load}:

$$P_{PV} + P_{WT} = P_{load} + P_{grid}; P_{ESS} = 0; \tag{23}$$

- Interconnected mode, DG = P_{load}:

$$P_{PV} + P_{WT} = P_{load}; P_{grid} = 0; P_{ESS} = 0; \tag{24}$$

- Interconnected mode, DG < P_{load}:

$$P_{PV} + P_{WT} + P_{grid} = P_{load}; P_{ESS} = 0. \tag{25}$$

For isolated conditions, no power can be transferred to or absorbed from the electricity grid. Therefore, if the balance between generation and load demand does not occur, the battery must supply or absorb the difference. However, the DG units must reduce their generated power to meet the load demand if the battery is completely charged. Thus, the following conditions must be verified:

- Isolated mode, DG > P_{load}, battery not fully charged:

$$P_{PV} + P_{WT} - P_{ESS} = P_{load}; P_{grid} = 0, \text{PV, and wind units under MPPT;} \tag{26}$$

- Isolated mode, DG > P_{load}, battery fully charged:

$$P_{PV} + P_{WT} = P_{load}; P_{grid} = 0; P_{ESS} = 0, \text{PV, and wind units under reduced power mode;} \tag{27}$$

- Isolated mode, DG = P_{load}:

$$P_{PV} + P_{WT} = P_{load}; P_{grid} = 0; P_{ESS} = 0, \text{PV, and wind units under MPPT;} \tag{28}$$

- Isolated mode, DG < P_{load}:

$$P_{PV} + P_{WT} + P_{ESS} = P_{load}; P_{grid} = 0, \text{PV, and wind units under MPPT.} \tag{29}$$

6. Results

A simulation was performed in which load steps were applied for evaluating the operation of the microgrid under all seven of the operational scenarios described.

The behavior of the microgrid DC bus voltage for the different control strategies and operating modes is shown in Figure 13. It can be seen how the DC bus voltage tends to the reference value, regardless of the interconnected or isolated condition of the microgrid, except when the microgrid is isolated, the battery is fully charged, and the load demanded power is lower than the DG available power. In this case, the power generated by each DG unit is reduced in approximate proportion to the increase in the DC bus voltage, as expected. As shown in Figure 14, a relatively smooth transition is achieved for each operating mode of the microgrid.

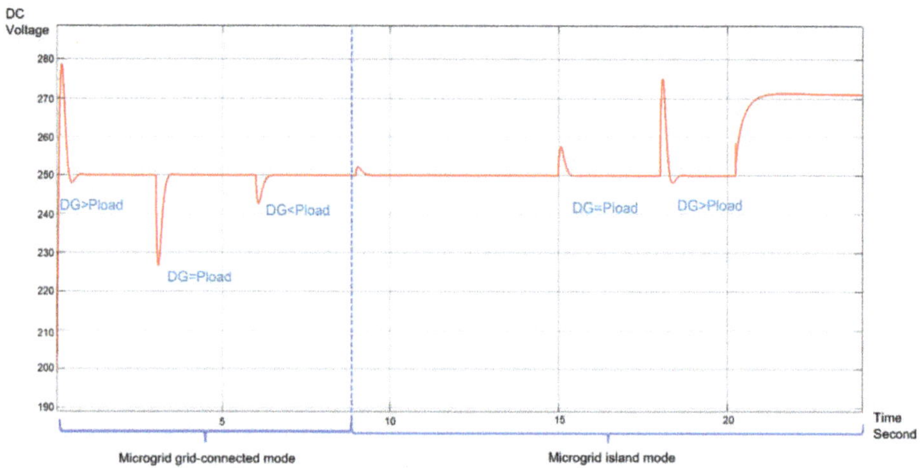

Figure 13. Voltage DC bus microgrid.

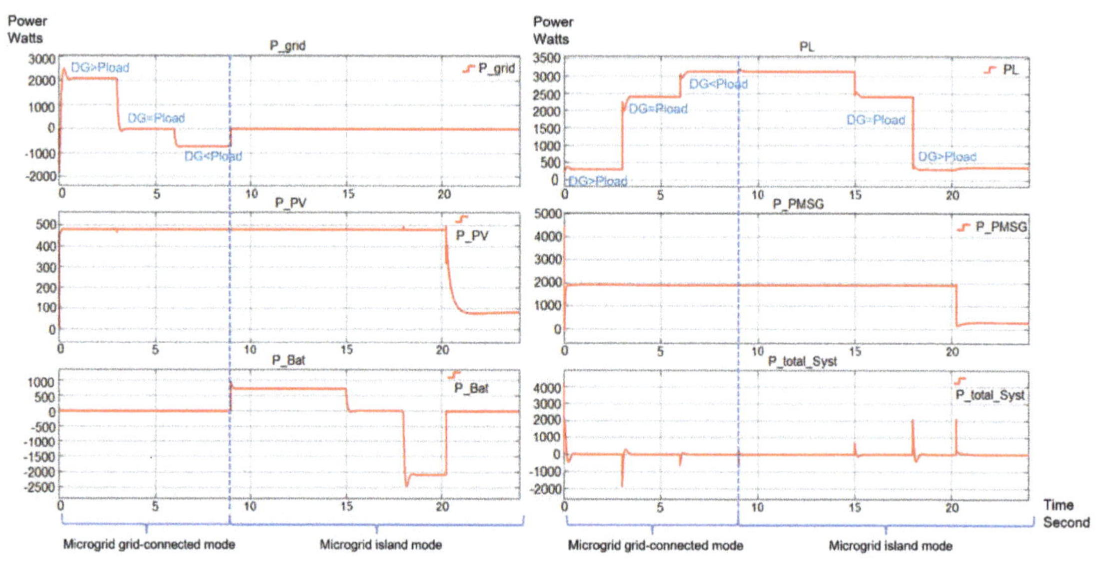

Figure 14. Relation of electrical power between the control strategies and modes of operation of the microgrid with P_{PV}, P_{WT}, P_{ESS}, P_{load}, and P_{grid}.

Initially, the microgrid is connected to the main power system, and three possible situations are considered. (i) DG > P_{load}: in this case, since the main microgrid DC/AC converter transfers to the grid the excess of generated power, the DC bus voltage is controlled to remain at the rated value, as shown in Figure 13. Additionally, the PV and wind generation units operate under MPPT mode, producing around 480 W and 1900 W, respectively, as observed in Figure 14. The battery is considered to be fully charged and therefore does not absorb any power. (ii) DG = P_{load}: at t = 3 s, the load power abruptly increases, becoming approximately equal to the total power produced by the DG units. Due to the net power (P_{net} = DG-P_{load}) reduction, the DC bus voltage starts to decrease, but the DC bus voltage controller asks for the DC/AC converter power (P_{grid}) to rise until the DC bus voltage returns to the rated value, which occurs when $P_{grid} = P_{net} = 0$.

(iii) At t = 6 s, the total load power increases again, making DG < P_{load}. In this case, the DC bus voltage controller is forced to absorb active power from the AC mains (P_{grid} < 0) to make V_{DC} = 250 V. The microgrid is connected to the main power system and the DG units generated powers are constant since they remain under MPPT operation. (iv) At t = 9 s, the microgrid interconnection circuit breaker is turned off, making P_{grid} = 0. At this moment, the battery interface converter assumes the responsibility for the DC bus voltage control and starts to deliver the necessary net power to regulate the DC bus voltage at its nominal value. (v) At t = 15 s, the total load power decreases so that DG = P_{load} again. Once the battery is delivering some power to the DC bus voltage, its value tends to increase right after t = 15 s, but the DC bus voltage controller commands a decrease in the power delivered by the battery. (vi) At t = 18 s, an additional decrease in the total load power makes DG > P_{load} again, and the DC bus voltage experiences a transient increase, but the battery converter again regulates this voltage and absorbs the excess power. However, at some instant after t = 20 s (vii), the battery becomes wholly charged (SoC = max), and the battery cannot absorb any more power. Since DG > P_{load}, the net power is delivered to the DC bus, the voltage of which naturally starts to increase. During phases (iv), (v), and (vi), the DG units operate under MPPT, but the energy management strategy commands off-MPPT behavior when SoC = max. The power generated by each DG unit decreases in inverse proportion to the DC bus voltage until the total generated power becomes equal to the power demanded by the microgrid loads. As can be seen in Figures 13 and 14, the proposed energy management strategy, together with the proposed DG units' control schemes, operate as theoretically predicted.

7. Conclusions

The control of a microgrid makes it possible to take full advantage of the distributed generation resources, and this, in turn, allows it to be used to satisfy the energy demand of low voltage electrical installations. These control strategies are necessary to achieve smooth transitions and interaction between the electrical network and the micro-network for each of the operating modes. This paper simulates and evaluates a control strategy that allows power transitions between the microgrid interconnected to the university campus and all possible operating scenarios. The converters have a low number of switches but allow adequate control of each element of the microgrid so that the GD units operate under maximum power point tracking (MPPT) mode or promotes the balance of the generated and load powers.

As observed in the simulations, the correct control strategies for all the elements that make up the distributed generation system, such as for PV MPPT control or load power-sharing, PMSG-based wind turbine MPPT control or load power sharing, the control of the microgrid inverter and the synchronization algorithm demonstrated that the proposed strategy is adequate to integrate electrical microgrids in the low voltage network. This type of technology allows a change in the current electrical energy distribution system, making end-users important actors in regulating, controlling, and decongesting electrical networks.

Currently, work is being done on the implementation phase of the proposed control system in the microgrid that is being developed at the Technological Institute of Santo Domingo as part of a project seeking to integrate electrical microgrids in distributed low voltage electrical networks. Future work will be dedicated to implementing the energy management and control system (EMCS) to allow the management of the microgrid in a testing center that will help to analyze the behavior of the control system proposed.

Author Contributions: Conceptualization, M.A.-M., L.L.-V. and F.N.; methodology, M.A.-M., L.L.-V. and F.N.; software, M.A.-M. and F.N.; validation, F.S., F.N. and V.A.G.; formal analysis, M.A.-M., L.L.-V. and F.N.; investigation, M.A.-M. and L.L.-V.; writing—original draft preparation, M.A.-M.; writing—review and editing, F.S., F.N., D.M.-H. and V.A.G.; visualization, M.A.-M. and D.M.-H.; supervision, L.L.-V. and F.N.; funding acquisition, M.A.-M. All authors have read and agreed to the published version of the manuscript.

Funding: This research was funded by FONDOCYT Grant No. 2018-2019-3C1-160 (055-2019 INTEC) in the Dominican Republic.

Institutional Review Board Statement: Not applicable.

Informed Consent Statement: Not applicable.

Data Availability Statement: Not applicable.

Acknowledgments: The authors gratefully acknowledge the National Background of Innovation and Scientific and Technological Development of the Dominican Republic (FONDOCYT) for the financial support for this research.

Conflicts of Interest: The authors declare no conflict of interest.

References

1. Impram, S.; Varbak Nese, S.; Oral, B. Challenges of renewable energy penetration on power system flexibility: A survey. *Energy Strateg. Rev.* **2020**, *31*, 100539. [CrossRef]
2. Čepin, M. Evaluation of the power system reliability if a nuclear power plant is replaced with wind power plants. *Reliab. Eng. Syst. Saf.* **2019**, *185*, 455–464. [CrossRef]
3. Aybar-Mejía, M.; Villanueva, J.; Mariano-Hernández, D.; Santos, F.; Molina-García, A. A Review of Low-Voltage Renewable Microgrids: Generation Forecasting and Demand-Side Management Strategies. *Electronics* **2021**, *10*, 2093. [CrossRef]
4. Ma, T.; Cintuglu, M.H.; Mohammed, O. Control of hybrid AC/DC microgrid involving energy storage, renewable energy and pulsed loads. In Proceedings of the 2015 IEEE Industry Applications Society Annual Meeting, Addison, TX, USA, 18–22 October 2015; pp. 1–8.
5. Hosseinzadeh, M.; Salmasi, F.R. Robust Optimal Power Management System for a Hybrid AC/DC Micro-Grid. *IEEE Trans. Sustain. Energy* **2015**, *6*, 675–687. [CrossRef]
6. Hu, J.; Shan, Y.; Xu, Y.; Guerrero, J.M. A coordinated control of hybrid ac/dc microgrids with PV-wind-battery under variable generation and load conditions. *Int. J. Electr. Power Energy Syst.* **2019**, *104*, 583–592. [CrossRef]
7. Silva, E.A.; Bradaschia, F.; Cavalcanti, M.C.; Nascimento, A.J. Parameter Estimation Method to Improve the Accuracy of Photovoltaic Electrical Model. *IEEE J. Photovolt.* **2016**, *6*, 278–285. [CrossRef]
8. Slootweg, J.G.; Polinder, H.; Kling, W.L. Dynamic modelling of a wind turbine with doubly fed induction generator. In Proceedings of the 2001 Power Engineering Society Summer Meeting. Conference Proceedings (Cat. No.01CH37262), Vancouver, BC, Canada, 15–19 July 2001; Volume 1, pp. 644–649.
9. Wijewardana, S.M. New Dynamic Battery Model for Hybrid Vehicles and Dynamic Model Analysis Using Simulink. *Eng. J. Inst. Eng. Sri Lanka* **2014**, *47*, 53. [CrossRef]
10. Zhu, C.; Li, X.; Song, L.; Xiang, L. Development of a theoretically based thermal model for lithium ion battery pack. *J. Power Sources* **2013**, *223*, 155–164. [CrossRef]
11. Mejias, M.A.; Landera, Y.G.; Viltre, L.L. Comparison of maximum power point tracking techniques used in photovoltaic system. *ITEGAM-JETIA* **2021**, *7*, 4–12. [CrossRef]
12. Akagi, H.; Watanabe, E.H.; Aredes, M. *Instantaneous Power Theory and Applications to Power Conditioning;* Wiley: Hoboken, NJ, USA, 2007; ISBN 9780470118924.
13. Limongi, L.R.; Bojoi, R.; Griva, G.; Tenconi, A. Digital current-control schemes. *IEEE Ind. Electron. Mag.* **2009**, *3*, 20–31. [CrossRef]
14. Neto, R.C.; Neves, F.A.S.; de Souza, H.E.P. Complex Controllers Applied to Space Vectors: A Survey on Characteristics and Advantages. *J. Control. Autom. Electr. Syst.* **2020**, *31*, 1132–1152. [CrossRef]
15. Bojrup, M.; Karlsson, P.; Alaküla, M.; Gertmar, L. Multiple Rotating Integrator Controller for Active Filters. In Proceedings of the EPE 99 Conference, Lausanne, Switzerland, 7–9 September 1999.
16. Yuan, X.; Allmeling, J.; Merk, W.; Stemmler, H. Stationary frame generalized integrators for current control of active power filters with zero steady state error for current harmonics of concern under unbalanced and distorted operation conditions. In Proceedings of the Conference Record of the 2000 IEEE Industry Applications Conference. Thirty-Fifth IAS Annual Meeting and World Conference on Industrial Applications of Electrical Energy (Cat. No.00CH37129), Rome, Italy, 8–12 October 2000; Volume 4, pp. 2143–2150.
17. Lascu, C.; Asiminoaei, L.; Boldea, I.; Blaabjerg, F. High Performance Current Controller for Selective Harmonic Compensation in Active Power Filters. *IEEE Trans. Power Electron.* **2007**, *22*, 1826–1835. [CrossRef]
18. Neves, F.A.S.; Arcanjo, M.A.C.; Azevedo, G.M.S.; de Souza, H.E.P.; Viltre, L.T.L. The SVFT-Based Control. *IEEE Trans. Ind. Electron.* **2014**, *61*, 4152–4160. [CrossRef]
19. Escobar, G.; Hernandez-Briones, P.G.; Martinez, P.R.; Hernandez-Gomez, M.; Torres-Olguin, R.E. A Repetitive-Based Controller for the Compensation of $6\ell\pm1$ Harmonic Components. *IEEE Trans. Ind. Electron.* **2008**, *55*, 3150–3158. [CrossRef]
20. Luo, Z.; Su, M.; Yang, J.; Sun, Y.; Hou, X.; Guerrero, J.M. A Repetitive Control Scheme Aimed at Compensating the 6k + 1 Harmonics for a Three-Phase Hybrid Active Filter. *Energies* **2016**, *9*, 787. [CrossRef]

21. Zimann, F.J.; Neto, R.C.; Neves, F.A.S.; de Souza, H.E.P.; Batschauer, A.L.; Rech, C. A Complex Repetitive Controller Based on the Generalized Delayed Signal Cancelation Method. *IEEE Trans. Ind. Electron.* **2019**, *66*, 2857–2867. [CrossRef]
22. Neto, R.C.; Neves, F.A.S.; de Souza, H.E.P. Complex nk+m Repetitive Controller Applied to Space Vectors: Advantages and Stability Analysis. *IEEE Trans. Power Electron.* **2021**, *36*, 3573–3590. [CrossRef]
23. Kaura, V.; Blasko, V. Operation of a phase locked loop system under distorted utility conditions. *IEEE Trans. Ind. Appl.* **1997**, *33*, 58–63. [CrossRef]
24. Chung, S.-K. A phase tracking system for three phase utility interface inverters. *IEEE Trans. Power Electron.* **2000**, *15*, 431–438. [CrossRef]
25. Golestan, S.; Monfared, M.; Freijedo, F.D. Design-Oriented Study of Advanced Synchronous Reference Frame Phase-Locked Loops. *IEEE Trans. Power Electron.* **2013**, *28*, 765–778. [CrossRef]
26. Neves, F.A.S.; Cavalcanti, M.C.; de Souza, H.E.P.; Bradaschia, F.; Bueno, E.J.; Rizo, M. A Generalized Delayed Signal Cancellation Method for Detecting Fundamental-Frequency Positive-Sequence Three-Phase Signals. *IEEE Trans. Power Deliv.* **2010**, *25*, 1816–1825. [CrossRef]
27. Neves, F.A.S.; de Souza, H.E.P.; Cavalcanti, M.C.; Peña, E. Low effort digital filters for fast sequence components separation of unbalanced and distorted three-phase signals. In Proceedings of the 2010 IEEE International Symposium on Industrial Electronics, Bari, Italy, 4–7 July 2010; pp. 2927–2932.

applied
sciences

Article

Linear Programming Coordination for Overcurrent Relay in Electrical Distribution Systems with Distributed Generation

Daniel Alcala-Gonzalez [1], Eva M. García del Toro [1], M. Isabel Más-López [2], Sara García-Salgado [1] and Santiago Pindado [3,*]

1   Departamento de Ingeniería Civil, Hidráulica y Ordenación del Territorio, ETSI Civil, Universidad Politécnica de Madrid, Alfonso XII, 3, 28014 Madrid, Spain; d.alcalag@upm.es (D.A.-G.); evamaria.garcia@upm.es (E.M.G.d.T.); sara.garcia@upm.es (S.G.-S.)
2   Departamento de Ingeniería Civil, Construcción Infraestructura y TransporteETSI Civil, Universidad Politécnica de Madrid Alfonso XII, 3, 28014 Madrid, Spain; mariaisabel.mas@upm.es
3   Instituto Universitario de Microgravedad "Ignacio Da Riva" (IDR/UPM), ETSI Aeronáutica y del Espacio, Universidad Politécnica de Madrid, Pza. del Cardenal Cisneros 3, 28040 Madrid, Spain
*   Correspondence: santiago.pindado@upm.es

Abstract: Electric power distribution networks are generally radial in nature, with unidirectional power flows transmitted from the highest voltage levels to the consumption levels. The protection system in these distribution networks is relatively simple and consists mainly of fuses, reclosers (RC) and overcurrent relays (OCRs). The installation of distributed generation (DG) in a network causes coordination problems between these devices, because the power flows are no longer unidirectional and can flow upstream to the substation. For this reason, the work proposed here analyzes the most significant impacts that DG has on the protection devices and proposes an adjustment method for the OCRs based on linear programming (LP) techniques with the aim of improving their response time to the different faults that may occur in the main feeder of the network. The distribution system selected for the study is the IEEE 34 bus system using DIgSILENT 14.1 software for its modeling and Matlab for the adjustment of the overcurrent devices. Results indicate that better coordination between protection devices are achieved if LP is used.

Keywords: adaptive protection; distributed power generation; power distribution; power system protection

Citation: Alcala-Gonzalez, D.; García del Toro, E.M.; Más-López, M.I.; García-Salgado, S.; Pindado, S. Linear Programming Coordination for Overcurrent Relay in Electrical Distribution Systems with Distributed Generation. *Appl. Sci.* **2022**, *12*, 4279. https://doi.org/10.3390/app12094279

Academic Editors: Luis Hernández-Callejo, Maria del Carmen Alonso García and Sara Gallardo Saavedra

Received: 27 March 2022
Accepted: 19 April 2022
Published: 23 April 2022

Publisher's Note: MDPI stays neutral with regard to jurisdictional claims in published maps and institutional affiliations.

1. Introduction

The current philosophy in the planning, management and control of a radial power distribution network is based on the assumption of the existence of unidirectional power flows, which is transmitted from the highest transport voltage levels to the distribution levels. We can assume that short-circuit currents behave similarly. These assumptions allow for the implementation of relatively simple and inexpensive protection schemes with which a selective operation of the protection system is achieved [1,2]. According to the principles of selectivity [3], only the protection device closest to the fault should operate to clear the fault, leaving the rest of the network energized.

The installation of DGs at medium and low voltage levels changes this fundamental basis. Power flows and short-circuit currents can now have different upstream directions and values [4]. As a consequence, the initial schemes implemented (e.g., main feeder protection) may no longer work or be less effective [1,5].

With the presence of DG the distribution system becomes more active, i.e., both load and generation significantly affect the state of the network [6]. The effects of DG on distribution network devices have been discussed in different literature [4,7]. To reduce them, several "mitigation methods" have been proposed, thus, several authors recommend acting directly on the DG [3,8], e.g., disconnecting it just before fault detection or limiting its power and therefore its effect on the protection system [9], others recommend reconfiguring the network topology [10,11], or the use of fault current limiters (FLCs) [12,13].

In recent years, mitigation methods based on coordination algorithms have been developed using linear programming (LP), non-linear programming (NPL), or genetic algorithms (GA) [14–17] that either modify the coordination interval time (CTI) between devices, determine the optimal location of the DG [15], or in combination, without compromising the protection system.

Although these methods can be effective, they have some disadvantages and limitations. Therefore, the disconnection of the DG immediately before fault detection proposed in [3,18], can cause asynchronous reconnections and cause severe damage to both the DG and the distribution network, especially at high penetration levels [3,10]. The limitation of the DG capacity proposed in [9] is also undesirable, since it also limits the penetration level and therefore the advantages of its installation close to the consumption points, (loss reduction, improvement of supply quality, etc.). The network reconfiguration suggested in [10,11] is costly and sometimes unfeasible. Finally, the use of current limiters [12,14] entails an additional installation cost that can be prohibitive.

Regarding the use of NLP methods for the determination of the coordination index mentioned in [15] can be very complex, particularly if the number of protection devices and DG's is high, since the calculation time required to determine their operating parameters will also be high [19–21], on the other hand, not always the optimal location point of DG is viability from the perspective of network operation, e.g., availability of power evacuation at the point of common connection (PCC).

The investigations described in [14,16,17], use GA-based optimization algorithms to determine optimal coordination between OCR and distance relays [14]. In [16] a hybrid GA and NLP algorithm is developed to choose the optimal value of DIAL or time multiplier setting (TMS) of all OCR, the use of a continuous genetic algorithm (GAC) faster than GA for the same purpose is developed in [17]. However, all the optimization algorithms described do not consider DG.

In the present work:

- the most significant impacts that DG causes on protection devices are analyzed,
- an adjustment method for OCRs based on LP techniques is proposed,
- the analysis is carried out by using the IEEE 34-node test feeder system modeled in DIgSILENT PowerFactory.

As far as the authors know, it seems that there is a lack of similar studies in the available literature.

2. Effect Caused by the DG on Protective Devices

2.1. Loss of Sensitivity

The problem related to the loss of sensitivity to main feeder protection is often referred to as protection blinding [8]. Installation of DGs in the distribution system can reduce the value of the short-circuit current detected by the protection of the main substation and affect the response time of the circuit breaker, which will depend, to a large extent, on the size and location of the DG within the distribution network.

The loss of sensitivity of the protection device can be analyzed as a function of where the short circuit is located in relation to the location of the DG, distinguishing between short circuits located upstream of the DG and short circuits located downstream of it. Figure 1 shows the topology of a radial distribution network where S.E is the substation, R is the main feeder protection, RC is the automatic recloser, and F is a protection fuse for the branch line.

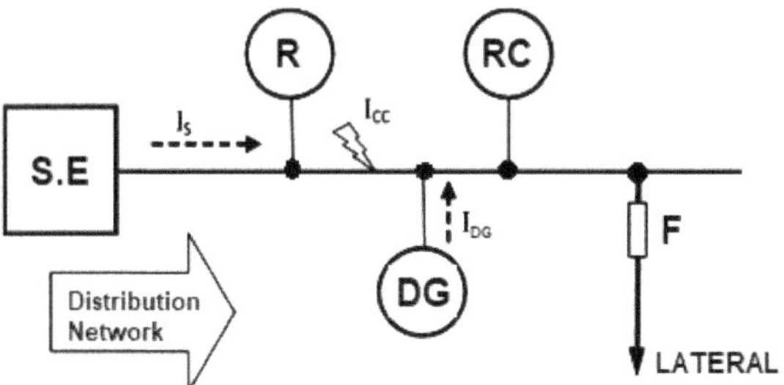

Figure 1. Sketch of a short circuit upstream of DG.

For faults located upstream (between the substation—S.E, and the DG), the contribution current of the S.E to the short-circuit is independent of the size of the DG. A three-phase short-circuit is also represented upstream of the DG, for different penetration levels: 17%, 33%, and 50%. In the operating characteristic curves of the time overcurrent (t/i) devices shown in Figure 2, this effect is observed, as the current detected by the relay remains constant at 1.59 kA, regardless of the of penetration level of the DG.

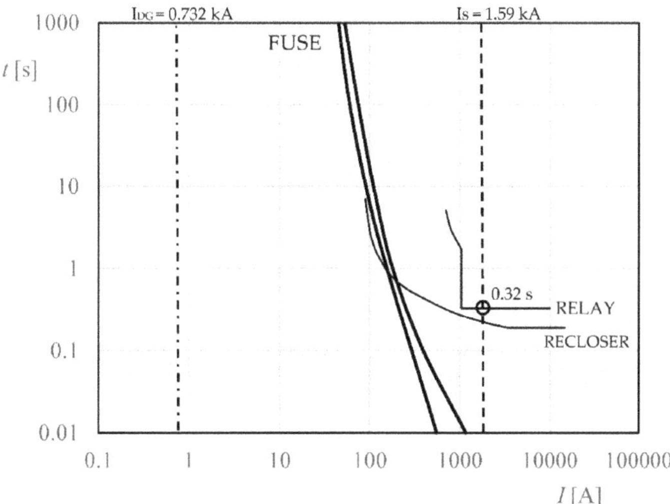

Figure 2. Short circuit upstream of DG. See also Figure 1.

In contrast, for faults located downstream of the substation and the DG (as shown in Figure 3), it is observed that the current contribution of the DG to the short circuit causes the current measured by the main relay, R, of the S.E to decrease from 1.21 with 17% penetration to 1.14 kA with 50% penetration level of the DG, which represents a 6.1% loss in the sensitivity of the protection device. This effect will cause a delay in the operating time of the protection device, as shown in Figure 4. The extreme case of this effect occurs when the main relay does not detect the fault current, as shown in Figure 5. The maximum value at which the relay stops detecting the fault current appears when the DG is installed close to the substation and under fault conditions at the main feeder [12].

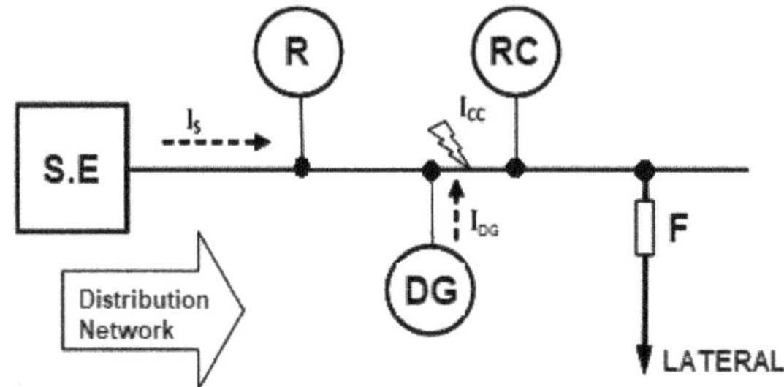

Figure 3. Sketch of a short circuit downstream of DG.

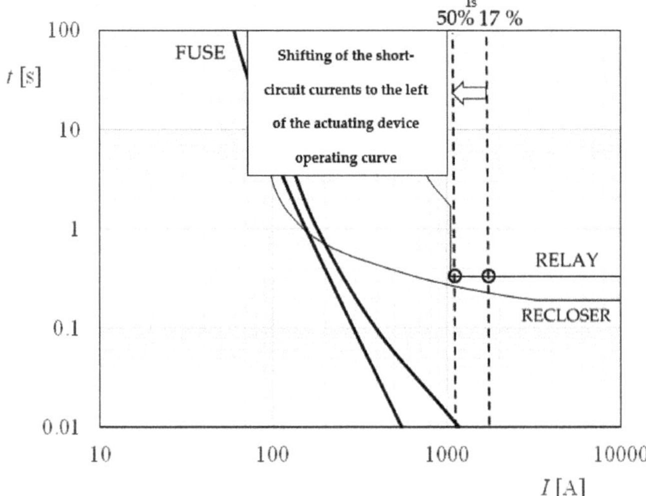

Figure 4. Short circuit downstream of DG. See also Figure 3.

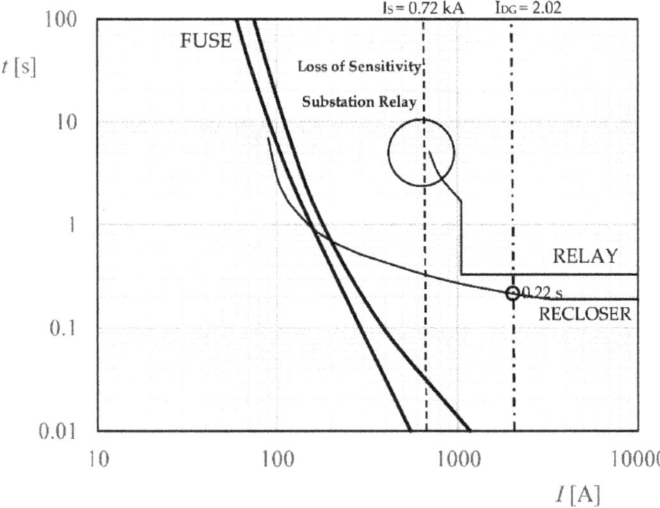

Figure 5. Short loss of sensitivity substation relay. Loss of sensitivity.

2.2. Loss of Coordination

Under normal operating conditions, protection devices are coordinated in such a way that, in the event of a fault in the network, the main protection acts before backup protection. A large majority of faults occurring in the network are of a temporary nature, the purpose of the recloser, RC, is to try to clear these faults and on the one hand avoid an unnecessary interruption of the electrical service, and on the other hand to safeguard the protection fuse of the branch line under its supervision. Depending on the initial settings of the coordination schemes, and taking into account the size, location, and type of DG installed, it may happen that a temporary fault in a branch line causes the loss of coordination between fuse and recloser, affecting their coordination time due to the additional current with which the DG contributes to the short-circuit. To observe this impact, the situation of temporary fault, I_{CC}, in a branch line (lateral) is shown in Figure 6.

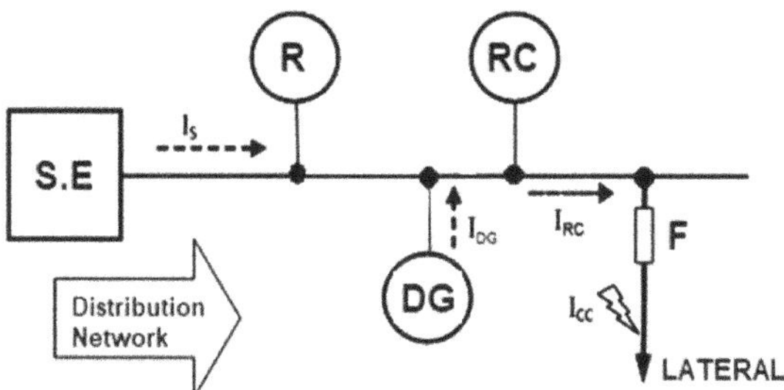

Figure 6. Short circuit in branch line.

The characteristic t/i trip curves are represented in Figure 7 where the impact called (fuse nuisance blowing) is shown. We can observe how a temporary fault in a branch line causes a loss of coordination between the fuse and the recloser. Thus, the fuse detects a fault current of 0.25 kA and trips in 0.15 s, while the recloser is crossed by a current of 0.650 kA, which would cause it to trip in 0.32 s. This time is higher than the fuse tripping.

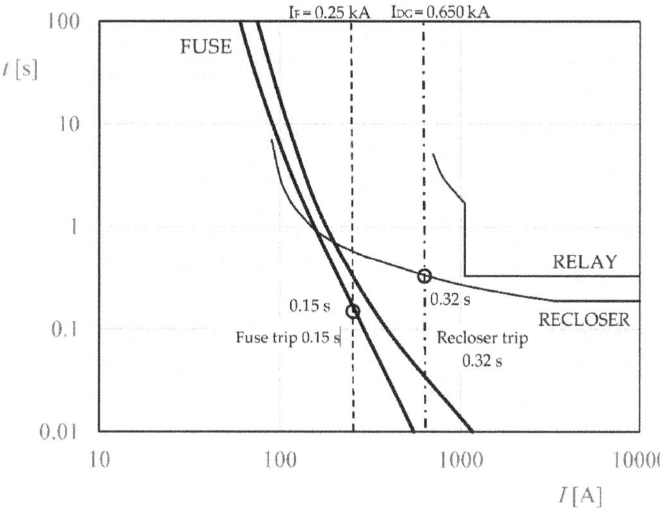

Figure 7. Failure coordination between recloser, RC, and fuse, F, see also Figure 6.

3. Coordination with Linear Programming

OCRs are currently the most widely used protection devices [13,22], this device generally being used as backup protection. However, in some situations, it may be the only protection. The OCR should act as primary protection by removing faults that occur in the area under its supervision. Only in the event that the primary protection fails to operate should the backup protection initiate tripping. A typical electrical distribution network may consist of hundreds of equipment protection relays. Each relay in the system must be coordinated with the adjacent equipment protection relay. If backup protection is not well coordinated, coordination failure may occur, and therefore, coordination of backup protections is one of the main concerns in the network protection system [13,23]. Generally, protection coordination can be performed by topology [24,25], by optimization methods [26,27], or by expert methods [28]. Topological analysis is used for relay tuning in multiterminal networks, graph theory, and functional approximation techniques are employed to provide the best solution that does not necessarily have to be the optimal one. In optimization methods, some researchers [26,29] use non-linear programming techniques to determine the optimal relay settings, subject to constraints due to coordination and limits of the relay settings themselves. In reference [30], the big M method is proposed to find the optimal value of the OCR time multiplier setting (TMS) in which the stated values of the plug setting are assumed to be known and fixed.

The problem of OCR coordination in the distribution system with DG presence can thus be defined as an optimization problem with constraints. The objective is to minimize the operating time of the relay closest to the place where the fault has occurred. The constraints imposed are due to limits on the operating time of the relay, coordination criteria, and relay characteristics. In this work, the OCR coordination problem in radial distribution systems is formulated as a linear programming (LP) problem with constraints, where the sum of the operating times of the relays in the system for different minimum fault points:

$$\min \frac{\alpha}{\left(k_{i,j}\right)^{\beta} - 1} \sum_{i=1}^{n} \left(\sum_{j=1}^{m} \mathrm{TMS}_{i,j} - \mathrm{TMS}_{i+1,j+1} \right), \tag{1}$$

where α and β are OCR shape constants according to Table 1, $k_{i,j}$ is the ratio between the short-circuit current, I_{CC}, and the relay setting current, I_{ar}, regarding relay i with fault in section j, and TMS is the time multiplier setting.

Table 1. Form constants for the exponential equation IEC [31].

Type of Curve	α	β
Standard inverse time	0.14	0.02
Very inverse	13.5	1
Extremely inverse	80	2
Long inverse time	120	1

The proposed optimization problem is subjected to the following group of constraints:

- Coordination criterion—The protection coordination criterion establishes the minimum time that must elapse between the operation of the primary protection and the operation of the backup protection. The fault is detected simultaneously by the primary protection, PP, and the secondary protection, PS. To avoid erroneous operation, the PS will only have to operate in case the PP fails. If we define R_i as the primary fault protection at a certain point, j, and R_{i+1} as the secondary or backup protection for the same fault. The constraint condition for coordination criteria is:

$$t_{i+1,j} - t_{i,j} > \Delta t, \tag{2}$$

where, $t_{i+1,j}$ is the operation time of the back-up, PS, operation for fault at point j, $t_{i,j}$ is the operation time of PP operation for the same fault, and Δt is the coordination time interval (CTI).

- Relay operating time limits—The time taken by a relay to detect and isolate a fault produced in its zone of influence must be bounded. This is the constraint imposed by the operating time of the relays:

$$t_{i,min} \leq t_{i,j} \leq t_{i,max} \tag{3}$$

where, $t_{i,min}$ and $t_{i,max}$ are the minimum and maximum operating times of relay i (at any point).

- Time multiplier setting, TMS, limits—The operating time of a relay is directly proportional to the TMS. Therefore:

$$\text{TMS}_{i,min} \leq \text{TMS}_{i,j} \leq \text{TMS}_{i,max}, \tag{4}$$

where $\text{TMS}_{i,min}$ is the minimum TMS value for relay i and, $\text{TMS}_{i,max}$ is maximum value of TMS for relay i. The TMS values usually taken are 0.025–1.2, respectively [32].

- Relay operating characteristics—To extend the work region of the overcurrent relay specified by the standard [33], the relay's parameters are optimized considering the maximum value of the relay and the minimum value of the relay, as expressed respectively. We consider all relays with the same characteristics:

$$t_{op} = \frac{\alpha}{\text{PSM}^{\beta} - 1}\text{TMS}, \tag{5}$$

where t_{op} is the relay operating time and PSM is the plug setting multiplier ($k_{i,j}$ in Equation (1)). For inverse characteristic curves it is usually $\alpha = 0.02$ and $\beta = 0.14$ [32,33].

The relay setting current, I_{ar}, is determined from the system requirements. Thus, the above equation can be rewritten as:

$$t_{op} = a \times \text{TMS}, \tag{6}$$

where:

$$a = \frac{\alpha}{\text{PMS}^{\beta} - 1}. \tag{7}$$

4. Case Study Results and Discussion

The distribution system selected for the simulations is the one taken from the test feeders of the Distribution System Analysis Subcommittee of the Institute of Electrical and Electronics Engineers (IEEE) [34], shown in Figure 8. The main characteristics of the distribution network on which the short circuits were simulated are the following:

- The voltage across the distribution network is 15 kV.
- The short-circuit power of the external network will be 100 MVA.
- The power demand of the electrical loads connected to the network is 12 MW and load flow analysis they have been considered balanced.
- Overhead network with double circuit with LA-110 conductor. The electrical characteristics of this conductor are shown in [35].

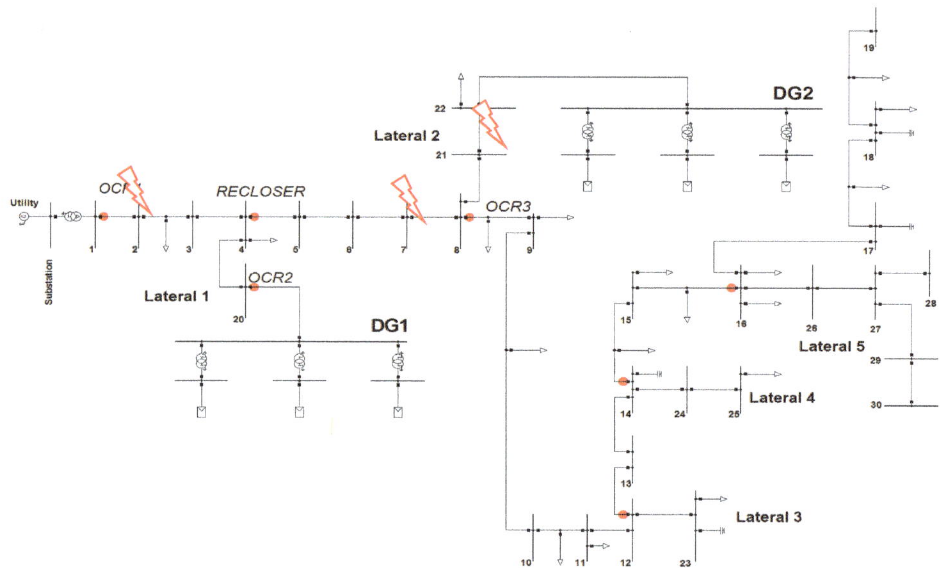

Figure 8. Modified IEEE 34-node test feeder system [34].

Three-phase and single-phase short circuits were simulated on the distribution network, at busbars 2, 7, and 21, as shown in Figure 8, and with different DG penetration levels ranging from 33% to 50%. A first distributed generation unit, DG1, was installed at busbar 20, whereas a second one, DG2, was installed at busbar 22.

The DG protection device should be properly coordinated with the other protection devices in the distribution network. When a fault occurs on the line to which the DG unit is connected, the DG unit must be disconnected before the protection device on that line is disconnected. This ensures that the DG does not cause disturbances in the operation of the line protections. Furthermore, when a fault occurs on a branch line, the DG must remain in service and will not disconnect before the fault line does [8]. According to this, the coordination criterion shown in Table 2 is proposed.

Table 2. Coordination criteria according fault location.

Bus (Node)	Location Faults	Main Protection	Backup Protection
2	F2	OCR2	OCR1
7	F7	RC	OCR1
21	F21	RC	OCR3

The transformation ratio of the current transformers used in the test is 400/5, chosen according to the full-load currents of the system and the short-circuit currents detected by each of the protection devices according to the location of the fault (shown in Table 3).

Table 3. Short circuit (expressed in amperes, A) detected by the protection devices, for different grounding resistor values, R_f, expressed in ohm.

Fault Type		F2 Penetration Level DG 17%	F2 Penetration Level DG 33%	F2 Penetration Level DG 50%	F7 Penetration Level DG 17%	F7 Penetration Level DG 33%	F7 Penetration Level DG 50%	F21 Penetration Level DG 17%	F21 Penetration Level DG 33%	F21 Penetration Level DG 50%
Three-phase	I_{OCR1}	1568	1568	1568	1193	1116	1104	321	325	328
	$I_{OCR2(DG)}$	625	894	1077	570	778	919	392	407	417
	I_{OCR3}	-	-	-	-	-	-	361	368	373
	I_{RC}	-	-	-	1717	1922	2052	667	681	689
Single-phase $R_f = 0$	I_{OCR1}	539	608	665	505	556	575	404	397	395
	$I_{OCR2(DG)}$	710	1076	1351	670	960	1165	480	465	474
	I_{OCR3}	30.8	30	29	31	30	29	230	238	241
	I_{RC}	644	627	619	1078	1445	1662	728	746	754
$R_f = 5$	I_{OCR1}	564	604	623	517	533	532	398	195	393
	$I_{OCR2(DG)}$	736	943	1059	669	827	918	446	461	469
	I_{OCR3}	32.45	33	33	31.6	32	32	211	216	219
	I_{RC}	677	697	706	1060	1263	1359	733	748	754
$R_f = 10$	I_{OCR1}	523	535	535	490	488	462	396	393	464
	$I_{OCR2(DG)}$	664	677	820	616	703	751	443	456	391
	I_{OCR3}	33.3	33	33.9	32	33	33	194	199	200
	I_{RC}	695	705	708	1000	1103	1146	735	747	753
$R_f = 15$	I_{OCR1}	488	491	488	466	459	452	394	391	389
	I_{OCR2}	605	667	699	572	628	658	440	452	459
	I_{OCR3}	33	33	33.6	32	33	33	180	184	195
	I_{RC}	698	703	699	944	1004	1028	735	746	751
$R_f = 20$	I_{OCR1}	464	464	461	448	440	434	392	448	387
	$I_{OCR2(DG)}$	563	607	629	540	580	602	437	389	454
	I_{OCR3}	33	33.5	33	33	33	33	168	171	172
	I_{RC}	697	699	699	900	040	956	734	744	748
$R_f = 25$	I_{OCR1}	447	446	443	435	427	422	390	387	385
	$I_{OCR2(DG)}$	534	567	598	517	548	565	434	444	456
	I_{OCR3}	33	33.3	33.3	33	33	33	158	160	161
	I_{RC}	696	697	696	869	897	908	733	742	745

4.1. Objective Function According to Fault Location

According to the location of the fault, and based on the protection criteria included in Table 2, three objective functions to minimize can be obtained, each of them will result in a different adjustment, an adaptive adjustment.

The scenario analyzed is the three-phase short-circuit scenario with a DG1 penetration level of 17%, the results obtained for the other scenarios can be found in Table A3 in the Appendix A.

Fault 2: OCR$_2$ with OCR$_1$

$$\min z = \sum_{i=1}^{m} a_{i,k}(\text{TMS})_i = 2.27\,\text{TMS}_1 + 3.33\,\text{TMS}_2 \tag{8}$$

Fault 7: Recloser with OCR$_2$

$$\min z = \sum_{i=1}^{m} a_{i,k}(\text{TMS})_i = 3.49\,\text{TMS}_2 + 2.21\,\text{TMS}_{RC2} \tag{9}$$

Fault 21: OCR$_3$ with recloser

$$\text{minz} = \sum_{i=1}^{m} a_{i,k}(\text{TMS})_i = 4.57\,\text{TMS}_3 + 3.23\,\text{TMS}_{RC} \tag{10}$$

Constraints:

- The minimum operating time of each relay is 0.1 s.
- The normal range of TMS from 0.025 to 1.2.
- Typical CTI setting is 0.3 s.

Coordination criterion
Fault 2: OCR_2 with OCR_1:

$$2.27\text{TMS}_1 - 3.33\text{TMS}_2 \geq 0.3 \tag{11}$$

Fault 7: Recloser with OCR_2:

$$3.49\text{TMS}_2 - 2.21\text{TMS}_{RC} \geq 0.3 \tag{12}$$

Fault 21: OCR_3 with Recloser:

$$4.57\text{TMS}_3 - 3.23\text{TMS}_{RC} \geq 0.3 \tag{13}$$

Relay operating time limits
Fault 2: OCR_2 with OCR_1:

$$\begin{aligned} 2.27\text{TMS}_1 \geq 0.1\ \text{TMS}_1 \leq 0.044 \\ 3.33\text{TMS}_2 \geq 0.1\ \text{TMS}_2 \leq 0.030 \end{aligned} \tag{14}$$

Fault 7: Recloser with OCR_2:

$$\begin{aligned} 3.49\text{TMS}_2 \geq 0.1\ \text{TMS}_2 \leq 0.038. \\ 2.21\text{TMS}_{RC} \geq 0.1\ \text{TMS}_{RC} \leq 0.045 \end{aligned} \tag{15}$$

Fault 21: OCR_3 with recloser:

$$\begin{aligned} 4.57\text{TMS}_3 \geq 0.1\ \text{TMS}_3 \leq 0.021. \\ 3.23\text{TMS}_{RC} \geq 0.1\ \text{TMS}_{RC} \leq 0.030 \end{aligned} \tag{16}$$

The normal range of TMS is from 0.025 to 1.2:

$$\begin{aligned} \text{TMS}_1 &\geq 0.025 \\ \text{TMS}_2 &\geq 0.025 \\ \text{TMS}_3 &\geq 0.025 \\ \text{TMS}_{RC} &\geq 0.025 \end{aligned} \tag{17}$$

4.2. Relay Operating Time Calculation

Using linear programming, the TMS value was calculated as a function of the fault location point.

Substituting the TMS value and the obtained value of the relay constant in Equation (4), we obtain the operating time of each protection relay.

Fault 2: OCR_2 with OCR_1:

$$\begin{aligned} \text{OCR1}\ t_{op1} &= a_1(\text{TMS}_1) = 2.27 \cdot 0.17 = 0.39\ \text{s} \\ \text{OCR2}\ t_{op2} &= a_2(\text{TMS}_2) = 3.33 \cdot 0.030 = 0.10\ \text{s} \end{aligned} \tag{18}$$

Fault 7: Recloser with OCR_2:

$$\begin{aligned} \text{OCR2}\ t_{op2} &= a_2(\text{TMS}_2) = 3.49 \cdot 0.038 = 0.13\ \text{s} \\ \text{RECLOSER}\ t_{opRC} &= a_{RC}(\text{TMS}_{RC}) = 2.21 \cdot 0.11 = 0.24\ \text{s} \end{aligned} \tag{19}$$

Fault 21: OCR_3 with Recloser

$$OCR3\ t_{op3} = a_3(TMS_3) = 4.57 \cdot 0.065 = 0.30\ s$$
$$RECLOSER\ t_{opRC} = a_{RC}(TMS_{RC}) = 3.23 \cdot 0.025 = 0.1\ s \tag{20}$$

The values of constants a and TMS and relay operation times *top* for the rest of the scenarios are shown in Appendix A Tables A1 and A2. Table A3, shows a comparison of the operation times between the coordination performed in a classical way and with the calculation of adjustments carried out by using LP. Implementing the results obtained in the simulation software, we obtain the following protection schemes shown below.

The classical coordination plots versus linear programming coordination in the case of three-phase short-circuits in the locations described with 17% DG1, are shown in Figures 9–11. It can be seen that the response of the OCR protection devices is optimized using the method proposed in this work.

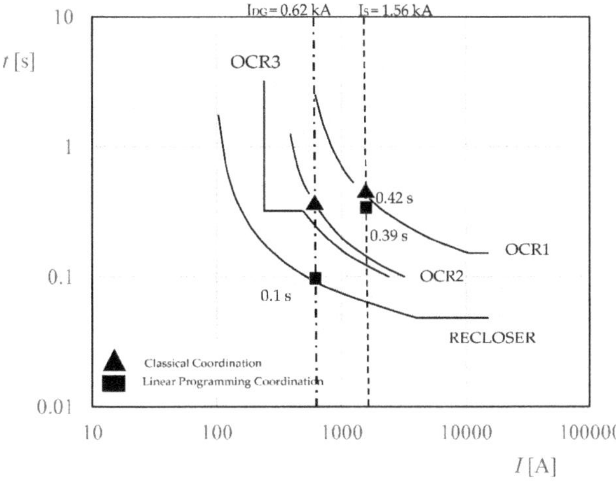

Figure 9. Short-circuit in F2 with penetration level of DG1 17 %. Classical coordination versus linear programming coordination.

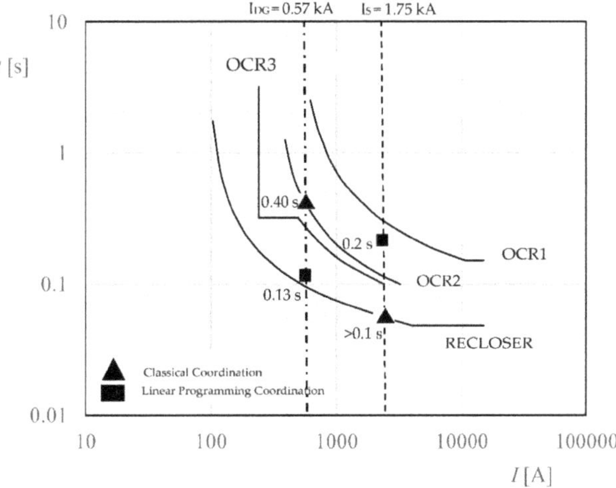

Figure 10. Short-circuit in F7 with penetration level of DG1 17%. Classical coordination versus al linear programming coordination.

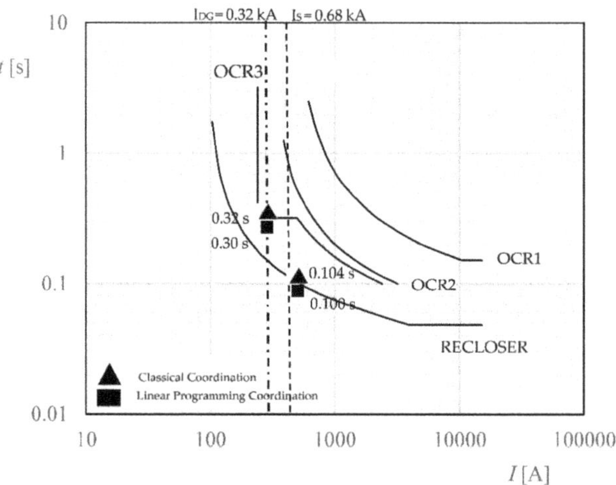

Figure 11. Short-circuit in F21 with penetration level of DG1 17%. Classical coordination versus linear programming coordination.

5. Conclusions

In this work, a unified protection system that offers double functionality is proposed. The system was proven to be capable of both optimizing the settings of the relays for each network situation.

The performance of the relay has been tested for different DG penetration levels, different types of short-circuit faults (single-phase to ground and three-phase), and different fault zones (faults in the main feeder and the laterals).

The proposed scheme can be implemented in distribution systems using the communication standard IEC61850 [31].

In all the cases studied, the system has proven capable of adapting, in real time, the performance parameters of the relays that protect the faulted line section depending on the operating conditions of the system at the time of the fault.

The proposed protection system has been applied to a distribution system in the presence of DG. The results demonstrate that the protection coordination proposed in this paper is capable of reducing relay operating times by more than 80% compared to classical coordination, which produces a more reliable operation of a network with DG.

Author Contributions: Conceptualization, E.M.G.d.T. and M.I.M.-L.; methodology, D.A.-G., E.M.G.d.T., M.I.M.-L., S.G.-S. and S.P.; software, D.A.-G.; validation, S.G.-S., E.M.G.d.T. and M.I.M.-L.; formal analysis, E.M.G.d.T., M.I.M.-L. and D.A.-G.; investigation, D.A.-G., E.M.G.d.T., M.I.M.-L., S.G.-S. and S.P.; data curation, E.M.G.d.T. and M.I.M.-L.; writing—original draft preparation, D.A.-G., E.M.G.d.T., S.G.-S., M.I.M.-L. and S.P.; writing—review and editing, E.M.G.d.T. and M.I.M.-L.; supervision, S.G.-S., M.I.M.-L. and S.P. All authors have read and agreed to the published version of the manuscript.

Funding: This research received no external funding.

Institutional Review Board Statement: Not applicable.

Informed Consent Statement: Not applicable.

Data Availability Statement: Not applicable.

Conflicts of Interest: The authors declare no conflict of interest.

Appendix A

Table A1. Values of constant relay ($a_{i,j}$) for the different scenarios analyzed.

Type of Fault			F2			F7			F21		
			$a_{17\%}$	$a_{33\%}$	$a_{50\%}$	$a_{17\%}$	$a_{33\%}$	$a_{50\%}$	$a_{17\%}$	$a_{33\%}$	$a_{50\%}$
Three-phase		I_{OCR1}	2.27	2.27	2.27	2.52	2.58	2.59	4.96	4.92	4.89
		$I_{OCR2(DG)}$	3.33	2.83	2.62	3.49	3.00	2.79	4.33	4.23	4.17
		I_{OCR3}	-	-	-	-	-	-	4.57	4.51	4.47
		I_{RC}	-	-	-	2.21	2.13	2.08	3.23	3.19	3.18
Single-phase	$R_f = 0$	I_{OCR1}	3.59	3.38	3.23	3.72	3.54	3.47	4.25	4.30	4.31
		$I_{OCR2(DG)}$	3.13	2.62	2.4	3.22	2.74	2.54	3.83	3.90	3.86
		I_{OCR3}	-	-	-	-	-	-	6.55	6.35	6.27
		I_{RC}	3.28	3.33	3.35	2.62	2.34	2.23	3.10	3.06	3.05
	$R_f = 5$	I_{OCR1}	3.51	3.39	3.34	3.68	3.63	3.62	4.29	7.78	4.32
		$I_{OCR2(DG)}$	3.08	2.76	2.64	3.22	2.92	2.79	4.00	3.92	3.88
		I_{OCR3}	-	-	-	-	-	-	7.14	6.97	6.88
		I_{RC}	3.2	3.16	3.14	2.63	2.46	2.40	3.09	3.06	3.05
	$R_f = 10$	I_{OCR1}	3.65	3.61	3.61	3.79	3.8	3.92	4.30	4.32	3.91
		$I_{OCR2(DG)}$	3.23	3.2	2.93	3.35	3.15	3.05	4.02	3.95	4.34
		I_{OCR3}	-	-	-	-	-	-	7.83	7.61	7.56
		I_{RC}	3.16	3.14	3.14	2.7	2.59	2.56	3.08	3.06	3.05
	$R_f = 15$	I_{OCR1}	3.8	3.78	3.8	3.9	3.93	3.97	4.32	4.34	4.35
		$I_{OCR2(DG)}$	3.39	3.23	3.15	3.48	3.32	3.25	4.03	3.97	3.93
		I_{OCR3}	-	-	-	-	-	-	8.56	8.33	7.78
		I_{RC}	3.16	3.15	3.15	2.76	2.69	2.67	3.08	3.06	3.05
	$R_f = 20$	I_{OCR1}	3.91	3.91	3.92	3.99	4.03	4.06	4.33	3.99	4.37
		$I_{OCR2(DG)}$	3.51	3.38	3.32	3.59	3.46	3.39	4.05	4.35	3.96
		I_{OCR3}	-	-	-	-	-	-	9.36	9.14	9.07
		I_{RC}	3.16	3.15	3.15	2.82	2.77	2.75	3.08	3.06	3.06
	$R_f = 25$	I_{OCR1}	3.99	4.00	4.02	4.06	4.11	4.13	4.34	4.37	4.38
		$I_{OCR2(DG)}$	3.61	3.5	3.41	3.68	3.56	3.51	4.06	4.01	3.95
		I_{OCR3}	-	-	-	-	-	-	10.2	10.0	9.93
		I_{RC}	3.16	3.16	3.16	2.86	2.82	2.81	3.09	3.07	3.06

Table A2. Relay TMS values for the different scenarios analyzed.

Type of Fault			F2			F7			F21		
			$TMS_{17\%}$	$TMS_{33\%}$	$TMS_{50\%}$	$TMS_{17\%}$	$TMS_{33\%}$	$TMS_{50\%}$	$TMS_{17\%}$	$TMS_{33\%}$	$TMS_{50\%}$
Three-phase		I_{OCR1}	0.17	0.17	0.17	-	-	-	-	-.	-
		$I_{OCR2(DG)}$	0.03	0.04	0.04	0.04	0.03	0.04	-	-	-
		I_{OCR3}	-	-	-	-	-	-	0.08	0.09	0.09
		I_{RC}	-	-	-	0.19	0.18	0.19	0.03	0.03	0.03
Single-phase	$R_f = 0$	I_{OCR1}	0.11	0.11	0.12	-	-	-	-	-	-
		$I_{OCR2(DG)}$	0.03	0.04	0.04	0.03	0.04	-	-	-	-
		I_{OCR3}	-	-	-	-	-	0.06	0.06	0.06	0.06
		I_{RC}	-	-	-	0.15	0.14	0.03	0.03	0.03	0.03
	$R_f = 5$	I_{OCR1}	0.11	0.11	0.11	-	-	-	-	-	-
		$I_{OCR2(DG)}$	0.03	0.04	0.04	0.03	0.03	-	-	-	-
		I_{OCR3}	-	-	-	-	-	0.06	0.06	0.07	0.07
		I_{RC}	-	-	-	0.15	0.01	0.03	0.03	0.03	0.03

Table A2. *Cont.*

			F2			**F7**			**F21**		
Type of Fault			$TMS_{17\%}$	$TMS_{33\%}$	$TMS_{50\%}$	$TMS_{17\%}$	$TMS_{33\%}$	$TMS_{50\%}$	$TMS_{17\%}$	$TMS_{33\%}$	$TMS_{50\%}$
Single-phase	$R_f = 10$	I_{OCR1}	0.10	0.11	0.11	-	-	-	-	-	-
		$I_{OCR2(DG)}$	0.03	0.03	0.04	0.03	0.03	-	-	-	-
		I_{OCR3}	-	-	-	-	-	0.05	0.05	0.06	0.06
		I_{RC}	-	-	-	0.14	0.01	0.03	0.03	0.03	0.03
	$R_f = 15$	I_{OCR1}	0.10	0.10	0.10	-	-	-	-	-	-
		$I_{OCR2(DG)}$	0.03	0.03	0.03	0.03	-	-	-	-	-
		I_{OCR3}	-	-	-	-	-	0.05	0.05	0.05	0.05
		I_{RC}	-	-	-	0.14	-	0.03	0.03	0.03	0.03
	$R_f = 20$	I_{OCR1}	0.10	0.10	0.10	-	-	-	-	-	-
		$I_{OCR2(DG)}$	0.03	0.03	0.03	0.03	-	-	-	-	-
		I_{OCR3}	-	-	-	-	-	0.04	0.04	0.05	0.05
		I_{RC}	-	-	-	0.14	-	0.03	0.03	0.03	0.03
	$R_f = 25$	I_{OCR1}	0.10	0.10	0.10	-	-	-	-	-	-
		$I_{OCR2(DG)}$	0.03	0.03	0.03	0.03	-	-	-	-	-
		I_{OCR3}	-	-	-	-	-	0.04	0.04	0.05	0.04
		I_{RC}	-	-	-	0.13	-	0.03	0.03	0.03	0.03

Table A3. Operation time PL coordination versus operation time from classical coordination.

			F2						**F7**						**F21**					
Type of Fault		Type Coordination	t_{op} 17%		t_{op} 33%		t_{op} 50%		t_{op} 17%		t_{op} 33%		t_{op} 50%		t_{op} 17%		t_{op} 33%		t_{op} 50%	
			C	PL	C	PL	C	PL	C	PL	C	PL	C	PL	C	PL	C	PL	C	PL
Three-phase		I_{OCR1}	0.42	**0.39**	0.42	**0.39**	0.42	**0.39**	-	-	-	-	-	-	-	-	-	-	-	-
		$I_{OCR2(GD)}$	0.20	**0.20**	0.19	**0.10**	0.15	**0.08**	0.20	0.13	0.20	0.10	0.20	0.10	-	-	-	-	-	-
		I_{OCR3}	-	-	-	-	-	-	-	-	-	-	-	-	0.42	**0.38**	0.42	**0.40**	0.42	**0.40**
		I_{RC}	-	-	-	-	-	-	0.45	**0.42**	0.45	**0.38**	0.45	**0.40**	0.12	**0.08**	0.12	**0.10**	0.12	**0.10**
Single-phase	$R_f = 0$	I_{OCR1}	0.42	**0.39**	0.43	**0.37**	0.42	**0.39**	-	-	-	-	-	-	-	-	-	-	-	-
		$I_{OCR2(DG)}$	0.14	**0.10**	0.14	**0.10**	0.14	**0.10**	0.14	**0.10**	0.14	**0.10**	0.14	**0.13**	-	-	-	-	-	-
		I_{OCR3}	-	-	-	-	-	-	-	-	-	-	-	-	0.40	**0.39**	0.14	**0.38**	0.14	**0.38**
		I_{RC}	-	-	-	-	-	-	0.40	**0.39**	0.40	**0.33**	0.40	**0.31**	0.40	**0.10**	0.40	**0.09**	0.40	**0.09**
	$R_f = 5$	I_{OCR1}	0.41	**0.39**	0.42	**0.37**	0.41	**0.37**	-	-	-	-	-	-	-	-	-	-	-	-
		$I_{OCR2(DG)}$	0.14	**0.10**	0.14	**0.10**	0.14	**0.10**	0.14	**0.10**	0.14	**0.10**	0.14	**0.08**	-	-	-	-	-	-
		I_{OCR3}	-	-	-	-	-	-	-	-	-	-	-	-	0.39	**0.43**	0.45	**0.49**	0.45	**0.48**
		I_{RC}	-	-	-	-	-	-	0.40	**0.39**	0.40	**0.3**	0.40	**0.40**	0.40	**0.09**	0.40	**0.09**	0.40	**0.09**
	$R_f = 10$	I_{OCR1}	0.41	**0.37**	0.42	**0.40**	0.41	**0.40**	-	-	-	-	-	-	-	-	-	-	-	-
		$I_{OCR2(DG)}$	0.14	**0.10**	0.14	**0.10**	0.14	**0.11**	0.14	**0.10**	0.14	**0.09**	0.14	**0.09**	-	-	-	-	-	-
		I_{OCR3}	-	-	-	-	-	-	-	-	-	-	-	-	-	**0.39**	-	**0.46**	-	**0.45**
		I_{RC}	-	-	-	-	-	-	0.40	**0.38**	0.40	**0.3**	0.40	**0.3**	0.40	**0.10**	0.4	**0.09**	0.40	**0.09**
	$R_f = 15$	I_{OCR1}	0.40	**0.38**	0.41	**0.38**	0.40	**0.38**	-	-	-	-	-	-	-	-	-	-	-	-
		$I_{OCR2(DG)}$	0.14	**0.10**	0.14	**0.10**	0.14	**0.09**	0.14	**0.10**	0.14	**0.10**	0.14	**0.10**	-	-	-	-	-	-
		I_{OCR3}	-	-	-	-	-	-	-	-	-	-	-	-	0.45	**0.39**	-	**0.42**	-	**0.39**
		I_{RC}	-	-	-	-	-	-	0.40	**0.39**	0.40	**0.3**	0.40	**0.3**	0.40	**0.09**	-	**0.09**	-	**0.09**
	$R_f = 20$	I_{OCR1}	0.40	**0.39**	0.41	**0.39**	0.40	**0.39**	-	-	-	-	-	-	-	-	-	-	-	-
		$I_{OCR2(DG)}$	0.14	**0.10**	0.14	**0.10**	0.14	**0.11**	0.14	**0.10**	0.14	**0.10**	0.14	**0.10**	-	-	-	-	-	-
		I_{OCR3}	-	-	-	-	-	-	-	-	-	-	-	-	-	**0.39**	-	**0.46**	-	**0.45**
		I_{RC}	-	-	-	-	-	-	0.40	**0.39**	0.40	**0.3**	0.40	**0.03**	0.40	**0.09**	0.40	**0.09**	0.40	**0.09**
	$R_f = 25$	I_{OCR1}	0.42	**0.40**	0.40	**0.40**	0.40	**0.40**	-	-	-	-	-	-	-	-	-	-	-	-
		$I_{OCR2(DG)}$	0.14	**0.10**	0.14	**0.10**	0.14	**0.11**	0.14	**0.10**	0.14	**0.11**	0.14	**0.11**	-	-	-	-	-	-
		I_{OCR3}	-	-	-	-	-	-	-	-	-	-	-	-	-	**0.40**	-	**0.50**	-	**0.40**
		I_{RC}	-	-	-	-	-	-	0.40	**0.37**	0.40	**0.3**	0.40	**0.03**	0.40	**0.09**	0.40	**0.09**	0.40	**0.09**

C—classical coordination; **PL**—lineal programming coordination.

References

1. Dugan, R.C.; McDermott, T.E. Distributed generation. *IEEE Ind. Appl. Mag.* **2002**, *8*, 19–25. [CrossRef]
2. Salman, S.; Rida, I. Investigating the impact of embedded generation on relay settings of utilities electrical feeders. *IEEE Trans. Power Deliv.* **2001**, *16*, 246–251. [CrossRef]
3. Anderson, P.M. *Power System Protection*; Sons, J.W., Ed.; IEEE Press: Piscataway, NJ, USA, 1998.
4. Girgis, A.; Brahma, S.M. Effect of distributed generation on protective device coordination in distribution system. In Proceedings of the Large Engineering Systems Conference of Power Engineering, Halifax, NS, Canada, 11–13 July 2001.
5. Scheepers, M.; van Werven, M.; Mutale, J.; Strbac, G.; Porter, D. Distributed generation in electricity markets, its impact on distribution system operators, and the role of regulatory and commercial arrangements. *Int. J. Distrib. Energy Resour.* **2006**, *2*, 35–38.
6. Jenkins, N. Embedded generation. *Power Eng. J.* **1995**, *9*, 145–150. [CrossRef]
7. Doyle, M.T. Reviewing the impacts of distributed generation on distribution system protection. *Power Eng. Soc. Summer Meet. IEEE* **2002**, *1*, 103–105.
8. Repo, S.; Jarventausta, P.; Maki, K. Protection requirement graph for interconnection of distributed generation on distribution level. *Int. J. Energy Issues* **2007**, *28*, 47–64.
9. Girgis, A.; Fallon, C.M.; Lubkeman, D.L. A fault location technique for rural distribution feeders. *IEEE Trans. Ind. Appl.* **1993**, *29*, 1170–1175. [CrossRef]
10. Das, R. Estimating locations of shunt faults on distribution lines. Master's Thesis, College of Graduate Studies and Research, University of Saskatchewan, Saskatoon, SK, Canada, 1995.
11. Das, R.; Sachdev, M.S.; Sidhu, T.S. A technique for estimating locations of shunt faults on distribution lines. In Proceedings of the IEEE WESCANEX 95. Communications, Power, and Computing. Conference Proceedings, Winnipeg, MB, Canada, 15–16 May 1995; Volume 1, pp. 6–11.
12. Tarek, K.; Girgis, A. Protection Coordination Planning with Distributed Generation. Master's Thesis, CETC Varennes—Energy Technology and Programs Sector, Varennes, QC, Canada, 2007.
13. Warrington, A.R.V.C. *Protective Relays: Their Theory and Practice*; Wiley, J., York, S.N., Eds.; Springer: Cham, Switzerland, 1969.
14. Sant, M.T.; Paithankar, Y.G. Online digital fault locator foroverhead transmission line. *Proc. Inst. Electr.* **1979**, *126*, 1181–1185. [CrossRef]
15. Sant, M.T.; Paithankar, Y. Fault locator for long ehv transmission lines. *Electr. Power Syst. Res.* **1983**, *6*, 305–310. [CrossRef]
16. Takagi, T.; Yamakoshi, Y.; Yamaura, M.; Kondow, R.; Matsushima, T. Development of a new type fault locator using the One-Terminalvoltage and current data. *IEEE Trans. Power Appar. Syst.* **1982**, *PAS-101*, 2892–2898. [CrossRef]
17. Edmond, S., III. Evaluation and development of transmission line fault-locating techniques which use sinusoidal steady-state. *Comput. Electr. Eng.* **1983**, *10*, 269–278.
18. Johns, A.T.; Lai, L.L.; El-Hami, M.; Daruvala, D.J. New approach to directional fault location for overhead power distribution feeders. *Gener. Transm. Distrib. IEE Proc. C* **1991**, *138*, 351–357. [CrossRef]
19. Jang, J.S. ANFIS: Adaptive-network-based fuzzy inference system. *IEEE Trans. Syst. Man Cybern.* **1993**, *23*, 665–685. [CrossRef]
20. Russell, S.; Norvig, P. *Artificial Intelligence: A Modern Approach*, 2nd ed.; Prentice Hall: Hoboken, NJ, UK, 2002.
21. Brand, C.B.K.; Wimmer, W. *Design of IEC 61850 Based Substation Automation Systems According to Customer Requirements*; Paper b5-103; CIGRE: Paris, France, 2004.
22. Paithankar, Y.G. *Transmission Network Protection: Theory and Practice*; Routledge: Oxfordshire, UK, 1997.
23. Noghabi, A.S.; Sadeh, J.; Mashhadi, H.R. Considering different network topologies in optimal overcurrent relay coordination using a hybrid GA. *IEEE Trans. Power Deliv.* **2019**, *24*, 1857–1863. [CrossRef]
24. Knable, A. A standardized approach to relay coordination. *IEEE Winter Power Meet.* **1969**, *69*, 58.
25. Jenkins, L.; Khicha, H.; Shivakumar, S.; Dash, P. An application of function dependencies to the topological analysis of protection schemes. *IEEE Trans. Power Deliv.* **1992**, *7*, 77–83. [CrossRef]
26. Urdaneta, A.J.; Nadira, R.; Perez Jimenez, L.G. Optimal coordination of directional overcurrent relays in interconnected power systems. *IEEE Trans. Power Deliv.* **1988**, *3*, 903–911. [CrossRef]
27. Chattopadhyay, B.; Sachdev, M.S.; Sidhu, T.S. An on-line relay coordination algorithm for adaptive protection using linear programming technique. *IEEE Trans. Power Deliv.* **1996**, *11*, 165–173. [CrossRef]
28. So, C.W.; Li, K.K. Time coordination method for power system protection by evolutionary algorithm. *IEEE Trans. Ind. Appl.* **2000**, *36*, 1235–1240. [CrossRef]
29. Laway, N.A.; Gupta, H.O. A method for adaptive coordination of overcurrent relays in an interconnected power system. In Proceedings of the Fifth International Conference on Developments in Power System Protection, York, UK, 30 March–2 April 1993; pp. 240–243.
30. Bedekar, P.; Bhide, S.; Kale, V. Optimum time coordination of overcurrent relays in distribution system using big-m (penalty) method. *WSEAS Trans. Power Syst.* **2019**, *4*, 341–350.
31. *IEC 61850-SER*; Communication Networks and Systems in Substations. All Parts. IEC Std: Geneva, Switzerland, 2005.
32. Soman, S. Lectures on Power System Protection. Module 5, Lecture 19. NPTEL Online. 2010. Available online: www.cdeep.iitb.ac.in (accessed on 26 March 2022).

33. Keil, T.; Jager, J. Advanced coordination method for overcurrent protection relays using nonstandard tripping characteristics. *IEEE Trans. Power Deliv.* **2018**, *23*, 52–57. [CrossRef]
34. IEEE. *34 Node Test Feeder*; The Institute of Electrical and Electronics Engineers, Inc.: Piscataway, NJ, USA, 2004.
35. *Proyecto Tipo Unión Fenosa Líneas Eléctricas Aéreas de Hasta 20 kV*; UFD: Madrid, Spain, 2017.

applied sciences

Article

Effect of Distributed Photovoltaic Generation on Short-Circuit Currents and Fault Detection in Distribution Networks: A Practical Case Study

Daniel Alcala-Gonzalez [1], Eva Maria García del Toro [1], María Isabel Más-López [2] and Santiago Pindado [3,*]

[1] Departamento de Ingeniería Civil: Hidráulica y Ordenación del Territorio ETSI Civil, Universidad Politécnica de Madrid Alfonso XII, 3, 28014 Madrid, Spain; d.alcalag@upm.es (D.A.-G.); evamaria.garcia@upm.es (E.M.G.d.T.)
[2] Departamento de Ingeniería Civil: Construcción, Infraestructura y Transporte ETSI Civil, Universidad Politécnica de Madrid Alfonso XII, 3, 28014 Madrid, Spain; mariaisabel.mas@upm.es
[3] Instituto Universitario de Microgravedad "Ignacio Da Riva" (IDR/UPM), ETSI Aeronáutica y del Espacio, Universidad Politécnica de Madrid, Pza. del Cardenal Cisneros 3, 28040 Madrid, Spain
* Correspondence: santiago.pindado@upm.es

Abstract: The increase in the installation of renewable energy sources in electrical systems has changed the power distribution networks, and a new scenario regarding protection devices has arisen. Distributed generation (DG) might produce artificial delays regarding the performance of protection devices when acting as a result of short-circuits. In this study, the preliminary research results carried out to analyze the effect of renewable energy sources (photovoltaic, wind generation, etc.) on the protection devices of a power grid are described. In order to study this problem in a well-defined scenario, a quite simple distribution network (similar to the ones present in rural areas) was selected. The distribution network was divided into three protection zones so that each of them had DG. In the Institute of Electrical and Electronic Engineers (IEEE) system 13 bus test feeder, the short-circuits with different levels of penetration were performed from 1 MVA to 3 MVA (that represent 25%, 50%, and 75% of the total load in the network). In the simulations carried out, it was observed that the installation of DG in this distribution network produced significant changes in the short-circuit currents, and the inadequate performance of the protection devices and the delay in their operating times (with differences of up to 180% in relation to the case without DG). The latter, that is, the impacts of photovoltaic DG on the reactions of protection devices in a radial distribution network, is the most relevant outcome of this work. These are the first results obtained from a research collaboration framework established by staff from ETSI Civil and the IDR/UPM Institute, to analyze the effect of renewable energy sources (as DG) on the protection devices of a radial distribution network.

Keywords: coordination protection; distributed generation; photovoltaic resources; DigSILENT

Citation: Alcala-Gonzalez, D.; García del Toro, E.M.; Más-López, M.I.; Pindado, S. Effect of Distributed Photovoltaic Generation on Short-Circuit Currents and Fault Detection in Distribution Networks: A Practical Case Study. *Appl. Sci.* **2021**, *11*, 405. https://doi.org/10.3390/app11010405

Received: 16 November 2020
Accepted: 10 December 2020
Published: 4 January 2021

Publisher's Note: MDPI stays neutral with regard to jurisdictional claims in published maps and institutional affiliations.

1. Introduction

The distributed generation (DG) based on renewable energy sources in some distribution networks has caused a major change in the traditional model of power supply. According to the new paradigm of DG, generation of electricity power should come closer to the user, in contradiction to the traditional generation in centralized plants [1,2]. Therefore, it is necessary to implement electrical systems with an infrastructure that allows the energy to be distributed and made available to users in optimal conditions for their use [3].

Although the presence of DG in power networks can be beneficial [4,5], it produces a structural transformation, failing to behave as a classical radial distribution network. Reversed power flows can occur both in steady state and in the presence of faults [6–10]. The impacts that DG may have on these distribution networks will depend, among other aspects, on the type of generation, technology used, installed power, and location in the network [11–15]. Furthermore, these new fault modes resulting from the introduction of DG, based on renewable energy such as photovoltaic (PV) generation, can affect the reliability of these PV systems and decrease the revenues foreseen [16,17].

Among the different problems arising with the increasing of DG within power networks, the protection relays coordination is one of the most relevant (other problems being harmonic distortion, frequency drop, and stability and reliability of the network [18–22]). Protection in traditional networks was designed for unidirectional power distribution. However, this situation has changed with the greater importance of renewable energy [23], and other ways to reduce power production from the traditional energy sources that have a negative impact on the climate (e.g., the use of electric vehicles as power sources [24–29]). Within the networks with DG, such as the ones with photovoltaic generation located in different points of the grid, the coordination between protection devices might fail [30–38]. Additionally, each kind of renewable energy source presents new fault modes in a network. In [39], there is thorough review of the fault modes that a photovoltaic installation can bring to a distribution network.

The management and control of a radial electric distribution network is based on the assumption of the existence of unidirectional flows of power, which is transmitted from the highest levels of transport voltage to the distribution levels. We can assume that the short-circuit currents behave in a similar way. These assumptions allow the implementation of relatively simple and economical protection schemes, in order to achieve selective operation of the protection system (according to the principles of selectivity, only the protection device closest to the defect must function to clear the fault, leaving the rest of the network energized). There are numerous literature reviews on the matter of the protection of power grids with DG; the recent works [40–44] being worthy of mention.

The installation of DG in medium and low voltage levels changes the fundamental basis of the aforementioned unidirectional power flow [45]. Both the power flows and the short-circuit currents can now have upstream addresses, these also being of different values than the ones initially foreseen [6]. As mentioned above, the initial schemes applied (that is, the main feeder protection) might start being less effective or even stop working [46]. This problem being caused as the value of the short-circuit current detected by the main protection of the substation may be altered, and therefore affect the response time of the protection devices that will depend, to a great extent, on the size and location of the DG within the distribution network [15,47,48].

As stated, one of the main problems of the presence of DG in distribution networks is the loss of coordination and untimely triggering of protection devices [49], as short-circuit currents can increase as a result of the contribution from those DG. The fault currents suffer variations that can run the fuse in both directions, which means that it can be traversed by currents generated in points downstream and upstream of its location. Therefore, it is necessary to reconfigure the protection devices of these distribution networks. The loss of sensitivity of the protection devices can then be analyzed according to the location of short-circuits in relation to the DG, differentiating among faults located upstream and downstream of the aforementioned DG.

The aim of this work is to describe the preliminary results obtained from a research collaboration framework established by researchers from ETSI Civil and the IDR/UPM Institute to analyze the effect of renewable energy sources (as DG) on the protection devices of a power grid. The aforementioned framework is based on the work already carried out on protection in power grids with distributed generation [50], and the work carried out in analytical modeling on solar panels and batteries [51–60]. In the present work, photovoltaic energy was selected as the example of DG. It should be underlined that this particular source of energy is associated with some specific problems when considered as DG [61–65], and should be analyzed while taking into account its particular performance in relation to temperature and irradiance [66]. Finally, it should be underlined that although there are many works in the available literature on the effect of DG on distribution grids [6,11,12,14,18,20,24,29,36,67,68], it appears that the effect of DG on the protection coordination of relays curves.

In order to obtain quick results that could lead to general but solid conclusions, a simple distribution network has been analyzed in the present work. This specific power distribution grid was selected as it represents a well-defined scenario in which relevant conclusions can be derived.

As a first work on protection for grids with DG based on renewable energy sources, we chose to analyze a simple distribution network similar to the ones typical of rural areas. With the objective of ensuring the electricity supply in these distribution networks when DG is connected, and maintaining the functionality of the protections, the following aspects were studied:

- How the installation of DG in the studied distribution network can cause significant changes in the fault currents;
- The untimely actuation of the protections; and
- The effect on the protection devices relating to the amount of DG power supplied to the system.

Finally, it is also fair to say that the present work analyzed the effect of photovoltaic distributed generation in a network, leaving aside other possible renewable sources, such as wind power generation. This was a drawback that will be overcome in future works.

The following paper is organized as follows: in Section 2, the distribution network and different cases studied are described, the results being included in Section 3. Finally, conclusions are summarized in Section 4.

2. Materials and Methodology

The protection of the distribution networks studied in this work was guaranteed by main protection relays located at the beginning of the line (that is, at the electric substation (S)), and cut-out fuses (F1, F2) at the laterals. The basic topology of this kind of network is shown in Figure 1, where a distributed generation (DG) source has been added. The main feeder is protected by a relay (R), whereas the laterals supplying the loads are protected by means of fuses.

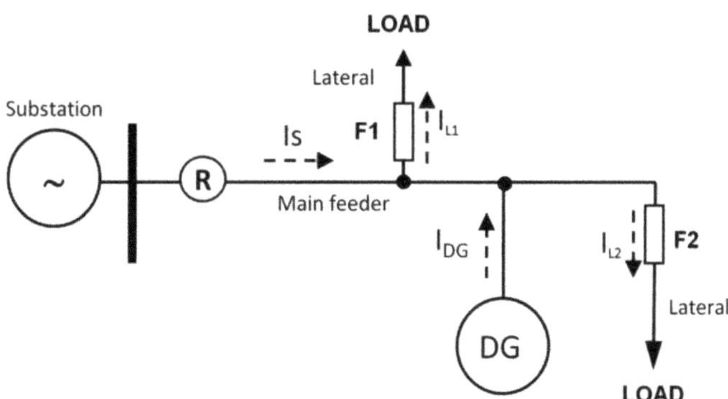

Figure 1. Basic topology of the simple power distribution network studied. A distributed generation (DG) source has been added.

Relays are normally equipped with inverse time overcurrent devices, whose performance is defined by the following equation [67]:

$$t(I) = \frac{a}{\left(\frac{I}{I_{pick-up}}\right)^b - 1} \text{TMS} \tag{1}$$

where t is the time of operation of the overcurrent device, I is the fault current detected by the device, $I_{pick\text{-}up}$ is the adjustment current (relay pick-up curve) of the device, a and b are the control parameters of the actuation curve, and TMS is the Time Multiplier Setting (expressed in seconds) [69]. The characteristic values of the relays used in the present work are included in Table 1.

Table 1. Parameters for the different relay characteristics used in the Institute of Electrical and Electronic Engineers (IEEE)-13 bus test feeder system used in the present work (see Equation (1)). These values were selected according to the IEC 60255-51 standard.

Time-Current Curve Type	Settings	
	a	*b*
Inverse	0.14	0.02
Very inverse	13.5	1
Extremely inverse	80	2

Each relay in a protection system has a set of relay settings that determine the primary and backup protection that the relay will provide. The relay settings for each relay are calculated so that the relay fulfills the primary and backup protection requirements of the network it is protecting. Calculations are based on the maximum load current, the maximum and minimum fault currents, and/or the impedance of feeders that the relay is protecting.

On the other hand, fuses have an inverse current-time characteristic that is usually plotted as a log-log curve, which is better approximated by a second-order polynomial function. However, it should be underlined that the most interesting part of this curve, in terms of its practical applications, can be approximated by a linear expression [70]. Therefore, it can be assumed that the general equation for the characteristic curve of an expulsion fuse can be expressed as [68,71–74]:

$$\log(t) = m \log(I) + n \tag{2}$$

where t is the actuation time of the ejection fuse, I is the current through it, and m and n are fuse constants to be determined as in [75].

The present work was carried out by using computer simulation of a rural/small grid with 3 possible photovoltaic DG sources, selected as a case study test network. To analyze the effect of the photovoltaic DG on the protection devices of such a distribution network, the IEEE-13 bus test feeder system was used (see Figure 2).

The simulation was performed with DIgSILENT Power Factory software, already used to study the protection of distribution systems with DG [72,76–79]. This distribution network design has been successfully used to study static short-circuit currents [80], the maximum possible photovoltaic power penetration into a network in relation to the demand response [81], voltage regulation strategies in DG [78,82,83], or fault ride-through in power networks related to renewable energy (wind and photovoltaic) [79]. Besides, it should be also said that DIgSILENT Power Factory software is a powerful simulation tool that integrates the following standard methodologies for short-circuit calculations: IEC 60909, IEEE 141/ANSI C37, VDE 0102/0103, G74, and IEC 61363.

The analyzed distribution network was divided into three Zones of Protection (Zones 1, 2, and 3), assigning a specific area for each DG (starting each one of these zones from the beginning of the output feeder from the electrical substation, see Figure 2). The total load of the distribution network was 4 MW, which was considered constant in all simulations. The connection points of the DG were chosen based on their distance from the main substation.

The three different case studies analyzed in the present work are indicated Figure 2 (each of them involves only one photovoltaic DG):

- N1: installation of a photovoltaic DG at bus 646, located at the middle of the distance to the substation (L = 2400 m), within Protection Zone 1;

- N2: installation of a photovoltaic DG at bus 633, closer to the substation (L = 1500 m), within Protection Zone 2;
- N3: installation of a photovoltaic DG at bus 652, farthest from the substation (L = 7000 m), within Protection Zone 3.

These case studies were analyzed with different DG penetration levels (1, 2, and 3 MW, representing 25%, 50%, and 75% of the power load, respectively). These levels represent normal variations in the behavior of the solar panels in relation to irradiance on the solar cells and their temperature (the use of Maximum Power Point Tracking is supposed). In recent works, we have reflected these variations (both measured and calculated by using implicit and explicit models) [59,60,84].

To the authors' knowledge and after a thorough review of the available literature, the present approach to analyze the performance of a protection scheme based on both relays and cut-out fuses in a well-known power distribution grid, and in relation to the power supplied from photovoltaic DG, represents a novelty.

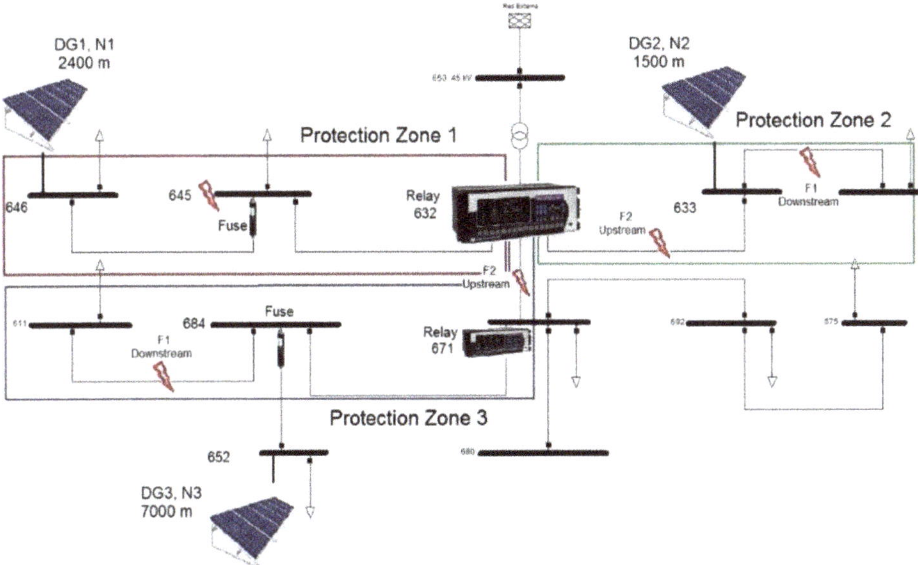

Figure 2. Example of the rural network topology studied in the present work. The position of the three DG included in the topology, N1, N2, and N3 (case studies), is indicated. The different protection zones are also indicated in the figure, together with the simulated short-circuits.

3. Results and Discussion

The results, as presented for all scenarios, show various ranges of DG capacity which provide different values of fault current. Adding distributed generation to the 20 kV distribution network creates a situation where networks that were designed primarily as tail, radial, or open radial networks become looped networks.

As a result, distributed generation can cause relays in a protection system to under-reach or over-reach. This has been illustrated in this paper, by sample calculations and an actual protection review (using DigSILENT PowerFactory software).

In Figure 3, the electric circuit corresponding to Protection Zone 1 with a 1 MW photovoltaic DG source (connected to bus 646) is shown (see also Figure 2). Protection elements in this area include an overcurrent relay at the beginning of the line (bus 632), and a fuse (F645) to protect the line connecting the DG. Only a 1 MW power supply through bus 646 was analyzed, as it represented the maximum power that was able to be injected by the DG in continuous operation (larger installed powers always tripped fuse F645).

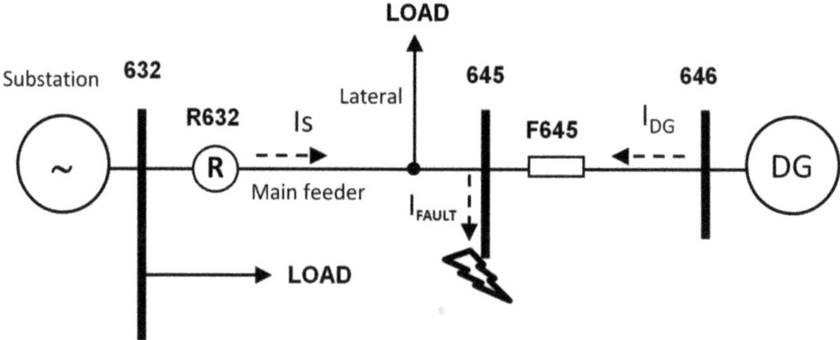

Figure 3. Protection Zone 1 electric circuit (see also Figure 2), in which a short-circuit in bus 645 is indicated.

A three-phase short-circuit at bus 645 was simulated; the characteristic curves being plotted in Figure 4. In the graph, it shows that in the case of a fault between the substation and the DG, the presence of the latter represented a decrease in the value of the short-circuit current detected by the relay (R632 in Figure 3), from 536 A to 513 A. Thus, a loss in relay sensitivity (binding), as relay R632 at the main feeder detected a lower value of the fault current, with the triggering time also being delayed from 0.3 s to 0.35 s. These results are also summarized in Figure 5, and Table 2 at the end of this paper.

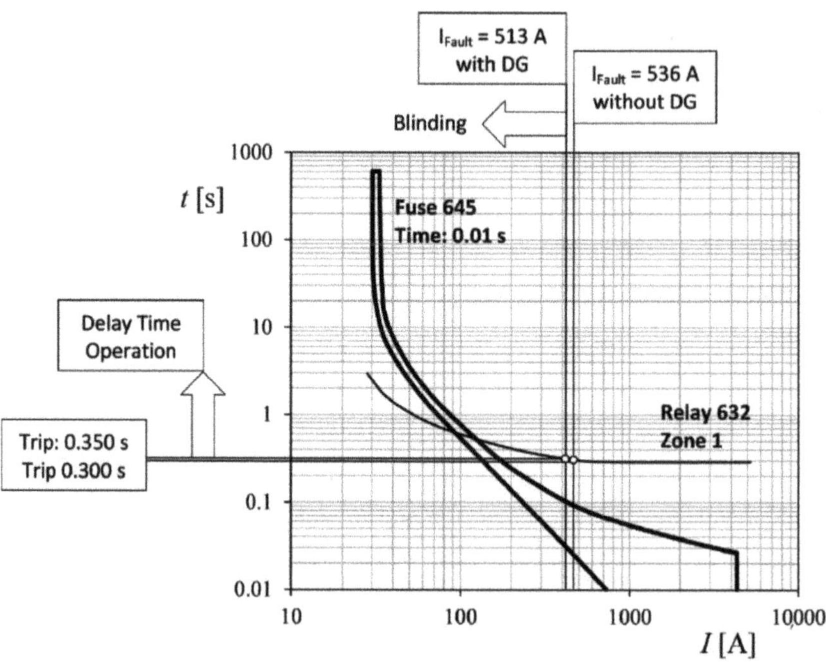

Figure 4. Protection Zone 1 current-time (*I-t*) performance in case of a three-phase short-circuit with 25% DG power (1 MW) at bus 645 (see Figure 3).

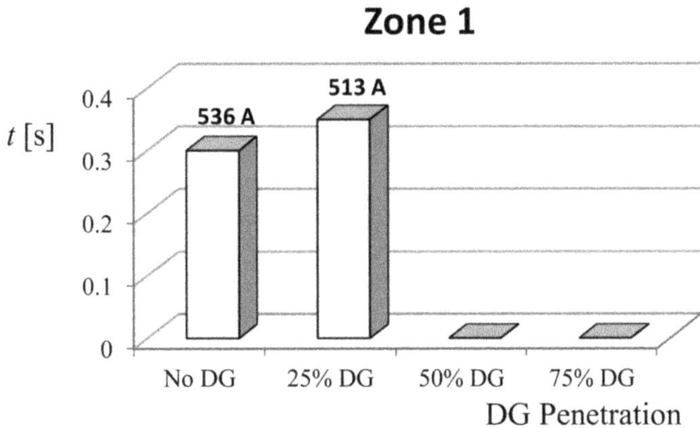

Figure 5. Triggering time, t, of protection relay R632 in Protection Zone 1 (see Figures 2 and 3), in relation to the power supply from DG. The currents detected by the relay when triggering were added to the graph in each case.

Table 2. Relay and fuse currents, I, and triggering times, t, in the protection devices from the defined Protection Zones 1, 2, and 3 (see Figure 2, Figure 3, Figure 6, and Figure 10), in relation to the power supplied by the DG (1, 2, and 3 MW, representing 25%, 50%, and 75% of the photovoltaic installed power).

Protection Zone	Fault	Protection Device	Without DG		25% DG (1 MW)		50% DG (2 MW)		75% DG (3 MW)	
			I_{sc} [A]	t [s]	I_{sc} [A]	t [s]	I_{sc} [A]	t [s]	I_{sc} [A]	t [s]
Zone 1	Bus 645	R632	536.3	0.300	513.6	0.350	-	-	-	-
		F645	-	-	726.4	0.010	-	-	-	-
Zone 2	Fault F1	R632	488.3	0.380	465.1	0.389	455.5	0.393	449.3	0.400
	Fault F2	R632	498.1	0.370	477.1	0.384	467.9	0.388	461.2	0.390
Zone 3	Fault F1	R632	1400.1	0.300	490.0	0.800	438.2	0.840	432.5	0.840
		R671	-	-	440.1	0.380	438.2	0.390	432.5	0.380
		F684	-	-	452.3	0.113	440.8	0.112	432.5	0.110
	Fault F2	R632	1113.2	0.560	420.3	0.850	490.3	0.860	438.3	0.870
		R671	1113.2	0.310	420.3	0.390	427.6	0.390	438.3	0.750
		F684	-	-	439.5	0.110	427.6	0.119	438.0	0.120

The Protection Zone 2 analyzed scenario (Figure 2) involved a 1, 2, and 3 MW DG power supply. A protection relay R632 was installed at the substation (see Figure 6), and two faults were studied, at points F1 (downstream from the DG source) and F2 (upstream from the DG source). In case of a short-circuit at point F1, the current detected by the relay decreased from 488 A (without DG) to 455 A (with DG), with a triggering delay of 13 ms (see in Figure 7 the case corresponding to 50% DG power at bus 633). Similarly, in case of a short-circuit at point F2, the currents decreased from 498 A to 467 A, with a triggering delay of 10 ms (see in Figure 8 the case corresponding to 50% DG power at bus 633). The results are summarized in Figure 9 and Table 2.

As performed in relation to the fault analysis carried out in Protection Zone 2, three different DG power supply scenarios (1, 2, and 3 MW) were studied in relation to Protection Zone 3 (Figure 2). The protection devices installed were two overcurrent relays (R671 and R632) within the main distribution line, and a fuse 684 DG protection, see Figure 10. Two three-phase short-circuits were simulated within the corresponding Protection Zone 3: at point F1, located downstream from the DG, and at point F2, upstream from the DG. The results are respectively included in the graphs in Figures 11 and 12. In case of short-circuits at F1, the DG installation implies a decrease of the current detected by the relay, from 1113 A to 420 A, which represents a 62% decrease of the relay sensitivity and a delay of 550 ms in relation to the triggering time (see Figure 11). Similarly, in the case of short-circuit at point F2 (see Figure 12), the connection of the DG source introduced a decrease of the detected current from 1400 A to 490 A, and a delay of the triggering time, from 300 ms to 800 ms. The results are summarized in Figure 13 and Table 2.

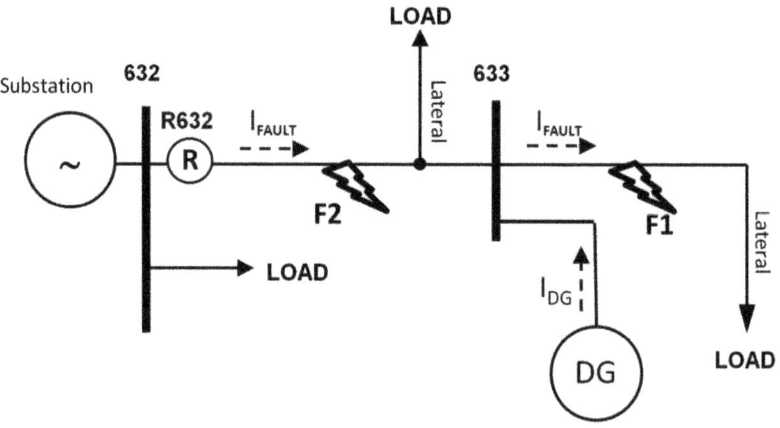

Figure 6. Protection Zone 2 electric circuit (see also Figure 2), in which short-circuits in points F1 and F2 are indicated.

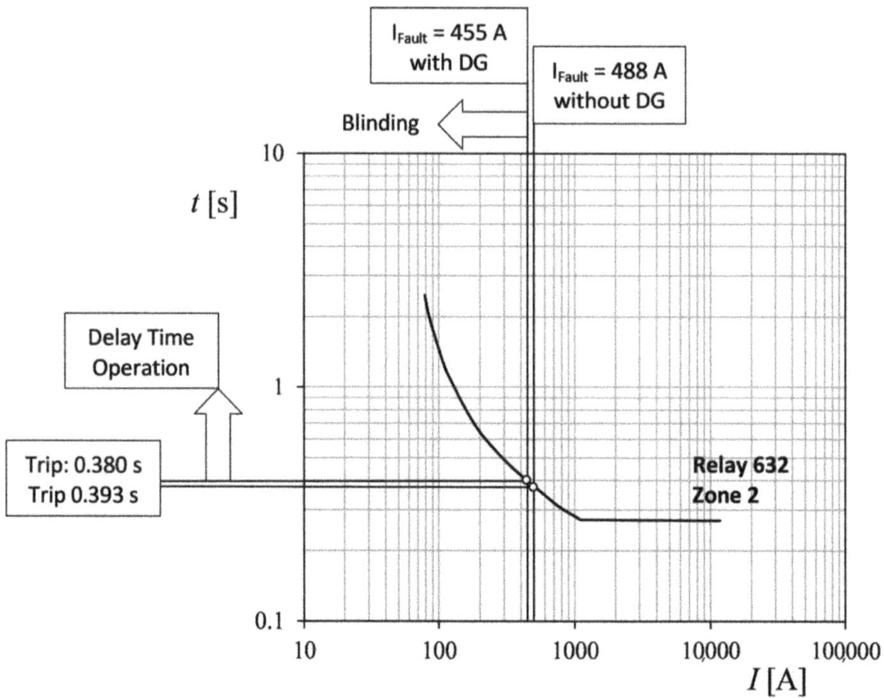

Figure 7. Protection Zone 2 current-time (*I-t*) performance of in case of a three-phase short-circuit at point F1, with 50% DG power (2 MW) at bus 633 (see Figure 6).

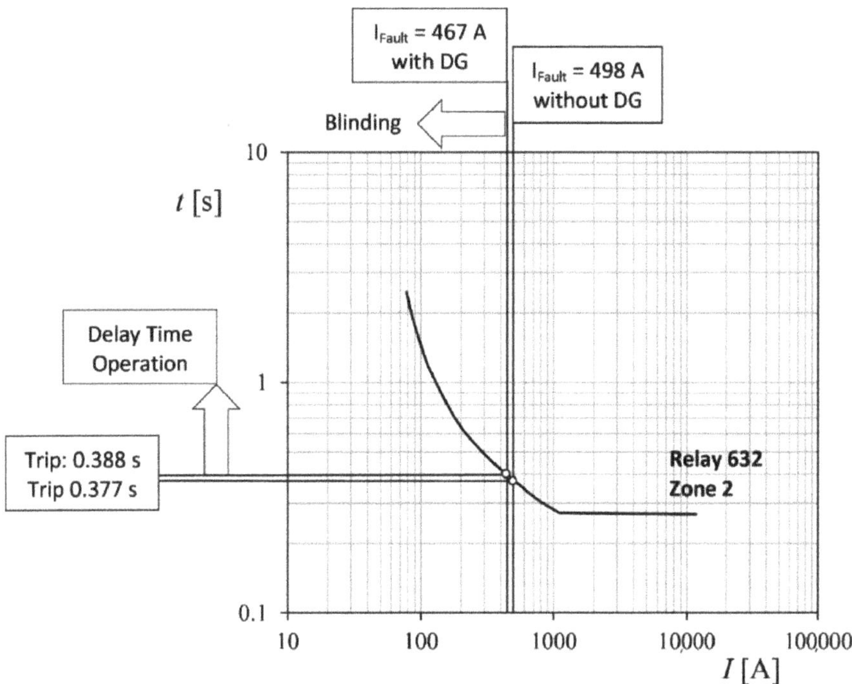

Figure 8. Protection Zone 2 current-time (*I-t*) performance in case of a three-phase short-circuit at point F2, with 50% DG power (2 MW) at bus 633 (see Figure 6).

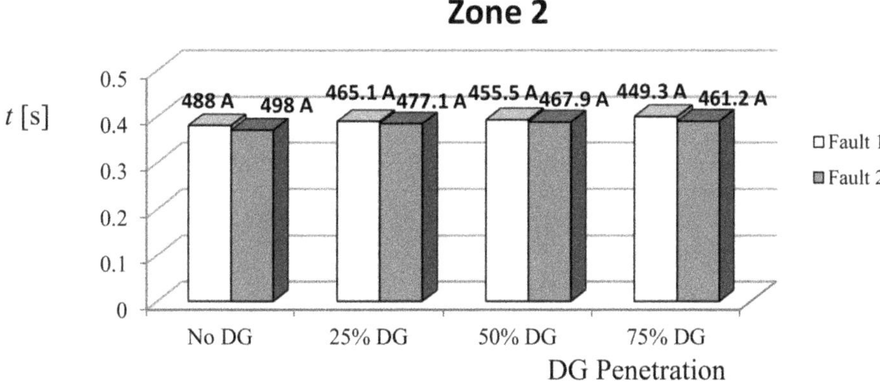

Figure 9. Triggering time, *t*, of the main feeder protection relay R632 in Protection Zone 2 (see Figures 2 and 6) for fault F1 downstream from the DG and fault F2 upstream from the DG, in relation to DG power supply. The currents detected by the relay when triggering were added to the graph in each case.

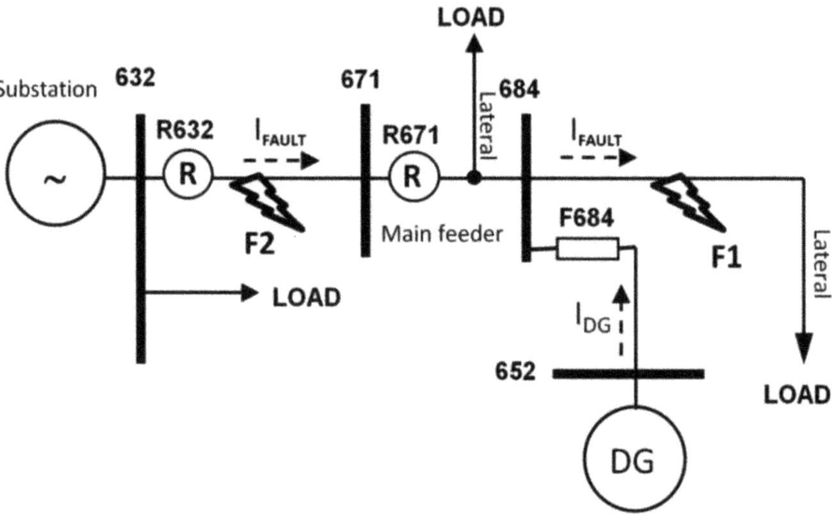

Figure 10. Protection Zone 3 electric circuit (see also Figure 2), in which short-circuits in points F1 and F2 are indicated.

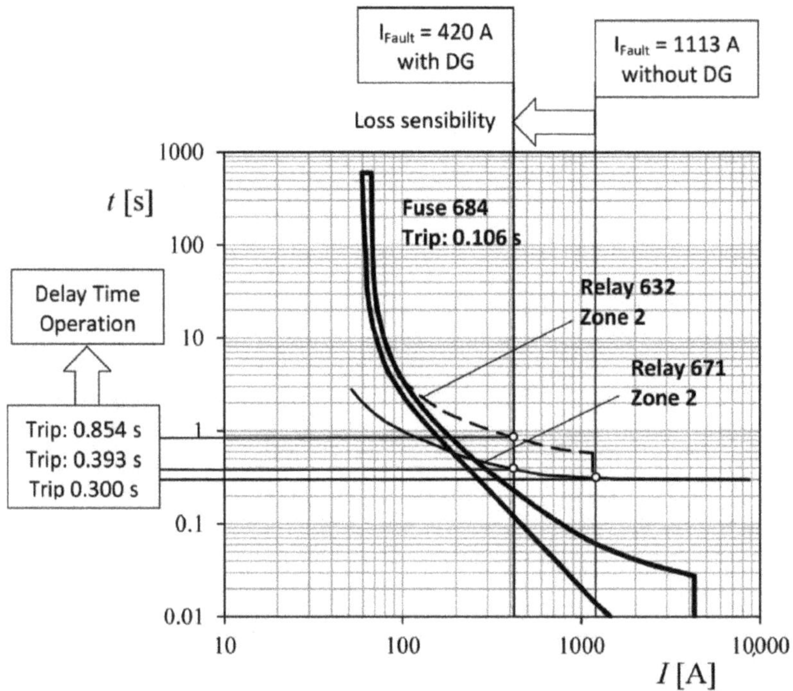

Figure 11. Protection Zone 3 current-time (*I-t*) performance in case of a three-phase short-circuit at point F1, with 25% DG power (1 MW) at bus 652 (see Figure 10).

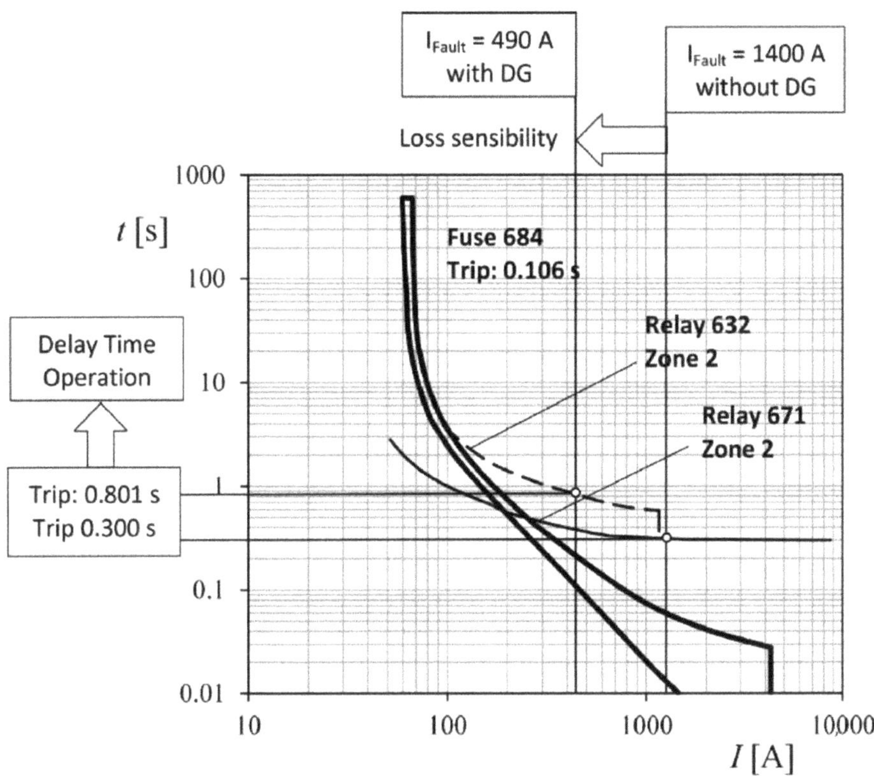

Figure 12. Protection Zone 3 current-time (*I-t*) performance in case of a three-phase short-circuit at point F2, with 25% DG power (1 MW) at bus 652 (see Figure 10).

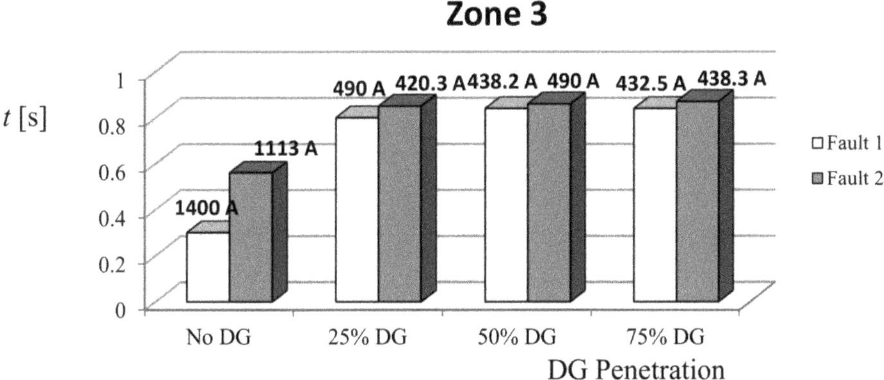

Figure 13. Triggering time, *t*, of the main feeder protection relay R632 in Protection Zone 3 (see Figures 2 and 10) for fault F1 upstream the DG and fault F2 downstream the DG, in relation to DG power supply. The currents detected by the relay when triggering were added to the graph in each case.

In Table 2, the results obtained in relation to each of the analyzed scenarios are included. The fault currents and the protection relay tripping/triggering times are shown with regard to:

- the Zone in which the fault was produced; and
- the penetration level (which is the ratio of the power supplied by the photovoltaic DG to the power demanded by the load, as indicated in the Section 2).

4. Conclusions

A coordination study was presented and applied to the IEEE 13-node test feeder to evaluate the effect of the PV DG penetration on the protection devices coordination. Different cases were studied by changing DG penetration levels and locations for each possible fault location. A protection coordination assessment was carried out by analyzing the location of the faults (devices see fault currents for downstream and upstream faults).

The most relevant conclusions from this work are:

- Faults upstream in relation to the last DG were not seen by the protection devices located downstream from this DG. Even for larger DG penetration these downstream devices did not have coordination problems between them (as they did not see any fault);
- On the other hand, although devices saw fault currents for upstream faults, coordination was lost if they saw the same fault current for a fault downstream as well as for a fault upstream.

Additionally, the present protection procedures need to be revised, as the presence of DG sources might represent a quite serious effect on the current and triggering time reaction of the protection relays. According to the results and depending on the distance to the substation, the tripping time with DG was increased up to 167% (for 25% PV DG penetration) and 180% (for 50% and 75% PV DG penetration).

The results from this work can be reasonably extrapolated to grid networks, bearing in mind the current direction effects on the protection relays, and the DG penetration.

Future works of the research team that carried out the present work are the development of simplified techniques to locate faults in distribution networks with DG, new methods to select the appropriate fuses in these networks (to avoid unexpected fuse tripping), and a procedure to successfully reprogram overcurrent relays.

Finally, it should also be noted that the present work will be followed by others that will include research on the dynamic modeling of power supplies from DG in distribution networks, and simplified methodologies on fault locations in grids with DG.

Author Contributions: Conceptualization, D.A.-G. and S.P.; methodology, D.A.-G., E.M.G.d.T., M.I.M.-L. and S.P.; software, D.A.-G.; validation, D.A.-G., E.M.G.d.T. and M.I.M.-L.; formal analysis, D.A.-G.; investigation, D.A.-G., E.M.G.d.T., M.I.M.-L. and S.P.; data curation, D.A.-G.; writing—original draft preparation, D.A.-G. and S.P.; writing—review and editing, D.A.-G. and S.P.; supervision, D.A.-G. and S.P. All authors have read and agreed to the published version of the manuscript.

Funding: This research received no external funding.

Acknowledgments: The authors are indebted to the IDR/UPM Institute Angel Sanz-Andres for its kind support regarding the research related to power systems at the research institute. The authors are also grateful to Fernando Gallardo and SIELEC for the kind supervision and advise in relation to power distribution networks and protection systems. Finally, the authors are grateful to the reviewers, whose comments helped in improving the present work.

Conflicts of Interest: The authors declare no conflict of interest.

References

1. Kezunovic, M. *Fundamentals of Power System Protection*; Elsevier BV: Amsterdam, The Netherlands, 2005; pp. 787–803.
2. Jennett, K.; Coffele, F.; Booth, C. Comprehensive and quantitative analysis of protection problems associated with increasing penetration of inverter-interfaced DG. In Proceedings of the 11th IET International Conference on Developments in Power Systems Protection (DPSP 2012), Institution of Engineering and Technology (IET), Birmingham, UK, 23–26 April 2012.

3. Cidrás, J.; Miñambres, J.F.; Alvarado, F.L. Fault Analysis and Protection Systems. Analysis and Operation. In *Electric Energy Systems. Analysis and Operation*; Gómez-Expósito, A., Conejo, A.J., Cañizares, C., Eds.; CRC Press: Boca Raton, FL, USA, 2009; pp. 303–354. ISBN 9781315221809.

4. Liu, Z.; Wen, F.; Ledwich, G. Potential benefits of distributed generators to power systems. In Proceedings of the 2011 IEEE 4th International Conference on Electric Utility Deregulation and Restructuring and Power Technologies (DRPT), Weihai, China, 6–9 July 2011; pp. 1417–1423.

5. Chamandoust, H.; Hashemi, A.; Derakhshan, G.; Abdi, B. Optimal hybrid system design based on renewable energy resources. In Proceedings of the IEEE Proceedings 2017 Smart Grid Conference (SGC), Tehran, Iran, 20–21 December 2017; pp. 1–5.

6. Bastiao, F.; Cruz, P.; Fiteiro, R. Impact of distributed generation on distribution networks. In Proceedings of the 2008 5th International Conference on the European Electricity Market, Institute of Electrical and Electronics Engineers (IEEE), Lisbon, Portugal, 25 July 2008; pp. 1–6.

7. Hsu, C.T.; Tsai, L.J.; Cheng, T.J.; Chen, C.S.; Hsu, C.W. Solar PV generation system controls for improving voltage in distribution network. In Proceedings of the 2013 International Symposium on Next-Generation Electronics, Institute of Electrical and Electronics Engineers (IEEE), Kaohsiung, Taiwan, 25–26 February 2013; pp. 486–489.

8. Zheng, S.; Xu, Z.H.; Liu, J.J.; Fu, Q. The Application of Electronic Communication Relay Protection in Distribution Network with Distributed Generation. *Adv. Mater. Res.* **2014**, *1070*, 938–942. [CrossRef]

9. Li, Y.; Ren, H.; Zhou, L.; Wang, F.; Li, J. Inverse-time protection scheme for active distribution network based on user-defined characteristics. In Proceedings of the 2017 IEEE PES Innovative Smart Grid Technologies Conference Europe (ISGT-Europe), Institute of Electrical and Electronics Engineers (IEEE), Torino, Italy, 26–29 September 2017; Volume 2018, pp. 1–6.

10. Zhang, Y.; Song, X.; Li, J.; Gao, F.; Deng, Z. Research on Voltage Rise Risk Assessment Model and Method for Distribution Network with Distributed Generation. In Proceedings of the 2018 2nd IEEE Conference on Energy Internet and Energy System Integration (EI2), Institute of Electrical and Electronics Engineers (IEEE), Beijing, China, 20–22 October 2018; pp. 1–5.

11. Deng, W.; Pei, W.; Qi, Z. Impact and improvement of Distributed Generation on voltage quality in Micro-grid. In Proceedings of the 2008 Third International Conference on Electric Utility Deregulation and Restructuring and Power Technologies, Institute of Electrical and Electronics Engineers (IEEE), Nanjing, China, 6–9 April 2008; pp. 1737–1741.

12. A review on power quality challenges in renewable Energy grid integration. *Int. J. Curr. Eng. Technol.* **2011**, *6*, 1573–1578. [CrossRef]

13. Hakimi, S.M.; Moghaddas-Tafreshi, S.M. Optimal Planning of a Smart Microgrid Including Demand Response and Intermittent Renewable Energy Resources. *IEEE Trans. Smart Grid* **2014**, *5*, 2889–2900. [CrossRef]

14. Sandhu, M.; Thakur, T. Issues, Challenges, Causes, Impacts and Utilization of Renewable Energy Sources - Grid Integration. *J. Eng. Res. Appl.* **2014**, *4*, 636–643.

15. Ehsan, A.; Yang, Q. Optimal integration and planning of renewable distributed generation in the power distribution networks: A review of analytical techniques. *Appl. Energy* **2018**, *210*, 44–59. [CrossRef]

16. Chiacchio, F.; Famoso, F.; D'Urso, D.; Brusca, S.; Aizpurua, J.I.; Cedola, L. Dynamic Performance Evaluation of Photovoltaic Power Plant by Stochastic Hybrid Fault Tree Automaton Model. *Energies* **2018**, *11*, 306. [CrossRef]

17. Chiacchio, F.; D'Urso, D.; Famoso, F.; Brusca, S.; Aizpurua, J.I.; Catterson, V.M. On the use of dynamic reliability for an accurate modelling of renewable power plants. *Energy* **2018**, *151*, 605–621. [CrossRef]

18. Kadir, A.A.; Mohamed, A.; Shareef, H. Harmonic impact of different distributed generation units on low voltage distribution system. In Proceedings of the 2011 IEEE International Electric Machines & Drives Conference (IEMDC), Institute of Electrical and Electronics Engineers (IEEE), Niagara Falls, ON, Canada, 15–18 May 2011; pp. 1201–1206.

19. Antonova, G.; Nardi, M.; Scott, A.; Pesin, M. Distributed generation and its impact on power grids and microgrids protection. In Proceedings of the 2012 65th Annual Conference for Protective Relay Engineers, Institute of Electrical and Electronics Engineers (IEEE), College Station, TX, USA, 2–5 April 2012; pp. 152–161.

20. Bignucolo, F.; Cerretti, A.; Coppo, M.; Savio, A.; Turri, R. Impact of Distributed Generation Grid Code Requirements on Islanding Detection in LV Networks. *Energies* **2017**, *10*, 156. [CrossRef]

21. Senarathna, S.; Hemapala, K.U. Review of adaptive protection methods for microgrids. *AIMS Energy* **2019**, *7*, 557–578. [CrossRef]

22. Adefarati, T.; Bansal, R.C. Reliability assessment of distribution system with the integration of renewable distributed generation. *Appl. Energy* **2017**, *185*, 158–171. [CrossRef]

23. Barik, M.A.; Pota, H.R. Complementary effect of wind and solar energy sources in a microgrid. In Proceedings of the IEEE PES Innovative Smart Grid Technologies, Tianjin, China, 21–24 May 2012; pp. 1–6.

24. Clement-Nyns, K.; Haesen, E.; Driesen, J. The impact of vehicle-to-grid on the distribution grid. *Electr. Power Syst. Res.* **2011**, *81*, 185–192. [CrossRef]

25. López, M.; Martín, S.; Aguado, J.; De La Torre, S. V2G strategies for congestion management in microgrids with high penetration of electric vehicles. *Electr. Power Syst. Res.* **2013**, *104*, 28–34. [CrossRef]

26. García-Villalobos, J.; Zamora, I.; Martín, J.I.; Asensio, F.J.; Aperribay, V. Plug-in electric vehicles in electric distribution networks: A review of smart charging approaches. *Renew. Sustain. Energy Rev.* **2014**, *38*, 717–731. [CrossRef]

27. López, M.A.; de la Torre, S.; Martín, S.; Aguado, J.A. Demand-side management in smart grid operation considering electric vehicles load shifting and vehicle-to-grid support. *Int. J. Electr. Power Energy Syst.* **2015**, *64*, 689–698. [CrossRef]

28. Tan, K.M.; Ramachandaramurthy, V.K.; Yong, J.Y. Integration of electric vehicles in smart grid: A review on vehicle to grid technologies and optimization techniques. *Renew. Sustain. Energy Rev.* **2016**, *53*, 720–732. [CrossRef]
29. Galiveeti, H.R.; Goswami, A.K.; Choudhury, N.B.D. Impact of plug-in electric vehicles and distributed generation on reliability of distribution systems. *Eng. Sci. Technol. Int. J.* **2018**, *21*, 50–59. [CrossRef]
30. Margossian, H.; Capitanescu, F.; Sachau, J. Distributed generator status estimation for adaptive feeder protection in active distribution grids. In Proceedings of the 22nd International Conference and Exhibition on Electricity Distribution (CIRED 2013), Institution of Engineering and Technology (IET), Stockholm, Sweden, 10–13 June 2013; p. 677.
31. Margossian, H.; Capitanescu, F.; Sachau, J. Feeder protection challenges with high penetration of inverter based distributed generation. In Proceedings of the IEEE Eurocon 2013, Zagreb, Croatia, 1–4 July 2013; pp. 1369–1374.
32. Neshvad, S.; Margossian, H.; Sachau, J.; Sachau, J. Topology and parameter estimation in power systems through inverter-based broadband stimulations. *IET Gener. Transm. Distrib.* **2016**, *10*, 1710–1719. [CrossRef]
33. Margossian, H.; Sachau, J.; Deconinck, G. Short circuit calculation in networks with a high share of inverter based distributed generation. In Proceedings of the 2014 IEEE 5th International Symposium on Power Electronics for Distributed Generation Systems (PEDG), Galway, Ireland, 24–27 June 2014; pp. 1–5.
34. Dewadasa, M.; Ghosh, A.; Ledwich, G. Fold back current control and admittance protection scheme for a distribution network containing distributed generators. *IET Gener. Transm. Distrib.* **2010**, *4*, 952. [CrossRef]
35. Dewadasa, M.; Ghosh, A.; Ledwich, G. Protection of distributed generation connected networks with coordination of overcurrent relays. In Proceedings of the IECON 2011—37th Annual Conference of the IEEE Industrial Electronics Society, Melbourne, VIC, Australia, 7–10 November 2011; pp. 924–929.
36. Sa'ed, J.A.; Favuzza, S.; Ippolito, M.G.; Massaro, F. An Investigation of Protection Devices Coordination Effects on Distributed Generators Capacity in Radial Distribution Systems. In Proceedings of the IEEE/ICCEP 2013, Alghero, Italy, 11–13 June 2013; pp. 686–692.
37. Fani, B.; Bisheh, H.; Sadeghkhani, I. Protection coordination scheme for distribution networks with high penetration of photovoltaic generators. *IET Gener. Transm. Distrib.* **2018**, *12*, 1802–1814. [CrossRef]
38. Brahma, S.; Girgis, A. Microprocessor-based reclosing to coordinate fuse and recloser in a system with high penetration of distributed generation. In Proceedings of the 2002 IEEE Power Engineering Society Winter Meeting Conference Proceedings (Cat. No.02CH37309), New York, NY, USA, 7–10 November 2011; pp. 453–458.
39. Pillai, D.S.; Blaabjerg, F.; Rajasekar, N. A Comparative Evaluation of Advanced Fault Detection Approaches for PV Systems. *IEEE J. Photovoltaics* **2019**, *9*, 513–527. [CrossRef]
40. Safaei, A.; Zolfaghari, M.; Gilvanejad, M.; Gharehpetian, G.B. A survey on fault current limiters: Development and technical aspects. *Int. J. Electr. Power Energy Syst.* **2020**, *118*, 105729. [CrossRef]
41. Kumar, R.; Saxena, D. A Literature Review on Methodologies of Fault Location in the Distribution System with Distributed Generation. *Energy Technol.* **2020**, *8*, 8. [CrossRef]
42. Barra, P.; Coury, D.; Fernandes, R. A survey on adaptive protection of microgrids and distribution systems with distributed generators. *Renew. Sustain. Energy Rev.* **2020**, *118*, 109524. [CrossRef]
43. Norshahrani, M.; Mokhlis, H.; Abu Bakar, A.H.; Jamian, J.J.; Sukumar, S. Progress on Protection Strategies to Mitigate the Impact of Renewable Distributed Generation on Distribution Systems. *Energies* **2017**, *10*, 1864. [CrossRef]
44. Manditereza, P.T.; Bansal, R.C. Renewable distributed generation: The hidden challenges—A review from the protection perspective. *Renew. Sustain. Energy Rev.* **2016**, *58*, 1457–1465. [CrossRef]
45. Abapour, S.; Zare, K.; Mohammadi-Ivatloo, B. Dynamic planning of distributed generation units in active distribution network. *IET Gener. Transm. Distrib.* **2015**, *9*, 1455–1463. [CrossRef]
46. Wilks, J.; Dip, E.E.; Dip, O. Developments in power system protection. In Proceedings of the Annual Conference of Electric Energy Association of Australia, Canberra, Australia, 9–10 August 2002.
47. Oree, V.; Hassen, S.Z.S.; Fleming, P.J. Generation expansion planning optimisation with renewable energy integration: A review. *Renew. Sustain. Energy Rev.* **2017**, *69*, 790–803. [CrossRef]
48. Huda, A.; Živanović, R. Large-scale integration of distributed generation into distribution networks: Study objectives, review of models and computational tools. *Renew. Sustain. Energy Rev.* **2017**, *76*, 974–988. [CrossRef]
49. Girgis, A.; Brahma, S. Effect of distributed generation on protective device coordination in distribution system. In Proceedings of the IEEE LESCOPE 01, 2001 Large Engineering Systems Conference on Power Engineering Conference Proceedings, Theme: Powering Beyond 2001 (Cat. No.01ex490), Halifax, NS, Canada, 11–13 July 2001; pp. 115–119.
50. Alcala, D.; Gonzalez-Juarez, L.G.; Valino, V.; García, J.L. A 'Binocular' Method for Detecting Faults in Electrical Distribution Networks with Distributed Generation. *Elektron. Elektrotechnika* **2016**, *22*, 3–8. [CrossRef]
51. Cubas, J.; Pindado, S.; Victoria, M. On the analytical approach for modeling photovoltaic systems behavior. *J. Power Sources* **2014**, *247*, 467–474. [CrossRef]
52. Cubas, J.; Pindado, S.; De Manuel, C. Explicit Expressions for Solar Panel Equivalent Circuit Parameters Based on Analytical Formulation and the Lambert W-Function. *Energies* **2014**, *7*, 4098–4115. [CrossRef]
53. Cubas, J.; Pindado, S.; Sanz-Andrés, Á. Accurate Simulation of MPPT Methods Performance When Applied to Commercial Photovoltaic Panels. *Sci. World J.* **2015**, *2015*, 914212. [CrossRef] [PubMed]

54. Pindado, S.; Cubas, J.; Sorribes-Palmer, F. On the Analytical Approach to Present Engineering Problems: Photovoltaic Systems Behavior, Wind Speed Sensors Performance, and High-Speed Train Pressure Wave Effects in Tunnels. *Math. Probl. Eng.* **2015**, *2015*, 897357. [CrossRef]
55. Cubas, J.; Pindado, S.; Sorribes-Palmer, F. Analytical Calculation of Photovoltaic Systems Maximum Power Point (MPP) Based on the Operation Point. *Appl. Sci.* **2017**, *7*, 870. [CrossRef]
56. Pindado, S.; Cubas, J. Simple mathematical approach to solar cell/panel behavior based on datasheet information. *Renew. Energy* **2017**, *103*, 729–738. [CrossRef]
57. Pindado, S.; Cubas, J.; Roibás-Millán, E.; Bugallo-Siegel, F.; Sorribes-Palmer, F. Assessment of Explicit Models for Different Photovoltaic Technologies. *Energies* **2018**, *11*, 1353. [CrossRef]
58. Porras-Hermoso, Á.; Pindado, S.; Cubas, J. Lithium-ion battery performance modeling based on the energy discharge level. *Meas. Sci. Technol.* **2018**, *29*, 117002. [CrossRef]
59. Cubas, J.; Gomez-Sanjuan, A.M.; Pindado, S. On the thermo-electric modelling of smallsats. In Proceedings of the 50th International Conference on Environmental Systems—ICES 2020, Lisbon, Portugal, 12–16 July 2020; pp. 1–12.
60. Porras-Hermoso, Á.; Cobo-Lopez, B.; Cubas, J.; Pindado, S. Simple solar panels/battery modeling for spacecraft power distribution systems. *Acta Astronaut.* **2021**, *179*, 345–358. [CrossRef]
61. Qu, L.; Zhao, D.; Shi, T.; Chen, N.; Ding, J. Photovoltaic Generation Model for Power System Transient Stability Analysis. *Int. J. Comput. Electr. Eng.* **2013**, *5*, 297–300. [CrossRef]
62. Christodoulou, C.A.; Papanikolaou, N.P.; Gonos, I.F. Design of Three-Phase Autonomous PV Residential Systems With Improved Power Quality. *IEEE Trans. Sustain. Energy* **2014**, *5*, 1027–1035. [CrossRef]
63. Perpinias, I.I.; Tatakis, E.C.; Papanikolaou, N. Optimum design of low-voltage distributed photovoltaic systems oriented to enhanced fault ride through capability. *IET Gener. Transm. Distrib.* **2015**, *9*, 903–910. [CrossRef]
64. Desai, J.V.; Dadhich, P.K.; Bhatt, P.K. Investigations on Harmonics in Smart Distribution Grid with Solar PV Integration. *Technol. Econ. Smart Grids Sustain. Energy* **2016**, *1*, 11. [CrossRef]
65. Kharrazi, A.; Sreeram, V.; Mishra, Y. Assessment techniques of the impact of grid-tied rooftop photovoltaic generation on the power quality of low voltage distribution network—A review. *Renew. Sustain. Energy Rev.* **2020**, *120*, 109643. [CrossRef]
66. Papanikolaou, N.P.; Tatakis, E.C.; Kyritsis, A.C. Analytical Model for PV—Distributed Generators, suitable for Power Systems Studies. In Proceedings of the IEEE 2009 13th European Conference on Power Electronics and Applications, Barcelona, Spain, 8–10 September 2009.
67. Zeineldin, H.H.; Mohamed, Y.A.-R.I.; Khadkikar, V.; Pandi, V.R. A Protection Coordination Index for Evaluating Distributed Generation Impacts on Protection for Meshed Distribution Systems. *IEEE Trans. Smart Grid* **2013**, *4*, 1523–1532. [CrossRef]
68. Javadian, S.A.M. Impact of distributed generation on distribution system's reliability considering recloser-fuse miscoordination-A practical case study. *Indian J. Sci. Technol.* **2011**, *4*, 1279–1284. [CrossRef]
69. Damchi, Y.; Mashhadi, H.R.; Sadeh, J.; Bashir, M. Optimal coordination of directional overcurrent relays in a microgrid system using a hybrid particle swarm optimization. In Proceedings of the IEEE 2011 International Conference on Advanced Power System Automation and Protection, Beijing, China, 16–20 October 2011; Volume 2, pp. 1135–1138.
70. Chaitusaney, S.; Yokoyama, A. Prevention of Reliability Degradation from Recloser–Fuse Miscoordination Due to Distributed Generation. *IEEE Trans. Power Deliv.* **2008**, *23*, 2545–2554. [CrossRef]
71. Costa, G.B.; Marchesan, A.C.; Morais, A.P.; Cardoso, G.; Gallas, M. Curve fitting analysis of time-current characteristic of expulsion fuse links. In Proceedings of the 2017 IEEE International Conference on Environment and Electrical Engineering and 2017 IEEE Industrial and Commercial Power Systems Europe (EEEIC/I&CPS Europe), Milan, Italy, 6–9 June 2017; pp. 1–6.
72. Supannon, A.; Jirapong, P. Recloser-fuse coordination tool for distributed generation installed capacity enhancement. In Proceedings of the 2015 IEEE Innovative Smart Grid Technologies—Asia (ISGT ASIA), Bangkok, Thailand, 3–6 November 2015; pp. 1–6.
73. Kim, I.; Regassa, R.; Harley, R.G. The modeling of distribution feeders enhanced by distributed generation in DIgSILENT. In Proceedings of the 2015 IEEE 42nd Photovoltaic Specialist Conference (PVSC), New Orleans, LA, USA, 14–19 June 2015; pp. 1–5.
74. Tamizkar, R.; Javadian, S.; Haghifam, M.-R. Distribution system reconfiguration for optimal operation of distributed generation. In Proceedings of the IEEE 2009 International Conference on Clean Electrical Power, Capri, Italy, 9–11 June 2009; pp. 217–222.
75. Fazanehrafat, A.; Javadian, S.A.M.; Bathaee, S.M.T.; Haghifam, M.R. Maintaining the recloser-fuse coordination in distribution systems in presence of DG by determining DG's size. In Proceedings of the IET 9th International Conference on Developments in Power Systems Protection (DPSP 2008), Glasgow, UK, 17–20 March 2008; pp. 132–137.
76. Sookrod, P.; Wirasanti, P. Overcurrent relay coordination tool for radial distribution systems with distributed generation. In Proceedings of the IEEE 2018 5th International Conference on Electrical and Electronic Engineering (ICEEE), Istanbul, Turkey, 3–5 May 2018; pp. 13–17.
77. Shobole, A.; Baysal, M.; Wadi, M.; Tur, M.R. Effects of distributed generations' integration to the distribution networks case study of solar power plant. *Int. J. Renew. Energy Res.* **2017**, *7*, 954–964.
78. Ahmad, I.; Palensky, P.; Gawlik, W. Multi-Agent System based voltage support by distributed generation in smart distribution network. In Proceedings of the IEEE 2015 International Symposium on Smart Electric Distribution Systems and Technologies (EDST), Vienna, Austria, 8–11 September 2015; pp. 329–334.

79. Alsokhiry, F.S.; Lo, K.L. Effect of distributed generations based on renewable energy on the transient fault—Ride through. In Proceedings of the IEEE 2013 International Conference on Renewable Energy Research and Applications (ICRERA), Madrid, Spain, 20–23 September 2013; pp. 1102–1106.
80. Chapi, F.; Fonseca, A.; Perez, F. Determination of Overcurrent Protection Settings Based on Estimation of Short-Circuit Currents Using Local Measurements. In Proceedings of the 2019 IEEE Fourth Ecuador Technical Chapters Meeting (ETCM), Guayaquil, Ecuador, 1–15 November 2019; pp. 1–6.
81. Rahman, M.; Arefi, A.; Shafiullah, G.M.; Hettiwatte, S. Penetration maximisation of residential rooftop photovoltaic using demand response. In Proceedings of the IEEE 2016 International Conference on Smart Green Technology in Electrical and Information Systems (ICSGTEIS), Bali, Indonesia, 6–8 October 2016; pp. 21–26.
82. Tengku Hashim, T.J.; Mohamed, A.; Shareef, H. Comparison of decentralized voltage control methods for managing voltage rise in active distribution networks. *Prz. Elektrotechniczny* **2013**, *89*, 214–218.
83. Abbott, S.R.; Fox, B.; Morrow, D.J. Distribution network voltage support using sensitivity-based dispatch of Distributed Generation. In Proceedings of the 2013 IEEE Power Energy Society General Meeting, Vancouver, BC, Canada, 21–25 July 2013; pp. 1–5. [CrossRef]
84. Roibás-Millán, E.; Alonso-Moragón, A.; Jiménez-Mateos, A.G.; Pindado, S. Testing solar panels for small-size satellites: The UPMSAT-2 mission. *Meas. Sci. Technol.* **2017**, *28*, 115801. [CrossRef]

applied sciences

Article

Detection Criterion for Progressive Faults in Photovoltaic Modules Based on Differential Voltage Measurements

Luis Diego Murillo-Soto [1,*,†] and Carlos Meza [1,2,†]

1. Costa Rica Institute of Technology, Electromechanical Engineering School, Cartago 159-7050, Costa Rica
2. Department of Electrical, Mechanical and Industrial Engineering, Anhalt University of Applied Sciences, 06366 Köthen, Germany; carlos.meza@hs-anhalt.de
* Correspondence: lmurillo@tec.ac.cr
† These authors contributed equally to this work.

Abstract: PV modules may experience degradation conditions that affect their power efficiency and affect the rest of the PV array. Based on the literature review, this paper links the parameter variation on a PV module with the six most common degradation faults, namely, series resistance degradation, optical homogeneous degradation, optical heterogeneous degradation, potential induced degradation, micro-cracks, and light-induced degradation. A Monte Carlo-based numerical simulation was used to study the effect of the faults mentioned above in the voltage of the modules in a PV array with one faulty module. A simple expression to identify faults was derived based on the obtained results. The simplicity of this expression allows integrating the fault detection technique in low-cost electronic circuits embedded in a PV module, optimizer, or microinverter.

Keywords: fault location in photovoltaic arrays; failure modes simulation; fault detection criterion

Citation: Murillo-Soto, L.D.; Meza, C. Detection Criterion for Progressive Faults in Photovoltaic Modules Based on Differential Voltage Measurements. *Appl. Sci.* **2022**, *12*, 2565. https://doi.org/10.3390/app12052565

Academic Editors: Luis Hernández-Callejo, Maria del Carmen Alonso García and Sara Gallardo Saavedra

Received: 1 February 2022
Accepted: 18 February 2022
Published: 1 March 2022

Publisher's Note: MDPI stays neutral with regard to jurisdictional claims in published maps and institutional affiliations.

1. Introduction

According to [1], a fault can be defined as "an abnormal condition that may cause a reduction in, or loss of, the capability of a functional unit to perform a required function". Faults in photovoltaic (PV) installations can significantly affect their energy yield, which is why the development of fault detection and diagnosis techniques has become an essential topic of research in recent years, Refs. [2–4]. Research works that analyze failure mode in solar modules, e.g., Refs. [5–11], have identified that the power reduction in PV modules under fault causes a deformation of the current-voltage curve, e.g., curves in [12].

Recently, Ref. [13] presented a performance analysis with 30 faulty modules of different brands. The study shows that, on average, the power loss after two years, with several faults, drops around 1.08%; also, the voltage variation at maximum power was, on average −1.17%. Similar research in India concluded that, on average, power degradation in PV modules is found to increase 1.4% per year over 25 years [14]. But this degradation rate is not a static value; it depends on the life stage of the PV installations as is shown in [15], the first 18 years around 0.1% per year; then the next period of 10 years, the rate raises around 1%/year, and the final life stage, the rate increases more than 1.2%/year.

It is important to highlight that failure modes at the module level affect the whole PV installation performance because the PV array is a symmetrical composition of PV modules as is demonstrated by Gokmen in [16]. Eighteenth types of faults in solar modules have been described in [17], their detectable effects sometimes overlap, so it is not easy to distinguish among them. Furthermore, the diagnoses of faults require combining knowledge of different domains (visual inspection, thermography, electrical, chemical, material analyses, so on), Refs. [18,19]. This means that the specific failure modes could have, for example, the same behavior in the electric domain.

The detection criterion used in this paper is based on the widely accepted five parameters model, which, according to [20], can represent a single solar cell or several cells connected in series (PV module). This model is mathematically represented as follows,

$$I = I_{ph} - I_s \left(e^{\frac{V + IR_s}{N_s \eta V_t}} - 1 \right) - \frac{(V + IR_s)}{R_p} \tag{1}$$

where I and V represent the current and voltage of the module, I_{ph} is the photo-current generated, I_s is the saturation current of the diode, R_s and R_p are the parasitic series and parallel resistances, N_s is the number of cells in series, η represent the ideality factor, and V_t is the thermal voltage.

An increase in the PV module series resistance can model several fault mechanisms, as it has been indicated in [21,22]. However, using other parameters to explain other failure modes is scarce in the literature. In this regard, one of the contributions of this work is to propose parameter ranges in the PV model that represent others failure modes not included in the typical models, e.g., only series resistance variation. This work will focus on the most common types of degrading faults listed next.

- Series resistance degradation (SRD) faults. The increase of the parasitic resistance R_s could be expected by thermal cycling, solder corrosion, broken ribbons, and homogeneous soldering disconnection, so on. For example, Ref. [23] indicates that normally the increment is 10 %/year. However, the variation of R_s could be up to ten times its original value, according to [24].
- Optical homogeneous degradation (OD) faults. The photocurrent I_{ph} is affected by homogeneous glass corrosion, homogeneous contamination, loss of transparency, homogeneous corrosion of the anti-reflection coating. For instance, the yellowing and the browning are faults that cause power losses up to 50% [25]. According to [23] this parameter decreases its value between $[0, 2]$ %/year, for simulation purpose a window of 20 years were simulated.
- Optical heterogeneous degradation (OHD) faults. This fault family is similar to the one presented before, but in this case, the level of affectation is not homogeneous in two ways, location and degree of degradation. According to [17] the parasitic parallel resistance R_p decreases, as well as I_{ph}. The variation of I_{ph} was taken as equivalent to OD faults, and the variation of R_p was taken as the worse case reported in PID faults.
- Potential induced degradation (PID) faults. This fault occurs when high potential voltages of the system cause a leakage current between the frame and the solar cells. According to [26], the shunt resistance R_p decreases as well as the short-circuit current I_{sc}, but for this work, we assume that the changes in I_{sc} are caused by I_{ph} because these variables are proportional $I_{sc} \propto I_{ph}$. In [27] has proposed that both parameter decrease proportionally, $\downarrow R_p \propto \downarrow I_{sc}$. Hence, the range of variation for these parameters are $R_p = [1, 100]\%$ and $I_{ph} = [80, 100]\%$. We call this variant as PID^1.

  Also, another PID fault variation, we call PID^2, is found in [28], and they show that the parameters R_p moves between $[0.1, 100]\%$, I_s varies between $[100, 300]\%$ and η moves between $[100, 112]\%$. These percentages were calculated by analyzing the maximum and minimum values reported in the graphical results.
- Micro cracks (MC) faults. Micro crack can reduce energy production up to 50% of a module with only 40% of the affected area in one cell [29]. For these faults, according to [30], I_s and I_{ph} decrease proportionally because both parameters are inversely proportional to the affected area; hence the range of variation is $[1, 100]\%$.
- Light induced degradation (LID) faults. According to [31] LID is defined as the increment of the recombination current in the base in P-type silicon wafers. The effect is also observed in the open-circuit voltage [17]. For a particular module, the saturation current I_s could be moved in a range of $[1, 50]$ times; this variation

can produce changes of open-circuit voltage V_{oc} in the range of $[0, 18]\%$ [20]. These percentages were calculated by analyzing graphical results.

A second contribution is the proposal of new detection criteria based on the measurements of the module voltages. Due to its simplicity, this criterion is suitable for implementation in monitoring strategies such as the one presented in [32]. In fact, several approaches use voltage measurements to detect and diagnose faults in the array, for example [33–38], but they require other variables like the string current, time, or even the ambient variables, to generate alarms adequately. Also, in those papers, the nature of faults analyzed is different because they focus mainly on faults such as diode short-circuit, open-circuits, partial shadows, and degradation faults represented only as an increase of the serial resistance. Furthermore, a similar indicator based on voltage is presented in [39], which is called ΔV, but this indicator requires knowing precisely the output voltage of the array and the theoretical voltage at its maximum power. This indicator is quite different from our approach since our criterion is calculated from the voltages of the modules in the PV string.

This work presents and analyzes new detection criteria to demonstrate that is capable to detect all the progressive faults only with the module voltage variable in multi-crystal PV modules. This numerical simulation study is divided into the following main sections: Section 2 describes the theoretical framework used to derive the detection criterion. Section 3 links specific fault conditions with changes in the variation of some parameters of the PV array. This section also explains the simulated experiment, circuits and software used, and the calibration process. Sections 4 and 5 show the main results and the discussion. Finally, the main conclusions are highlighted in section 6.

2. Theoretical Framework: Detection Criterion Fundamentals

The explicit expression of (1) for the module voltage is obtained using Lambert-$\mathcal{W}$ transformation, if we have an equation with the form $y = xe^x$, the term x could be obtained as $x = \mathcal{W}(y)$ for any $x > -1/e$. Therefore, the voltage is expressed as follows,

$$V = (I_{ph} + I_s)R_p - I(R_p + R_s) - N_s\eta V_t \mathcal{W}\left(\frac{R_p I_s}{N_s \eta V_t}e^{\left(\frac{(I_{ph} + I_s - I)R_p}{N_s \eta V_t}\right)}\right). \tag{2}$$

It is clear that if a failure affects the PV module performance, the parameters in (1) and (2) change their values and hence the voltage V in (2) also varies [28]. If the initial values of all the parameters are known it is possible to identify the degradation condition of one PV module using (2). Nevertheless, the aforementioned is seldom the case. Not only the initial values are unknown but also a PV installation is comprised of several PV modules connected in series and parallel and the relationships between the parameter variation in one PV module will affect the others.

In a typical photovoltaic plant, PV modules are connected using a configuration known as series-parallel (SP). In an SP configuration, the PV modules are first connected in series forming strings which are then connected in parallel as depicted in Figure 1, forming a PV array of $m \times n$ modules. The voltages in the j-string, formed by m modules, are governed by Kirchhoff's voltage law [20], as follows;

$$\sum_{i=1}^{m} V(i,j) - V_{blk}(j) - V_{op} = 0 \quad , \tag{3}$$

where $V(i,j)$ is the differential voltage of the module i in the string j as in Equation (2), $V_{blk}(j)$ is the voltage of the blocking diode, and V_{op} is the operational voltage point of the array.

When all the modules are identical (3) can be simplified as follows,

$$mV(j) - V_{blk}(j) - V_{op} = 0. \tag{4}$$

Now consider the case in which at least one module is degraded or faulty. Due to the SP configuration the sum of all the voltages in the string with the faulty PV module does not change, see Equation (3). Therefore, the PV modules that are part of the string in which a faulty module is located will change their electrical operating point. We call the aforementioned elements *affected* PV modules, as it can be seen in Figure 2.

If we denote m_f as the number of modules with the same fault, (3) can be rewritten as,

$$(m - m_f) \cdot V_a(j) + m_f \cdot V_f(j) = V_{op} + V_{blk}(j) \tag{5}$$

where $V_f(j)$ is the voltage of the faulty module (*degradeted*) and $V_a(j)$ is the voltage of the affected modules.

Notice that the sum of the voltages at the faulty and at the affected modules in a given string should be equal to the sum of the voltages of non-faulty modules in a string parallel to it. Hence, the variation in the voltages of the affected modules and the faulty modules must behave oppositely, i.e., when the voltage at the faulty decreases the voltage at the affected modules increases. Moreover, the difference between the voltages at the affected and the faulty modules can be expressed as follows,

$$\Delta_V = \frac{V_a - V_f}{V_a} 100, \qquad \forall I > 0. \tag{6}$$

Given that the current trough the string, I, is shared by the modules, (6) can be rewritten as

$$\Delta_V = \frac{IV_a - IV_f}{IV_a} 100 = \frac{P_a - P_f}{P_a} 100, \qquad \forall I > 0. \tag{7}$$

where P_a and P_f are the power produced by the affected and faulty modules, and I is the current of the j-string (j). Therefore (6) represents a measure proportional to the power reduction in a PV module.

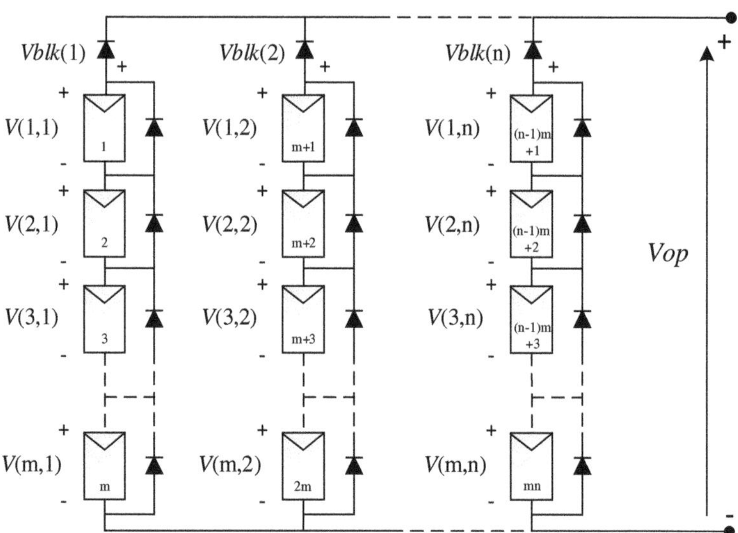

Figure 1. Nomeclature for the SP PV array.

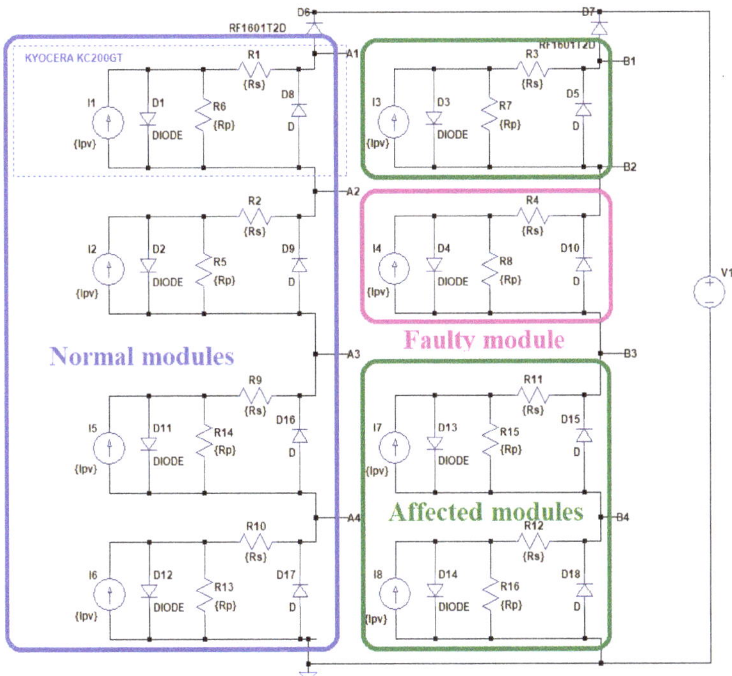

Figure 2. Circuit simulated in LtSpice software with seven types of faults.

3. Materials and Methods

3.1. Parameter Variation in Failure Modes

The way in which photovoltaic modules degrade their power production is called failure mode, in other words, the process of how a PV module is going to fail is also known as a fault. The numerical simulation study considers variations in the parameters for 20 years according to the measured values of parameters reported in [17,20,23–30] for the faults mentioned in the Introduction. In this regard, Table 1 presents the range of variation of these parameters.

Table 1. Parameter variations for different fault families.

Fault Type:	$R_s(\%)$	$R_p(\%)$	$I_s(\%)$	$I_{ph}(\%)$	$\eta(\%)$
SRD	$[100, 1000]$				
OD				$[60, 100]$	
OHD		$[0.1, 100]$		$[60, 100]$	
PID[1]		$[1, 100]$		$[80, 100]$	
PID[2]		$[0.1, 100]$	$[100, 300]$		$[100, 112]$
MC			$[1, 100]$	$[1, 100]$	
LID			$[100, 5000]$		

3.2. Test Bench for Simulation

The PV array was simulated using LtSpice [40] and was composed of 4×2 solar modules; each module uses the parameters of the Kyocera KC200GT PV module, as shown in Table 2. The simulated circuit is presented in Figure 2. The parameters used for the module are shown in Table 3 and were derived using the methodology proposed in [41]. Theoretical operative points are calculated and compared with the computing simulation to validate the general model of the 4×2 PV modules. The results are shown in Table 4 and this comparison is made with the relative error (*R. Error*) calculated as shown in (8).

$$R.\ Error = \frac{\text{Theoretical value} - \text{Experimental value}}{\text{Theoretical value}} \times 100. \tag{8}$$

Table 2. Electrical Performance at Standard Test Conditions (STC).

Specification for KC200GT	Value
Maximum Power ($Pmax$)	200.0 W
Maximum Power Voltage ($Vmpp$)	26.3 V
Maximum Power Current ($Impp$)	7.61 A
Open Circuit Voltage (Voc)	32.9 V
Short Circuit Current (Isc)	8.21 A
Temperature Coefficient of Voc (K_V)	-1.23×10^{-1} V/°C
Temperature Coefficient of Isc (K_I)	3.18×10^{-3} A/°C
Number of series cell (N_s)	54

Table 3. Values of the five-parameter model of the KC200GT [41].

Parameter	Value
Saturation current (I_s)	9.825×10^{-8} A
Photo current(I_{ph})	8.214 A
Series resistance (R_s)	0.221 Ω
Parallel resistance (R_p)	415.405 Ω
Ideally factor (η)	1.3

Table 4. Electrical performance of the 4×2 array.

Specification	Theoretical Values	Simulation Values	R. Error
$Pmax$	1600 W	1599.4 W	0.04%
$Vmpp$	105.20 V	105.40 V	−1.19%
$Impp$	15.22 A	15.18 A	0.26%
Voc	131.6 V	132 V	−1.30%
Isc	16.42 A	16.42 A	0.00%

The simulations consist of selecting a PV module in the array and simulating each fault type 500 times. The selected PV module is called a faulty one, as shown in Figure 2. Furthermore, the numerical simulation changes the parameter values randomly according to Table 1 using a flat distribution over their predefined ranges. A sweep of voltage is made by the controlled voltage source for every new parameter setup, generating new curves. Also, it is important to mention that the voltage sweep varies from 0 to $4Voc$ V in steps of 0.5 V. In summary, every fault simulation has 500 parameter variations generating the following curves:

(a) the current of the string where is located the faulty module versus the operational voltage,

(b) the differential voltages of the modules in the faulty string versus the operational voltage,

(c) the relative percentage difference versus the operational voltage,

(d) the output power of the array versus the operational voltage.

To simplify the experiments and for readability issues, the results for the seven simulated fault types are shown in Figure 3, and in the Appendix A from Figures A1–A6. These charts correspond only to faults located in the position $(2,2)$ in the array as indicated by the magenta rectangle in Figure 2. These curves can be extrapolated and are valid for faults in any position inside the array. All the simulations of the circuit were performed at standard test conditions, i.e., at an irradiance of 1000 W/m^2 and a cell temperature of 25 ° C, which are the main input references to get the electrical measurements reported in data-sheets [42]. Also, the datasheets give data to contrast with the simulation. The translation equations were implemented according to [20], but no variations for these experiments were made on the temperature or the irradiance variables.

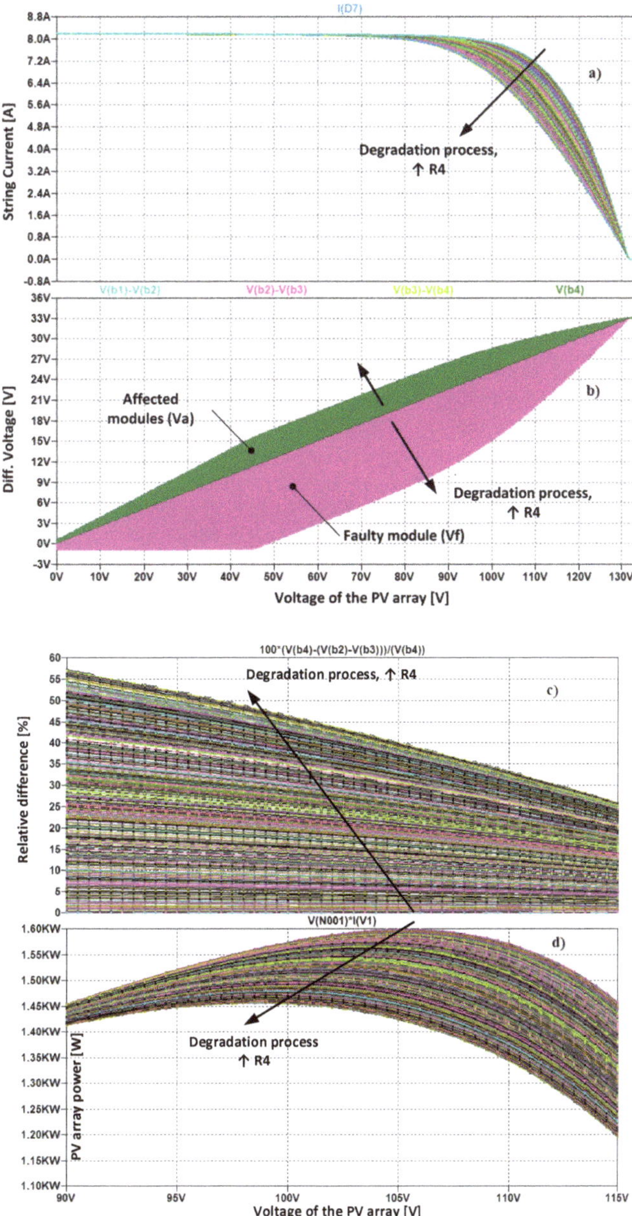

Figure 3. Series resistance degradation faults. This figure shows four charts as follows: (**a**) the current of the affected string versus the operating voltage of the array, (**b**) the differential voltage for all the modules in the affected string, (**c**) the relative percentage difference versus the operating voltage of the array, (**d**) peak behavior of output power in the array versus the operating voltage of the array.

4. Results

The results of the SRD fault are presented in Figure 3. The SRD fault was simulated by varying the resistance R4 up to ten times its original value. Figure 3a shows how the current of the faulty string changes as the degradation process caused by resistor R4 continues. This means that the differential voltages V_a and V_f are separating as the resistor R4 is increasing; see Figure 3b. The relative percentage difference calculated with Equation (6) is presented in Figure 3c. In this case, it is shown that there is a power loss by the faulty module concerning the non-faulty modules in the same string; the arrow shows the degradation process. The array's output power is shown in Figure 3d, and it goes down according to the increment of the serial resistance R4; also, the degradation process is indicated with the arrow.

Table 5 shows how the resistor value affects the voltage at the maximum power-point ($Vmpp$), the maximum power point ($Pmax$), and the total power loss (TPL). The TPL indicates how much power is lost in the whole PV array with respect to an expected non-faulty value of 1600 W, as indicated in Equation (8). Additionally, the table shows the value of voltage for the affected (V_a) and faulty (V_f) PV modules as well as the relative percentage difference (Δ_V) between them. The last column in the table shows the module power loss (MPL) by the faulty module with respect to its maximum power as reported in the datasheet (200 W).

The graphical results for the following six fault types are presented in the Appendix A, each figure belongs to one fault type, and it contains the same four charts explained before for the SRD fault. In the Appendix A, the Figure A1 presents OD fault, and the results were obtained moving I4 values randomly. Figure A2 presents OHD fault; this simulation changes the values for R8 and I4 randomly and separately. Figure A3 presents PID[1] fault; here, the elements R8 and I4 move randomly but proportionally. Next, Figure A4 presents PID[2] fault; for this simulation, R8, I4, and the parameter η in diode D4 move freely and randomly. Further, the results for MC fault are presented in Figure A5, here I4, and the parameter I_s in the diode D4 move randomly but in a proportional way. Finally, Figure A6 shows the results for LID fault when the parameter I_s moves in the diode D4. It is interesting to see that all fault types present similar patterns and behaviors in all the charts.

Table 5. Numeric results for the increment of R4 in the 4×2 array.

R4	$Vmpp$ (V)	$Pmax$ (W)	TPL (%)	V_a (V)	V_f (V)	Δ_V (%)	MPL (%)
$1.0R_s$	105.5	1599.4	0.04	26.59	26.59	0	0
$1.5R_s$	105.0	1592.9	0.44	26.67	25.84	3.13	3.14
$2.0R_s$	104.5	1586.1	0.86	26.75	25.09	6.21	6.26
$2.5R_s$	104.0	1579.1	1.30	26.83	24.35	9.27	9.35
$3.0R_s$	104.0	1571.9	1.75	27.03	23.75	12.15	12.40
$3.5R_s$	103.5	1564.5	2.21	27.11	23.02	15.09	15.40
$4.0R_s$	103.0	1556.9	2.69	27.19	22.29	17.99	18.39
$4.5R_s$	102.5	1549.1	3.42	27.44	21.02	23.42	24.17

5. Discussion

Only the PV module under examination has one type of progressive fault in the previous circuit simulations. The rest of the elements, such as PV modules, wires, bypass diodes, were assumed to work in a non-faulty condition. The aforementioned does not mean that the results are specific to the selected module location; if the faulty module is located at any other place in the array, the obtained charts would be equivalent to the graphs presented in this work.

Based on the simulation results it is confirmed that when a progressive fault affects a photovoltaic module, its differential voltage changes according to the severity of the failure mode. Figure 3b confirms that the progressive fault unbalances the voltage in the string; this means that as the progressive fault progress, the faulty module's differential voltage decrease, and the differential voltage of the affected modules increase. Hence, the relative percentage difference (Δ_V) increases, and the output power of the array decreases; these facts are easily checked in Figure 3c,d. For all the fault types analyzed, the same performance was observed in sub-charts in Figures A1–A6.

The degree of fault affectation in the module is calculated with the relative percentage difference (6), and this indicator is equivalent to the percentage of power loss if the modules work at the same string current. Table 5 presents the numerical results for the SRD fault. These results correspond to the evolution of the maximum power-point. In these maximum power points, it seems that the power loss indicator (Δ_V) is highly correlated with the power loss of the module calculated at standard test conditions, MPL. Actually, for this fault condition, the estimated power Δ_V and the MPL are highly correlated, the Pearson Coefficient [43] is, in this case, $r = 0.9999$; therefore, it is possible to calculate a linear regression as,

$$MPL = 1.0311\Delta_V - 0.1082, \; \forall \; \Delta_V \; at \; Pmax. \tag{9}$$

Moreover, the obtained linear regression has a very low variability, given a determination coefficient R^2 of 99.99%. This result is not unique for this failure mode, the correlation between Δ_V and MPL appears for all the fault types analyzed. Graphically, it can be appreciated in Figure 4 and the parameters for the best fit line equation presented in Table 6.

In this PV array model of 1.6 kWp, as the progressive fault increase, the $Vmpp$ moves affecting all the modules' power production; this means generating small losses in all the modules in the array. These small changes in power production caused by the progressive fault could be hidden because they are too small to be detected by the monitoring system. For the 1.6 kWp example, if R_s double its value in whatever PV module, the total power loss of the array drops about 0.86%, which could be negligible. However, this percentage represents one faulty module losing around 6.26% of its energy (See the third row in Table 5). This hidden effect is even more drastic in large PV installations; let's suppose that a faulty module has a constant power loss, and if the number of PV modules in the array is high enough, the total power loss tends to zero. However, the faulty module is still losing the same percentage of energy and may evolve into a more severe fault condition.

Table 6. Parameters of the best fit line equation for all fault types.

Fault Type	Slope m	Intersection b	r	$R^2(\%)$
SRD	1.0311	−1.1082	0.9999	99.99
OD	1.3257	−1.5509	0.9946	98.94
OHD	1.3539	−1.9737	0.9942	98.85
PID[1]	1.2502	−1.7063	0.9987	99.75
PID[2]	1.1499	−1.2365	0.9997	99.94
MC	1.3309	−1.9012	0.9959	99.18
LID	1.0191	−1.0461	0.9999	99.99

The correlation $MPL \propto \Delta_V$ can be used to detect abnormal behavior. Moreover, it is possible to develop a simple criterion for fault detection and location in SP arrays. The Equation (10) is based on (6), and it checks if the relative percentage difference between two voltages is more than a threshold value called δ. Here V_{max} represents the maximum differential voltage of all the modules in the string j, and $V(i)$ is the differential voltage of the analyzed module. It should be noticed that V_{max} is the higher voltage in the string, and it belongs to the module with less degradation. This criterion is simple to incorporate into other fault detection proposals that use differential voltages as input signals such as [44,45].

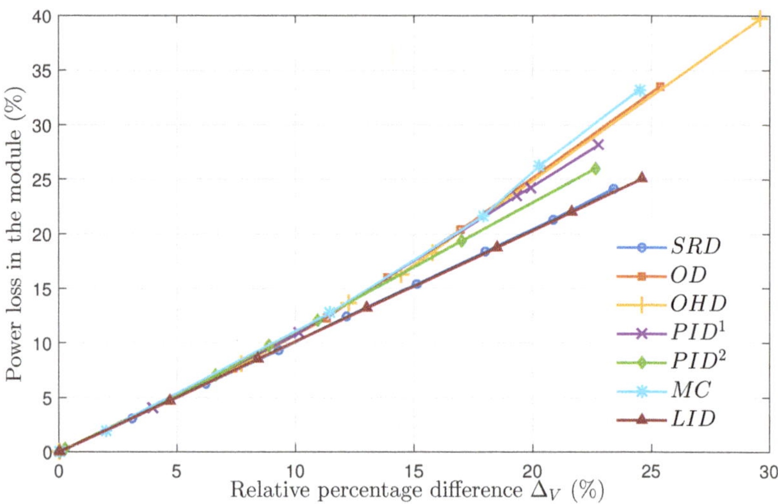

Figure 4. Power loss of the modules versus Δ_V at mpp for the studied faults.

$$Detect(V_{max}, V(i), \delta) = \begin{cases} True & if & \dfrac{V_{max} - V(i)}{V_{max}} > \delta \\[3mm] False & if & \dfrac{V_{max} - V(i)}{V_{max}} \leq \delta \end{cases}, \forall i = \{1, 2, ..., m\}. \tag{10}$$

To define the fault threshold, it is crucial to know the environment in which the solar array is placed. For instance, soiling may not be a problem in tropical regions benefiting from rainfall cleaning. On the contrary deserts or dry places, the PV array could be affected by dust storms or air pollution that could reduce the general performance [46,47]. In [48] it has been reported that power losses in outdoor conditions could be reduced between [5, 6]% and for laboratory conditions, it is possible to reduce it up to 40%. For instance, in tropical weather like Phitsanulok Tayland, it has been reported in [49] a decrease in solar radiation of [3.71, 11.15]% when the dust deposition rate (DDR) is 425 mg/m^{2}d in 60 days. On the contrary dry cities like Mexico City for also 60 days, the DDR reported is 102 g/m^{2}d to reduce the performance ratio up to 15%.

Detecting permanent faults such as PID, hotspots, or micro-cracks with online real-time methods is always a challenge. For instance, if a hotspot is considered mild, its temperature is just 10 °C higher than the other parts of the cells; however, if the hot spot is considered severe, it presents a temperature higher by approximately 18 °C. These facts mean that power production could be reduced between 4% and 10% [50]. A similar analysis can be done with micro-crack; for instance, in a PV module formed with 60 cells, if just one cell has an inactive area of 25%, the power loss in the whole module is about 10% [30]. Therefore, a rule to avoid false faults due to normal soiling or other temporal issues should use a δ value between 5 and 10 % as a threshold.

6. Conclusions

This work has analyzed several progressive faults presented in Table 1 and has been able to conclude the following:

- The differential voltage in the affected modules will always be higher than the differential voltage in the faulty module.
- The relative percentage difference (Δ_V) always increases in proportion to the severity (power loss) of the fault wherever operational points of the array.

- The power loss per module is proportional to the Δ_V if the PV system works at maximum power-point.

The criteria proposed to detect permanent and progressive faults are based on estimating the power degradation in the faulty PV module. This is done with the differential voltages of the modules. This new detection criterion is suitable for real-time online analysis in PV arrays, which is part of the additional work, to experimentally demonstrate the simplicity of this fault detection criterion, running on a real-time system with several faulty modules.

Author Contributions: Conceptualization, L.D.M.-S.; methodology, L.D.M.-S.; software, L.D.M.-S.; validation, L.D.M.-S. and C.M.; formal analysis, L.D.M.-S.; investigation, L.D.M.-S. and C.M.; resources C.M.; data curation, L.D.M.-S.; writing—original draft preparation, L.D.M.-S.; writing review and editing, C.M.; visualization, L.D.M.-S.; supervision, C.M.; project administration, C.M.; funding acquisition, C.M. All authors have read and agreed to the published version of the manuscript.

Funding: This work was supported by the scholarship program of the Costa Rica Institute of Technology and the VIE project 5402-1360-4201.

Data Availability Statement: Not applicable.

Acknowledgments: Special thanks to Giovanni Spagnuolo from the University of Salerno, Italy, for his comments on this work.

Conflicts of Interest: The authors declare no conflict of interest.

Abbreviations

The following abbreviations are used in this manuscript:

PV	PhotoVoltaic
SRD	Series Resistor Degradation
OD	Optical Homogeneous Degradation
OHD	Optical Heterogeneous Degradation
PID^1	Potential Induced Degradation variant one
PID^2	Potential Induced Degradation variant two
MC	Micro-cracks
LID	Light Induced Degradation
TPL	Total Power loss
MPL	Module Power loss

Appendix A. Curves for the Simulated Failure Modes

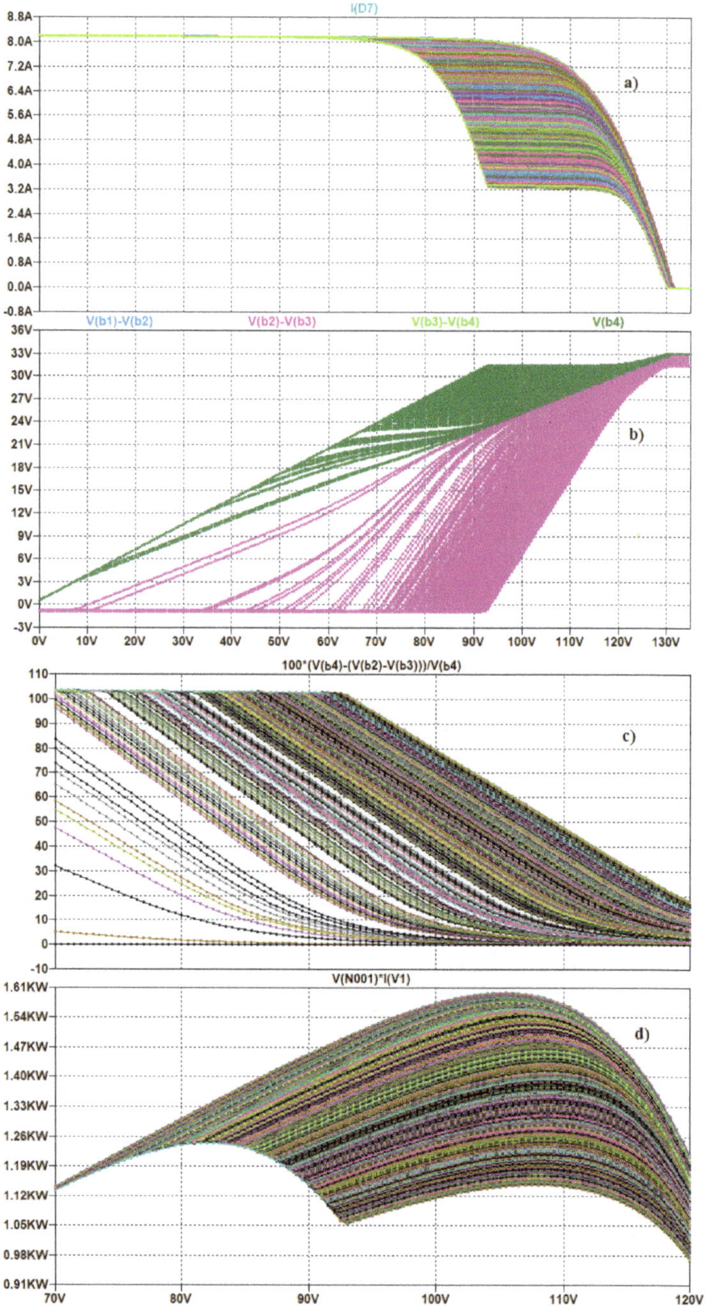

Figure A1. Optical homogeneous degradation faults. This figure shows four charts as following: (**a**) the current of the affected string versus the operation voltage of the array, (**b**) the differential voltage for all the modules in the affected string, (**c**) the relative percentage difference versus the operating voltage of the array, (**d**) peak behavior of output power in the array.

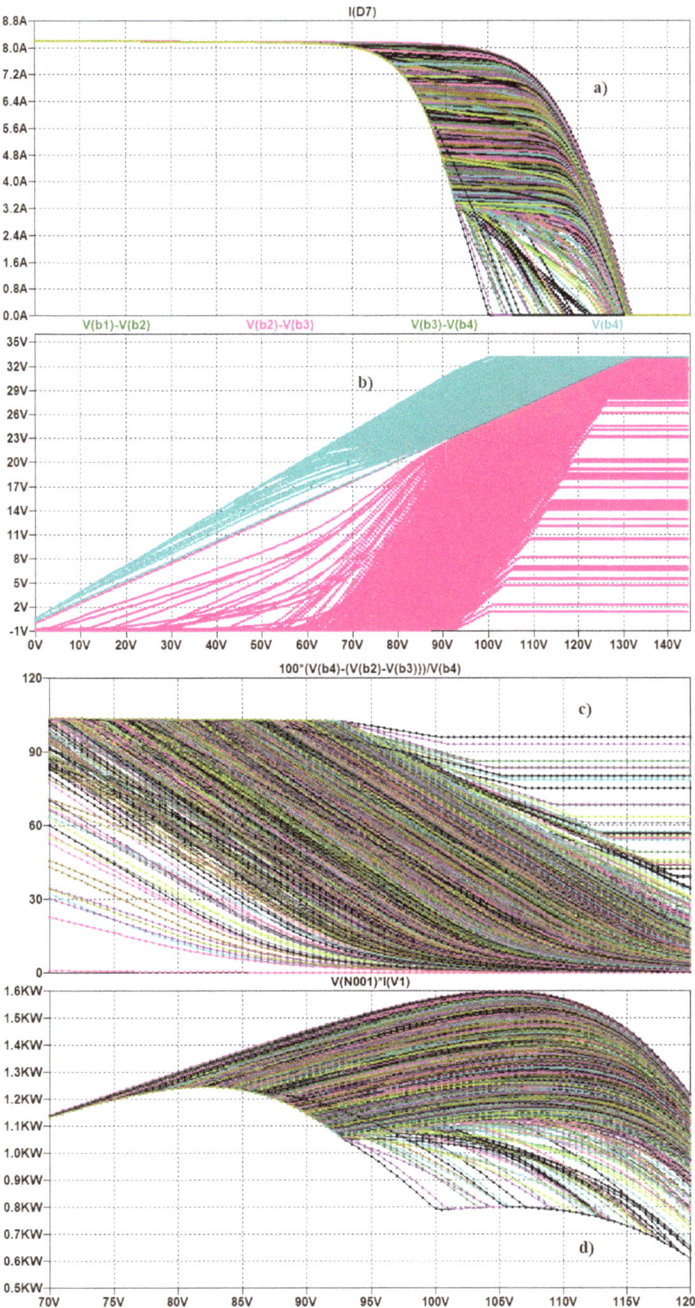

Figure A2. Optical heterogeneous degradation faults. This figure shows four charts as following: (**a**) the current of the affected string versus the operation voltage of the array, (**b**) the differential voltage for all the modules in the affected string, (**c**) the relative percentage difference versus the operating voltage of the array, (**d**) peak behavior of output power in the array.

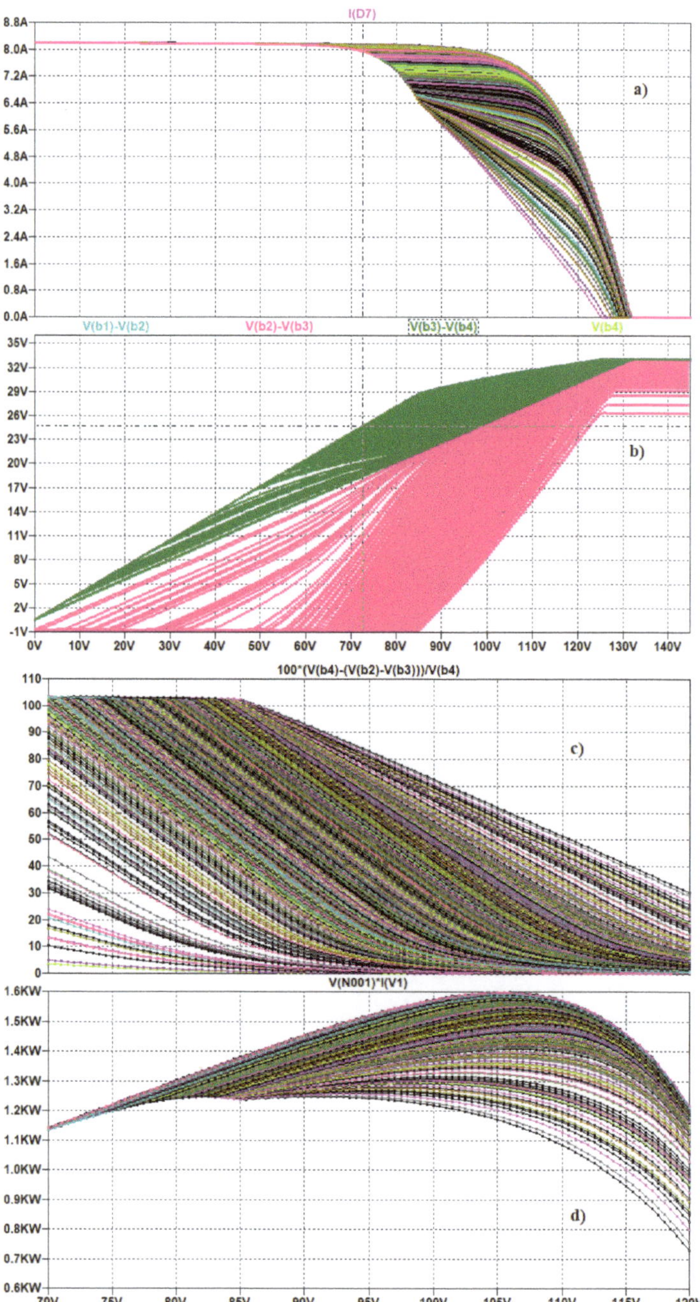

Figure A3. Potential induced degradation faults type 1. This figure shows four charts as following: (**a**) the current of the affected string versus the operation voltage of the array, (**b**) the differential voltage for all the modules in the affected string, (**c**) the relative percentage difference versus the operating voltage of the array, (**d**) peak behavior of output power in the array.

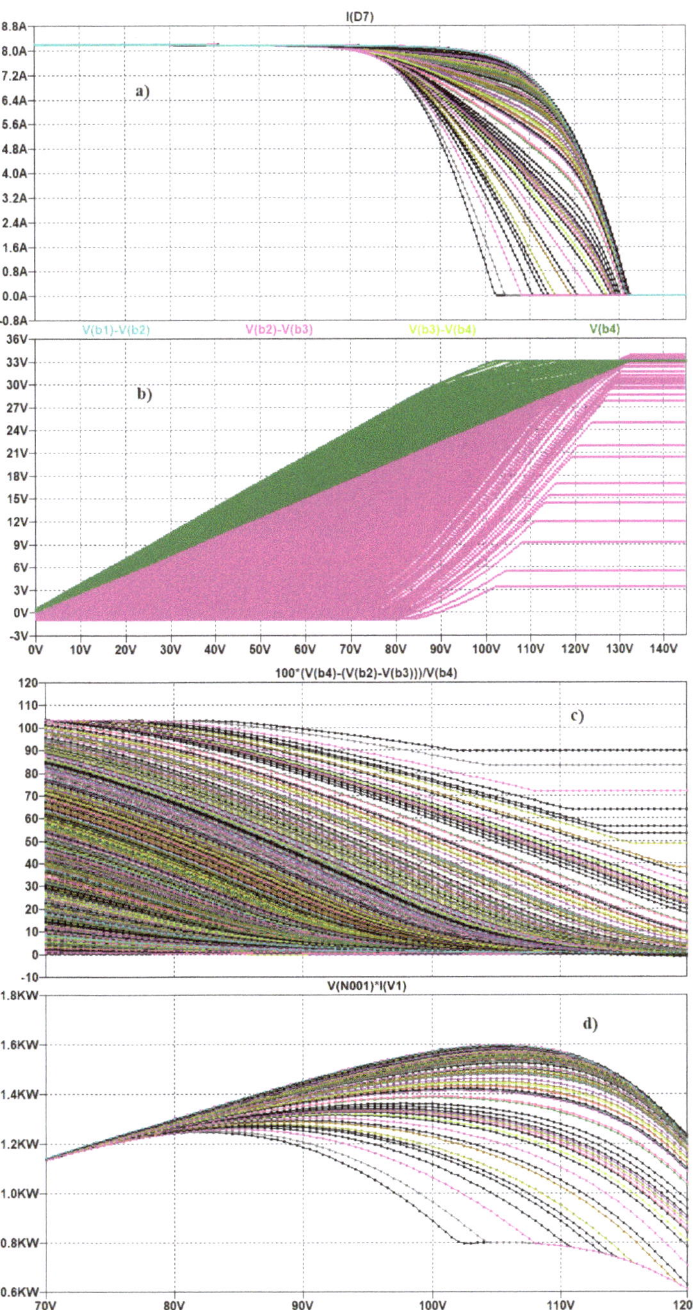

Figure A4. Potential induced degradation faults type 2. This figure shows four charts as following: (**a**) the current of the affected string versus the operation voltage of the array, (**b**) the differential voltage for all the modules in the affected string, (**c**) the relative percentage difference versus the operating voltage of the array, (**d**) peak behavior of output power in the array.

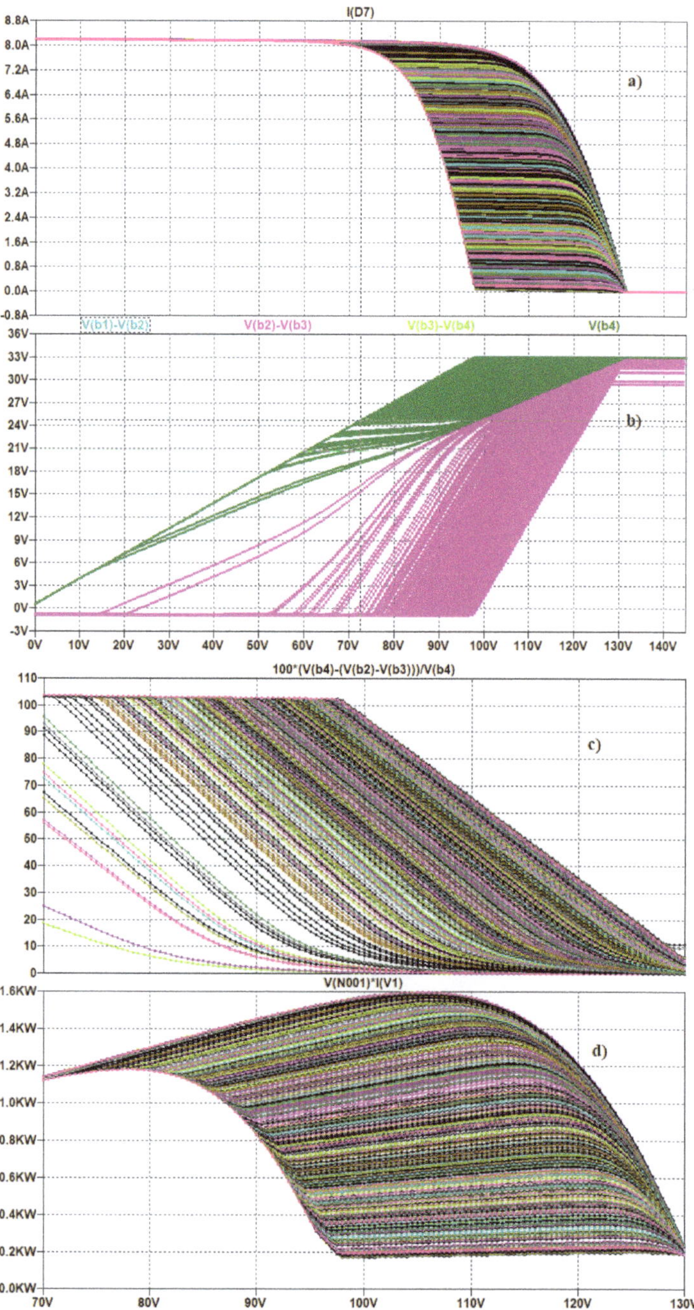

Figure A5. Micro cracks faults. This figure shows four charts as following: (**a**) the current of the affected string versus the operation voltage of the array, (**b**) the differential voltage for all the modules in the affected string, (**c**) the relative percentage difference versus the operating voltage of the array, (**d**) peak behavior of output power in the array.

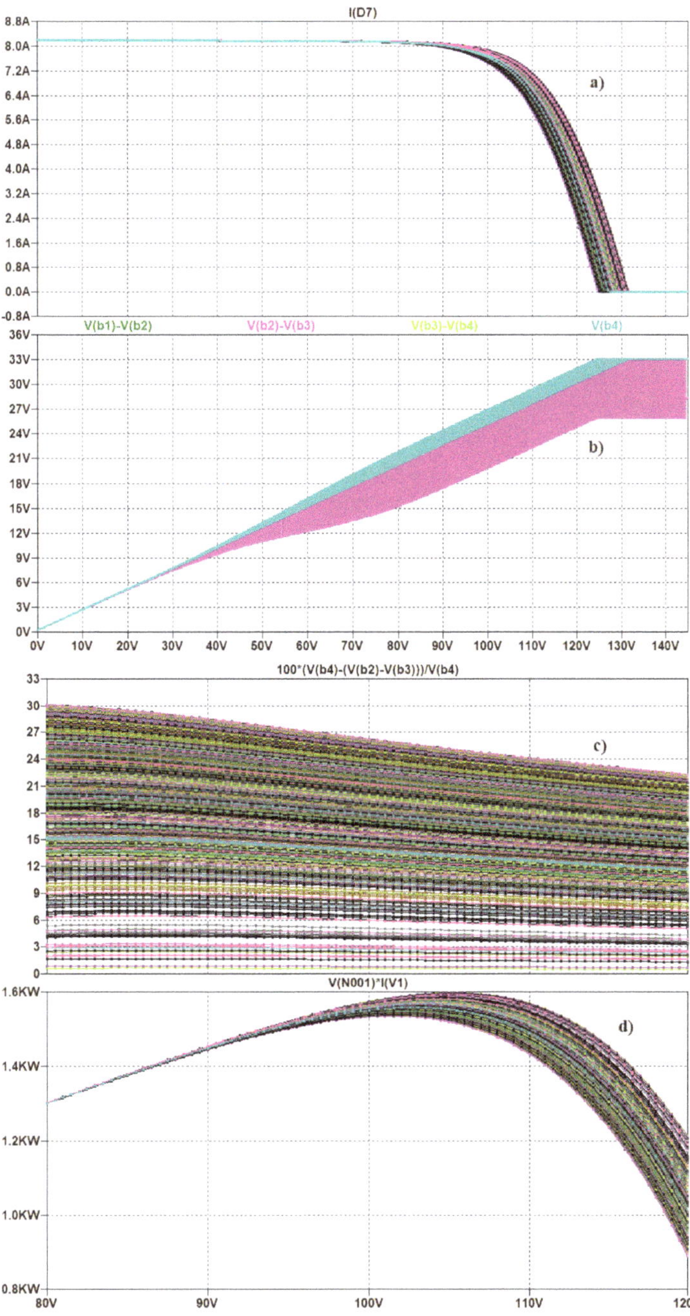

Figure A6. Light induced degradation faults. This figure shows four charts as follows: (**a**) the current of the affected string versus the operation voltage of the array, (**b**) the differential voltage for all the modules in the affected string, (**c**) the relative percentage difference versus the operating voltage of the array, (**d**) peak behavior of output power in the array.

References

1. ISO/IEC 2382-14. Reliability, Maintainability and Availability: Part 14—Vocabulary. 1997. Available online: https://www.iso.org/obp/ui/#iso:std:iso-iec:2382:-14:ed-2:v1:en (accessed on 31 January 2022).
2. Bastidas-Rodriguez, J.; Petrone, G.; Ramos-Paja, C.; Spagnuolo, G. Photovoltaic modules diagnostic: An overview. In Proceedings of the IECON 2013—39th Annual Conference of the IEEE Industrial Electronics Society, Vienna, Austria, 10–13 November 2013; pp. 96–101. [CrossRef]
3. Madeti, S.R.; Singh, S. A comprehensive study on different types of faults and detection techniques for solar photovoltaic system. *Sol. Energy* **2017**, *158*, 161–185. [CrossRef]
4. Mellit, A.; Tina, G.; Kalogirou, S. Fault detection and diagnosis methods for photovoltaic systems: A review. *Renew. Sustain. Energy Rev.* **2018**, *91*, 1–17. [CrossRef]
5. Manganiello, P.; Balato, M.; Vitelli, M. A Survey on Mismatching and Aging of PV Modules: The Closed Loop. *IEEE Trans. Ind. Electron.* **2015**, *62*, 7276–7286. [CrossRef]
6. Alam, M.K.; Khan, F.; Johnson, J.; Flicker, J. A Comprehensive Review of Catastrophic Faults in PV Arrays: Types, Detection, and Mitigation Techniques. *IEEE J. Photovolt.* **2015**, *5*, 982–997. [CrossRef]
7. Jordan, D.C.; Silverman, T.J.; Sekulic, B.; Kurtz, S.R. PV degradation curves: Non-linearities and failure modes. *Prog. Photovolt. Res. Appl.* **2017**, *25*, 583–591. [CrossRef]
8. Köntges, M.; Oreski, G.; Jahn, U. *Assessment of PV Module Failures in the Field*; Technical Report; International Energy Agency Photovoltaic Power Systems Programme: IEA PVPS Task 13. ISBN: 978-3-906042-54-1. 2017. Available online: https://iea-pvps.org/key-topics/report-assessment-of-photovoltaic-module-failures-in-the-field-2017/ (accessed on 31 January 2022).
9. Abdulmawjood, K.; Refaat, S.S.; Morsi, W.G. Detection and prediction of faults in photovoltaic arrays: A review. In Proceedings of the 2018 IEEE 12th International Conference on Compatibility, Power Electronics and Power Engineering (CPE-POWERENG 2018), Doha, Qatar, 10–12 April 2018; pp. 1–8. [CrossRef]
10. Gallardo-Saavedra, S.; Hernández-Callejo, L.; Duque-Pérez, O. Analysis and characterization of PV module defects by thermographic inspection. *Rev. Fac. Ing. Univ. Antioq.* **2019**, *93*, 92–104. [CrossRef]
11. Gallardo-Saavedra, S.; Hernández-Callejo, L.; Duque-Pérez, O. Quantitative failure rates and modes analysis in photovoltaic plants. *Energy* **2019**, *183*, 825–836. [CrossRef]
12. Li, B.; Delpha, C.; Migan-Dubois, A.; Diallo, D. Fault diagnosis of photovoltaic panels using full I–V characteristics and machine learning techniques. *Energy Convers. Manag.* **2021**, *248*, 114785. [CrossRef]
13. Ustun, T.S.; Nakamura, Y.; Hashimoto, J.; Otani, K. Performance analysis of PV panels based on different technologies after two years of outdoor exposure in Fukushima, Japan. *Renew. Energy* **2019**, *136*, 159–178. [CrossRef]
14. Chandel, S.; Naik, M.N.; Sharma, V.; Chandel, R. Degradation analysis of 28 year field exposed mono-c-Si photovoltaic modules of a direct coupled solar water pumping system in western Himalayan region of India. *Renew. Energy* **2015**, *78*, 193–202. [CrossRef]
15. Virtuani, A.; Caccivio, M.; Annigoni, E.; Friesen, G.; Chianese, D.; Ballif, C.; Sample, T. 35 years of photovoltaics: Analysis of the TISO-10-kW solar plant, lessons learnt in safety and performance—Part 1. *Prog. Photovolt. Res. Appl.* **2019**, *27*, 328–339. [CrossRef]
16. Gokmen, N.; Karatepe, E.; Celik, B.; Silvestre, S. Simple diagnostic approach for determining of faulted PV modules in string based PV arrays. *Sol. Energy* **2012**, *86*, 3364–3377. [CrossRef]
17. Köntges, M.; Kurtz, S.; Packard, C.; Jahn, U.; Berger, K.; Kato, K.; Friesen, T.; Liu, H.; Van Iseghem, M. *Review of Failures of Photovoltaic Modules*; Technical Report; International Energy Agency Photovoltaic Power Systems Programme: IEA PVPS Task 13. ISBN: 978-3-906042-16-9. 2014. Available online: https://iea-pvps.org/wp-content/uploads/2020/01/IEA-PVPS_T13-01_2014_Review_of_Failures_of_Photovoltaic_Modules_Final.pdf (accessed on 31 January 2022).
18. Škvarada, P.; Tománek, P.; Koktavý, P.; Macků, R.; Šicner, J.; Vondra, M.; Dallaeva, D.; Smith, S.; Grmela, L. A variety of microstructural defects in crystalline silicon solar cells. *Appl. Surf. Sci.* **2014**, *312*, 50–56. [CrossRef]
19. Papež, N.; Sobola, D.; Škvarenina, L.; Škarvada, P.; Hemzal, D.; Tofel, P.; Grmela, L. Degradation analysis of GaAs solar cells at thermal stress. *Appl. Surf. Sci.* **2018**, *461*, 212–220. [CrossRef]
20. Petrone, G.; Ramos-Paja, C.A.; Spagnuolo, G.; Xiao, W. *Photovoltaic Sources Modeling*; Wiley Online Library: Hoboken, NJ, USA, 2017.
21. Bastidas-Rodriguez, J.D.; Franco, E.; Petrone, G.; Ramos-Paja, C.A.; Spagnuolo, G. Model-Based Degradation Analysis of Photovoltaic Modules Through Series Resistance Estimation. *IEEE Trans. Ind. Electron.* **2015**, *62*, 7256–7265. [CrossRef]
22. Pei, T.; Hao, X. A Fault Detection Method for Photovoltaic Systems Based on Voltage and Current Observation and Evaluation. *Energies* **2019**, *12*, 1712. [CrossRef]
23. Friesen, T. *WP5 Deliverable 5.2 Lifetime Degradation Mechanisms Lifetime Degradation Mechanisms*; Technical Report 308991, Performance Plus Project: European Union's Seventh Programme No 308991. 2015. Available online: http://businessdocbox.com/Green_Solutions/106497963-Deliverable-5-2-lifetime-degradation-mechanisms-thomas-friesen-supsi-16-03-2015-version-final-checked-by-johannes-stockl-ait.html (accessed on 31 January 2022).
24. van Dyk, E.E.; Meyer, E.L. Analysis of the effect of parasitic resistances on the performance of photovoltaic modules. *Renew. Energy* **2004**, *29*, 333–344. [CrossRef]
25. Pern, F.; Czanderna, A.; Emery, K.; Dhere, R. Weathering degradation of EVA encapsulant and the effect of its yellowing on solar cell efficiency. In Proceedings of the Conference Record of the Twenty—Second IEEE Photovoltaic Specialists Conference-1991, Las Vegas, NV, USA, 7–11 October 1991; pp. 557–561. [CrossRef]

26. Pingel, S.; Frank, O.; Winkler, M.; Daryan, S.; Geipel, T.; Hoehne, H.; Berghold, J. Potential Induced Degradation of solar cells and panels. In Proceedings of the 2010 35th IEEE Photovoltaic Specialists Conference, Honolulu, HI, USA, 20–25 June 2010; pp. 002817–002822. [CrossRef]

27. Schutze, M.; Junghanel, M.; Koentopp, M.B.; Cwikla, S.; Friedrich, S.; Muller, J.W.; Wawer, P. Laboratory study of potential induced degradation of silicon photovoltaic modules. In Proceedings of the 2011 37th IEEE Photovoltaic Specialists Conference, Seattle, WA, USA, 19–24 June 2011; pp. 000821–000826. [CrossRef]

28. Slamberger, J.; Schwark, M.; Van Aken, B.B.; Virtič, P. Comparison of potential-induced degradation (PID) of n-type and p-type silicon solar cells. *Energy* **2018**, *161*, 266–276. [CrossRef]

29. Kontges, M.; Kunze, I.; Kajari-Schröder, S.; Breitenmoser, X.; Bjørneklett, B. Quantifying the risk of power loss in PV modules due to micro cracks. In Proceedings of the 25th European Photovoltaic Solar Energy Conference and Exhibition, Valencia, Spain, 6–10 September 2010. [CrossRef]

30. Kontges, M.; Kunze, I.; Kajari-Schrder, S.; Breitenmoser, X.; Bjørneklett, B. The risk of power loss in crystalline silicon based photovoltaic modules due to micro-cracks. *Sol. Energy Mater. Sol. Cells* **2011**, *95*, 1131–1137. [CrossRef]

31. Nehme, B.F.; Akiki, T.K.; Naamane, A.; M'Sirdi, N.K. Real-Time Thermoelectrical Model of PV Panels for Degradation Assessment. *IEEE J. Photovolt.* **2017**, *7*, 1362–1375. [CrossRef]

32. Spertino, F.; Ahmad, J.; Di Leo, P.; Ciocia, A. A method for obtaining the I-V curve of photovoltaic arrays from module voltages and its applications for MPP tracking. *Sol. Energy* **2016**, *139*, 489–505. [CrossRef]

33. Silvestre, S.; Chouder, A.; Karatepe, E. Automatic fault detection in grid connected PV systems. *Sol. Energy* **2013**, *94*, 119–127. [CrossRef]

34. Silvestre, S.; da Silva, M.A.; Chouder, A.; Guasch, D.; Karatepe, E. New procedure for fault detection in grid connected PV systems based on the evaluation of current and voltage indicators. *Energy Convers. Manag.* **2014**, *86*, 241–249. [CrossRef]

35. Nehme, B.; Msirdi, N.K.; Namaane, A.; Akiki, T. Analysis and Characterization of Faults in PV Panels. *Energy Procedia* **2017**, *111*, 1020–1029. [CrossRef]

36. Murillo-Soto, L.D.; Meza, C. Fault detection in solar arrays based on an efficiency threshold. In Proceedings of the 2020 IEEE 11th Latin American Symposium on Circuits & Systems (LASCAS), San Jose, Costa Rica, 25–28 February 2020; pp. 1–4.

37. Abd el Ghany, H.A.; ELGebaly, A.E.; Taha, I.B. A new monitoring technique for fault detection and classification in PV systems based on rate of change of voltage-current trajectory. *Int. J. Electr. Power Energy Syst.* **2021**, *133*, 107248. [CrossRef]

38. Pei, T.; Zhang, J.; Li, L.; Hao, X. A fault locating method for PV arrays based on improved voltage sensor placement. *Sol. Energy* **2020**, *201*, 279–297. [CrossRef]

39. Silvestre, S.; Kichou, S.; Chouder, A.; Nofuentes, G.; Karatepe, E. Analysis of current and voltage indicators in grid connected PV (photovoltaic) systems working in faulty and partial shading conditions. *Energy* **2015**, *86*, 42–50. [CrossRef]

40. Analog Devices. LTspice Sofware XVII. Available online: https://www.analog.com/en/design-center/design-tools-and-calculators/ltspice-simulator.html (accessed on 31 January 2022).

41. Villalva, M.G.; Gazoli, J.R.; Filho, E.R. Modeling and circuit-based simulation of photovoltaic arrays. In Proceedings of the 2009 Brazilian Power Electronics Conference, Bonito-Mato Grosso do Sul, Brazil, 27 September–1 October 2009; pp. 1244–1254. [CrossRef]

42. International Electrotechnical Commission. *IEC 61215-1: 2021 Terrestrial Photovoltaic (PV) Modules—Design Qualification and Type Approval—Part 1: Test Requirements*; International Electrotechnical Commission: Geneva, Switzerland, 2021.

43. Walpole, R.E.; Myers, R.H.; Myers, S.L.; Ye, K. *Probability and Statistics for Engineers and Scientists*; Macmillan: New York, NY, USA, 1993; Volume 5.

44. Murillo-Soto, L.D.; Meza, C. Diagnose Algorithm and Fault Characterization for Photovoltaic Arrays: A Simulation Study. In *ELECTRIMACS 2019*; Zamboni, W., Petrone, G., Eds.; Springer International Publishing: Cham, Switzerland, 2020; pp. 567–582.

45. Murillo-Soto, L.D.; Meza, C. Automated Fault Management System in a Photovoltaic Array: A Reconfiguration-Based Approach. *Energies* **2021**, *14*, 2397. [CrossRef]

46. Mejia, F.; Kleissl, J.; Bosch, J. The Effect of Dust on Solar Photovoltaic Systems. *Energy Procedia* **2014**, *49*, 2370–2376. [CrossRef]

47. Guo, B.; Javed, W.; Figgis, B.W.; Mirza, T. Effect of dust and weather conditions on photovoltaic performance in Doha, Qatar. In Proceedings of the 2015 First Workshop on Smart Grid and Renewable Energy (SGRE), Doha, Qatar, 22–23 March 2015; pp. 1–6. [CrossRef]

48. Rao, A.; Pillai, R.; Mani, M.; Ramamurthy, P. Influence of Dust Deposition on Photovoltaic Panel Performance. *Energy Procedia* **2014**, *54*, 690–700. [CrossRef]

49. Ketjoy, N.; Konyu, M. Study of Dust Effect on Photovoltaic Module for Photovoltaic Power Plant. *Energy Procedia* **2014**, *52*, 431–437. [CrossRef]

50. Acciani, G.; Falcone, O.; Vergura, S. Typical Defects of PV-cells. In Proceedings of the 2010 IEEE International Symposium on Industrial Electronics, Bari, Italy, 4–7 July 2010; pp. 2745–2749.

Article

Distribution Grid Stability—Influence of Inertia Moment of Synchronous Machines

Tomáš Petrík, Milan Daneček, Ivan Uhlíř, Vladislav Poulek and Martin Libra *

Department of Physics, Faculty of Engineering, Czech University of Life Sciences Prague, Kamýcká 129, 16500 Prague, Czech Republic; tom.petr11@seznam.cz (T.P.); milan.danecek@gmail.com (M.D.); uhliri@tf.czu.cz (I.U.); poulek@tf.czu.cz (V.P.)
* Correspondence: libra@tf.czu.cz

Received: 19 November 2020; Accepted: 14 December 2020; Published: 18 December 2020

Abstract: This paper shows the influence of grid frequency oscillations on synchronous machines coupled to masses with large moments of inertia and solves the maximum permissible value of a moment of inertia on the shaft of a synchronous machine in respect to the oscillation of grid frequency. Grid frequency variation causes a load angle to swing on the synchronous machines connected to the grid. This effect is particularly significant in microgrids. This article does not consider the effects of other components of the system, such as the effects of frequency, voltage, and power regulators.

Keywords: angle swinging; grid frequency oscillations; electromechanical system; inertial masses; microgrids

1. Introduction

The idea of grid voltage having a coherent sinusoidal course with constant frequency does not correspond with the reality of phenomena occurring in the distribution grid. Accurate measurements disclose oscillations of grid frequency and an instantaneous phase of distribution grid voltage. Grid frequency oscillation has a random noise trend because it is only the response of an energy system to a casually changing daily demand energy diagram [1]. This effect is particularly significant in microgrids [2–4]. The issue of network stability has been studied in a number of other works (e.g., [5]).

The phenomenon of grid oscillations may be described in terms of statistical dynamics using the power spectral density $S_\omega(\Omega)$ of a distribution grid's angular frequency fluctuation. An example of this function characterizing the frequency spectrum of grid frequency deviations is shown in Figure 1.

A spectrum of grid frequency deviations has two components: (i) A slow component for $\Omega < 2\text{s}^{-1}$ (mean $\Delta F < 0.32$ Hz), which is determined by the properties of rotational regulation in power plants as well as the properties of superior central frequency regulation. The interconnection of originally regional districts of the distribution grid into larger complex grids and, later, into the European central grid caused a reduction in the amplitude of very slow fluctuations. Then, this phenomenon continued, leading to the current state of zero steady-state error in the distribution grid frequency. (ii) A fast component for $\Omega > 2\text{s}^{-1}$ (mean $\Delta F > 0.32$ Hz), which is determined by an angular elasticity of the supply system in the case of power changes. A fast component originates from the magnitude and phase changes of instant voltage as it decreases on a power line, and the resulting fast load changes cause transformer impedances in the distribution grid. The angular elasticity refers to the angular elasticity of the phasors of voltage.

Figure 1 shows some samples of the spectrograms of fluctuation in the angular frequency of the distribution grid. Modern distribution grids with digital control of frequency have zero difference on average from central frequency, but there are visible fluctuations of frequency and phasor angle in an area up to $\Omega > 2\text{ s}^{-1}$ (mean $\Delta F > 0.32$ Hz).

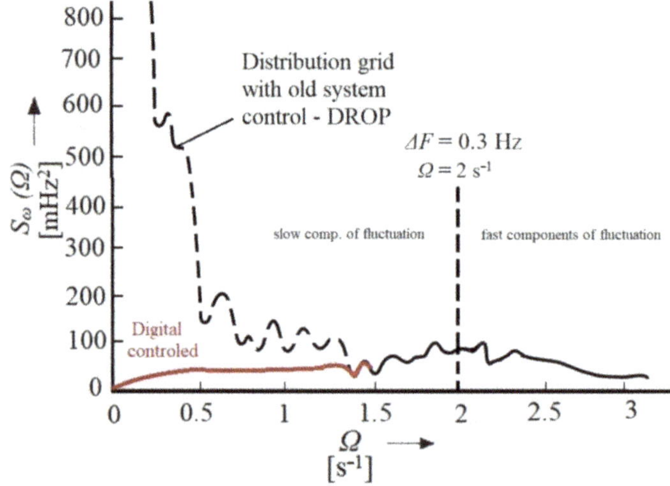

Figure 1. Examples of power spectral density $S\omega(\Omega)$ at frequency fluctuation Ω.

If a synchronous machine driving inertial masses is connected to the grid, the mechanical movement of a machine shaft must follow frequency and phase changes of a voltage phasor of the grid. A shaft follows the instantaneous grid voltage phase with an angle deviation, producing a load angle β in a synchronous machine. The instantaneous power consumed or supplied by a machine to a grid depends on the magnitude and orientation of the load angle β. The greater the inertial masses on the shaft are in comparison to the size of a synchronous machine, the greater the dynamic deviation of the following fast grid phase changes and load angle oscillations will be [6].

Since an electromechanical system (a synchronous machine and its inertial masses) has very low oscillation dumps, the load angles, excited by distribution grid frequency fluctuations, have a harmonic swinging character. In addition, they are accompanied by an undesirable overflow of energy between the grid and the inertial masses on the shaft.

2. Description of a Modeled System

The electromechanical swinging system consists of the rigidity of a synchronous machine magnetic field, the damping effects, and the moment of inertia, which are connected with a shaft as presented in Figure 2. The system in Figure 2 is described by the differential equation

$$I\ddot{\beta} + B\dot{\beta} + k\beta = \frac{I}{P_d}\dot{\omega}, \tag{1}$$

where β is the load angle of a synchronous machine, ω is the grid frequency (s^{-1}), I is the overall moment of inertia connected to a shaft (kg·m^2), B is the torsion damping constant (N·m·s), K is the torsion rigidity of a synchronous machine (N·m), and P_d is the number of pole pairs in a synchronous machine.

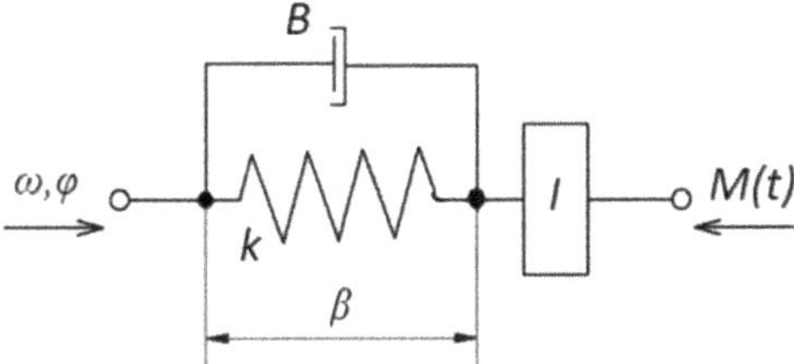

Figure 2. Electromechanical vibrating set of a synchronous machine.

For simplification, the following will be assumed:

- An unloaded synchronous machine;
- Linearity of the magnetic circuit;
- The exclusion of the influence of external voltage, power, and frequency control circuits;
- The possession of torque characteristics for small $|\beta|$ (< 0.1 rad) in linearized solutions.

The precision of the system is described by Equation (1) in Laplace's transformation [7,8]:

$$G(p) = \frac{pI}{P_d \ (p^2 I + pB + k)}. \tag{2}$$

The power spectral density of the load angle fluctuation is

$$S_\beta(\Omega) = \lim_{n \to \infty} (G(p))^2 \ S_\omega(\Omega), \tag{3}$$

where $S_\beta(\Omega)$ is the power spectral density of the load angle fluctuation and $S_\omega(\Omega)$ is the power spectral density of fluctuation for angular frequency fluctuations (s^{-2}).

$$S_\beta(\Omega) = \frac{\Omega^2 I^2 S_\omega(\Omega)}{P_d\left[(k - \Omega^2 I)^2 + \Omega^2 B^2\right]}. \tag{4}$$

The natural frequency of the system is Ω_0:

$$\Omega_0 = \sqrt{\frac{k}{I}}. \tag{5}$$

The coefficient of a relative system damping

$$a = \frac{B}{2\sqrt{kI}} \tag{6}$$

is very low for most synchronous machines; it is usually < 0.1, in which case the resonance is very selective. The system transmits just the natural frequency. A system response has almost a sinusoidal course of load density swinging with an amplitude β_A. From Equation (4), the results for $\Omega = \Omega_0$ can be written as

$$S_\beta(\Omega_0) = \beta_A^2 = \frac{I^2}{P_d^2 \ B^2} = S_\omega(\Omega_0). \tag{7}$$

3. Maximum Admissible Grid Frequency Oscillations

If the response of a load angle's swing has to have an amplitude A as the maximum, the power spectral density $S_\omega(\Omega_0)$ of the grid angle frequency must not be greater than $S_{\omega,max}(\Omega_0)$

$$S_{\omega,max}(\Omega_0) = \frac{P_d^2 \, B^2}{I^2} \, \beta_A^2,$$

(8)

where $S_{\omega,max}(\Omega_0)$ is the upper limit of the power spectral density of a grid angle frequency (s^{-2}).

Substituting the variable Ω_0 from Equation (5) into Equation (8) for the parameter I yields

$$S_{\omega,max}(\Omega_0) = \frac{P_d^2 \, B^2}{k^2} \, \beta_A^2 \, \Omega_0^4.$$

(9)

The maximum size of a flying wheel for which the amplitude of the load angle's swing does not exceed the given value β_A is thus determined.

Courses for $S_{\omega,max}(\Omega_0)$ according to Equation (9) for the chosen β_A (e.g., 0.1, 0.15, and 0.2 rad) are drawn in Figure 3. These courses constitute the upper limits of power spectral density in the grid angle frequency fluctuations $S_{\omega,max}(\Omega_0)$ for the permitted β_A. The courses of the upper limits are fitted with a scaled I according to Equation (5):

$$I = \frac{k}{\Omega_0^2}$$

(10)

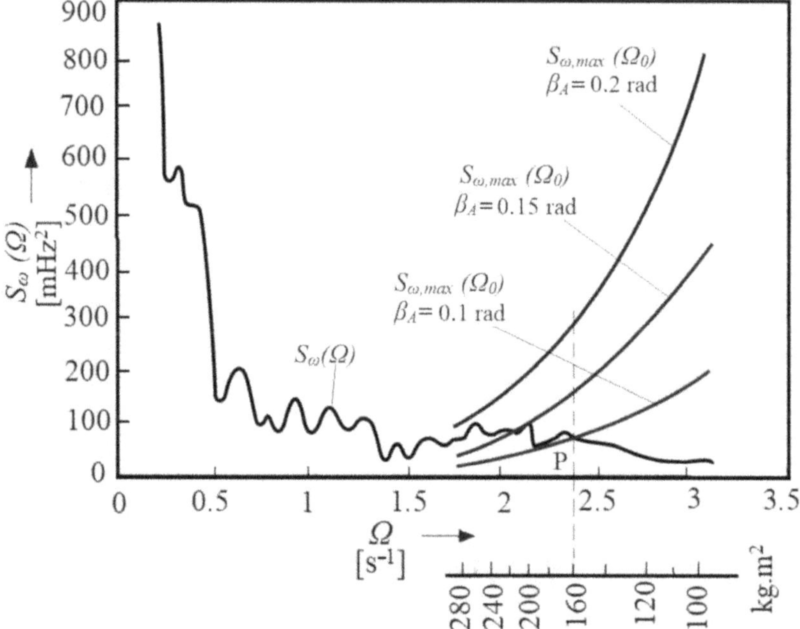

Figure 3. Influence of the fluctuation spectrum of the grid and inertial momentum I on the amplitude of angular oscillation.

The maximum value of the moment of inertia being placed on the shaft of a synchronous machine is given by the first intersection point (from the right) of the function $S_{\omega,max}(\Omega_0)$ for the permissible maximum of the load angle oscillation amplitude β_A with the course $S_\omega(\Omega)$ measured in the grid section at the considered time.

4. Practical Verifying

The synchronous machine MEZ-A 225 MO 4 with the parameters $P_s = 50$ kVA, $n = 1500$ min^{-1}, $k = 1140$ N·m, and $B = 14$ N·m·s was chosen as an example. MEZ-A 225 MO 4 is a synchronous hydroalternator with four poles, without a dumper, with a brushless exciter powered by an external electronic source with a constant current of 1.2 A. This corresponds to excitation at a nominal voltage of 3×400 V in the idle state.

This excitation current of the brushless exciter was kept constant throughout the experiment.

The shaft of the machine was connected by a rigid coupling with a steel flywheel with a diameter of 1300 mm mounted in its own bearings.

The moment of inertia $I = 160$ kg·m^2 was determined by calculation from the geometric dimensions and the catalog data. The damping constant $B = 14$ N·m·s was calculated from the attenuation of the transient during phasing into the distribution network.

The unit was started at synchronous velocity by using a friction coupling from a diesel engine.

During the experiment, the synchronous machine ran as a motor on the distribution network without any additional load. It was loaded by mechanical losses in the bearings and the mechanical power consumption of its own ventilation and exciter. The total load was estimated at 3 kW, which is less than 8% of the nominal power of the synchronous machine. This state was close to the idling state, which is theoretically discussed above.

Under the permissible load angle oscillation amplitude $\beta_A = 0.1$ rad (equal to the overflow of active power $P_a = 20$ kW), the intersection point P presented the maximum value $I = 160$ kg·m^2.

Figure 4 gives evidence of conformity to the calculated value of an average amplitude ($I = 160$ kg·m^2) with a recording of grid frequency fluctuation and the corresponding response of the load angle. This recording was measured from the machine during the experiment.

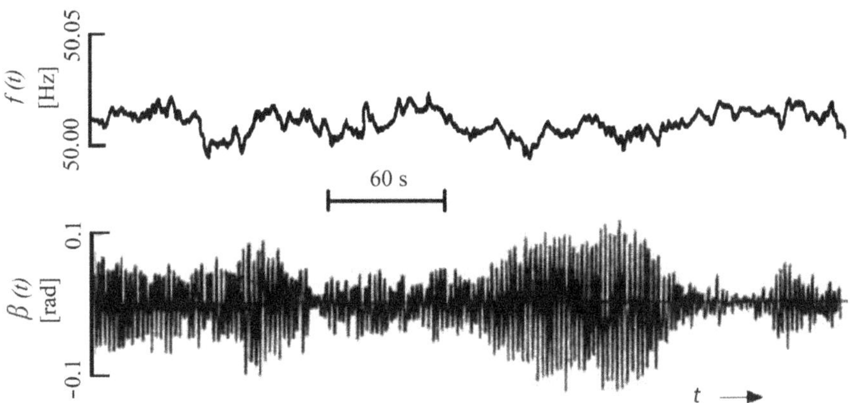

Figure 4. Random changes in grid frequency (top) and changes influenced by the oscillation of the load angle of a synchronous machine (bottom).

5. Conclusions

Grid frequency variation causes a load angle to swing in the synchronous machines connected to the grid. The grid frequency variation spectrum shape grows the inertial moment on the shaft of the synchronous machine while also growing the swinging amplitude. This is visible from the shown equations. The maximum inertia moment value was thus determined for safe swinging amplitude. The evidence of conformity to the calculated value of the moment of inertia is shown in Figure 4 as a recording of grid frequency and the corresponding response of the load angle. The average value of the load angle is present in the required interval. The instantaneous value of the load angle is mainly present in the required interval; however, the average value is more significant.

The load angle in this safe interval enables the machine to be driven in a safe and reliable manner by external controllers. These controllers are not considered in the article. This limitation should be respected both in drives with tens and hundreds of kW as well as in drives with small synchronous machines and stepping motors, where large inertial masses have to be driven.

As was written before, this paper did not consider external controllers. However, they are often a major part of frequency grid stability issues. Frequency grid fluctuations cause load angle fluctuations, which in turn cause frequency grid fluctuations. This interaction is a closed circle. In closing, it is necessary to mention trends and possibilities in frequency stabilization. A few studies last year discussed renewable sources of energy. These renewable sources can reduce the quality of grid services [9]. However, with the appropriate combination, along with batteries, they can be used to stabilize the properties of the grid [9]. A small battery source connected to photovoltaic panels was described in [10]. These battery sources, or even larger ones, would therefore be appropriate to include and use in the grid. This is especially true for intelligent buildings that combine many energy sources such as synchronous generators and photovoltaic panels; these batteries can be used as more than just backup sources.

Author Contributions: Conceptualization, T.P., I.U.; investigation, T.P., I.U., M.L.; methodology, M.D., V.P.; supervision, M.L.; validation, V.P., M.D.; visualization, V.P., M.L.; writing—original draft, I.U., T.P.; writing—review and editing, T.P., M.L. All authors have read and agreed to the published version of the manuscript.

Funding: This research received no external funding.

Conflicts of Interest: The authors declare no conflict of interest.

Nomenclature

$S_\omega(\Omega)$	power spectral density of distribution grid's angular frequency fluctuation
Ω	angular frequency of fluctuations
B	load angle of a synchronous machine
ω	grid frequency
I	overall moment of inertia connected to a shaft
B	torsion damping constant
K	torsion rigidity of a synchronous machine (N·m)
P_d	number of pole pairs of a synchronous machine
Ω_0	natural frequency of the system
S_β	(Ω) power spectral density of the load angle fluctuation β_A
β_A	load angle of a synchronous machine at the natural frequency of the system
a	coefficient of a relative system damping
$S_{\omega,max}$	(Ω_0) upper limit of the power spectral density of a grid angle frequency
P_s	power of a unit used for experimentation
n	nominal revolutions of a unit
P_a	active power

References

1. Amin, W.T.; Montoya, O.D.; Garrido, V.M.; Gil-González, W.; Garces, A. Voltage and Frequency Regulation on Isolated AC Three-phase Microgrids via s-DERs. In Proceeding of the 2019 IEEE Green Technologies Conference (GreenTech), Lafayette, LA, USA, 3–6 April 2019; pp. 1–6.
2. Ferro, G.; Robba, M.; Sacile, R. A Model Predictive Control Strategy for Distribution Grids: Voltage and Frequency Regulation for Islanded Mode Operation. *Energies* **2020**, *13*, 2637. [CrossRef]

3. Delfino, F.; Ferro, G.; Robba, M.; Rossi, M. An architecture for the optimal control of tertiary and secondary levels in small-size islanded microgrids. *Int. J. Electr. Power Energy Syst.* **2018**, *103*, 75–88. [CrossRef]
4. Delfino, F.; Rossi, M.; Ferro, G.; Minciardi, R.; Robba, M. MPC-based tertiary and secondary optimal control in islanded microgrids. In Proceedings of the 2015 IEEE International Symposium on Systems Engineering (ISSE), Rome, Italy, 28–30 September 2015; pp. 23–28.
5. Wu, Y.-K.; Tang, K.-T.; Lin, Z.K.; Tan, W.-S. Flexible Power System Defense Strategies in an Isolated Microgrid System with High Renewable Power Generation. *Appl. Sci.* **2020**, *10*, 3184. [CrossRef]
6. Serra, F.M.; Fernandez, M.L.; Montoya, O.D.; Gil-Gonzalez, W.J.; Hernandez, J.C. Nonlinear Voltage Control for Three-Phase DC-AC Converters in Hybrid Systems: An Application of the PI-PBC Method. *Electronics* **2020**, *9*, 847. [CrossRef]
7. Han, J.; Liu, Z.; Liang, N.; Song, Q.; Li, P. An Autonomous Power Frequency Control Strategy Based on Load Virtual Synchronous Generator. *Processes* **2020**, *8*, 433. [CrossRef]
8. Haque, M.E.; Negnevitsky, M.; Muttaqi, K.M. A Novel Control Strategy for a Variable Speed Wind Turbine with a Permanent-Magnet Synchronous Generator. *IEEE Trans. Ind. Appl.* **2010**, *46*, 331–339. [CrossRef]
9. Yoo, Y.; Jung, S.; Kang, S.; Song, S.; Lee, J.; Han, C.; Jang, G. Dispatchable Substation for Operation and Control of Renewable Energy Resources. *Appl. Sci.* **2020**, *10*, 7938. [CrossRef]
10. Poulek, V.; Dang, M.Q.; Libra, M.; Beránek, V.; Šafránková, J. PV Panel with Integrated Lithium Accumulators for BAPV Applications—One Year Thermal Evaluation. *IEEE J. Photovolt.* **2020**, *10*, 150–152, ISSN 2156-3403. [CrossRef]

Publisher's Note: MDPI stays neutral with regard to jurisdictional claims in published maps and institutional affiliations.

Article

Analytical Modeling of Current-Voltage Photovoltaic Performance: An Easy Approach to Solar Panel Behavior

José Miguel Álvarez [1], Daniel Alfonso-Corcuera [1,2], Elena Roibás-Millán [1,2], Javier Cubas [1,2], Juan Cubero-Estalrrich [1], Alejandro Gonzalez-Estrada [1], Rocío Jado-Puente [1], Marlon Sanabria-Pinzón [1] and Santiago Pindado [1,2,*]

[1] Instituto Universitario de Microgravedad "Ignacio Da Riva" (IDR/UPM), ETSI Aeronáutica y del Espacio, Universidad Politécnica de Madrid, Plaza del Cardenal Cisneros 3, 28040 Madrid, Spain; jm.alvarez@upm.es (J.M.Á.); daniel.alfonso.corcuera@upm.es (D.A.-C.); elena.roibas@upm.es (E.R.-M.); j.cubas@upm.es (J.C.); jl.cubero@alumnos.upm.es (J.C.-E.); alejandro.gonzalez.estrada@alumnos.upm.es (A.G.-E.); rocio.jado.puente@alumnos.upm.es (R.J.-P.); ms.sanabria@alumnos.upm.es (M.S.-P.)
[2] Departamento de Sistemas Aeroespaciales, Transporte Aéreo y Aeropuertos (SATAA), ETSI Aeronáutica y del Espacio, Universidad Politécnica de Madrid, Plaza del Cardenal Cisneros 3, 28040 Madrid, Spain
*   Correspondence: santiago.pindado@upm.es

Abstract: In this paper, we propose very simple analytical methodologies for modeling the behavior of photovoltaic (solar cells/panels) using a one-diode/two-resistor (1-D/2-R) equivalent circuit. A value of $a = 1$ for the ideality factor is shown to be very reasonable for the different photovoltaic technologies studied here. The solutions to the analytical equations of this model are simplified using easy mathematical expressions defined for the Lambert W-function. The definition of these mathematical expressions was based on a large dataset related to solar cells and panels obtained from the available academic literature. These simplified approaches were successfully used to extract the parameters from explicit methods for analyzing the behavior of solar cells/panels, where the exact solutions depend on the Lambert W-function. Finally, a case study was carried out that consisted of fitting the aforementioned models to the behavior (that is, the I-V curve) of two solar panels from the UPMSat-1 satellite. The results show a fairly high level of accuracy for the proposed methodologies.

Keywords: solar cell; solar panel; parameter extraction; analytical; Lambert W-function; spacecraft solar panels; I-V curve; modeling

Citation: Álvarez, J.M.; Alfonso-Corcuera, D.; Roibás-Millán, E.; Cubas, J.; Cubero-Estalrrich, J.; Gonzalez-Estrada, A.; Jado-Puente, R.; Sanabria-Pinzón, M.; Pindado, S. Analytical Modeling of Current-Voltage Photovoltaic Performance: An Easy Approach to Solar Panel Behavior. *Appl. Sci.* **2021**, *11*, 4250. https://doi.org/10.3390/app11094250

Academic Editor: Luis Hernández-Callejo

Received: 13 March 2021
Accepted: 3 May 2021
Published: 7 May 2021

Publisher's Note: MDPI stays neutral with regard to jurisdictional claims in published maps and institutional affiliations.

1. Introduction

Based on the installation of power plants over the last few decades, it can be seen that photovoltaic energy has emerged as a very important factor in policies relating to renewable energy sources [1–7]. As an energy source, solar panels have a significant competitive edge over other renewable energy sources due to their potential for dual-use at either the industrial or the domestic scale. Both possibilities have provided an impetus for the increasing growth in demand for photovoltaic systems [5,8–11].

Photovoltaic energy has also been demonstrated to be crucial for stand-alone power systems such as glow-in-the-dark lettering and illuminated signs on highways. One of the first such stand-alone industrial applications of photovoltaic technologies was in power generation in spacecraft. According to Rauschenbach, "The first solar cell array that successfully operated in space was launched on 17 March 1958, on board Vanguard I, the second U.S. earth satellite" [12].

Modeling the performance of power sub-systems for space missions is essential at the predesign stage. At these early configuration stages of a space mission, simple, fast simulations are required that are as accurate as possible. This trend has also been amplified by the increasing importance of Concurrent Design (CD) in industrial processes. According to the European Cooperation for Space Standardization (ECSS), CD is especially important in the early predesign phase of space missions. It is based on the parallelization of processes associated with different sub-systems and is carried out in a special working environment organized into Concurrent Design Facilities (CDFs), where design parameters are shared, and the information flow is organized thus that stable solutions can be reached as quickly as possible [13–17].

Research carried out at the IDR/UPM Institute on solar panels has been driven by the need for simple procedures to calculate the performance of these panels. This need has arisen in the context of mission predesign at the CDF [18,19], and particularly in relation to sub-system coupling effects, such as those that can be found when analyzing thermal control and power distribution in spacecraft.

The simulation of photovoltaic devices (solar cells/panels) is normally carried out by means of equivalent circuit models, which are defined by implicit mathematical expressions. However, problems arise in terms of:

- The extraction of the parameters appearing in this equation, which define the performance of the photovoltaic device at a certain temperature and irradiance level, and;
- The calculation of the output current as a function of the output voltage, or vice versa, since this equation is implicit.

There are multiple ways to fit the equation for the photovoltaic equivalent circuit models to experimental data [20–29]. All of these approaches can essentially be grouped into two different types: Numerical (in which the equation is fitted to a large dataset that represents the *I-V* curve of the photovoltaic device) and analytical (in which the parameters of the equation are extracted based on three points of the *I-V* curve: short circuit, maximum power, and open voltage).

The second issue described above, i.e., the calculation of the output current as a function of the output voltage once the parameters of the equation have been extracted, can only be solved by numerical approaches [30–46] (which include iterative solutions [47–49]), or by using of the Lambert W-function [50–55] (whose exact calculation requires a numerical approach).

Possible solutions to the aforementioned problems include explicit methods of photovoltaic modeling, which are based on explicit equations with parameters defined based on the characteristic points of the *I-V* curve. Some of these methods also require the use of the Lambert W-function to define the parameters [56].

In the present work, we explore two important aspects of the analytical approach to the one diode/two resistor (1-D/2-R) equivalent circuit model for solar panels:

- Selection of a proper ideality factor for extraction of the model parameters based on analytical methods, as this forms the cornerstone for extracting the other four parameters [57], and;
- Development of a simplified approach to the Lambert W-function, which is required in this analytical methodology in order to:
  - Obtain the value of the first parameter of the 1-D/2-R equivalent circuit model, i.e., the value of the resistance of the series-connected resistor, and;
  - Solve the implicit equation of this model to derive the output current in relation to the output voltage.

Our simplified approach to the Lambert W-function is also proven to be a relatively powerful tool for extracting the parameters of some of the most accurate explicit equations for photovoltaic modeling that can be found in the literature: the models of Kalmarlar and Haneefa [58,59] and Das [60].

The aim of this paper is to define very simple methodologies to measure the performance of photovoltaic devices (solar cells/panels), thus that these can be implemented as part of more complex simulations. One example of this type of simulation is the coupled thermo-electrical modeling of space systems with ESATAN$^©$ [61]. It should also be emphasized that these methodologies may be useful tools for professionals within small and medium-sized enterprises in the solar energy sector to allow them to estimate the performance of the systems they design. Additionally, the increasing use of photovoltaic generation in small grids [62,63] could increase the worth of the easy approximations to solar panels performance included in this paper.

This paper is organized as follows. A simple model of solar panels is described in Section 2, and simplified equations for solving the 1-D/2-R equivalent circuit model and the Lambert W-function are given in Section 3. A case study is also described and solved in Section 3, using the procedures described in the preceding sub-sections. Finally, conclusions are presented in Section 4.

2. Modeling of Photovoltaic Systems

2.1. The 1-D/2-R Equivalent Circuit Model

The most widely used method of modeling the performance of a solar cell/panel (based on its *I-V* curve, where *I* is the output current and *V* the output voltage) is an equivalent circuit based on one current source, one diode, and two resistors (series and shunt resistors), as shown in Figure 1. The equation that describes the performance of this model is as follows [57,64–66]:

$$I = I_{pv} - I_0 \left[\exp\left(\frac{V + IR_s}{naV_T} \right) - 1 \right] - \frac{V + IR_s}{R_{sh}} \tag{1}$$

where I_{pv} is the photocurrent, I_0 is the reverse saturation current of the diode, R_{sh} is the resistance of the shunt resistor, R_s is the resistance of the series-connected resistor, and V_T is the thermal voltage:

$$V_T = \frac{\kappa T}{q} \tag{2}$$

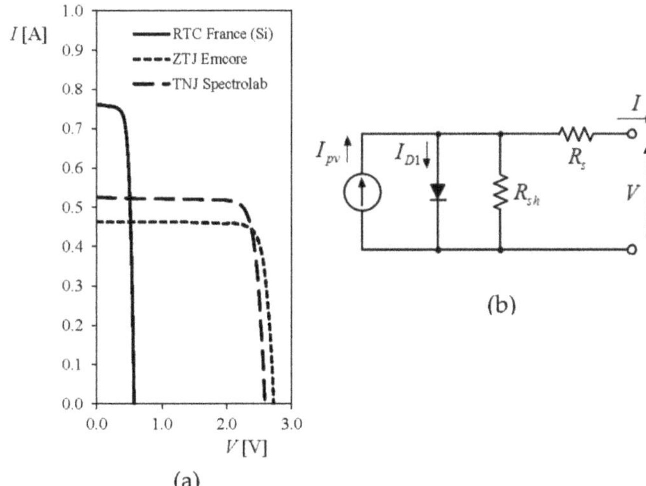

Figure 1. (a) *I-V* curves for three different solar cells; (b) 1-D/2-R equivalent circuit model for analyzing the behavior of a solar cell/panel.

In the above equation, $\kappa = 1.38064852 \cdot 10^{-23}$ m^2 kg s^{-2} K^{-1} is the Boltzmann constant, T is the temperature expressed in K, and $q = 1.60217662 \cdot 10^{-19}$ C is the electron charge. In Equation (1), *a* is the ideality factor of the diode (in principle, the effect of temperature on this parameter can be left aside [64]), and *n* is the number of series-connected cells in the solar panel (obviously, $n = 1$ when studying the performance of a single cell).

The first problem of modeling a photovoltaic device with the 1-D/2-R equivalent circuit model lies in extracting the five parameters of the model: I_{pv}, I_0, a, R_s, and R_{sh}. Depending on the available information, which might be either:

- The *I-V* curve with an enough large number of points, or;
- The three characteristic points of the *I-V* curve (short circuit output current, I_{sc}, open circuit output voltage, V_{oc}, and output current and voltage levels at the Maximum Power Point (MPP), I_{mp} and V_{mp},

It is possible to extract these five parameters either by numerically fitting Equation (1) to the *I-V* curve, or by using the three characteristic points, which in practice represent four conditions [67]: the first three of which force the equation to match the points $(0, I_{sc})$, (V_{mp}, I_{mp}) and $(V_{oc}, 0)$, and the fourth is related to the maximum power condition:

$$-\frac{\partial I}{\partial V}\bigg|_{(V_{mp},I_{mp})} = \frac{I_{mp}}{V_{mp}}. \tag{3}$$

Modeling the performance of photovoltaic systems with the equivalent circuit described here can be very advantageous, as these circuits somehow preserve the physical processes of a pair formed of a current source and a p-n junction [68]. The effect of the series-connected resistor, R_s, is mainly associated with power losses in the solder bonds and interconnections between cells, whereas the effect of the shunt resistor, R_{sh}, is associated with current leakages across the p-n junction [69,70]. In recent years, several reviews of the different procedures and techniques for the parameter extraction problem related to the 1-D/2-R model have been published [20,21,23,24,67,71–73]. The present work focuses on an analytical approach based on the use of information from the characteristic points to calculate the five parameters in Equation (1). The following equations can be derived [57]:

$$\frac{naV_T V_{mp}\left(2I_{mp} - I_{sc}\right)}{\left(V_{mp}I_{sc} + V_{oc}\left(I_{mp} - I_{sc}\right)\right)\left(V_{mp} - I_{mp}R_s\right) - naV_T\left(V_{mp}I_{sc} - V_{oc}I_{mp}\right)} = \exp\left(\frac{V_{mp} + I_{mp}R_s - V_{oc}}{naV_T}\right), \tag{4}$$

$$R_{sh} = \frac{\left(V_{mp} - I_{mp}R_s\right)\left(V_{mp} - R_s\left(I_{sc} - I_{mp}\right) - naV_T\right)}{\left(V_{mp} - I_{mp}R_s\right)\left(I_{sc} - I_{mp}\right) - naV_T I_{mp}}, \tag{5}$$

$$I_{pv} = \frac{R_{sh} + R_s}{R_{sh}}I_{sc}, \tag{6}$$

$$I_0 = \frac{\left(R_{sh} + R_s\right)I_{sc} - V_{oc}}{R_{sh}\exp\left(\frac{V_{oc}}{naV_T}\right)}. \tag{7}$$

Three problems arise at this point:

- A sufficiently accurate estimation of the ideality factor a is required;
- Equation (4) is an implicit mathematical expression for solving R_s, and an iterative process is, therefore, required to extract this parameter; and
- Equation (1), which defines the performance of the solar cell/panel, is also an implicit expression. As a consequence, once all the parameters of this equation have been extracted, an additional iterative process will be required to solve it (i.e., to derive the value of the output current, I, for a given value of the output voltage level, V).

Fortunately, the ideality factor can be estimated by taking a value within the range from $a = 1$ to $a = 1.5$ [74,75], and the value of the resistance of the series-connected resistor, R_s, can be derived from [64]:

$$R_s = A\left(W_{-1}\left(B\exp(C)\right) - (D + C)\right), \tag{8}$$

where W_{-1} is the negative branch of the Lambert W-function (see Section 2.1). The variables A, B, C, and D are defined as:

$$A = \frac{naV_T}{I_{mp}}, \tag{9}$$

$$B = -\frac{V_{mp}(2I_{mp} - I_{sc})}{V_{mp}I_{sc} + V_{oc}(I_{mp} - I_{sc})},$$

(10)

$$C = -\frac{2V_{mp} - V_{oc}}{naV_T} + \frac{V_{mp}I_{sc} - V_{oc}I_{mp}}{V_{mp}I_{sc} + V_{oc}(I_{mp} - I_{sc})},$$

(11)

$$D = \frac{V_{mp} - V_{oc}}{naV_T}.$$

(12)

However, the problem of solving the implicit equation in (1) still remains after the extraction of parameters a and R_s. Authors such as Peng et al. [50] have shown that this equation can also be solved using the Lambert W-function:

$$I = \frac{R_{sh}(I_{pv} + I_0) - V}{R_{sh} + R_s} - \frac{naV_T}{R_s}W_0\left(\frac{R_{sh}R_sI_0}{naV_T(R_{sh} + R_s)}\exp\left(\frac{R_{sh}R_s(I_{pv} + I_0) + R_{sh}V}{naV_T(R_{sh} + R_s)}\right)\right),$$

(13)

where W_0 is the positive branch of the Lambert W-function (see Section 2.1).

2.2. Explicit Equations/Models as Alternative to the 1-D/2-R Equivalent Circuit Model

The difficulties associated with solving the 1-D/2-R model implicit equation drove researchers to develop explicit equations for modeling the performance of solar cells/panels [58,59,76–82]. These photovoltaic models give the output current as a function of the output voltage by using simple mathematical models or equations that can be easily solved without the need for advanced mathematical tools. Of these, three models appear to be fairly accurate in relation to the *I-V* curve. The model proposed by Kalmarkar and Haneefa [58,59] is:

$$\frac{I}{I_{sc}} = 1 - (1 - \gamma)\frac{V}{V_{oc}} - \gamma\left(\frac{V}{V_{oc}}\right)^m,$$

(14)

where [56]:

$$\gamma = \frac{2\left(\frac{I_{mp}}{I_{sc}}\right) - 1}{(m - 1)\left(\frac{V_{mp}}{V_{oc}}\right)^m},$$

(15)

$$m = \frac{W_{-1}\left(-\left(\frac{V_{oc}}{V_{mp}}\right)^{\frac{1}{K}}\left(\frac{1}{K}\right)\ln\left(\frac{V_{mp}}{V_{oc}}\right)\right)}{\ln\left(\frac{V_{mp}}{V_{oc}}\right)} + \frac{1}{K} + 1,$$

(16)

and:

$$K = \frac{1 - \left(\frac{I_{mp}}{I_{sc}}\right) - \left(\frac{V_{mp}}{V_{oc}}\right)}{2\left(\frac{I_{mp}}{I_{sc}}\right) - 1},$$

(17)

the model proposed by Das [60] is:

$$\frac{I}{I_{sc}} = \frac{1 - \left(\frac{V}{V_{oc}}\right)^k}{1 + h\left(\frac{V}{V_{oc}}\right)},$$

(18)

where [56]:

$$k = \frac{W_{-1}\left(\left(\frac{I_{mp}}{I_{sc}}\right)\ln\left(\frac{V_{mp}}{V_{oc}}\right)\right)}{\ln\left(\frac{V_{mp}}{V_{oc}}\right)},$$

(19)

$$h = \left(\frac{V_{oc}}{V_{mp}}\right)\left(\frac{I_{sc}}{I_{mp}} - \frac{1}{k} - 1\right),$$

(20)

while the model of Pindado and Cubas [56,80] is:

$$
I = \begin{cases}
I_{sc}\left(1 - \left(1 - \frac{I_{mp}}{I_{sc}}\right)\left(\frac{V}{V_{mp}}\right)^{\frac{I_{mp}}{I_{sc}-I_{mp}}}\right) & ; V \leq V_{mp} \\
I_{mp}\left(\frac{V_{mp}}{V}\right)\left(1 - \left(\frac{V-V_{mp}}{V_{oc}-V_{mp}}\right)^{\phi}\right) & ; V \geq V_{mp}
\end{cases}
\tag{21}
$$

where:

$$
\phi = \left(\frac{I_{sc}}{I_{mp}}\right)\left(\frac{I_{sc}}{I_{sc}-I_{mp}}\right)\left(\frac{V_{oc}-V_{mp}}{V_{oc}}\right).
\tag{22}
$$

In the graph in Figure 2, these models are fitted to the experimental *I-V* curve for an RTC solar cell, taken from the well-known work by Easwarakhanthan et al. [83]. The 1-D/2-R equivalent circuit model is also fitted to these data, and it can be seen that the accuracy of the explicit models is quite high, as stated in [56,81]. The problem with using some of these models again lies in the need to solve the Lambert W-function, which requires some mathematical ability.

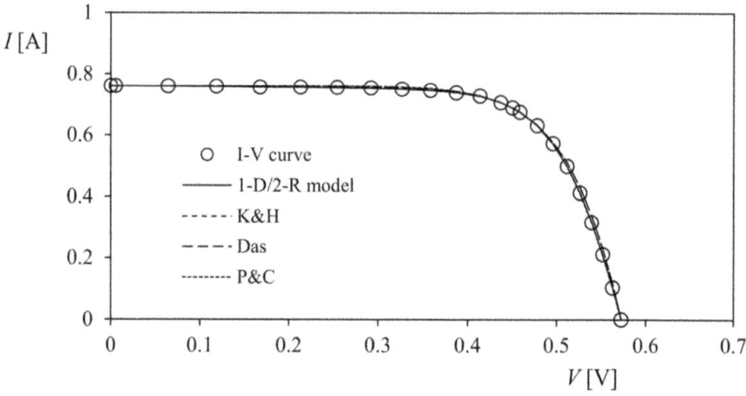

Figure 2. *I-V* curve of the RTC solar cell [83], with the results from the 1-D/2-R equivalent circuit model, and the explicit models of Kalmarkar and Haneefa, Das, and Pindado and Cubas.

2.3. The Lambert W-Function

From the methodologies described above, it is clear that in order to work with the equations of the implicit 1-D/2-R model and some of the explicit models, some knowledge of the Lambert W-function is required. The Lambert W-function, $W(z)$ (plotted in Figure 3), is defined as:

$$
z = W(z)\exp(W(z)),
\tag{23}
$$

In the above equation, z is a complex number. If a real variable x is considered, the Lambert function is then defined within the range $[-1/e, \infty]$. It should also be noted that this function gives a double value within the range $[-1/e, 0]$. Two different branches, called the positive and negative branches, are normally defined for this function, as follows:

- $W_0(x)$, for $W(x) \geq -1$, and
- $W_{-1}(x)$, for $W(x) \leq -1$.

This function is a useful tool for solving equations that involve exponentials since if $X = Y\exp(Y)$, then $Y = W(X)$. However, as stated above, a certain level of expertise is required to solve these [84,85]. Barry et al. developed an interesting explicit approach, although this may not be sufficiently direct to obtain a solution [86].

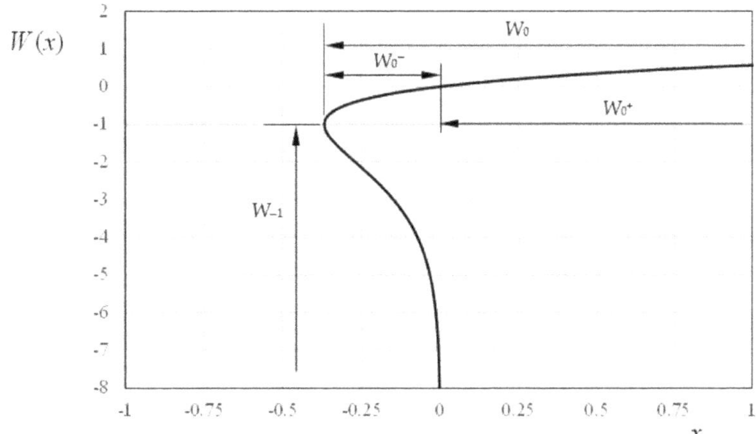

Figure 3. The Lambert W-function, with its two different branches.

3. Results

In this section, a method that facilitates the use of the 1-D/2-R equivalent circuit model is presented. It is organized into three sub-sections, as follows:

- Sub-Section 3.1 describes the problem of selecting a suitable value for the ideality factor, a, in order to obtain the values for the rest of the parameters.
- Very simple equations for the Lambert W-function, for use in solving the equations described in the section above, are included in Sub-Section 3.2.
- Finally, a case study in which our methodologies are applied to the modeling of spacecraft solar panels is included in Sub-Section 3.3.

3.1. On the Best Value for the Ideality Factor

In a paper published in the 2nd International Conference on Renewable Energy Research and Applications (ICRERA 2013), our research group demonstrated a hyperbolic relationship between the non-dimensional RMSE, ζ, of the 1-D/2-R equivalent circuit model fitted to the *I-V* curve of a solar cell, and the value of the ideality factor, a [87]. The non-dimensional (or normalized) RMSE is defined as:

$$\zeta = \frac{\text{RMSE}}{I_{sc}} = \frac{1}{I_{sc}}\sqrt{\frac{1}{N}\sum_{i=1}^{N}\left(I - I_{ref}\right)^2}, \tag{24}$$

where N is the number of point of the *I-V* curve, I is the current obtained from modeling the corresponding solar cell/panel, and I_{ref} is the measured current associated with each one of those points. The aforementioned relationship between ζ and a is plotted in Figure 4. It can be seen that the curve has a hyperbolic shape, with a minimum value of the non-dimensional RMSE at $a = 1.18$. This hyperbolic shape was also confirmed in a recent work by Elkholy and Abou El-Ela [88].

The normalized RMSE can be used to compare the accuracy of different models applied to different photovoltaic technologies. However, it should also be taken into account that variations in temperature and irradiance could somehow affect the accuracy of the model's accuracy. In the present work, the current at the short circuit from the reference *I-V* curve has been selected as it was in previous works such as [57]. This way to normalize the RMSE has been used by other authors [89]. However, it was not the only one, as the average current value from the measured dataset that represents the *I-V* curve [88,90], or the difference between the current at two points [91], has also been proposed as current values to normalize the RMSE.

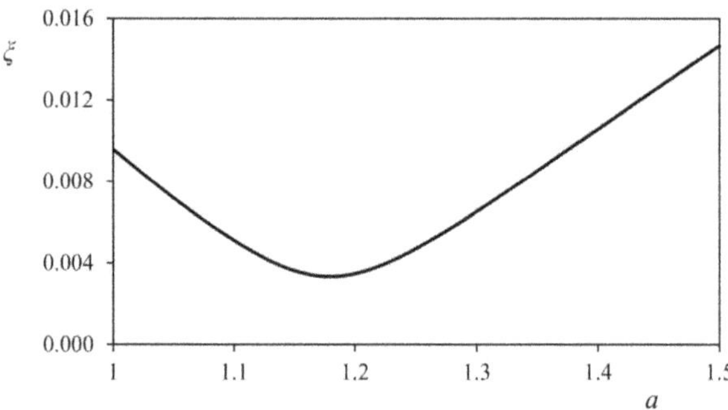

Figure 4. Non-dimensional RMSE, ζ, for a 1-D/2-R equivalent circuit model fitted to the *I-V* curve of an Emcore ZTJ solar cell (see Figure 1) versus the ideality factor, *a*.

The value of the ideality factor, *a*, was studied with the aim of providing future researchers with an accurate but much simpler fitting of the 1-D/2-R equivalent circuit model. To carry out this analysis, the *I-V* curves from four solar cells and three solar panels were selected from previous works [56,81], as shown in Table 1. The parameters extracted by fitting Equation (1) to these *I-V* curves using Matlab$^{©}$ are given in Table 2. These fits were used as a reference for the values calculated analytically since it could be reasonably assumed that they were the most accurate ones for the *I-V* curves. It should also be remarked that the equations for the 1-D/2-R equivalent circuit model resulting from these fits may not exactly meet the requirements for the short circuit, MPP, and open-circuit conditions, as they represent the best fits not only to the three characteristic points but to a large number of additional points.

Table 1. Solar cell/panel *I-V* curves used in the present work (where *n* is the number of series-connected cells for each of these photovoltaic devices; *T* is the temperature; and the characteristic points are I_{sc}, I_{mp}, V_{mp}, and V_{oc}).

Solar Cell/Panel	Technology	n	T [°C]	I_{sc} [A]	I_{mp} [A]	V_{mp} [V]	V_{oc} [V]
RTC France [1]	Si	1	33	0.7605	0.6894	0.4507	0.5727
TNJ Spectrolab [2]	GaInP2/GaAs/Ge	3	28	0.5259	0.4969	2.273	2.592
ZTJ Emcore [2]	InGaP/InGaAs/Ge	3	28	0.4634	0.4424	2.398	2.726
Azur Space 3G30C [3]	GaInP/GaAs/Ge	3	28	0.5270	0.5023	2.468	2.711
Photowatt PWP 201 [1]	Si	36	45	1.032	0.9255	12.493	16.778
Kyocera KC200GT-2 [2]	Si polycrystalline	54	25	8.182	7.605	26.90	32.92
Selex Galileo SPVS X5 [4]	GaInP/GaAs/Ge	15	20	0.5029	0.4783	12.406	13.603

[1] Taken from Easwarakhantan et al. [83]. [2] Graphically extracted from the manufacturer's datasheet. [3] Supplied by Azur Space. [4] Measured at CIEMAT (Spain).

Table 2. Parameters for the 1-D/2-E equivalent circuit models, numerically fitted to *I-V* curves based on data from RTC France, TNJ Spectrolab, ZTJ Emcore, and Azrur Space 3G30C solar cells, and Photowatt PWP 201, Kyocera KC200GT-2, and Selex Galileo SPVS X5 solar panels (where ζ is the non-dimensional RMSE for the fit).

Solar Cell/Panel	I_{pv} [A]	I_0 [A]	a	R_s [Ω]	R_{sh} [Ω]	ζ
RTC France	$7.617 \cdot 10^{-01}$	$2.746 \cdot 10^{-07}$	1.466	$3.697 \cdot 10^{-02}$	$4.421 \cdot 10^{+01}$	$8.49 \cdot 10^{-04}$
TNJ Spectrolab	$5.261 \cdot 10^{-01}$	$7.543 \cdot 10^{-15}$	1.045	$1.033 \cdot 10^{-01}$	$2.315 \cdot 10^{+02}$	$3.80 \cdot 10^{-03}$
ZTJ Emcore	$4.640 \cdot 10^{-01}$	$5.863 \cdot 10^{-14}$	1.180	$5.918 \cdot 10^{-02}$	$4.669 \cdot 10^{+02}$	$3.00 \cdot 10^{-03}$
Azur Space 3G30C	$5.274 \cdot 10^{-01}$	$8.522 \cdot 10^{-19}$	0.850	$8.580 \cdot 10^{-02}$	$2.202 \cdot 10^{+03}$	$1.57 \cdot 10^{-03}$
Photowatt PWP 201	$1.033 \cdot 10^{+00}$	$1.732 \cdot 10^{-06}$	1.282	$1.312 \cdot 10^{+00}$	$6.832 \cdot 10^{+02}$	$1.79 \cdot 10^{-03}$
Kyocera KC200GT-2	$8.179 \cdot 10^{+00}$	$2.693 \cdot 10^{-09}$	1.089	$2.280 \cdot 10^{-01}$	$1.449 \cdot 10^{+02}$	$3.53 \cdot 10^{-03}$
Selex Galileo SPVS X5	$4.994 \cdot 10^{-01}$	$6.165 \cdot 10^{-18}$	0.921	$1.885 \cdot 10^{-01}$	$1.422 \cdot 10^{+03}$	$6.50 \cdot 10^{-03}$

The 1-D/2-R equivalent circuit model was also analytically fitted to the aforementioned *I-V* curves. Using specific values for the ideality factor, a, and the characteristic points of the curve, I_{sc}, V_{mp}, I_{mp} and V_{oc}, the different parameters of the model were found using the following procedure:

- R_s was calculated using Equation (8) to (12);
- R_{sh} was calculated using Equation (5);
- I_{pv} was calculated using Equation (6);
- I_0 was calculated using Equation (7).

Once all the parameters were known, the *I-V* curve was found by obtaining the values of the current, *I*, for each value of the voltage, *V*, using Equation (13). The ideality factor was initially varied from $a = 0.5$ to $a = 1.8$ for each photovoltaic device (although this range was extended for some devices), and the non-dimensional RMSE, *x*, was calculated in each case. The results are plotted in Figure 5. Table 3 shows the values of the parameter a at which the minimum value of *x* is reached, with the rest of the parameters for the 1-D/2-R equivalent circuit.

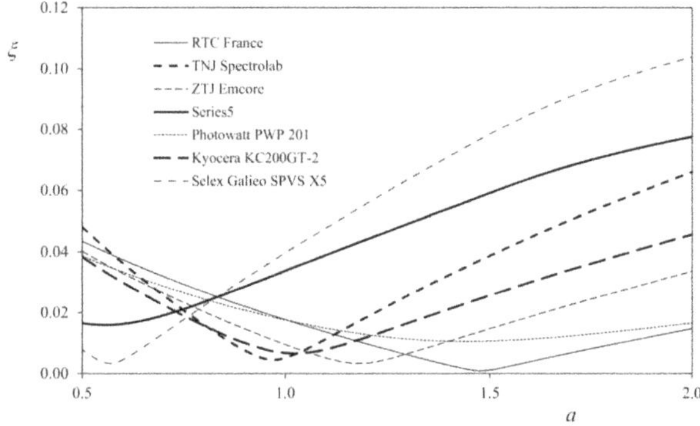

Figure 5. Non-dimensional RMSE, ξ, corresponding to the analytical approximations for extracting the 1-D/2-R equivalent circuit parameters in relation to the ideality factor, a, selected, for each one of the analyzed solar cells and panels.

Table 3. Parameters of the 1-D/2-R equivalent circuit models analytically fitted to *I-V* curves corresponding to data from RTC France, TNJ Spectrolab, ZTJ Emcore, and Azrur Space 3G30C solar cells, and Photowatt PWP 201, Kyocera KC200GT-2, and Selex Galileo SPVS X5 solar panels (where ξ is the non-dimensional RMSE for the fit).

Solar Cell/Panel	I_{pv} [A]	I_0 [A]	a	R_s [Ω]	R_{sh} [Ω]	ξ
RTC France	$7.610 \cdot 10^{-01}$	$3.200 \cdot 10^{-07}$	1.48	$3.621 \cdot 10^{-02}$	$5.195 \cdot 10^{+01}$	$9.33 \cdot 10^{-03}$
TNJ Spectrolab	$5.262 \cdot 10^{-01}$	$9.045 \cdot 10^{-16}$	0.98	$1.157 \cdot 10^{-01}$	$1.845 \cdot 10^{+02}$	$4.52 \cdot 10^{-03}$
ZTJ Emcore	$4.634 \cdot 10^{-01}$	$5.954 \cdot 10^{-14}$	1.18	$5.683 \cdot 10^{-02}$	$6.026 \cdot 10^{+02}$	$3.32 \cdot 10^{-03}$
Azur Space 3G30C	$5.275 \cdot 10^{-01}$	$5.012 \cdot 10^{-28}$	0.56	$1.322 \cdot 10^{-01}$	$1.554 \cdot 10^{+02}$	$1.60 \cdot 10^{-02}$
Photowatt PWP 201	$1.031 \cdot 10^{+00}$	$7.808 \cdot 10^{-06}$	1.44	$1.273 \cdot 10^{+00}$	$-1.121 \cdot 10^{+03}$	$1.06 \cdot 10^{-02}$
Kyocera KC200GT-2	$8.194 \cdot 10^{+00}$	$6.298 \cdot 10^{-10}$	1.02	$2.447 \cdot 10^{-01}$	$1.720 \cdot 10^{+02}$	$6.57 \cdot 10^{-03}$
Selex Galileo SPVS X5	$5.034 \cdot 10^{-01}$	$7.007 \cdot 10^{-29}$	0.56	$6.884 \cdot 10^{-01}$	$7.523 \cdot 10^{+02}$	$3.44 \cdot 10^{-03}$

Some important conclusions can be drawn based on these results, as follows:

- The analytical methodology suggested in the present work may give unacceptable results when the ideality factor exceeds a certain value, as complex numbers start to emerge in the calculations.
- One of the parameters obtained analytically for the best fit in terms of the non-dimensional RMSE has no physical meaning: this is the negative value of the resistance

of the shunt resistor shown in Table 3, which is obtained for the Photowatt PWP 201 solar panel.

- The best analytical fit gave remarkably low values of a in two cases: the Azur Space 3G30C solar cell, and the Selex Galileo SPVS X5 solar panel (composed of five Azur Space 3G28C series-connected cells), as their values were outside the range [1,1.5] suggested by Villalva et al. [74,75].

In order to compare both types of fit, the values of the ideality factor that gave the lowest values of x for the analytical fittings, $a_{bf\text{-}a}$, were plotted against the values obtained from the numerical fittings, $a_{bf\text{-}n}$, in Figure 6. It can be seen that the correlation between the fits is high, with the exception of only two devices: the Azur Space 3G30C solar cell, and the Selex Galileo SPVS X5 solar panel (which are both based on the same photovoltaic technology, as shown in Table 1). However, this correlation does not allow us to identify any rule that would suggest a reasonable value for the ideality factor. We, therefore, plotted the average values of the non-dimensional RMSE, ζ_{av}, obtained from the analytical fits versus the ideality factor, a. It can be seen from Figure 6 that the minimum value of ζ_{av} was obtained at $a = 0.98$. As a consequence, the value suggested for all devices is $a = 1$.

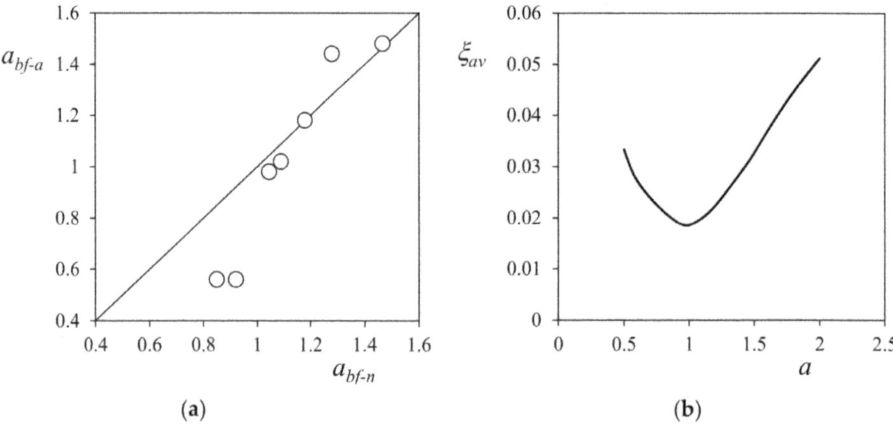

(a)   **(b)**

Figure 6. (a) Ideality factor corresponding to the best analytical fit to the 1-D/2-R equivalent circuit model, $a_{bf\text{-}a}$, for the studied *I-V* curves, versus ideality factors obtained from the numerical fittings, $a_{bf\text{-}n}$. The line corresponding to perfect correlation between both factors is shown as a reference. (b) Averaged non-dimensional RMSE for all analytical fittings, ζ_{av}, versus the ideality factor, a, used in the fits.

Figure 7 shows a comparison of the non-dimensional RMSE values obtained from the numerical fit, the analytical fit, and the analytical fit for $a = 1$ (see Table 4) for all of the photovoltaic devices. Although the differences may appear high, it should be taken into account that photovoltaic systems are normally designed to operate at the MPP, which is exactly reproduced by all of the analytical approximations for solving the parameter extraction of the 1-D/2-R equivalent circuit model. In addition, we reviewed the results from several works published over the last five years on parameter extraction for 1-D/2-R equivalent circuit models, based on the same experimental data from the RTC France solar cell and the Photowatt PWP 201 solar panel. The values obtained in these works for the non-dimensional RMSE for the fits were lower than the results from our method, with average values of $\zeta = 1.20\cdot10^{-03}$ (RTC France solar cell) and $\zeta = 4.03\cdot10^{-03}$ (Photowatt PWP 201 solar panel). Although these more accurate values were based on numerical rather than analytical procedures, it should be pointed out that numerical approaches do not always reach a more accurate solution [92].

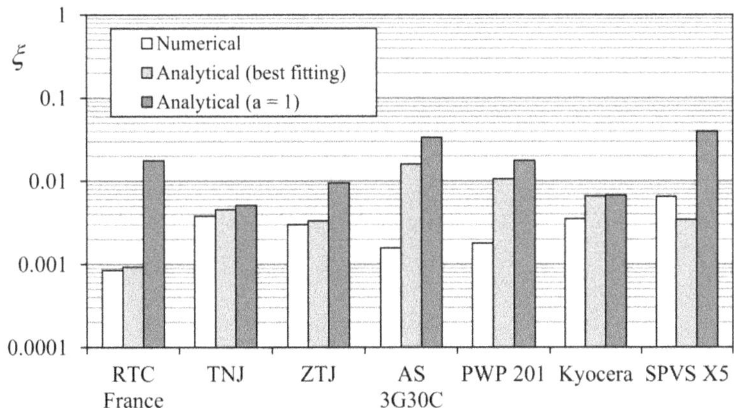

Figure 7. Values for the non-dimensional RMSE, x, obtained from a numerical fit of the 1-D/2-R equivalent circuit model to the studied *I-V* curves (see Table 2), an analytical fit (see Table 3), and an analytical fit with $a = 1$ (see Table 4).

Table 4. Analytically fitted parameters for 1-D/2-R equivalent circuit models, with $a = 1$, for *I-V* curves corresponding to RTC France, TNJ Spectrolab, ZTJ Emcore, and Azrur Space 3G30C solar cells, and Photowatt PWP 201, Kyocera KC200GT-2, and Selex Galileo SPVS X5 solar panels (where ζ is the non-dimensional RMSE for the fit).

Solar Cell/Panel	I_{pv} [A]	I_0 [A]	a	R_s [Ω]	R_{sh} [Ω]	ζ
RTC France	$7.638 \cdot 10^{-01}$	$2.721 \cdot 10^{-10}$	1	$6.912 \cdot 10^{-02}$	$1.615 \cdot 10^{+01}$	$1.77 \cdot 10^{-02}$
TNJ Spectrolab	$5.262 \cdot 10^{-01}$	$1.786 \cdot 10^{-15}$	1	$1.078 \cdot 10^{-01}$	$1.895 \cdot 10^{+02}$	$5.08 \cdot 10^{-03}$
ZTJ Emcore	$4.636 \cdot 10^{-01}$	$2.835 \cdot 10^{-16}$	1	$1.348 \cdot 10^{-01}$	$3.723 \cdot 10^{+02}$	$9.51 \cdot 10^{-03}$
Azur Space 3G30C	$5.269 \cdot 10^{-01}$	$3.881 \cdot 10^{-16}$	1	$-6.068 \cdot 10^{-02}$	$2.643 \cdot 10^{+02}$	$3.38 \cdot 10^{-02}$
Photowatt PWP 201	$1.036 \cdot 10^{+00}$	$4.163 \cdot 10^{-08}$	1	$1.982 \cdot 10^{+00}$	$5.430 \cdot 10^{+02}$	$1.76 \cdot 10^{-02}$
Kyocera KC200GT-2	$8.195 \cdot 10^{+00}$	$3.950 \cdot 10^{-10}$	1	$2.522 \cdot 10^{-01}$	$1.638 \cdot 10^{+02}$	$6.71 \cdot 10^{-03}$
Selex Galileo SPVS X5	$5.028 \cdot 10^{-01}$	$1.261 \cdot 10^{-16}$	1	$-3.091 \cdot 10^{-01}$	$1.185 \cdot 10^{+03}$	$3.96 \cdot 10^{-02}$

3.2. Lambert W-Function Simplified Equations for Solar Cell/Panel Modeling

A thorough review of the available literature from between 2000 and 2020 was carried out to find relevant data on photovoltaic devices, including the five parameters of the 1-D/2-R equivalent circuit model (I_{pv}, I_0, a, R_s, and R_{sh}), the characteristic points of the *I-V* curve (I_{sc}, V_{mp}, I_{mp}, and V_{oc}), the temperature of the cells, T, which is related to the performance curve, and the number of series-corrected cells, n. Information was found for 90 photovoltaic devices, most of which were solar panels.

The positive branch of the Lambert W-function needs to be solved for in Equation (13), and must be evaluated at certain points x for a given value of the output voltage, V (within the range $[0, V_{oc}]$):

$$x = f\left(I_{pv}, I_0, a, R_s, R_{sh}, n, T, V\right). \tag{25}$$

After post-processing the data from the 90 solar cells/panels, it was clear that Equation (13) refers to the right-side section of the Lambert's W-function positive branch, W_0^+. The following expressions were proposed:

$$W_0^+(x) = x \exp\left(0.71116 x^2 - 0.98639 x\right), \quad x \in \left[2 \cdot 10^{-16}, 2 \cdot 10^{-1}\right], \tag{26}$$

$$W_0^+(x) = -1.6579 + 0.1396\left(2.9179 \cdot 10^5 - (x - 22.8345)^4\right)^{0.25}, \quad x \in [0.2, 1.2], \tag{27}$$

$$W_0^+(x) = -1.2216 + 3.4724 \cdot 10^{-2}\left(1.7091 \cdot 10^8 - (x - 114.146)^4\right)^{0.25}, \; x \in [1.2, 10]. \quad (28)$$

The error for these equations was below 0.15% (Equation (26)), 0.16% (Equation (27)), and 0.08% (Equation (28)). For larger values of x, we recommend using the approximation proposed by Barry et al. [86] (see Appendix A). In Figure 8, the positive branch of Lambert's W-function, W_0^+, calculated at the points x (Equation (25)) corresponding to the data for the solar cells/panels, evaluated at $V = 0$ and $V = V_{oc}$, is shown together with Equation (26) to (28). Depending on the magnitude of x (that is, depending on how closely it approaches 0^+), W_0^+ can be reasonably estimated as:

$$W_0^+ = x - 0.98639x^2 + 1.1976x^3 \approx x - 0.98639x^2 \approx x, \quad (29)$$

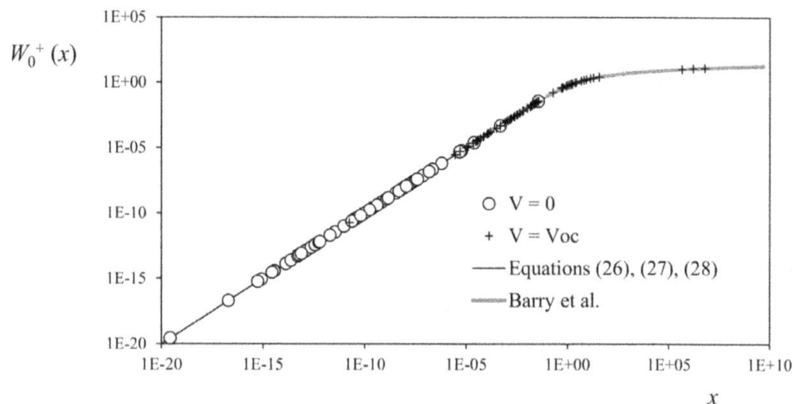

Figure 8. Positive branch of Lambert's W-function (right-side), W_0^+, calculated at $V = 0$ and $V = V_{oc}$ for points x (Equation (25)) corresponding to the 1-D/2-R equivalent circuit model, based on the data for the solar cells/panels. Our approximation for W_0^+ (Equation (26) to (28)) is also shown.

The negative branch of the Lambert W-function, W_{-1}, needs to be calculated at:

$$x = f\left(I_{pv}, I_0, a, R_s, R_{sh}, n, T, V\right) \quad (30)$$

when estimating the series resistance, R_s, for the 1-D/2-R equivalent circuit model (Equation (8) to (12)). Since the above equation for the different solar cells/panels from the database gives values of x within the range $[-2.77 \cdot 10^{-3}, -10^{-35}]$, two expressions were defined:

$$W_{-1}(x) = 9.7117 \cdot 10^{-5} \ln(-x)^3 + 6.8669 \cdot 10^{-3} \ln(-x)^2 + 1.2 \ln(-x) - 1.1102, \\ x \in \left[-10^{-2}, -5 \cdot 10^{-13}\right] \quad (31)$$

$$W_{-1}(x) = 1.6705 \cdot 10^{-6} \ln(-x)^3 + 4.4514 \cdot 10^{-4} \ln(-x)^2 + 1.0511 \ln(-x) - 2.3364, \\ x \in \left[-5 \cdot 10^{-13}, -10^{-40}\right] \quad (32)$$

The error for these equations was below 0.42% (Equation (31)) and 0.02% (Equation (32)) (see Figure 9). Equation (32) can be applied for values of up to $x = -10^{-50}$, with errors of below 0.07%.

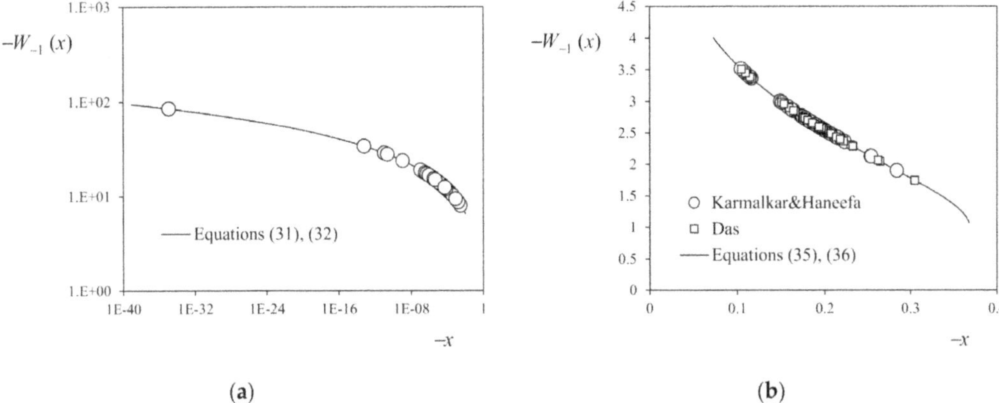

Figure 9. (a) Negative branch of Lambert's W-function, W_{-1}, calculated for points x (Equation (30)) corresponding to the series resistance, R_s, of the 1-D/2-R equivalent circuit model, based on data for solar cells/panels. Our approximate expression for W_{-1} (Equations (31) and (32)) is also shown. (b) Negative branch of Lambert's W-function, W_{-1}, calculated for points x (Equation (33)) corresponding to the explicit models proposed by Kalmarkar and Haneefa and Das, based on data for solar cells/panels. Our approximate expression for W_{-1} (Equations (35) and (36)) is also shown.

Finally, if the explicit models proposed by Kalmarkar and Haneefa, and Das are used, W_{-1} needs to be calculated at points x defined by the characteristic points of the *I-V* curve (I_{sc}, V_{mp}, I_{mp}, and V_{oc} (see Equations (16) and (19)):

$$x = f\left(I_{sc}, I_{mp}, V_{oc}, V_{mp}\right) \tag{33}$$

as shown in Figure 9. Based on the post-processed data for the 90 solar cells/panels, the values of x calculated using these explicit methods were within the range $[-0.304, -0.1]$. The following equation for W_{-1} was initially suggested, which was characterized by an error of less than 1.6%22 within this range:

$$W_{-1}(x) = 248.42x^4 + 134.24x^3 + 4.4258x^2 - 14.629x - 4.9631 \tag{34}$$

In order to extend both the range of validity and the accuracy of the above equation, a splitting into two sub-branches at the point where the curvature of the function changes its sign ($x = -0.27$), the following equations were suggested:

$$W_{-1}(x) = -1 - \sqrt{42.949x^2 + 37.694x + 8.0542}, \ x \in [-0.36785, -0.27] \tag{35}$$

$$W_{-1}(x) = 0.14279\ln(-x)^3 + 1.04416\ln(-x)^2 + 3.92\ln(-x) + 1.65795, \\ x \in [-0.27, -0.0732] \tag{36}$$

The error for the above equations was below 0.23% (Equation (35)) and 0.02% (Equation (36)) (see Figure 9). It should also be emphasized that caution needs to be used with regard to Equation (35), as it gives complex solutions when $x \rightarrow -1/e^+$ (that is, when $x \in [-1/e, -0.36785]$).

3.3. Case Study: The Solar Panels of the UPMSat-1

The general characteristics from the *I-V* curves for the UPM-5 and UPM-6 solar panels for UPMSat-1 (see Figure 10) are shown in Tables 1, 2 and 5 (Appendix B). These panels are based on two different technologies: Si (UPM-5) and Ga-As (UPM-6). In the latter, the number of series-connected cells n was 32; however, since they were based on double-junction technology, this number must be multiplied by two, unlike in the

1-D/2-R equivalent circuit model (the Selex Galileo SPVS X5, formed from five Azur Space triple-junction technology series-connected solar cells, where $n = 15$, as shown in Table 1).

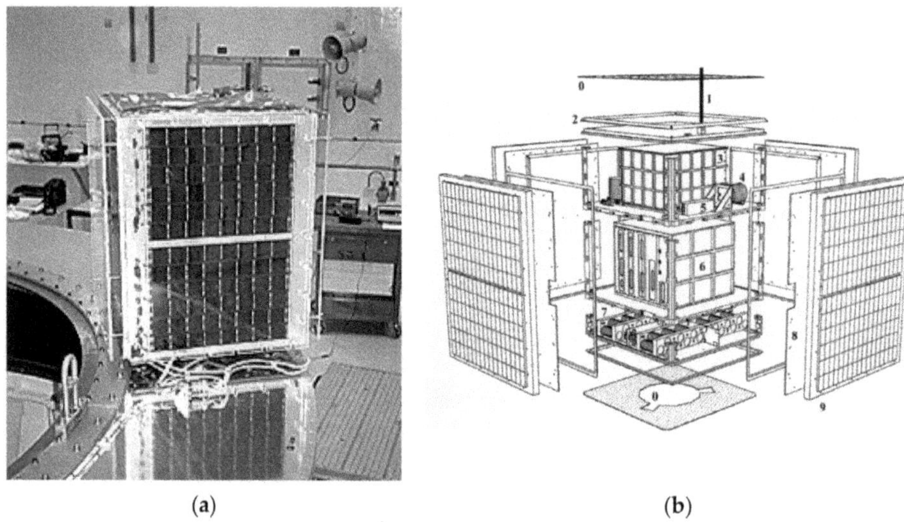

(a)   (b)

Figure 10. The UPMSat-1 satellite; launched in 1995, this was the tenth university-developed space mission in history [93]. (a) Satellite during the integration in the Ariane IV-40 launcher (V75 flight). (b) Exploded view drawing.

Table 5. Characteristics of the measured *I-V* curves corresponding to the UPM-5 and UPM-6 solar panels of UPMSat-1 (see Figure 10).

Solar Panel	Technology	n	T [°C]	I_{sc} [A]	I_{mp} [A]	V_{mp} [V]	V_{oc} [V]
UPM-5	Si	51	25	1.431	1.329	25.139	30.513
UPM-6	Ga-As	64	25	1.423	1.318	25.806	31.351

The *I-V* curves for the UPM-5 and UPM-6 solar panels are plotted in Figures 11 and 12, respectively, with several other curves corresponding to:

- The best fit of the 1-D/2-R equivalent circuit model (see Table 6);
- The analytical fit for $a = 1$ (Equations (5–12)), and our approximations to the Lambert W-function (Equations (26)–(28),(31,32));
- The explicit models proposed by Kalmarkar and Haneefa, Das (Equations (14)–(20), (35,36)), and Pindado and Cubas (Equations (21) and (22)), see Table 7.

Table 6. Parameters for the 1-D/2-R equivalent circuit model (where Num. represents the numerical best fit, and An. represents the analytical fit with $a = 1$) for the UPM-5 and UPM-6 solar panels of the UPMSat-1 satellite. ξ is the non-dimensional RMSE for the experimental data.

Solar Panel	Fitting	I_{pv} [A]	I_0 [A]	a	R_s [Ω]	R_{sh} [Ω]	ξ
UPM-5	Num.	1.4314	$1.0495 \cdot 10^{-09}$	1.105	1.0368	$4.3761 \cdot 10^{+03}$	$1.80 \cdot 10^{-03}$
	An.	1.4323	$1.0855 \cdot 10^{-10}$	1	1.2045	$1.3023 \cdot 10^{+03}$	$5.63 \cdot 10^{-03}$
UPM-6	Num.	1.4295	$1.0285 \cdot 10^{-09}$	0.902	0.8483	$9.9115 \cdot 10^{+02}$	$9.50 \cdot 10^{-03}$
	An.	1.4238	$7.3448 \cdot 10^{-09}$	1	0.7207	$1.3524 \cdot 10^{+03}$	$1.70 \cdot 10^{-02}$

Table 7. Parameters for the explicit models proposed by Kalmarkar and Haneefa, Das and Pindado and Cubas, for the UPM-5 and UPM-6 solar panels of the UPMSat-1 satellite. ζ is the non-dimensional RMSE for the experimental data.

Model	Parameters; ζ	UPM-5	UPM-6
Kalmarkar and Haneefa	γ	$9.942 \cdot 10^{-01}$	$9.912 \cdot 10^{-01}$
	m	$1.401 \cdot 10^{-01}$	$1.392 \cdot 10^{-01}$
	ζ	$3.35 \cdot 10^{-02}$	$3.71 \cdot 10^{-02}$
Das	k	$1.401 \cdot 10^{-01}$	$1.393 \cdot 10^{-01}$
	h	$6.526 \cdot 10^{-03}$	$9.568 \cdot 10^{-03}$
	ζ	$3.33 \cdot 10^{-02}$	$3.69 \cdot 10^{-02}$
Pindado and Cubas	ϕ	2.6605	2.5879
	ζ	$2.04 \cdot 10^{-02}$	$2.12 \cdot 10^{-02}$

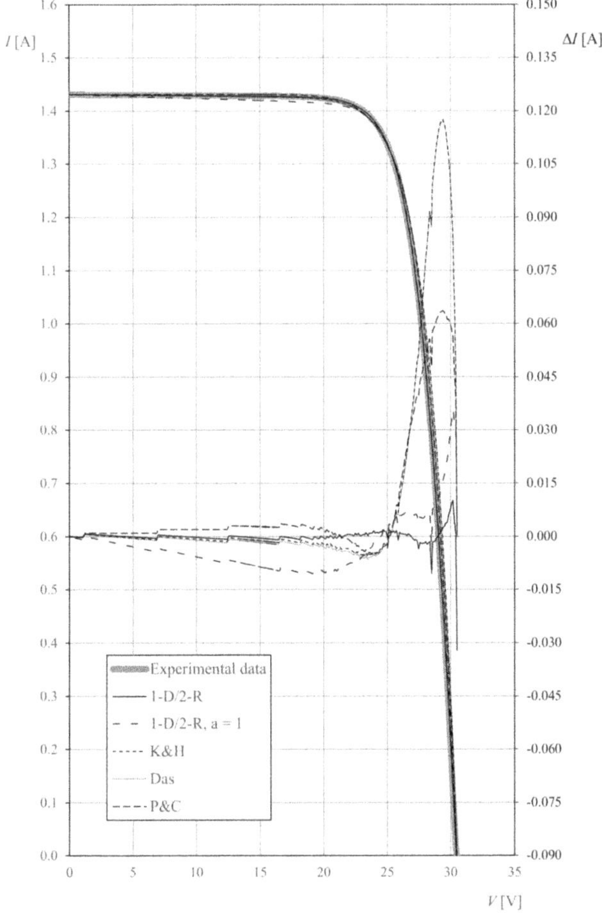

Figure 11. *I-V* experimental curve (left axis) for the UPM-5 solar panel. Curves for the 1-D/2-R equivalent circuit model (obtained using two procedures: best fit to the parameters obtained numerically and analytical extraction) and the explicit methods proposed by Kalmarkar and Haneefa, Das and Pindado and Cubas are shown. The differences in the current compared to the experimental data are indicated on the right axis.

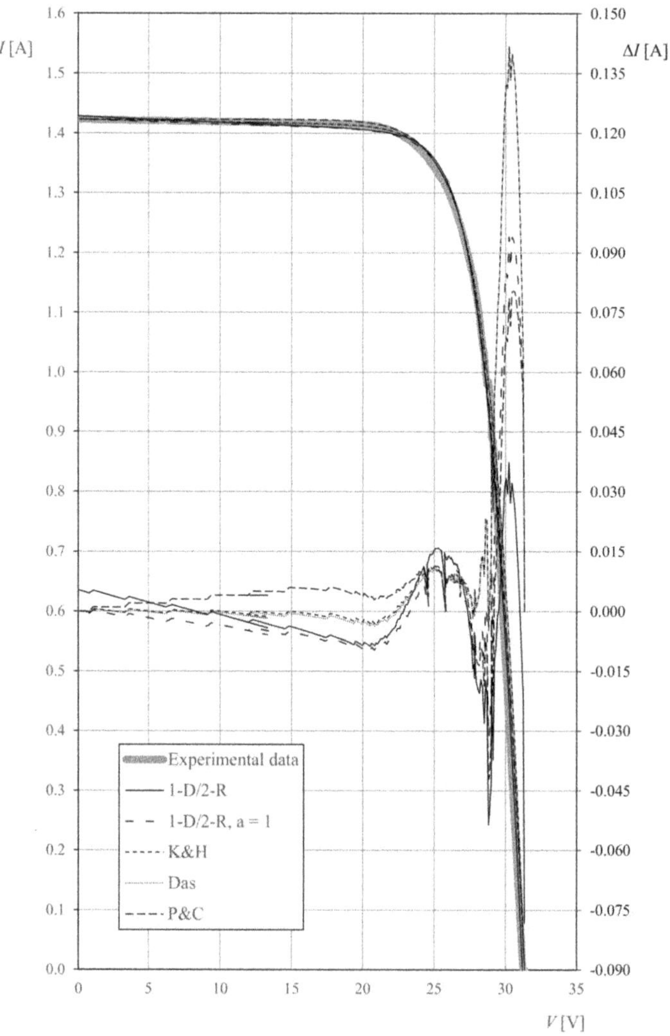

Figure 12. *I-V* experimental curve (left axis) for the UPM-6 solar panel. Curves for the 1-D/2-R equivalent circuit model (obtained using two procedures: best fit to the parameters obtained numerically and analytical extraction) and the explicit methods of Kalmarkar and Haneefa, Das and Pindado and Cubas are shown. The differences in the current compared to the experimental data are indicated on the right axis.

The differences in the current for these models with regard to the experimental *I-V* curves are plotted in Figures 11 and 12.

The differences were relatively small, with the maximum always located in a reasonably small range between the MPP and the open voltage point. The values of the non-dimensional RMSE, ζ, for the fits are plotted in Figure 13. The results show reasonably good agreement with the mathematical approaches proposed in the present work.

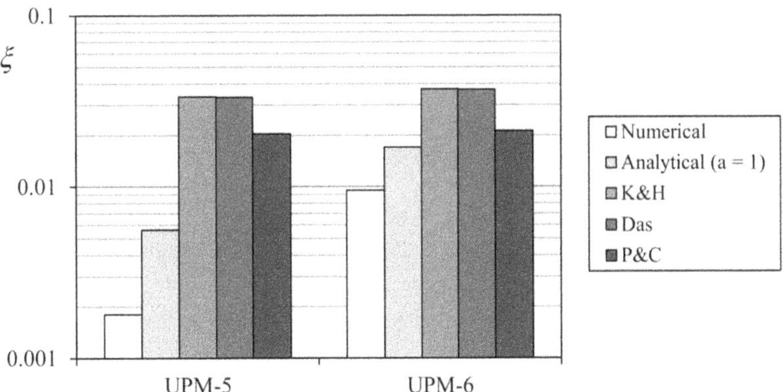

Figure 13. Non-dimensional RMSE, ζ, for the numerical fit of the 1-D/2-R equivalent circuit model, and the analytical fit with the proposed method, for the UPM-5 and UPM-6 solar panels of UPMSat-1. Values derived from the explicit methods studied here are also shown.

4. Conclusions

In this study, we have reviewed analytical approaches for extracting the five parameters of the 1-D/2-R equivalent circuit model from the characteristic points of the *I-V* curve (short circuit, MPP, and open circuit points). Five different problems were identified, as follows:

- An initial estimation of the ideality factor, a, is required.
- The equation for the value of the resistance of the series-connected resistor, R_s, is an implicit expression, meaning that either an iterative process or the Lambert W-function is required.
- When all the parameters for the 1-D/2-R equivalent circuit model have been extracted, an implicit equation must be solved (or the Lambert W-function must be used) to derive the value of the output current for a given output voltage.
- The use of the Lambert W-function requires some numerical and calculation resources and skills.
- The use of explicit models rather than the 1-D/2-R equivalent circuit model, in order to avoid the problems described above, may not be possible, as some of them require the Lambert W-function to derive their parameters, based on the characteristic points of the *I-V* curve.

I-V curves from seven different solar cells/panels were used to test our proposed approach. *I-V* curves from two of the solar panels of the UPMSat-1 spacecraft were also analyzed as a case study.

The most important conclusions of this work are as follows:

- A value of $a = 1$ for the 1-D/2-R equivalent circuit model was shown to be reasonable for most photovoltaic technologies.
- The analytical procedure for extracting the parameters for the 1-D/2-R equivalent circuit model may give values for the resistance of one resistor (or even both) that are negative. However, this does not affect the results (i.e., the modeled performance of the photovoltaic device).
- The Lambert W-function can be simplified for use in modeling the performance of photovoltaic devices. Accurate simplified versions of the Lambert W-function are proposed here for three cases, depending on the specific need: (i) calculation of R_s; (ii) calculation of the output current using the equation for the 1-D/2-R equivalent circuit; or (iii) calculation of the parameters for certain explicit models.
- Explicit models are also accurate alternatives to the 1-D/2-R equivalent circuit model.

Our approach was carefully verified in a case study in which the fits of the1-D/2-R equivalent circuit model and the explicit methods to the measured data (*I-V* curve) for two solar panels from the UPMSat-1 satellite were compared with the results of the proposed method.

Finally, it should be highlighted that the results from this work open up new possibilities for coupled calculations, both for the performance of photovoltaic systems and in other disciplines such as thermodynamics and power distribution in grids. The simple but accurate solutions to the 1-D/2-R equivalent circuit model and the explicit methods described here can easily be implemented in software packages such as ESATAN$^©$.

Author Contributions: Project administration, S.P., J.C., E.R.-M.; methodology, S.P.; conceptualization, S.P.; formal analysis, S.P.; supervision S.P., J.C., E.R.-M.; investigation, J.M.Á., D.A.-C., J.C.-E., M.S.-P., R.J.-P., A.G.-E.; software, J.M.Á., D.A.-C., J.C.-E., M.S.-P., R.J.-P., A.G.-E.; validation, visualization, J.M.Á., D.A.-C., J.C.-E., M.S.-P., R.J.-P., A.G.-E.; writing—original draft, J.M.Á., D.A.-C., E.R.-M., J.C.-E., M.S.-P., R.J.-P., A.G.-E.; writing—review and editing, S.P. All authors have read and agreed to the published version of the manuscript.

Funding: This research received no external funding.

Institutional Review Board Statement: Not applicable.

Informed Consent Statement: Not applicable.

Data Availability Statement: Not applicable.

Acknowledgments: This work was carried out within the framework of the Educational Innovation Project PIE IE1920.1402 'PIRAMIDE: *Proyectos de Investigación Realizados por Alumnos de Máster/Grado para la Innovación y el Desarrollo Espacial'*. This Academic Innovation Project (AIP) was funded by *Universidad Politécnica de Madrid* (UPM) in the 2019/2020 call '*Ayudas a la innovación educativa y a la mejora en la calidad de la enseñanza'*. The authors are indebted to Ángel Sanz-Andrés for his support with regard to the research program on the performance of solar panels at *Instituto Universitario de Microgravedad "Ignacio Da Riva"* (IDR/UPM), *Universidad Politécnica de Madrid* (UPM). The authors are grateful to the reviewers for their kind help in improving the present paper.

Conflicts of Interest: The authors declare no conflict of interest.

Appendix A. Approximation to the Right-Side of the Lambert W-Function Positive Branch by Larry et al. (2000) (see References)

$$W_0^+(x) = 1.4586887 \ln\left(\frac{1.2x}{\ln(2.4x/\ln(1+2.4x))}\right) - 0.4586887 \ln\left(\frac{2x}{\ln(1+2x)}\right). \quad (A1)$$

Appendix B. *I-V* Curves of the UPMSat-1 Solar Panels.

Table 1. *I-V* curve of the UPM-5 solar panel of the UPMSat-1.

V [V]	I [A]	V [V]	I [A]	V [V]	I [A]	V [V]	I [A]	V [V]	I [A]	V [V]	I [A]
−0.950	1.431	5.189	1.430	10.813	1.429	16.204	1.428	21.510	1.418	26.428	1.213
−0.801	1.431	5.330	1.430	10.944	1.429	13.338	1.428	21.632	1.417	26.545	1.198
−0.660	1.431	5.471	1.430	11.077	1.429	16.459	1.428	21.758	1.416	26.657	1.182
−0.510	1.431	5.613	1.430	11.217	1.429	16.589	1.427	21.879	1.415	26.775	1.165
−0.370	1.431	5.744	1.430	11.348	1.429	16.723	1.427	22.001	1.414	26.887	1.148
−0.221	1.431	5.888	1.430	11.480	1.429	16.848	1.427	22.128	1.413	27.006	1.129
−0.079	1.431	6.026	1.430	11.614	1.429	16.980	1.427	22.248	1.412	27.118	1.110
0.067	1.431	6.167	1.430	11.745	1.429	17.101	1.427	22.372	1.411	27.226	1.090
0.210	1.431	6.309	1.430	11.885	1.429	17.234	1.427	22.493	1.409	27.343	1.068
0.360	1.431	6.440	1.430	12.019	1.429	17.388	1.427	22.617	1.408	27.453	1.048

Table 1. *Cont.*

V [V]	I [A]	V [V]	I [A]	V [V]	I [A]	V [V]	I [A]	V [V]	I [A]	V [V]	I [A]
0.600	1.431	6.580	1.430	12.149	1.429	17.489	1.427	22.741	1.406	27.572	1.021
0.660	1.431	6.727	1.430	12.282	1.429	17.621	1.427	22.861	1.405	27.681	0.997
0.782	1.431	6.863	1.430	12.413	1.429	17.744	1.427	22.987	1.403	27.789	0.972
0.924	1.431	6.983	1.429	12.545	1.428	17.999	1.427	23.099	1.401	27.908	0.943
1.071	1.431	7.136	1.429	12.678	1.428	18.122	1.427	23.221	1.399	28.013	0.916
1.214	1.430	7.277	1.429	12.810	1.428	18.253	1.426	23.343	1.396	28.122	0.888
1.364	1.430	7.407	1.429	12.942	1.428	18.278	1.426	23.485	1.394	28.230	0.858
1.503	1.430	7.546	1.429	13.075	1.428	18.510	1.426	23.588	1.391	28.345	0.825
1.643	1.430	7.689	1.429	13.206	1.428	18.635	1.426	23.712	1.388	28.483	0.793
1.786	1.430	7.822	1.429	13.336	1.428	18.756	1.426	23.827	1.385	28.558	0.761
1.936	1.430	7.963	1.429	13.470	1.428	18.889	1.426	23.948	1.381	28.665	0.727
2.076	1.430	8.094	1.429	13.601	1.428	19.011	1.426	24.071	1.378	28.780	0.689
2.216	1.430	8.235	1.429	13.734	1.428	19.138	1.426	24.183	1.374	28.884	0.653
2.367	1.430	8.374	1.429	13.884	1.428	19.267	1.425	24.305	1.370	28.992	0.615
2.508	1.430	8.507	1.429	13.997	1.428	19.390	1.425	24.427	1.365	29.096	0.577
2.648	1.430	8.648	1.429	14.129	1.428	19.514	1.425	24.550	1.360	29.200	0.538
2.709	1.430	8.780	1.429	14.262	1.428	19.648	1.425	24.671	1.354	29.314	0.494
2.930	1.430	8.923	1.429	14.392	1.428	19.769	1.424	24.784	1.349	29.420	0.452
3.068	1.430	9.053	1.429	14.515	1.428	19.892	1.424	24.905	1.343	29.524	0.410
3.219	1.430	9.194	1.429	14.649	1.428	20.016	1.424	25.027	1.338	29.629	0.366
3.359	1.430	9.325	1.429	14.780	1.428	20.146	1.424	25.139	1.329	29.738	0.320
3.500	1.430	9.456	1.429	14.911	1.428	20.263	1.423	25.261	1.322	29.847	0.271
3.642	1.430	9.598	1.429	15.042	1.428	20.395	1.423	25.377	1.314	29.960	0.220
4.064	1.430	9.730	1.429	15.168	1.428	20.518	1.422	25.496	1.305	30.074	0.167
4.204	1.430	9.870	1.429	16.297	1.428	20.639	1.422	25.617	1.296	30.189	0.113
4.347	1.430	10.003	1.429	15.430	1.428	20.762	1.422	25.729	1.286	30.299	0.067
4.485	1.430	10.136	1.429	15.561	1.428	20.896	1.421	25.848	1.278	30.409	0.018
4.628	1.430	10.275	1.429	15.666	1.428	21.018	1.421	25.961	1.265	30.513	0.000
4.767	1.430	10.407	1.429	15.815	1.428	21.140	1.420	26.082	1.253		
4.909	1.430	10.540	1.429	15.949	1.428	21.265	1.419	26.193	1.241		
5.047	1.430	10.661	1.429	16.082	1.428	21.388	1.419	26.314	1.227		

Table 2. *I-V* curve of the UPM-6 solar panel of the UPMSat-1.

V [V]	I [A]	V [V]	I [A]	V [V]	I [A]	V [V]	I [A]	V [V]	I [A]	V [V]	I [A]
−0.638	1.423	9.402	1.419	19.166	1.416	22.469	1.399	25.814	1.310	28.142	1.093
−0.518	1.423	9.722	1.419	19.269	1.416	22.593	1.398	25.844	1.309	28.249	1.069
−0.198	1.423	10.042	1.419	19.373	1.416	22.687	1.396	25.917	1.305	28.367	1.047
0.122	1.423	10.362	1.419	19.476	1.415	22.801	1.395	25.990	1.300	28.485	1.031
0.442	1.423	10.882	1.419	19.560	1.415	22.905	1.393	26.063	1.296	28.581	0.999
0.782	1.423	11.002	1.419	19.684	1.415	23.009	1.391	26.136	1.292	28.680	0.980
1.082	1.422	13.322	1.419	19.788	1.415	23.113	1.390	26.209	1.287	28.795	0.984
1.402	1.422	11.842	1.419	19.892	1.415	23.217	1.388	26.282	1.283	28.990	0.923
1.722	1.422	11.962	1.419	19.986	1.415	23.320	1.386	26.365	1.278	29.007	0.892
2.042	1.422	12.282	1.418	20.100	1.414	23.424	1.384	26.428	1.273	29.112	0.883
2.362	1.422	12.602	1.418	20.204	1.414	23.528	1.382	26.500	1.268	29.216	0.828
2.882	1.422	12.922	1.418	20.308	1.414	23.632	1.380	26.573	1.262	29.321	0.793
3.002	1.422	13.242	1.418	20.411	1.414	23.736	1.377	26.646	1.257	29.424	0.753
3.322	1.422	13.682	1.418	20.515	1.414	23.840	1.375	26.718	1.251	29.527	0.718
3.642	1.421	13.882	1.418	20.619	1.413	23.944	1.372	26.792	1.248	29.630	0.678
3.882	1.421	14.202	1.418	20.723	1.413	24.048	1.370	26.865	1.240	29.732	0.632

Table 2. *Cont.*

V [V]	I [A]	V [V]	I [A]	V [V]	I [A]	V [V]	I [A]	V [V]	I [A]	V [V]	I [A]
4.282	1.421	14.522	1.418	20.827	1.413	24.152	1.367	26.938	1.234	29.834	0.589
4.602	1.421	14.842	1.417	20.931	1.412	24.265	1.364	27.011	1.227	29.935	0.543
4.922	1.421	15.162	1.417	21.035	1.412	24.590	1.362	27.084	1.221	30.035	0.502
5.242	1.421	15.482	1.417	21.139	1.411	24.483	1.359	27.157	1.214	30.138	0.465
5.562	1.421	15.802	1.417	21.243	1.410	24.567	1.356	27.230	1.207	30.239	0.416
5.882	1.421	16.122	1.417	21.248	1.410	24.671	1.352	27.303	1.200	30.344	0.382
6.202	1.421	16.442	1.417	21.450	1.409	24.776	1.349	27.378	1.193	30.455	0.329
6.622	1.420	16.762	1.417	21.554	1.408	24.879	1.346	27.448	1.186	30.568	0.282
6.842	1.420	17.082	1.417	21.668	1.408	24.983	1.342	27.521	1.178	30.682	0.234
7.182	1.420	17.402	1.417	21.762	1.407	25.087	1.338	27.594	1.170	30.797	0.187
7.482	1.420	17.722	1.416	21.868	1.406	25.180	1.335	27.687	1.163	30.911	0.139
7.802	1.420	18.042	1.416	21.970	1.405	25.284	1.331	27.740	1.154	31.021	0.094
8.122	1.420	18.362	1.416	22.074	1.404	25.398	1.327	27.813	1.146	31.131	0.048
8.442	1.420	18.682	1.416	22.178	1.403	25.502	1.323	27.805	1.141	31.241	0.002
8.762	1.420	19.002	1.416	22.282	1.402	25.806	1.318	27.916	1.130	31.351	0.000
9.082	1.420	19.061	1.416	22.386	1.400	25.710	1.314	28.032	1.112		

References

1. IEA. *Technology Roadmap-Solar Photovoltaic Energy 2014*; IEA: Paris, France, 2014.
2. Jäger-Waldau, A. Snapshot of photovoltaics February 2018. *EPJ Photovolt.* **2018**, *9*, 6. [CrossRef]
3. Zafrilla Rodríguez, J.E.; Arce Gonzalez, G.; Cadarso Vecina, M.Á.; Córcoles Fuentes, C.; Gómez Sanz, N.; López Santiago, L.A.; Monsalve Serrano, F.; Tobarra Gómez, M.Á. *El Desarrollo Actual de la Energía Solar Fotovoltaica en España*; Universidad de Murcia: Murcia, Spain, 2018.
4. Pandey, A.; Tyagi, V.; Selvaraj, J.A.; Rahim, N.; Tyagi, S. Recent advances in solar photovoltaic systems for emerging trends and advanced applications. *Renew. Sustain. Energy Rev.* **2016**, *53*, 859–884. [CrossRef]
5. Sommerfeld, J.; Buys, L.; Vine, D. Residential consumers' experiences in the adoption and use of solar PV. *Energy Policy* **2017**, *105*, 10–16. [CrossRef]
6. Hussain, A.; Arif, S.M.; Aslam, M. Emerging renewable and sustainable energy technologies: State of the art. *Renew. Sustain. Energy Rev.* **2017**, *71*, 12–28. [CrossRef]
7. Ueckerdt, F.; Brecha, R.; Luderer, G. Analyzing major challenges of wind and solar variability in power systems. *Renew. Energy* **2015**, *81*, 1–10. [CrossRef]
8. Kylili, A.; Fokaides, P.A.; Ioannides, A.; Kalogirou, S. Environmental assessment of solar thermal systems for the industrial sector. *J. Clean. Prod.* **2018**, *176*, 99–109. [CrossRef]
9. Kannan, N.; Vakeesan, D. Solar energy for future world—A review. *Renew. Sustain. Energy Rev.* **2016**, *62*, 1092–1105. [CrossRef]
10. Abolhosseini, S.; Heshmati, A. The main support mechanisms to finance renewable energy development. *Renew. Sustain. Energy Rev.* **2014**, *40*, 876–885. [CrossRef]
11. Zahedi, A. Solar photovoltaic (PV) energy; latest developments in the building integrated and hybrid PV systems. *Renew. Energy* **2006**, *31*, 711–718. [CrossRef]
12. Rauschenbach, H.S. *Solar Cell Array Design Handbook*; J.B. Metzler: Berlin, Germany, 1980.
13. Bandecchi, M.; Melton, B.; Ongaro, F. Concurrent engineering applied to space mission assessment and design. *ESA Bull. Sp. Agency* **1999**, *99*, 34–40.
14. Wall, S.D. Use of Concurrent Engineering in Space Mission Design. In Proceedings of the 2nd European Systems Engineering Conference (EuSEC), Munich, Germany, 13–15 September 2000; pp. 1–6.
15. Braukhane, A.; Quantius, D. Interactions in space systems design within a Concurrent Engineering facility. In Proceedings of the 2011 International Conference on Collaboration Technologies and Systems (CTS), Philadelphia, PA, USA, 23–27 May 2011; pp. 381–388.
16. Ivanov, A.B.; Masson, L.; Belloni, F. Operation of a Concurrent Design Facility for university projects. In Proceedings of the 2016 IEEE Aerospace Conference, Big Sky, MT, USA, 5–12 March 2016; pp. 1–9.
17. Loureiro, G.; Panades, W.; Silva, A. Lessons learned in 20 years of application of Systems Concurrent Engineering to space products. *Acta Astronaut.* **2018**, *151*, 44–52. [CrossRef]
18. Ballesteros, J.B.; Alvarez, J.M.; Arcenillas, P.; Roibas, E.; Cubas, J.; Pindado, S. CDF as a tool for space engineering master's student collaboration and concurrent design learning. In Proceedings of the 8th International Workshop on System & Concurrent Engineering for Space Applications (SECESA 2018), Glasgow, UK, 26–28 September 2018.
19. Roibás-Millán, E.; Sorribes-Palmer, F.; Chimeno-Manguán, M. The MEOW lunar project for education and science based on concurrent engineering approach. *Acta Astronaut.* **2018**, *148*, 111–120. [CrossRef]

20. Cotfas, D.; Cotfas, P.; Kaplanis, S. Methods to determine the dc parameters of solar cells: A critical review. *Renew. Sustain. Energy Rev.* **2013**, *28*, 588–596. [CrossRef]
21. Ortiz-Conde, A.; García-Sánchez, F.J.; Muci, J.; Sucre-González, A. A review of diode and solar cell equivalent circuit model lumped parameter extraction procedures. *Facta Univ. Ser. Electron. Energetics* **2014**, *27*, 57–102. [CrossRef]
22. Jena, D.; Ramana, V.V. Modeling of photovoltaic system for uniform and non-uniform irradiance: A critical review. *Renew. Sustain. Energy Rev.* **2015**, *52*, 400–417. [CrossRef]
23. Chin, V.J.; Salam, Z.; Ishaque, K. Cell modelling and model parameters estimation techniques for photovoltaic simulator application: A review. *Appl. Energy* **2015**, *154*, 500–519. [CrossRef]
24. Humada, A.M.; Hojabri, M.; Mekhilef, S.; Hamada, H.M. Solar cell parameters extraction based on single and double-diode models: A review. *Renew. Sustain. Energy Rev.* **2016**, *56*, 494–509. [CrossRef]
25. Ibrahim, H.; Anani, N. Evaluation of Analytical Methods for Parameter Extraction of PV modules. *Energy Procedia* **2017**, *134*, 69–78. [CrossRef]
26. Abbassi, R.; Abbassi, A.; Jemli, M.; Chebbi, S. Identification of unknown parameters of solar cell models: A comprehensive overview of available approaches. *Renew. Sustain. Energy Rev.* **2018**, *90*, 453–474. [CrossRef]
27. Bader, S.; Ma, X.; Oelmann, B. One-diode photovoltaic model parameters at indoor illumination levels-A comparison. *Sol. Energy* **2019**, *180*, 707–716. [CrossRef]
28. Yang, X.; Gong, W.; Wang, L. Comparative study on parameter extraction of photovoltaic models via differential evolution. *Energy Convers. Manag.* **2019**, *201*, 112113. [CrossRef]
29. Oulcaid, M.; El Fadil, H.; Ammeh, L.; Yahya, A.; Giri, F. Parameter extraction of photovoltaic cell and module: Analysis and discussion of various combinations and test cases. *Sustain. Energy Technol. Assess.* **2020**, *40*, 100736. [CrossRef]
30. Zhang, Y.; Ma, M.; Jin, Z. Comprehensive learning Jaya algorithm for parameter extraction of photovoltaic models. *Energy* **2020**, *211*, 118644. [CrossRef]
31. Li, S.; Gu, Q.; Gong, W.; Ning, B. An enhanced adaptive differential evolution algorithm for parameter extraction of photovoltaic models. *Energy Convers. Manag.* **2020**, *205*. [CrossRef]
32. Ben Messaoud, R. Extraction of uncertain parameters of a single-diode model for a photovoltaic panel using lightning attachment procedure optimization. *J. Comput. Electron.* **2020**, *19*, 1192–1202. [CrossRef]
33. Ridha, H.M.; Gomes, C.; Hizam, H. Estimation of photovoltaic module model's parameters using an improved electromagnetic-like algorithm. *Neural Comput. Appl.* **2020**, *32*, 12627–12642. [CrossRef]
34. Luu, T.V.; Nguyen, N.S. Parameters extraction of solar cells using modified JAYA algorithm. *Optics* **2020**, *203*, 164034. [CrossRef]
35. Abdulrazzaq, A.K.; Bognár, G.; Plesz, B. Evaluation of different methods for solar cells/modules parameters extraction. *Sol. Energy* **2020**, *196*, 183–195. [CrossRef]
36. Gude, S.; Jana, K.C. Parameter extraction of photovoltaic cell using an improved cuckoo search optimization. *Sol. Energy* **2020**, *204*, 280–293. [CrossRef]
37. Yousri, D.; Thanikanti, S.B.; Allam, D.; Ramachandaramurthy, V.K.; Eteiba, M. Fractional chaotic ensemble particle swarm optimizer for identifying the single, double, and three diode photovoltaic models' parameters. *Energy* **2020**, *195*, 116979. [CrossRef]
38. Diab, A.A.Z.; Sultan, H.M.; Do, T.D.; Kamel, O.M.; Mossa, M.A. Coyote Optimization Algorithm for Parameters Estimation of Various Models of Solar Cells and PV Modules. *IEEE Access* **2020**, *8*, 111102–111140. [CrossRef]
39. Liang, J.; Ge, S.; Qu, B.; Yu, K.; Liu, F.; Yang, H.; Wei, P.; Li, Z. Classified perturbation mutation based particle swarm optimization algorithm for parameters extraction of photovoltaic models. *Energy Convers. Manag.* **2020**, *203*, 112138. [CrossRef]
40. Liang, J.; Qiao, K.; Yuan, M.; Yu, K.; Qu, B.; Ge, S.; Li, Y.; Chen, G. Evolutionary multi-task optimization for parameters extraction of photovoltaic models. *Energy Convers. Manag.* **2020**, *207*, 112509. [CrossRef]
41. Long, W.; Cai, S.; Jiao, J.; Xu, M.; Wu, T. A new hybrid algorithm based on grey wolf optimizer and cuckoo search for parameter extraction of solar photovoltaic models. *Energy Convers. Manag.* **2020**, *203*, 112243. [CrossRef]
42. Ridha, H.M.; Heidari, A.A.; Wang, M.; Chen, H. Boosted mutation-based Harris hawks optimizer for parameters identification of single-diode solar cell models. *Energy Convers. Manag.* **2020**, *209*, 112660. [CrossRef]
43. Xiong, G.; Zhang, J.; Shi, D.; Zhu, L.; Yuan, X.; Tan, Z. Winner-leading competitive swarm optimizer with dynamic Gaussian mutation for parameter extraction of solar photovoltaic models. *Energy Convers. Manag.* **2020**, *206*, 112450. [CrossRef]
44. Zhang, Y.; Jin, Z.; Mirjalili, S. Generalized normal distribution optimization and its applications in parameter extraction of photovoltaic models. *Energy Convers. Manag.* **2020**, *224*, 113301. [CrossRef]
45. Montano, J.; Tobón, A.F.; Villegas, J.P.; Durango, M. Grasshopper optimization algorithm for parameter estimation of photovoltaic modules based on the single diode model. *Int. J. Energy Environ. Eng.* **2020**, *11*, 367–375. [CrossRef]
46. Ramadan, A.; Kamel, S.; Korashy, A.; Yu, J. Photovoltaic Cells Parameter Estimation Using an Enhanced Teaching–Learning-Based Optimization Algorithm. *Iran. J. Sci. Technol. Trans. Electr. Eng.* **2019**, *44*, 767–779. [CrossRef]
47. Benahmida, A.; Maouhoub, N.; Sahsah, H. An Efficient Iterative Method for Extracting Parameters of Photovoltaic Panels with Single Diode Model. In Proceedings of the 2020 5th International Conference on Renewable Energies for Developing Countries (REDEC), Marrakech, Morocco, 29–30 June 2020; pp. 1–6.
48. Ndegwa, R.; Simiyu, J.; Ayieta, E.; Odero, N. A Fast and Accurate Analytical Method for Parameter Determination of a Photovoltaic System Based on Manufacturer's Data. *J. Renew. Energy* **2020**, *2020*, 1–18. [CrossRef]

49. Arabshahi, M.; Torkaman, H.; Keyhani, A. A method for hybrid extraction of single-diode model parameters of photovoltaics. *Renew. Energy* **2020**, *158*, 236–252. [CrossRef]

50. Peng, L.; Sun, Y.; Meng, Z.; Wang, Y.; Xu, Y. A new method for determining the characteristics of solar cells. *J. Power Sources* **2013**, *227*, 131–136. [CrossRef]

51. Batzelis, E.I.; Anagnostou, G.; Chakraborty, C.; Pal, B.C. Computation of the Lambert W Function in Photovoltaic Modeling. In *Lecture Notes in Electrical Engineering*; J.B. Metzler: Berlin, Germany, 2020; pp. 583–595.

52. Yu, F.; Huang, G.; Xu, C. An explicit method to extract fitting parameters in lumped-parameter equivalent circuit model of industrial solar cells. *Renew. Energy* **2020**, *146*, 2188–2198. [CrossRef]

53. Ćalasan, M.; Aleem, S.H.A.; Zobaa, A.F. On the root mean square error (RMSE) calculation for parameter estimation of photovoltaic models: A novel exact analytical solution based on Lambert W function. *Energy Convers. Manag.* **2020**, *210*, 112716. [CrossRef]

54. Panchal, A.K. A per-unit-single-diode-model parameter extraction algorithm: A high-quality solution without reduced-dimensions search. *Sol. Energy* **2020**, *207*, 1070–1077. [CrossRef]

55. Nassar-eddine, I.; Obbadi, A.; Errami, Y.; El fajri, A.; Agunaou, M. Parameter estimation of photovoltaic modules using iterative method and the Lambert W function: A comparative study. *Energy Convers. Manag.* **2016**, *119*, 37–48. [CrossRef]

56. Pindado, S.; Cubas, J.; Roibas-Millan, E.; Bugallo-Siegel, F.; Sorribes-Palmer, F. Assessment of Explicit Models for Different Photovoltaic Technologies. *Energies* **2018**, *11*, 1353. [CrossRef]

57. Cubas, J.; Pindado, S.; Victoria, M. On the analytical approach for modeling photovoltaic systems behavior. *J. Power Sour.* **2014**, *247*, 467–474. [CrossRef]

58. Karmalkar, S.; Haneefa, S. A Physically Based ExplicitModel of a Solar Cell for Simple Design Calculations. *IEEE Electron. Device Lett.* **2008**, *29*, 449–451. [CrossRef]

59. Saleem, H.; Karmalkar, S. An Analytical Method to Extract the Physical Parameters of a Solar Cell From Four Points on the Illuminated. *J. Curve. IEEE Electron. Device Lett.* **2009**, *30*, 349–352. [CrossRef]

60. Das, A.K. An explicit J–V model of a solar cell using equivalent rational function form for simple estimation of maximum power point voltage. *Sol. Energy* **2013**, *98*, 400–403. [CrossRef]

61. Cubas, J.; Gomez-Sanjuan, A.M.; Pindado, S. On the thermo-electric modelling of smallsats. In Proceedings of the 50th International Conference on Environmental Systems—ICES 2020, Lisbon, Portugal, 16–20 July 2020; pp. 1–12.

62. Alcala-Gonzalez, D.; Del Toro, E.M.G.; Más-López, M.I.; Pindado, S. Effect of Distributed Photovoltaic Generation on Short-Circuit Currents and Fault Detection in Distribution Networks: A Practical Case Study. *Appl. Sci.* **2021**, *11*, 405. [CrossRef]

63. Pindado, S.; Alcala-Gonzalez, D.; Alfonso-Corcuera, D.; del Toro, E.G.; Más-López, M. Improving the Power Supply Performance in Rural Smart Grids with Photovoltaic DG by Optimizing Fuse Selection. *Agronomy* **2021**, *11*, 622. [CrossRef]

64. Cubas, J.; Pindado, S.; De Manuel, C. Explicit Expressions for Solar Panel Equivalent Circuit Parameters Based on Analytical Formulation and the Lambert W-Function. *Energies* **2014**, *7*, 4098–4115. [CrossRef]

65. Cubas, J.; Pindado, S.; Sanz-Andres, A. Accurate Simulation of MPPT Methods Performance When Applied to Commercial Photovoltaic Panels. *Sci. World J.* **2015**, *2015*, 1–16. [CrossRef]

66. Cubas, J.; Pindado, S.; De Manuel, C. New Method for Analytical Photovoltaic Parameters Identification: Meeting Manufacturer's Datasheet for Different Ambient Conditions. In *Proceedings of the International Congress on Energy Efficiency and Energy Related Materials (ENEFM2013)*; Springer Proceedings in Physics 155; Antalya, Turkey, 9–12 October 2014, Oral, A.Y., Bahsi, Z.B., Ozer, M., Eds.; Springer: Berlin, Germany, 2014; Volume 155, pp. 161–169.

67. Toledo, F.; Blanes, J.M.; Galiano, V.; Laudani, A. In-depth analysis of single-diode model parameters from manufacturer's datasheet. *Renew. Energy* **2021**, *163*, 1370–1384. [CrossRef]

68. Sze, S.M.; Mattis, D.C. Physics of Semiconductor Devices. *Phys. Today* **1970**, *23*, 75. [CrossRef]

69. Van Dyk, E.; Meyer, E. Analysis of the effect of parasitic resistances on the performance of photovoltaic modules. *Renew. Energy* **2004**, *29*, 333–344. [CrossRef]

70. Wolf, M.; Rauschenbach, H. Series resistance effects on solar cell measurements. *Adv. Energy Convers.* **1963**, *3*, 455–479. [CrossRef]

71. Yadir, S.; Bendaoud, R.; El-Abidi, A.; Amiry, H.; BenHmida, M.; Bounouar, S.; Zohal, B.; Bousseta, H.; Zrhaiba, A.; Elhassnaoui, A. Evolution of the physical parameters of photovoltaic generators as a function of temperature and irradiance: New method of prediction based on the manufacturer's datasheet. *Energy Convers. Manag.* **2020**, *203*. [CrossRef]

72. Nguyen-Duc, T.; Nguyen-Duc, H.; Le-Viet, T.; Takano, H. Single-Diode Models of PV Modules: A Comparison of Conventional Approaches and Proposal of a Novel Model. *Energies* **2020**, *13*, 1296. [CrossRef]

73. Humada, A.M.; Darweesh, S.Y.; Mohammed, K.G.; Kamil, M.; Mohammed, S.F.; Kasim, N.K.; Tahseen, T.A.; Awad, O.I.; Mekhilef, S. Modeling of PV system and parameter extraction based on experimental data: Review and investigation. *Sol. Energy* **2020**, *199*, 742–760. [CrossRef]

74. Villalva, M.G.; Gazoli, J.R.; Filho, E.R. Comprehensive Approach to Modeling and Simulation of Photovoltaic Arrays. *IEEE Trans. Power Electron.* **2009**, *24*, 1198–1208. [CrossRef]

75. Villalva, M.G.; Gazoli, J.R.; Filho, E.R. Modeling and circuit-based simulation of photovoltaic arrays. In Proceedings of the 2009 Brazilian Power Electronics Conference, Bonito-Mato Grosso do Sul, Brazil, 27 September–1 October 2009; pp. 1244–1254.

76. Akbaba, M.; Alattawi, M.A. A new model for I–V characteristic of solar cell generators and its applications. *Sol. Energy Mater. Sol. Cells* **1995**, *37*, 123–132. [CrossRef]

77. El Tayyan, A. An Empirical model for Generating the IV Characteristics for a Photovoltaic System. *J. Al Aqsa Univ.* **2006**, *10*, 214–221.

78. El Tayyan, A.A. A simple method to extract the parameters of the single-diode model of a PV system. *Turk. J. Phys.* **2013**, *37*, 121–131. [CrossRef]

79. Das, A.K. An explicit J–V model of a solar cell for simple fill factor calculation. *Sol. Energy* **2011**, *85*, 1906–1909. [CrossRef]

80. Saetre, T.O.; Midtgård, O.-M.; Yordanov, G.H. A new analytical solar cell I–V curve model. *Renew. Energy* **2011**, *36*, 2171–2176. [CrossRef]

81. Pindado, S.; Cubas, J. Simple mathematical approach to solar cell/panel behavior based on datasheet information. *Renew. Energy* **2017**, *103*, 729–738. [CrossRef]

82. Oulcaid, M.; El Fadil, H.; Ammeh, L.; Yahya, A.; Giri, F. One shape parameter-based explicit model for photovoltaic cell and panel. *Sustain. EnergyGrids Netw.* **2020**, *21*, 100312. [CrossRef]

83. Easwarakhanthan, T.; Bottin, J.; Bouhouch, I.; Boutrit, C. Nonlinear Minimization Algorithm for Determining the Solar Cell Parameters with Microcomputers. *Int. J. Sol. Energy* **1986**, *4*, 1–12. [CrossRef]

84. Caillol, J.-M. Some applications of the Lambert W function to classical statistical mechanics. *J. Phys. A Math. Gen.* **2003**, *36*, 10431–10442. [CrossRef]

85. Veberič, D. Lambert W function for applications in physics. *Comput. Phys. Commun.* **2012**, *183*, 2622–2628. [CrossRef]

86. Barry, D.A.; Parlange, J.-Y.; Li, L.; Prommer, H.; Cunningham, C.J.; Stagnitti, F. Analytical approximations for real values of the Lambert W-function. *Math. Comput. Simul.* **2000**, *53*, 95–103. [CrossRef]

87. Cubas, J.; Pindado, S.; Farrahi, A. New method for analytical photovoltaic parameter extraction. In Proceedings of the 2013 International Conference on Renewable Energy Research and Applications (ICRERA), Madrid, Spain, 20–23 October 2013; pp. 873–877.

88. Elkholy, A.; El-Ela, A.A. Optimal parameters estimation and modelling of photovoltaic modules using analytical method. *Heliyon* **2019**, *5*, e02137. [CrossRef] [PubMed]

89. Jadli, U.; Thakur, P.; Shukla, R.D. A New Parameter Estimation Method of Solar Photovoltaic. *IEEE J. Photovolt.* **2018**, *8*, 239–247. [CrossRef]

90. Maouhoub, N. Photovoltaic module parameter estimation using an analytical approach and least squares method. *J. Comput. Electron.* **2018**, *17*, 784–790. [CrossRef]

91. Oliva, D.; El Aziz, M.A.; Hassanien, A.E. Parameter estimation of photovoltaic cells using an improved chaotic whale optimization algorithm. *Appl. Energy* **2017**, *200*, 141–154. [CrossRef]

92. Chin, V.J.; Salam, Z. A New Three-point-based Approach for the Parameter Extraction of Photovoltaic Cells. *Appl. Energy* **2019**, *237*, 519–533. [CrossRef]

93. Swartwout, M. Reliving 24 Years in the Next 12 Minutes: A Statistical and Personal History of University-Class Satellites. In Proceedings of the 32nd AIAA/USU Conference on Small Satellites, Salt Lake City, UT, USA, 4–9 August 2018; pp. 1–20.

 *applied sciences*

Article

Coverage Path Planning with Semantic Segmentation for UAV in PV Plants

Andrés Pérez-González *, Nelson Benítez-Montoya , Álvaro Jaramillo-Duque and Juan Bernardo Cano-Quintero

Research Group in Efficient Energy Management (GIMEL), Electrical Engineering Department, Universidad de Antioquia, Calle 67 No. 53-108, Medellín 050010, Colombia; nelson.benitez@udea.edu.co (N.B.-M.) alvaro.jaramillod@udea.edu.co (Á.J.-D.); bernardo.cano@udea.edu.co (J.B.C.-Q.)
* Correspondence: afernando.perez@udea.edu.co; Tel.: +57-322-639-7758

Citation: Pérez-González, A.é.; Benítez-Montoya, N.; Jaramillo-Duque, Á.; Cano-Quintero, J.B. coverage path planning with Semantic Segmentation for UAV in PV Plants. *Appl. Sci.* **2021**, *11*, 12093. https://doi.org/10.3390/app112412093

Academic Editors: Luis Hernández-Callejo, Maria del Carmen Alonso García and Sara Gallardo Saavedra

Received: 16 November 2021
Accepted: 15 December 2021
Published: 19 December 2021

Abstract: Solar energy is one of the most strategic energy sources for the world's economic development. This has caused the number of solar photovoltaic plants to increase around the world; consequently, they are installed in places where their access and manual inspection are arduous and risky tasks. Recently, the inspection of photovoltaic plants has been conducted with the use of unmanned aerial vehicles (UAV). Although the inspection with UAVs can be completed with a drone operator, where the UAV flight path is purely manual or utilizes a previously generated flight path through a ground control station (GCS). However, the path generated in the GCS has many restrictions that the operator must supply. Due to these restrictions, we present a novel way to develop a flight path automatically with coverage path planning (CPP) methods. Using a DL server to segment the region of interest (RoI) within each of the predefined PV plant images, three CPP methods were also considered and their performances were assessed with metrics. The UAV energy consumption performance in each of the CPP methods was assessed using two different UAVs and standard metrics. Six experiments were performed by varying the CPP width, and the consumption metrics were recorded in each experiment. According to the results, the most effective and efficient methods are the exact cellular decomposition boustrophedon and grid-based wavefront coverage, depending on the CPP width and the area of the PV plant. Finally, a relationship was established between the size of the photovoltaic plant area and the best UAV to perform the inspection with the appropriate CPP width. This could be an important result for low-cost inspection with UAVs, without high-resolution cameras on the UAV board, and in small plants.

Keywords: deep learning (DL); unmanned aerial vehicle (UAV); photovoltaic (PV) plants; semantic segmentation; coverage path planning (CPP)

1. Introduction

According to REN21, over the past two years, global photovoltaic (PV) plants capacities and annual additions have grown and expanded rapidly. For instance, 621 GW were installed in the year 2019 and 760 GW in the year 2020 [1], despite the reduction in electricity consumption and shifted daily demand patterns due to the COVID-19 pandemic [2]. Additionally, it has become one of the most profitable options and is an energy resource that has recently decreased in cost. As a result, solar electricity generation has grown in residential, commercial, and utility-scale projects [3]. The future of PV generation will focus on optimizing hybrid systems [4] and improving the performance of each element of the system, as well as reducing their cost due to large-scale production [5]. Furthermore, the sector trend has been asking for low prices, and the competitive market has encouraged investment in solar PV technologies across the entire value chain, particularly in solar cells and modules, to improve efficiencies and reduce the levelized cost of energy (LCOE) [1]. As a result, PV power plants could grow almost sixfold over the next ten years, reaching a cumulative capacity of 2840 GW globally by 2030 and rising to 8519 GW by 2050, according to [6].

Thus, it is established that the number of PV plants and generated power have increased around the world. This implies specific technical challenges in their maintenance and operation (O and M) [7]. Some of these threats affect their production, which increases cost, and decreases profitability. The most common threats are the failures in the inverters and PV modules [8]. Through dirty equipment, the state of the environment, or manufacturing defects in the PV modules, PV plant energy generation can be curtailed by 31% in the worst cases [9–11].

One must consider that PV plants are commonly installed on roofs, rooftops, canopies, or facades for urban environments. Likewise, solar farms utilize rural environments, such as deserts, plains, and hills [7,12,13]. Depending on the location of the PV plant, the manual inspection tasks could be exhausting and take up to 8 h/MW, if the number of modules is considerable. The amount of inspection time increases for solar PV plants on rooftops or canopies, by virtue of aspects of their installation [14]. In addition, inspections to detect threats in the panels must be conducted by trained personnel. In some cases, problems occur in elevated installations, for which special training and certification is required. These jobs could put people and facilities at risk [15].

In recent decades, unmanned aerial vehicles (UAVs) have been increasingly used in inspection and patrol tasks [16–18]. UAV-based applications for PV plant inspections have many advantages in comparison with the manual inspection methods. The main advantages are flexibility, lower cost, larger area coverage, faster detection, higher precision, and the capacity to perform a superior and automatic inspections [11,17–21].

There are many approaches to performing an inspection with UAVs in PV modules. One of them used UAVs with a thermal imaging camera to take photos in the infrared spectrum to evaluate the UAVs parameters, such as height, speed, viewing angle, sun reflection, irradiance and temperature, all of which are necessary to perform defect inspection [22]. In a second approach, UAVs were implemented to inspect different solar PV plants, wherein analysis of the correlation between altitude and the pixel resolution was used to detect PV panel defects and features like shape, size, location, and color, among others, of a particular defect were also detected [7,23–26]. In a third approach, the authors proposed PV plant fault inspection with UAVs using image mosaics combined with orthophotography techniques to create a digital map; image mosaics were combined with multiple visible and/or infrared range images into a single mosaic image covering a large area [14,27,28]. Whereas the orthophotography technique is a vertical photograph that shows images of objects in true planimetric position [29]. Thus, these two techniques were integrated with previous works to achieve an advanced tool that allows monitoring and taking actions in the operation and maintenance (O and M) of PV plants [7]. In short, this tool has been used to detect defects and dust or dirt in PV modules [28]. Apart from this, some approaches have been developed for the detection of defects in PV modules using artificial vision techniques, machine learning, deep learning, and the integration therewith of the previous approaches [7,24,30–33]. Additionally, these approaches to planning the UAV's flight path were configured from a ground control station (GCS) program. Irrespective of these approaches, an important aspect is that the UAV should automatically follow the path to cover important points in PV plants. Many research efforts have been made to calculate the paths of and solve the waypoint planning problem for UAV inspection of solar PV plant applications [34–37], but none of them propose the coverage path planning (CPP) as a method to complete this task.

The CPP, given a region of interest (RoI) in a 2D environment, consists of calculating the path that passes through each one of the points that make up the desired environment and must be found considering the limitations of movement [38,39]. CPP is classified as a classical NP-hard problem in the field of computational complexity. These problems were initially analyzed for indoor environments with mobile robots. However, with the development of GPS, CPPs began to be used for missions with UAVs. Due to the environment in which the task is performed and the obstacles present, precise localization in the environment is an arduous task, which makes the CPP a difficult problem [40]. Additionally, it is classified as a motion planning subtopic in robotics, and has two approaches, heuristic and complete. In heuristic approaches, the robots follow a set of rules defining their behavior, but do not present a guarantee for successful full coverage. These guarantee using the cellular decomposition of the area, which involves space discretization into cells to simplify the coverage in each sub-region, unlike complete methods, which cannot afford such processing. Another important issue mentioned by the authors of reference [39] is the flight time to fully cover the area, which can be reduced using multiple robots and by reducing the number of turning maneuvers. Finally, the available RoI information is important; several approaches accept previous knowledge of the robot's respecting the search RoI (offline), while sensor-based approaches collect such information in real-time along the coverage (online) [41].

In the literature, CPP approaches are needed in several application areas, such as floor cleaning [42], agriculture [43,44], wildfire tracking [45], bush trimming [46], power line inspection [47], photogrammetry [39], visual inspection [48], and many more. Additionally, many surveys regarding CPP present several approaches and techniques for performing missions with, mostly, land vehicles [41,49,50]. The research interest in aerial robots (indoor and outdoor) has surely motivated the research of CPP [51]. This can be implemented in many UAVs platforms, such as fixed-wing, rotary-wing, and hybrid UAV (VTOL) [39]. Rotary-wing UAVs are inexpensive and have good maneuverability, and their small payload capacity limits the weight of on-board sensors and flight time. Hence, they are more suitable for CPP missions on a small scale. Additionally, the increasing usage of UAVs in applications with complicated missions has led to CPP methods being a very active research area for single and multiple UAVs, especially recently [39,52–54], As evidenced in a previous work [43], the classic taxonomy of coverage paths in UAVs are classified into no decomposition, exact cellular decomposition, and approximate cellular decomposition. The first performs the coverage with a single UAV, for which no decomposition technique is required, because the shape of the RoI has a non-complex geometry. The second divides the free space into simple, non-overlapping regions called cells. The union of all cells fills the free space completely. The cells in which there are no obstacles are easy to cover and can be covered by the robot with simple movements. The third is based on grids. They use a representation of the environment decomposed into a collection of uniformly squared cells [55], considering rectangular, concave, and convex polygons for RoIs. In addition to this, CPP performances were assessed with applied metrics according to [39]. The elemental approach most used to solve offline CPP problem sis the area decomposition into non-overlapping sub-regions [56], to determine the appropriate visiting sequence of each sub-region and to cover each decomposed region in a back-and-forth movement to secure a complete coverage path. As a result, the methods for obtaining complete coverage of an RoI are the exact and approximate cellular decomposition methods [57–59].

On the other hand, image processing helps to obtain a map of the robot or RoI. Robots, such as UAVs, need to know the RoI before commencing CPP [60], which represents where the PV plants are and can be determined in a process called boundary extraction [61,62]. Then, the deep learning (DL) image segmentation technique, also known as semantic segmentation [63], is achieved by applying deep convolutional neural networks (CNN), such as the U-Net network model [64,65] or the FCN model [66], which dramatically enhance the segmentation results. Once the segmentation is done and the mask is obtained or the RoI is identified, the GCS calculates the CPP that guides the UAV in the automatic plant inspection, during which it captures images of PV plants [67]. Most of the failures occur at the centimeter or millimeter level, and that poses a challenge for the inexpensive sensors available today [68]. Tests of coverage paths it can be conducted with drones in a real or simulated environment. The most viable option for this stage of the work is simulation, as verified by other research [69–72]. Simulation has been recognized as an important research tool; initially, simulation was an academic research tool, but with the advancement of computers, simulation has reached new levels. It is a remarkable tool that guarantees support in design, planning, analysis, and making decisions in different areas of research and development. Of course, robotics, as modern technology, has not been an exception [73,74].

This work is focused on implementing the best strategy of coverage path planning (CPP) over PV plants with UAVs using semantic segmentation in a deep learning server to obtain the RoI. The experimental results were obtained by simulating the CPP methods and using UAVs, such as the 3DR Iris and Typhoon aerial robots.

The key contributions of this work are as follows:

- This work proposes CPPs as a novel strategy for conducting an inspection flight over a PV plant with an UAV, since there are no previous reports of such work.
- This work used three CPP methods over three PV plants, which were modeled in a simulation environment to evaluate metrics and parameters. As a result, a relationship was found between the CPP width and energy consumption, and according to that, the best CPP method to implement.
- This work proposes a hybrid CPP method that uses image processing and a DL server to find the RoI quickly and accurately, becoming a semi-automatic process.
- A free simulation tool is provided, with an interface to simulate the inspection of PV plants with UAVs.

This paper is structured as follows. In Section 2, necessary definitions and the techniques used to obtain the results are described. In Section 3, the three CPP methods implemented are compared to show relationships among the CPP width, number of maneuvers, and energy consumption, with the aim of finding the best CPP method to implement. Finally, in Section 4, some conclusions are given.

2. Materials and Methods

In this work, three PV power plants were selected that met the image requirements of no light distortions, non-complex geometry, and grouped panels. These plants are in different parts of the world. The first PV plant has an area of $35,975$ square meters, located in Brazil (-22.119948323621525, -51.44666423682603), known as Usina Solar Unioeste 1 [75]. The second PV plant has an area of $25,056$ square meters, located in Iran (34.0504329771808, 49.796635265177294), known as Arak power plant [37]. The third plant is in the United States (38.55989816199527, -121.42374978032115); the plant has 1344 square meters of area, and it is part of a PV plant located on the roof of the California State University Sacramento Library (CSUSL) [76].

These plants were subjected to a series of processing stages, as shown in Figure 1. The RoIs were obtained from Google Maps satellite images with a predefined altitude, and limits, from which the image (input image) of the desired PV system was obtained as the first stage, shown in Figure 1. In the second stage, the image was entered into a DL server that was developed in this work, taking into account the previous work [65]. The DL server was launched with TensorFlow [77] and Flask [78]. This stage obtains a mask (image segmentation) of the PV plant, as shown in Figure 1. In the third stage it was necessary to apply a series of OpenCV functions (post-processing), as referenced in Algorithm A1 in the Appendix A, to adjust the mask (output mask) to the PV plant area, as seen in Figure 1. Then, the output mask or RoI was introduced in the GUI interface, where the CPP method was selected to internally execute. Later, the path GPS positions (CPP computed) were sent to the UAV through MAVlink commands [79,80]. The UAV executed these commands in the Gazebo platform (CPP simulation) and, at the same time, the simulation data was fed back the GUI interface. Simultaneously all GPS points reached by the UAV were drawn in the GCS platform (CPP). In this stage, the trajectory was validated. Each stage is described in greater detail in the following sections.

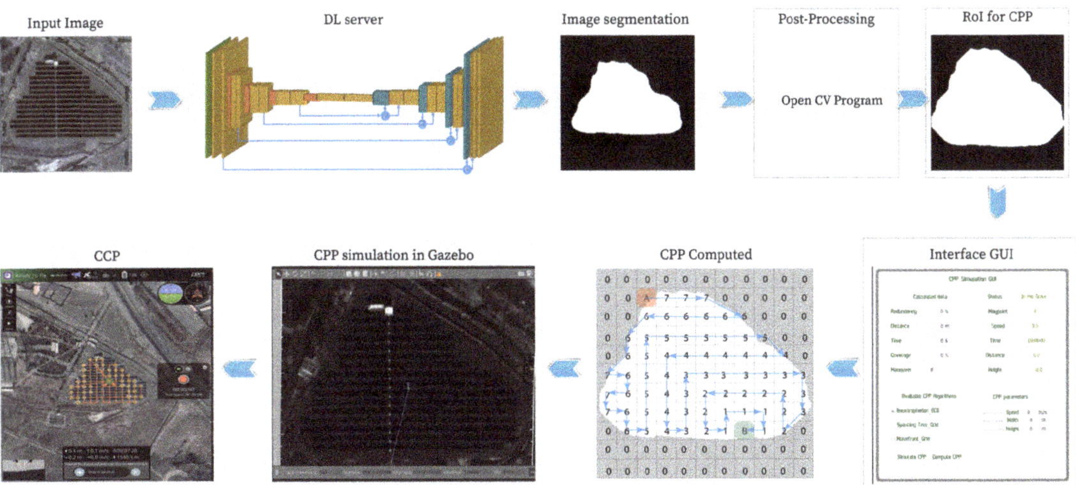

Figure 1. Stages for CPP with semantic segmentation for UAV in PV plants.

2.1. Deep Learning (DL) Server for Segmentation

A deep learning (DL) server for segmentation was necessary to extract the RoIs from the Google Maps images. Additionally, the server was used to achieve this task automatically with the process called semantic segmentation, wherein each pixel is labeled with the class of its enclosing object [65,66]. In previous work, a convolutional neural network was proposed, wherein a public database was used; this data was prepared, resized for training, and assessed with two network structures. The U-net network had the best performance, in terms of metrics, in the semantic segmentation task [65]. Then a DL server was employed to perform image segmentation and obtain the RoI [81].

2.2. Post Processing

In this step, a set of OpenCV functions were applied with Python 3.7. For example, in the first function, morphological operators like "Erode" and "Dilate" were applied to the images, then the "FindCountours" function was applied to help extract their contours. The contour can be defined as a curve that joins all the continuous points at the boundary of the PV installation. So, the "ContourArea" function was then used to find the area of the previous contour. Following this pattern, the area was compared with 400 others to filter the bigger area and eliminate the little areas belonging to false positives. Then the "ApproxPolyDP" function was used to approximate a shape of the contour to another shape with fewer vertices. Subsequently, the "DrawContour" function was used to draw the resulting contour [82,83]. Finally, the "Erode" morphological operators were used again, to expand the known area and compensate for the limitations of the mask with regard to the CPP method and some faulty occurrences caused by the false positives of the DL server. The pseudocode of the openCV functions used is shown in Algorithm A1 in the Appendix A.

2.3. 2D Coverage Path Planning Method in the GUI Interface

In previous studies of CPPs, there were many existing methods from which to select to solve the CPP problem. In this work, three methods based on CPP were selected, considering the following criteria: time of execution, ease of implementation, and more, which were used to cover the RoI. The methods were selected according to [38,39].

The boustrophedon exact cellular decomposition (BECD), which was proposed by [84], was the first selected. The CPP method is delineated in Figure 2.

The second was grid-based spanning tree coverage (GBSTC), which works with cellular decomposition, first proposed by [85], and depicted in Figure 3.

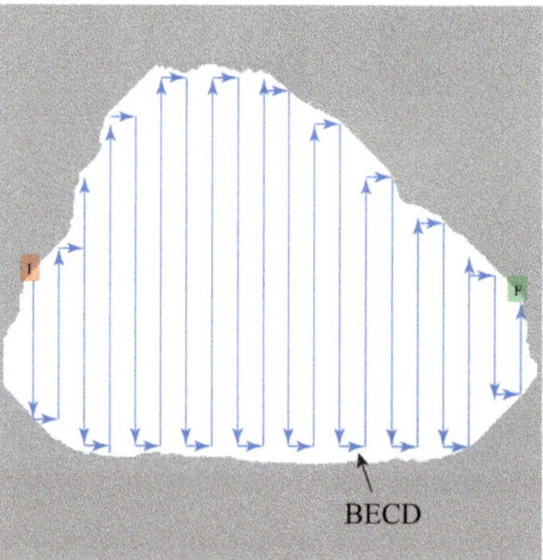

Figure 2. The RoI and the path generated from the initial point (I) to the final point (F) by BECD.

The third method selected for this project was grid-based wavefront coverage (GBWC), first proposed by [86]. The method is illustrated in Figure 4. Each of these CPP methods is explained in more detail in the following sections.

(a). The boustrophedon exact cellular decomposition (BECD) Method: This method takes the robot's free space and obstacles and splits them into cells. These cells are covered by the robot using a back-and-forth pattern from the initial point to the final point, using maneuvers of 90 degrees to change direction from south to north or vice versa, as shown in Figure 2. This method improves the trapezoid decomposition technique, as it exploits the structure of the polygon to determine the start and end of an obstacle, and thus is able to divide the free space into a few cells that do not require a redundant step, and it permits the coverage of curved areas [84].

(b). Grid-Based Spanning Tree Coverage (GBSTC): This method is based on approximate cell decomposition and differs from the previous method in that the following postulates were considered. First, the method divides the space into grids of side L. Second, the robot only moves in perpendicular directions to the sides of the grid. Third, every grid is subdivided into four grids of side $L/2$. Finally, GBSTC discards space that is partly occupied by obstacles. Consequently, considering these previous postulates, the method consists of several stages: in the first stage, a graphic structure is defined, $S(N, E)$, where N is nodes, defined as the central point of each grid, and E is edges, defined as the line segments connecting the centers of adjoining grids, as shown in Figure 3a. In the second stage, the method builds a spanning tree for S, and employs this tree to plan a cover path as follows. Starting in grid I with a sub-grid of side $L/2$, the robot begins by travelling between adjoining sub-grids along a path that moves around the spanning tree at a constant distance and in a counterclockwise direction, finishing when the initial sub-grid, I, is found again, which means it is also in the final point, F [85]. An example of this method is illustrated in Figure 3b. The approximation depends on the side length, L, of the grid.

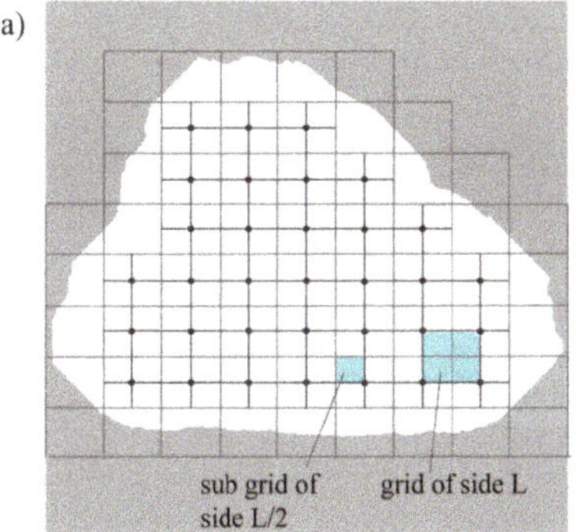
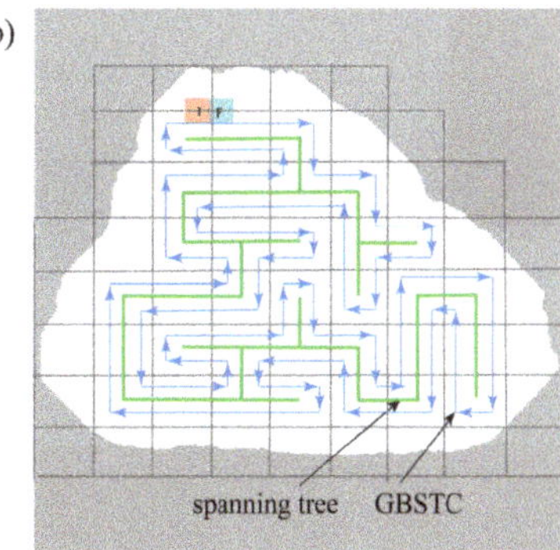

Figure 3. (**a**) Approximate cell decomposition in grids, and sub grids. (**b**), coverage path generated with the GBSTC method.

(c). grid-based wavefront coverage (GBWC): The first grid-based method proposed for CPP, GBWC is an offline method that uses a grid representation, in addition to applying a full CPP method. The method requires an initial grid, I, and a final grid, F. A distance transformation that propagates a wavefront from the final to the initial point is used to assign a specific number to each item on the grid. That is, the method first assigns a zero to the final item, and then a one to all its surrounding grids. Then all unmarked grids adjoining those marked one are numbered two. The process is incrementally repeated until the wavefront reaches the initial grid [86], as illustrated in Figure 4a.

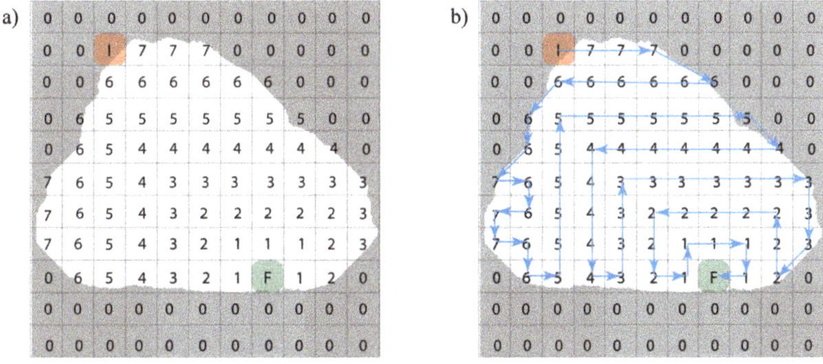

Figure 4. (**a**) Wavefront distance transforms for the selection of the initial position (I) and final position (F). (**b**) Coverage path generated using the wavefront distance transforms with the selection of the initial position (I).

Once the distance transformation is determined, a coverage path can be found by starting at the initial grid, I, and selecting the adjoining grid with the highest number that has not been explored. If two or more are unexplored, and adjoining grids share the same number, one of them is randomly selected, as shown in Figure 4b.

2.3.1. Metrics

The metrics evaluate the performance of the three CPP methods. Such assessment can be performed by considering five commonly used metrics to evaluate the effectiveness of the proposed CPP methods both theoretically and in simulation (dynamically) [7,41,43,62]. These five metrics are covered path length, flight time, energy consumption, redundancy of points traveled, and percentage of coverage of the total area. Each of these metrics is described below.

The L_{ct} metric (covered path length) is the length of the entire path covered by the UAV from the initial to the final point. For a trajectory in a 2D plane composed of n points, assuming the initial point as $(x_1, f(x_1))$ and the end point as $(x_n, f(x_n))$, the L_{ct} can be computed as shown in Equation (1):

$$L_{ct} = \sum_{i=1}^{n-1} \sqrt{(x_i + 1 - x_i)^2 + (f(x_1 + 1) - f(x_i))^2} \tag{1}$$

where $(x_i, f(x_i))$, in which $i = 1, 2, \ldots, n$, representing all the n points of the UAV flight path in 2D coordinates. More details can be found in [87].

The flight time metric of the PX4 SITL Gazebo model is the time required to travel the total flight path (takeoff, path travel, and landing) with a dynamic speed that considers the inertia and the variation in speed due to UAV turns angles. These data are collected through sensors in the Gazebo plugins [88].

The redundancy of points traveled, $R_\%$, corresponds to the number of points that are visited more than once from the total number of points that the path contains, Equation (2).

$$R_\% = \frac{P_{pc}}{P_{vmo}} \tag{2}$$

where P_{pc}, and P_{vmo} correspond to the number of points the trajectory contains, and the number of points visited more than once.

The percentage of coverage of the total area, $C_\%$, measures the number of effectively covered points within the total number thereof by the points the area to be covered contains, given by Equation (3).

$$C_\% = \frac{P_v}{P_T} \tag{3}$$

where P_v, and P_T correspond, respectively, to the total number of points visited and the total number of points in the area.

Moreover, the number of maneuvers metric, which is the number of turning maneuvers the UAV performs on a path, is often used as the main performance metric in coverage [89,90].

The energy consumption metric is computed from the voltage and current data from the power module of the PX4 SITL model [91]. Its value depends on parameters, such as the CPP width between lines and the speed and the height of the UAV at the time of implementing the CPP, which were configured in the interface and were simulated in Gazebo.

To validate the results of the methods described above, the BECD, GBSTC, and GBWC methods were implemented in two UAVs, simulated in Gazebo, and their performances were assessed by the metrics presented above [39]. The next section describes the results and compares the models in detail.

2.4. Simulation and Validation Platform

Based on [92], the Gazebo platform was selected to execute the simulation experiments, as it has extensive documentation on its webpage. In addition, it is the most mentioned and used simulation platform in previous work [73,74] that implemented path planning or CPP, used UAV sensors, or deployed several UAVs [92]. Gazebo also allows the modeling of commercial UAVs using the PX4 autopilot software [91], as shown in Figure 5. In addition, this figure shows one of the experiments conducted with the Typhoon UAV, flown over the CSUSL plant. The UAV sonar sensor, represented by blue lines, is also shown, as is the UAV camera, in the box at the upper right.

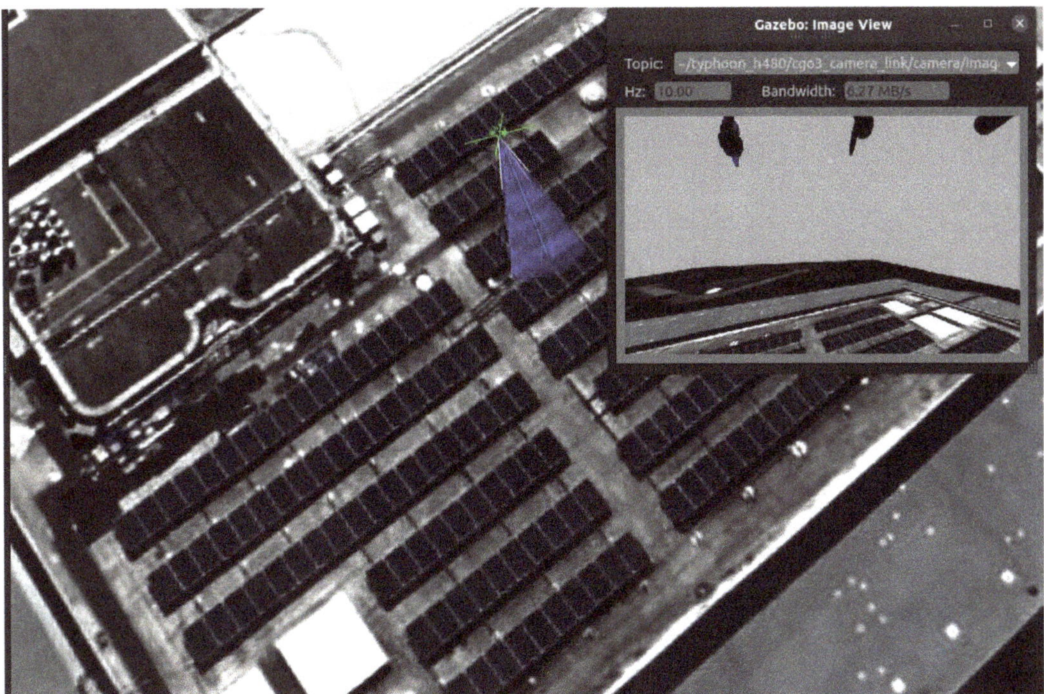

Figure 5. UAV over A PV plant, simulated in Gazebo.

The integration of Gazebo [92], PX4 [91], Python, and the CPP methods was implemented using ROS (Robot Operating System) as middleware [93]. This tool allows communication among nodes. The nodes are processes, and each node has a task associated with it, such as sending a Mavlink command to control the UAV trajectory and permitting reading messages from the UAV to discern its status in flight, using a simulation mode referred to as software in the loop (SITL). This simulator provides the ability to run different vehicles, such as a plane, copter, or rover, without a need for any microcontrollers or hardware [94]. In addition, two PX4 autopilot rotary wing UAVs, the 3DR Iris and Typhoon UAV, were chosen because they have good maneuverability and are more suitable for small-scale CPP missions. Furthermore, these UAVs have been widely used in other research [67,70,95].

The GCS software (QGroundControl) was selected to validate the CPP calculated for each UAV and PV plant obtained, because it is the most compatible tool with the PX4 autopilot. It is also recommended on the PX4 webpage. On other hand, another compatible tool, namely a GUI, was designed using Qt to modify and vary the parameters and to convert the way points from RoI pixels to geo-referenced points [96].

3. Results and Discussion

The procedure described above—DL-server, post-processing, CPP, and metrics evaluation—are represented in Figure 1, was applied to three different PV plant images. The results obtained are described in the following sections.

3.1. Results with Deep Server, and OpenCV Functions

OpenCV functions and a DL server were combined to accurately extract the mask and to trade off the DL server results of the PV plant area. Then, the CPP methods used the mask as a region of interest (RoI).

The stages to obtain the results of the RoI in the three images are shown in Figures 6–8. In the first stage, a high-resolution image from Google Maps, of predefined height and width, was obtained, and used as an input image in the DL server, depicted in Figure 6a. In the second stage, the output image of the DL server was the mask, shown in Figure 6b. In the third stage, the opening function was used in the mask, Figure 6c. In the fourth stage, the RoI was obtained using the draw contour method, as seen in Figure 6d. Finally, the RoI was blended with the input image with the aim of comparing the results in Figure 6e. The results were satisfactory and can be adapted, depending on the environment.

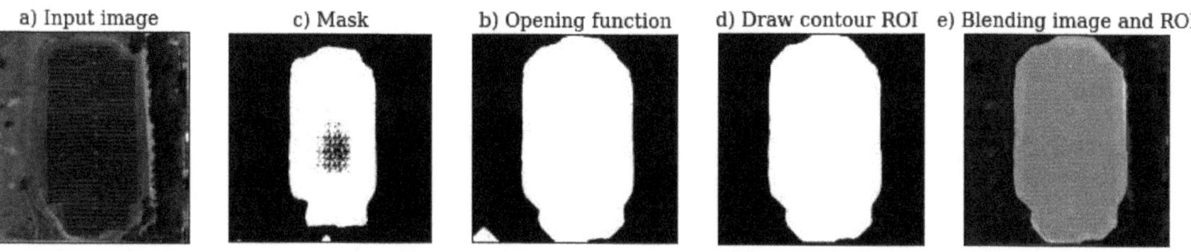

Figure 6. Steps of boundary extraction by the DL server and OpenCV functions in the Unioeste 1 PV plant.

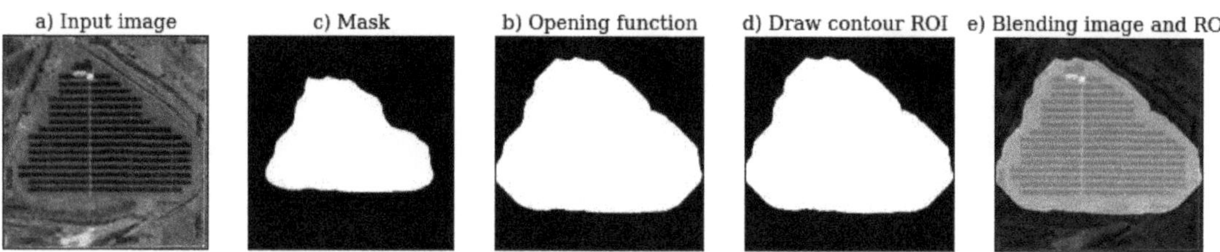

Figure 7. Steps of boundary extraction by DL server, and OpenCV functions in Arak PV plant.

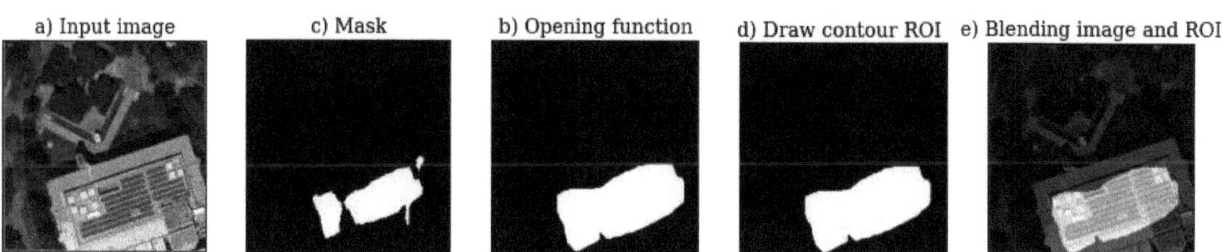

Figure 8. Steps of boundary extraction by DL server, and OpenCV functions in CSUSL PV plant.

3.2. Results of the CPP Method

The CCP method results were obtained from six experiments that implemented the five metrics previously described. Each experiment was conducted by selecting a PV plant and a UAV, then choosing the CPP method and the width, speed, and height of the UAV over the flight path. All of this was done in the implemented GUI interface. The six tests were conducted by varying the CPP width parameter, while other parameters were not varied, and the same experimental conditions were maintained throughout.

For these experiments two UAVs were selected, the Typhoon [97] and 3DR Iris [98], as mentioned in Section 2.4. Three simulated PV plants were also chosen, as highlighted in Section 2. The metrics referenced in Section 2.3.1 were assessed for each test. Battery consumption was obtained from the GCS software with the SITL parameter activated. The rest of the data was collected from the Gazebo simulations. The experiments are explained in detail in the following sections, and some results are discussed.

3.2.1. The Three First Experiments with a Typhoon UAV

The three first CPP experiments were conducted with a Typhoon UAV, varying the CPP width between 0 and 20 m, depending on the PV plant being covered. For example, in the first and second PV plants, Unioeste 1, and Arak, respectively, the CPP widths were varied between 5 and 20 m. In the third CSUSL PV plant, the width was varied between 1 and 8 m, due to the method's restrictions on running in small areas with a large width. Then, in each of these experiments, the resulting metrics were annotated in Tables, as shown in Appendix B.

The logged metrics were utilized to draw a clustered columns and lines chart for experiments 1, 2 and 3, performed with the Typhoon UAV, to highlight the most important information and to determine their correlations with each other. In these graphs the vertical axis was scaled logarithmically and the horizontal axis was labeled with the type of CPP method in use; in this way, it is possible to visualize the correlation of flight time, number of maneuvers, and path length covered with the energy consumption by each CPP method, as shown in Figures 9–11. In addition, the relationship between the redundancy metric and the CPP width is identified, showing that this metric is higher in BECD, but does not affect energy consumption, which is the primary metric of interest.

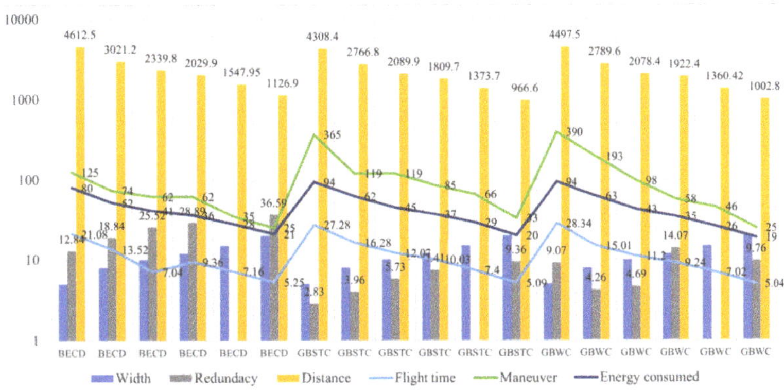

Figure 9. Performance metrics of Typhoon UAV on Unioeste PV plant (Experiment 1).

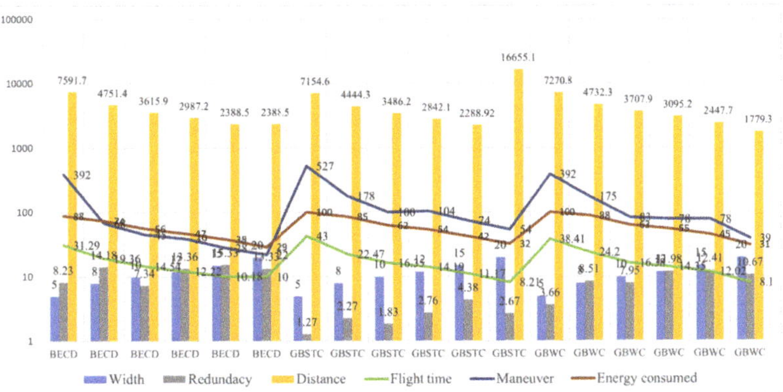

Figure 10. Performance metrics of Typhoon UAV on Arak PV plant (Experiment 2).

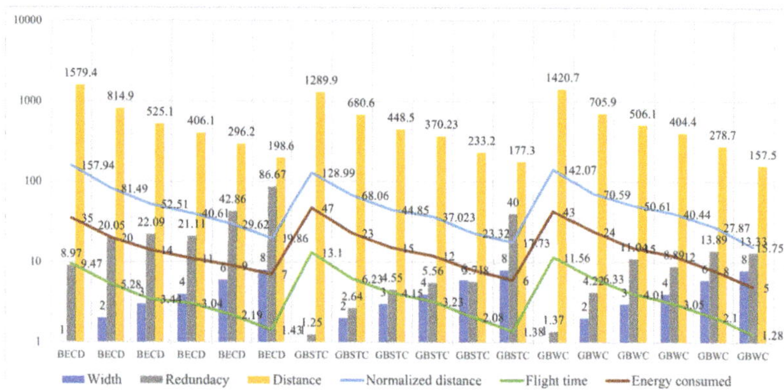

Figure 11. Performance metrics of Typhoon UAV on CSUSL PV plant (Experiment 3).

The UAVs' energy consumption, in conjunction with the CPP widths, were obtained from the energy consumption results logs, then a new Table 1 was composed of three large columns, corresponding to all the assessed PV plants, and each PV plant column is made up of three CPPs with their respective values of consumed energy. The rows of the table contain the CPP widths in increasing order from top to bottom, as shown in Table 1. Where it can be observed, some parameters, such as the energy consumption with regard to CPP width, were due to the size of the RoI. Additionally, it can be observed that the greater the CPP width, the lower the energy consumption, as seen in the columns of Table 1. For example, for the BECD CPP in Unioeste 1, with a width of 5 m, the percentage of energy consumed was 98%, while, for a width of 20, the energy consumed was 29%. It is also observed that the larger the PV plant, the UAV consumed all its energy in CPPs with narrow widths; on the other hand, if the plant is small, with the same width, the UAV does not have energy consumption problems. As can be seen in the row with a width of 8, where, for the Unioeste 1 and Arak PV plants, a lot of energy was consumed, between 88% and 52%, unlike the CSUSL plant, in which where the energy consumed was very little, between 7% and 5%. In the experiments for which energy consumption, with respect to CPP width, cannot be observed, (N/A) was used to annotate these results, which happened when the CPP width was very large with respect to the RoI, as this does not allow the generation of the route, or when the CPP width was very small with regard to the RoI causing a flight path in which the UAV consumes all its energy. In short, the UAV used can be undersized or oversized with respect to its intended PV plant.

The graphs in Figure 12 were constructed from Table 1 by quintic polynomial interpolations, which requires six data points to form a curve that passes through all given data points [99], where the abscissa axis is the CPP width and the ordinate axis is energy consumption. For the experiment conducted at the PV plant Unioeste 1, the graph in Figure 12a shows that the BECD method had the lowest energy consumption when the CPP width was in the range of 5 to 16 m, while GBSTC and GBWC had the lower energy consumption when the CPP width was in the range of 16 to 20 m. For the experiment tested at the Arak PV plant, the graph in Figure 12b shows that the BECD method had the lowest energy consumption when the CPP width was in the range of 5 to 10 m, while the GBSTC and GBWC had similar energy consumption when the CPP width was in the range of 10 to 15 m, and GBWC had the lowest energy consumption when the CPP width was in the range of 15 and 20 m. For tests conducted at the CSUSL PV plant, the graph in Figure 12c shows that the BECD method had the lowest energy consumption when the CPP width was in the range of 1 to 5.5 m, while the method GBSTC had the lowest energy consumption when the CPP width was in the range of 5.5 to 8 m. All plants were recreated in a simulation environment with their real dimensions.

Table 1. Comparison of three CPP methods with regard to CPP width, for a simulated Typhoon UAV at three PV plants.

| | Energy Consumption | | | | | | | | |
| PV Plant | Unioeste 1 (35,975 m^2) | | | Arak (25,056 m^2) | | | CSUSL (1344 m^2) | | |
CPP Width (m)	BECD	GBSTC	GBWC	BECD	GBSTC	GBWC	BECD	GBSTC	GBWC
1	*N/A*	*N/A*	*N/A*	*N/A*	*N/A*	*N/A*	35%	47%	43%
2	*N/A*	*N/A*	*N/A*	*N/A*	*N/A*	*N/A*	20%	23%	24%
3	*N/A*	*N/A*	*N/A*	*N/A*	*N/A*	*N/A*	14%	15%	15%
4	*N/A*	*N/A*	*N/A*	*N/A*	*N/A*	*N/A*	11%	12%	12%
5	98%	100%	100%	80%	94%	94%	*N/A*	*N/A*	*N/A*
6	*N/A*	*N/A*	*N/A*	*N/A*	*N/A*	*N/A*	9%	8%	8%
8	74%	85%	88%	52%	62%	63%	7%	6%	5%
10	56%	62%	63%	41.5%	45%	43%	5%	*N/A*	5%
12	47%	54%	55%	36%	37%	35%	*N/A*	*N/A*	*N/A*
15	38%	43%	45%	27%	29%	26%	*N/A*	*N/A*	*N/A*
20	29%	32%	31%	21%	20%	19%	*N/A*	*N/A*	*N/A*

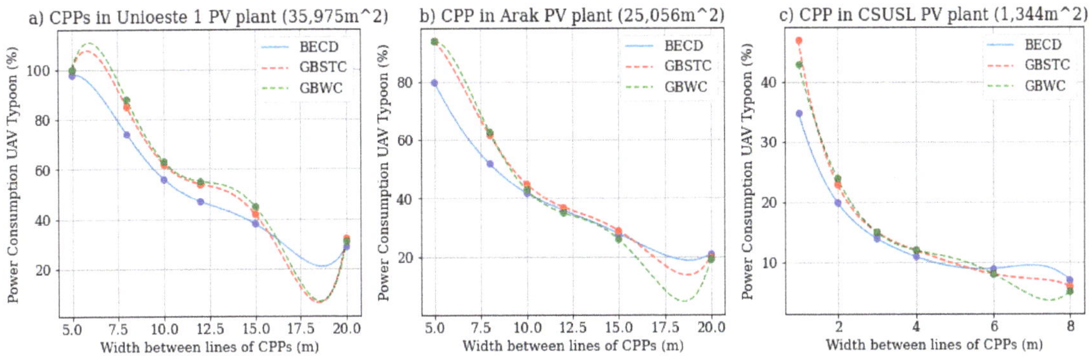

Figure 12. Performance test of Typhoon UAV varying the CPP width.

3.2.2. The Last Three Experiments with 3DR Iris UAV

The last three CPP experiments were conducted with a 3DR Iris UAV, varying the CPP width between 1 and 20 m, depending on the PV plant covered. For example, in the first and second PV plants, Unioeste 1 and Arak, respectively, the CPP widths were varied between 8 and 20 m; in the third PV plant, CSUSL, the width was varied between 1 and 8 m, due to the restrictions of the method requiring it be run in small areas with a large width. Then, in each of these experiments, the resulting metrics were annotated in tables, as shown in Appendix C.

As in the previous chapter, the recorded metrics were used to create bar and line graphs for Experiments 4, 5 and 6, which are those conducted with the 3DR Iris UAV, to highlight the most important information and to determine correlations among the metrics. In these graphs, the vertical and horizontal axes were scaled logarithmically and labeled in the same manner as above, with the aim of visualizing the correlations between flight time, number of maneuvers and path length covered, with regard to energy consumption by each CPP method, as shown in Figure 13–15 . Furthermore, the relationship between the redundancy metric and CPP width is identified, showing behavior similar to the previous UAV's data.

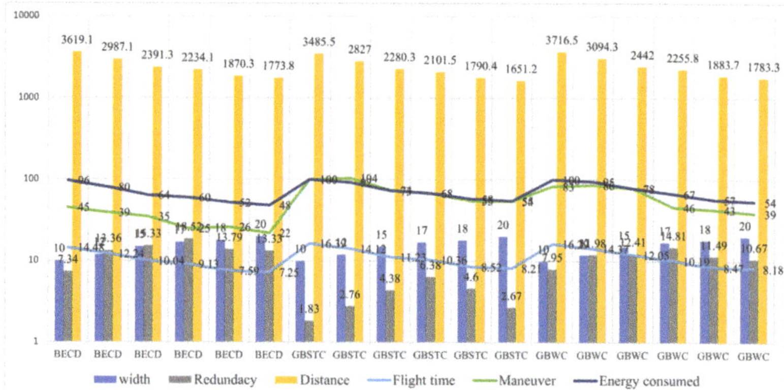

Figure 13. Performance metrics of 3DR Iris UAV on Unioeste PV plants (Experiment 4).

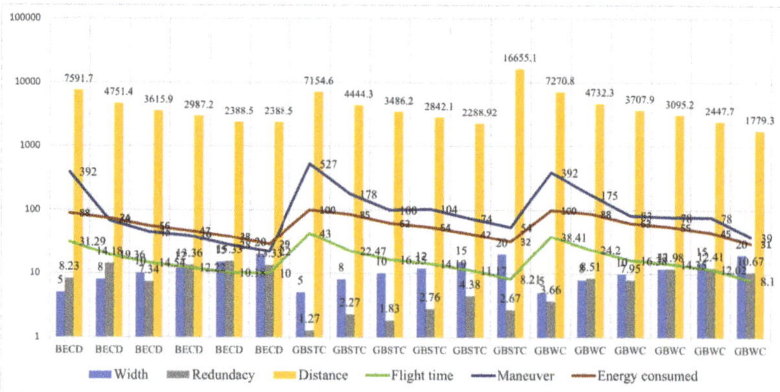

Figure 14. Performance metrics of 3DR Iris UAV on Arak PV plants (Experiment 5).

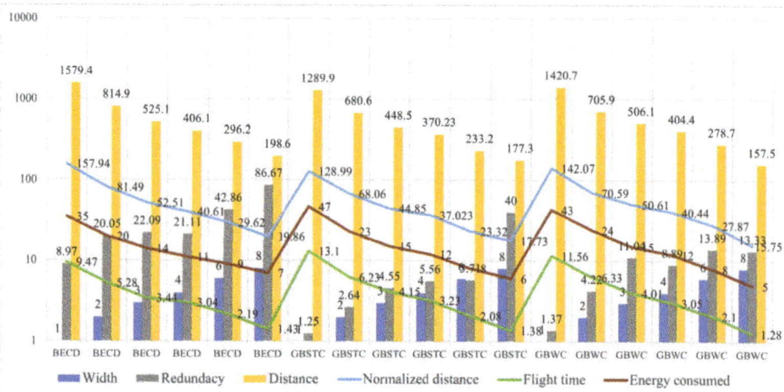

Figure 15. Performance metrics of 3DR Iris UAV on CSUSL PV plants (Experiment 6).

The UAV energy consumption and CPP widths were obtained from the energy consumption results logs, then a new Table 2 was established, in which three large columns correspond to the PV plants, where each PV plant column contains three CPPs with their respective values of consumed energy, and the rows of the table contain the CPP widths in an increasing direction from top to bottom. Here it can be observed, as in the last table, that some parameters' values, such as the energy consumption by CPP width, are due to the size of the RoI. It can also be observed that the greater the CPP width, the lower the energy consumption, as seen in the columns of Table 2. For example, using BECD at Unioeste 1 with a width of 10 m, the percentage of energy consumed was 100%, while for a width of 20 m, the energy consumed was 48%. It is also observed that, for larger PV plants, the UAV consumed all its energy in a CPP with a small width, whereas, for small plants of the same width, the UAV did not have energy waste problems. As can be seen in the row corresponding to a CPP width of 8 m, at the Unioeste 1 and Arak plants a lot of energy was consumed, between 100%, and 88%, unlike at the CSUSL plant, where the energy consumed was very little, between 9%, and 8%. Similarly, in the previous experiments, wherein the UAV's energy was exhausted and the CPP width did not allow the generation of the route or when the CPP width was large with regard to the RoI, the results were scored with (N/A). To summarize, the UAV used is suitable for the last PV plant.

On the other hand, the 3DR Iris UAV, with its design, size, autonomy, and performance is suitable when the PV plant to be inspected is small, of an approximate size of 5000 square meters or less, such as the CSUSL PV plant, at 1344 square meters. Owing to such PV plant sizes, any coverage method used in this work can be used to cover an area with a CPP width of 1 meter, while minimizing the metrics of path length covered, flight time and maneuverability, to obtain lower energy consumption, as shown in the Table 2. In addition, this UAV did not perform well at large or medium-sized plants, since, to cover them using the CPP method, the width that must be provided is greater than 8 m; therefore, though the UAV's inspection of the PV plant had no high-resolution cameras on board, they were needed; without them, the inspection cannot be guaranteed.

Table 2. Comparison of three CPP methods with regard to CPP width, simulated in a 3DR Iris UAV on three PV plants.

PV Plant	Unioeste 1 (35,975 m^2)			Arak (25,056 m^2)			CSUSL (1344 m^2)		
CPP Width (m)	BECD	GBSTC	GBWC	BECD	GBSTC	GBWC	BECD	GBSTC	GBWC
1	N/A	N/A	N/A	N/A	N/A	N/A	52%	72%	72%
2	N/A	N/A	N/A	N/A	N/A	N/A	30%	35%	38%
3	N/A	N/A	N/A	N/A	N/A	N/A	21%	24%	22%
4	N/A	N/A	N/A	N/A	N/A	N/A	17%	17%	15%
5	N/A	N/A	N/A	N/A	N/A	N/A	N/A	N/A	N/A
6	N/A	N/A	N/A	N/A	N/A	N/A	12%	12%	11%
8	100%	100%	100%	88%	98%	98%	9%	9%	8%
10	96%	100%	100%	70%	76%	71%	7%	N/A	7%
12	80%	92%	95%	60%	63%	59%	N/A	N/A	N/A
15	64%	73%	78%	45%	48%	43%	N/A	N/A	N/A
17	60%	68%	67%	N/A	N/A	N/A	N/A	N/A	N/A
20	48%	55%	54%	26%	26%	25%	N/A	N/A	N/A

The graphs in Figure 16 were constructed from Table 2 using quintic polynomial interpolations, wherein the abscissa axis is CPP width and the ordinate axis is energy consumption.

Additionally, the graph in Figure 16a shows that the BECD method had the lowest energy consumption; when the CPP width spanned the entire test range, the GBSTC and GBWC methods showed a higher energy consumption at the Unioeste 1 PV plant. The graph in Figure 16b shows that the BECD method had the lowest energy consumption when the CPP width was in the range of 8 to 10 m, whereas BECD and GBWC had similar energy consumption, leaving the GBWC method with the lowest consumption, when the

CPP width was in the range of 10 to 15 m. On the other hand, the GBWC method had the lowest energy consumption when the CPP width was in the range of 15 and 20 m; this test was conducted at the Arak PV plant. Finally, the graph in Figure 16c shows that the BECD method had the lowest energy consumption when the CPP width was in the range of 1 to 3 m, and the GBWC method had the lowest consumed energy when the CPP width was in the range of 3 to 8 m. The test was performed at CSUSL's PV plant.

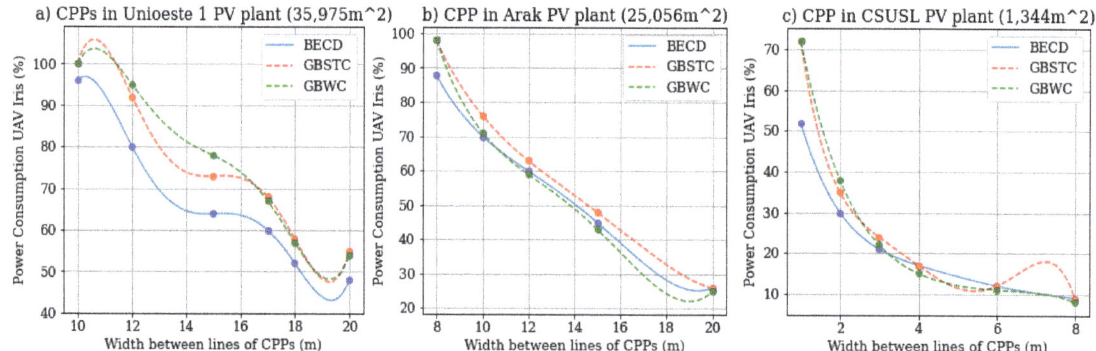

Figure 16. Performance test of UAV 3DR Iris varying the CPP width.

Summary Tables of the results of the metrics for each of the tests with the Typhoon UAV and 3DR Iris UAV are shown. All files, and logs for the experiments are available on GitHub at [100].

3.3. Discussion

The proposed strategy allows a semi-automatic and faster solution for achieving effective results when inspecting PV plants in geometrically simple areas, with certain limitations, although some results obtained in this work are theoretical results, such as from CPP width, for example; in practice, a camera of 14 megapixels is not adequate to inspect a PV plant with a CPP width of 20 m [37,101].

The results obtained in this work indicate that the most adequate method is BECD for a specific range of CPP widths, although it also shows adequate performance for various CPP widths in some RoIs. The GBWC showed a good performance when using a CPP width greater than 7 m in some RoIs, and all showed a relationship with the area to be covered, as shown in Tables 1 and 2.

An analysis of the metrics used in this work, such as redundancy, $R_\%$, which does not represent a significant factor when comparing these three CPP methods, showed that, although the BECD was the method with the highest redundancy, it was also the method with the lowest consumption of energy. On the other hand, the other metrics, such as the L_{ct} metric, flight time metric, and the number of maneuvers, are directly related to energy consumption, as can be seen in Appendix B and C. It also helps to conclude that the BECD method is more suitable for widths in a range between 0 and 7 m, due to the lower number of maneuvers in these ranges, as other authors have also mentioned [102,103]. The percentage of coverage of the total area, $C_\%$, always showed that coverage was total

The L_{ct} metric, flight time and number of maneuvers were directly related to the consumption of energy for all the experiments performed, as seen in Figures 9–11 and 13–15, and as confirmed by other authors [89,90], in addition to helping to reinforce that the most appropriate CPP for certain ranges is the BECD.

The BECD method is the best method among the three methods tested, in a specific width range, from 0 to 7 m, for all the RoIs tested, which means that for a 10 Megapixel camera with a horizontal field of view of 7 m, the CCP method's good images, in terms of what to inspect of a PV plant can be obtained [23].

The implemented BECD method divided the RoI into small regions, and then, over these regions, it implemented the round-trip coverage pattern, called boustrophedon [38,39]. This pattern allows spending less energy consumption with small widths, due to its low number of maneuvers compared with the other methods, as shown in Appendix B and C.

On the other hand, the GBWC method allowed lower energy consumption, according to the results of the simulations, in widths greater than a certain value, but depending on RoI size; this method also allows an approximate coverage outside the RoI, a characteristic that is very important in this type of application, and that, perhaps, is not very attractive for terrestrial robotics, from which this type of method originates [86].

A RoI with many concave points is a great challenge for performance metrics, since they increase with the distance of travel, and also increase the number of maneuvers and therefore increase energy consumption, according to these results in the following works [67,103], a more detailed analysis on these characteristics was made.

In future works, the methods (BECD, GBWC) could be implemented in UAVs with characteristics similar to those used in this work; these characteristics are shown in Tables A1 and A5. Additionally, with these UAVs, an inspection could be conducted in at least one PV plant. Where one can think about the implementation of an expert system that selects the coverage path between the two types of methods (BECD, GBWC), according to the CPP width size, required for a camera with a certain resolution, and focal length.

Finally, the results obtained with the BECD, and GBWC methods differ from the results obtained by other authors [34–37]. Who did not implement CPPs to solve these types of problems; they used other types of solutions that have restrictions when inspecting PV plants with UAVs. On the contrary, this work considers the CPPs, and obtained interesting results for future real implementation. The proposed CPP will increase the possibility of using inexpensive UAV systems for the inspection of PV systems on roofs of houses, and commercial buildings, and also, of the use of CPP with small widths to complete inspections at centimeter scales of the panel where the flaws can be better seen.

4. Conclusions

In this work, a method for implementing CPP in UAV for PV plant inspection was presented. The method consisted of a series of steps, one of these was the deployment of a DL-based U-net model to establish a DL service system from which to extract the limits of PV plants by extracting the boundaries of PV plants from an image. To summarize, the method was accurate, and fast without depending on the image, with low request latency and response.

This experiment focused on three novel path planning methods in the PV inspection missions in order to find the best path for covering each of the three PV power plants with less energy consumption. A GUI interface was used to order the UAV's maneuvers in the inspection of the simulated PV plants. The results of each CPP method in the simulation were compared. The best CPPs was the BECD, for a range of CPP widths of 0 to 7 m. These path planning algorithms can be performed by any multirotor UAV that receives Mavlink commands, can carry a camera sensor, and transmit real-time video to GCS.

Performance on the CPP tasks was measured using two different types of flying robots, a Typhoon UAV and a 3DR Iris UAV. It was demonstrated that the Typhoon UAV (or one with similar characteristics) is better suited for large or medium-sized PV plants (such as those in deserts, plains, and hills); instead, the drone-like 3DR Iris UAV is more suitable for small-sized PV generation (such as on roofs and rooftops, canopies, and facades). The proposed strategy allows such comparisons to be made and enables the selection of the most suitable UAV for each type of installation.

The values obtained for the metrics collected from each of the tests show a correlation between covered path length, flight time, number of maneuvers as regards energy consumption, and ensuring that the CPPs implemented in UAVs to inspect photovoltaic plants could be similar when implemented in real plants. The results also help to predict the energy consumption of a given UAV when performing a plant inspection.

Author Contributions: Conceptualization, A.é.P.-G., N.B.-M., Á.J.-D. and J.B.C.-Q.; methodology, A.é.P.-G.; software, A.é.P.-G., N.B.-M.; validation, A.é.P.-G.; formal analysis, A.é.P.-G.; investigation, A.é.P.-G., N.B.-M.; resources, A.é.P.-G., Á.J.-D. and J.B.C.-Q.; data curation, A.é.P.-G.; writing–original draft preparation, A.é.P.-G.; writing–review and editing, A.é.P.-G., N.B.-M., Á.J.-D. and J.B.C.-Q.; visualization, A.é.P.-G.; supervision, Á.J.-D. and J.B.C.-Q.; project administration, A.é.P.-G., Á.J.-D. and J.B.C.-Q.; funding acquisition, A.é.P.-G., Á.J.-D. and J.B.C.-Q. All authors have read and agreed to the published version of the manuscript.

Funding: This research was funded by the Colombia Scientific Program within the framework of the so-called Ecosistema Científico (Contract No. FP44842-218-2018).

Institutional Review Board Statement: Not applicable.

Informed Consent Statement: Not applicable.

Data Availability Statement: The models used in the computational experiment are available at GitHub in [100].

Acknowledgments: The authors gratefully acknowledge the support from the Colombia Scientific Program within the framework of the call Ecosistema Científico (Contract No. FP44842-218-2018). The authors also want to acknowledge Universidad de Antioquia for its support through the project "estrategia de sostenibilidad".

Conflicts of Interest: The authors declare no conflict of interest.

Abbreviations

The following abbreviations are used in this manuscript:

BECD	boustrophedon exact cellular decomposition
GBSTC	grid-based spanning tree coverage
GBWC	grid-Based wavefront coverage
DL	deep learning
UAV	unmanned aerial vehicle
PV	photovoltaic
O and M	operation and maintenance
CAGR	compound annual growth rate
LCOE	levelized cost of energy
List of Symbols	
L_{ct}	covered path length
$R_\%$	redundancy of points traveled
$C_\%$	percentage of coverage of the total area

Appendix A. Algorithm

In this step, a series of OpenCV functions, in python 3.7 and in jupyter-lab, were applied.

Algorithm A1: OpenCV functions.

1. **input** : A image I_r of size $w \times l$
2. **output**: A Map for robot
3. initialization
4. import cv 2, np, flask, tensorflow, matplotlib
5. **do**
6.   Im $\leftarrow$ cv2.imread($Mask$)
7.   th $\leftarrow$ cv2.Threshold(I_m,128,255,$THRESH_BINARY,THRESH_OTSU$)
8.   k $\leftarrow$ cv2.getStructuringElement($MORPH_RECT$, $(1,1)$)
9.   j $\leftarrow$ cv2.getStructuringElement($MORPH_RECT$, $(5,5)$)
10.   Ierosion $\leftarrow$cv 2.erode($th2[w,l]$,k,$iteration = 2$)
11.   Idilation $\leftarrow$ cv 2.dilate(Ierosion$[w,l]$,j,$iteration = 20$)
12.   ThImage $\leftarrow$ $Idilation.astype(np.uint8)$
13.   cnt $\leftarrow$ cv2.findCountours(
    ThImage$[w,l]$,$RETRE_XTER,CHAIN_APX_NONE$)
14.   **for** i **in** cnt:
15.   area $\leftarrow$ cv2.contourArea(cnt)
16.   **if** (area >400):
17.   apx $\leftarrow$ cv2.approxPolyDP(cnt,0.0010$\ast$ cv 2.$arcLength(cnt, True)$,$True$)
18.   cv2.drawContours($image_copy$, $[apx]$,-1 , $(0,0,255)$,7)
19.   mask $\leftarrow$ cv2.zeros($[w,l]$)
20.   cc $\leftarrow$ cv2.drawContours(mask,cnt,-1,$(255,255,255)$,-1,$FILLED$)
    while $True$

Appendix B. Tables of Typhoon UAV

Table A1. Yuneec Typhoon UAV technical specification.

Typhoon UAV	
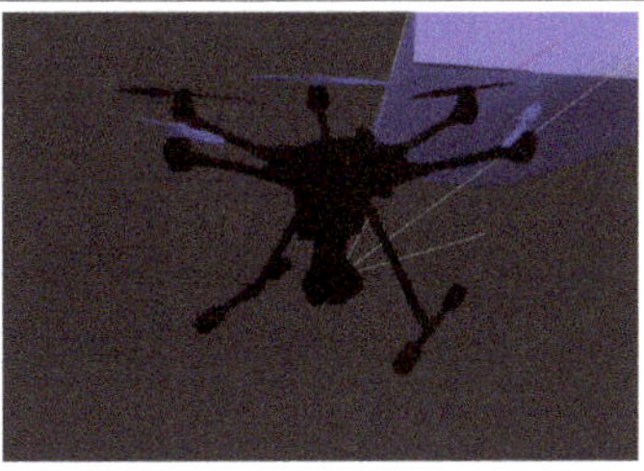	
Dimensions	$520 \times 457 \times 310$ mm ($20.5 \times 18 \times 12.2$ inches)
Weight	1980 g (69.8 ounces)
Battery	5400 mAh 4S/ 14.8 V (79.9 Wh)
Camera	12.4 megapixels, 14mm/F2.8
Flight Time	up to 25 min
Flight Speed	20o m/s
Payload Capacity	10.400 g
Motor	AC YUNH520120

Table A2. Experiment 1 with three CPP methods implemented using Typhoon UAV over Unioeste 1 PV Plant.

Width	CPP	Redundacy	Distance	Flight Time	Maneuver	Energy Consumed
5		8.23	7591.7	31.29	392	88
8		14.18	4751.4	19.36	68	74
10	BECD	7.34	3615.9	14.54	45	56
12		13.36	2987.2	12.22	39	47
15		15.33	2388.5	10.18	28	38
20		13.33	2388.5	10	22	29
5		1.27	7154.6	43	527	100
8		2.27	4444.3	22.47	178	85
10	GBSTC	1.83	3486.2	16.35	100	62
12		2.76	2842.1	14.19	104	54
15		4.38	2288.92	11.17	74	42
20		2.67	16,655.1	8.21	54	32
5		3.66	7270.8	38.41	392	100
8		8.51	4732.3	24.2	175	88
10	GBWC	7.95	3707.9	16.38	83	63
12		11.98	3095.2	14.35	78	55
15		12.41	2447.7	12.02	78	45
20		10.67	1779.3	8.1	39	31

Table A3. Experiment 2 with three CPP methods implemented using Typhoon UAV over Arak PV Plant.

Width	CPP	Redundacy	Distance	Flight Time	Maneuver	Energy Consumed
5		12.84	4612.5	21.08	125	80
8		18.84	3021.2	13.52	74	52
10	BECD	25.52	2339.8	7.4	62	41
12		28.89	2029.9	9.36	62	36
15		12.84	1547.95	7.16	35	28
20		36.59	1126.9	5.25	25	21
5		2.83	4308.4	27.28	365	94
8		3.96	2766.8	16.28	119	62
10	GBSTC	5.73	2089.9	12.07	119	45
12		7.41	1809.7	10.03	85	37
15		8.75	1373.7	7.4	66	29
20		9.36	966.6	5.09	33	20
5		9.07	4497.5	28.34	390	94
8		4.26	2789.6	15.01	193	63
10	GBWC	4.69	2078.4	11.2	98	43
12		14.07	1922.4	9.24	58	35
15		6.25	1360.42	7.02	46	26
20		9.76	1002.8	5.04	25	19

Table A4. Experiment 3 with three CPP methods implemented using Typhoon UAV over CSUSL PV Plant.

Width	CPP	Redundacy	Distance	Flight Time	Maneuver	Energy Consumed
1		8.97	1579.4	9.47	192	35
2	BECD	20.05	814.9	5.28	86	20
3		22.09	525.1	3.44	62	14
4		21.11	406.1	3.04	38	11
6		42.86	296.2	2.19	27	9
8		86.67	198.6	1.43	20	7
1		1.25	1289.9	13.1	542	47
2	GBSTC	2.64	680.6	6.23	154	23
3		4.55	448.5	4.15	88	15
4		5.56	370.23	3.23	58	12
6		5.71	233.2	2.08	30	8
8		40	177.3	1.38	19	6
1		1.37	1420.7	11.56	462	43
2	GBWC	4.22	705.9	6.33	168	24
3		11.04	506.1	4.01	69	15
4		8.89	404.4	3.05	44	12
6		13.89	278.7	2.1	24	8
8		13.33	157.5	1.28	14	5

Appendix C. Tables of 3DR Iris UAV

Table A5. 3D Robotics UAV 3DR Iris technical specification.

3D Robotics 3DR Iris	
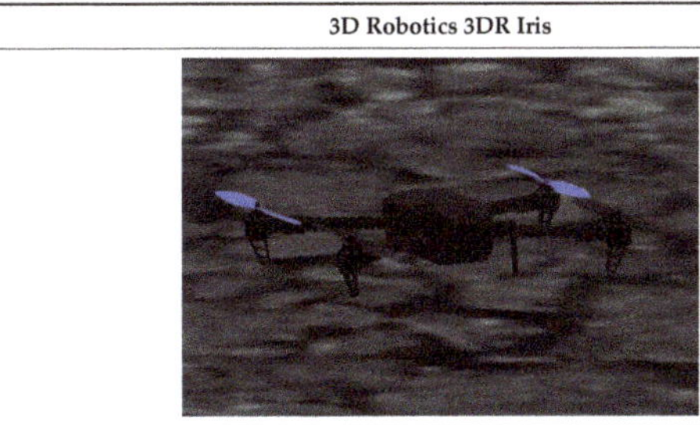	
Dimensions	10 cm in height, 55 cm motor-to-motor
Weight	1282 g
Battery	5100 mAh 3S
Camera	N/A
Flight Time	15–20 mins
Flight Speed	11 m/s
Payload Capacity	400 g
Motor	AC 2830, 950 kV

Table A6. Experiment 4 with three CPP methods implemented using UAV 3DR Iris over Unioeste 1 PV Plant.

Width	CPP	Redundacy	Distance	Flight Time	Maneuver	Energy Consumed
10	BECD	7.34	3619.1	14.48	45	96
12	BECD	13.36	2987.1	12.24	39	80
15	BECD	15.33	2391.3	10.04	35	64
17	BECD	18.52	2234.1	9.13	25	60
18	BECD	13.79	1870.3	7.59	26	52
20	BECD	13.33	1773.8	7.25	22	48
10	GBSTC	1.83	3485.5	16.39	100	100
12	GBSTC	2.76	2827	14.12	104	92
15	GBSTC	4.38	2280.3	11.23	74	73
17	GBSTC	6.38	2101.5	10.36	68	68
18	GBSTC	4.6	1790.4	8.52	55	58
20	GBSTC	2.67	1651.2	8.21	54	55
10	GBWC	7.95	3716.5	16.29	83	100
12	GBWC	11.98	3094.3	14.37	86	95
15	GBWC	12.41	2442	12.05	78	78
17	GBWC	14.81	2255.8	10.19	46	67
18	GBWC	11.49	1883.7	8.47	43	57
20	GBWC	10.67	1783.3	8.18	39	54

Table A7. Experiment 5 with three CPP methods implemented using the UAV 3DR Iris over Arak PV Plant.

Width	CPP	Redundacy	Distance	Flight Time	Maneuver	Energy Consumed
7	BECD	15.42	3292.7	15.31	80	95
8	BECD	18.84	3020.2	14.2	74	88
10	BECD	25.52	2339.3	12.52	62	70
12	BECD	28.89	2030.4	9.57	50	60
15	BECD	31.25	1548.5	7.27	35	45
20	BECD	36.59	1133.3	5.21	25	26
7	GBSTC	5.73	2090.4	43	527	100
8	GBSTC	3.96	2757.8	17.06	179	98
10	GBSTC	5.73	2090.4	12.36	119	76
12	GBSTC	7.41	1809.3	10.18	85	63
15	GBSTC	8.75	13,674.2	8	66	48
20	GBSTC	9.76	971	5.07	33	26
5	GBWC	3.66	7270.8	38.41	392	100
8	GBWC	4.69	2078.8	11.47	98	2
10	GBWC	4.69	2078.8	11.47	98	71
12	GBWC	14.07	1922.7	9.43	58	59
15	GBWC	6.25	1360.08	7.07	46	43
20	GBWC	6.25	1360.08	7.07	46	25

Table A8. Experiment 6 with three CPP methods implemented using the UAV 3DR Iris over CSUSL PV Plant.

Width	CPP	Redundacy	Distance	Flight Time	Maneuver	Energy Consumed
1		10.06	1724.8	10.39	211	52
2	BECD	15.38	978	6.12	95	30
3		19.3	597.2	4.18	62	21
4		26.67	441.5	3.33	47	17
6		47.22	316.7	2.3	30	12
8		66.67	197.4	1.47	20	9
1		1.77	1416.4	15.31	626	72
2	GBSTC	3.61	697.4	7.2	220	35
3		5.85	508.6	5	108	24
4		5.56	373.4	3.28	65	17
6		19.44	270.1	2.22	32	12
8		40	198.8	1.52	19	9
1		3.94	1414.9	15.3	626	72
2	GBWC	5.53	817.4	7.36	218	38
3		7.6	553	4.36	86	22
4		8.89	404.4	3.05	44	15
6		13.89	274.7	2.13	24	11
8		20	184.9	1.41	16	8

References

1. REN21®. GLOBAL STATUS REPORTS. 2021. Available online: https://www.ren21.net/reports/global-status-report/ (accessed on 14 August 2021).
2. Alkhraijah, M.; Alowaifeer, M.; Alsaleh, M.; Alfaris, A.; Molzahn, D.K. The Effects of Social Distancing on Electricity Demand Considering Temperature Dependency. *Energies* **2021**, *14*, 473. [CrossRef]
3. Mey, A. Most U.S. Utility-Scale Solar Photovoltaic Power Plants Are 5 Megawatts or Smaller-Today in Energy-U.S. Energy Information Administration (EIA). Available online: https://www.eia.gov/todayinenergy/detail.php?id=38272# (accessed on 3 October 2021).
4. Maghrabie, H.M.; Abdelkareem, M.A.; Al-Alami, A.H.; Ramadan, M.; Mushtaha, E.; Wilberforce, T.; Olabi, A.G. State-of-the-Art Technologies for Building-Integrated Photovoltaic Systems. *Buildings* **2021**, *11*, 383. [CrossRef]
5. Hao, P.; Zhang, Y.; Lu, H.; Lang, Z. A novel method for parameter identification and performance estimation of PV module under varying operating conditions. *Energy Convers. Manag.* **2021**, *247*, 114689. [CrossRef]
6. IRENA. Future of Solar Photovoltaic: Deployment, Investment, Technology, Grid Integration, and Socio-Economic Aspects. 2019. Available online: https:https://www.irena.org (accessed on 24 October 2021).
7. Grimaccia, F.; Leva, S.; Dolara, A.; Aghaei, M. Survey on PV modules common faults after an O and M flight extensive campaign over different plants in Italy. *IEEE J. Photovoltaics* **2017**, *7*, 810–816. [CrossRef]
8. Poulek, V.; Šafránková, J.; Černá, L.; Libra, M.; Beránek, V.; Finsterle, T.; Hrzina, P. PV Panel and PV Inverter Damages Caused by Combination of Edge Delamination, Water Penetration, and High String Voltage in Moderate Climate. *IEEE J. Photovoltaics* **2021**, *11*, 561–565. [CrossRef]
9. Jamil, W.J.; Rahman, H.A.; Shaari, S.; Salam, Z. Performance degradation of photovoltaic power system: Review on mitigation methods. *Renew. Sustain. Energy Rev.* **2017**, *67*, 876–891. [CrossRef]
10. Kaplani, E. PV cell and module degradation, detection and diagnostics. In *Renewable Energy in the Service of Mankind Vol II, Selected Topics from the World Renewable Energy Congress WREC, London, UK, 3–8 August 2014*; Springer: Cham, Switzerland, 2016; pp. 393–402.
11. Di Lorenzo, G.; Araneo, R.; Mitolo, M.; Niccolai, A.; Grimaccia, F. Review of O and M Practices in PV Plants: Failures, Solutions, Remote Control, and Monitoring Tools. *IEEE J. Photovoltaics* **2020**, *10*, 914–926. [CrossRef]
12. Donovan, C.W. *Renewable Energy Finance: Funding the Future of Energy*; World Scientific Publishing Co. Pte. Ltd.: Singapore, 2020.
13. Narvarte, L.; Fernández-Ramos, J.; Martínez-Moreno, F.; Carrasco, L.; Almeida, R.; Carrêlo, I. Solutions for adapting photovoltaics to large power irrigation systems for agriculture. *Sustain. Energy Technol. Assess.* **2018**, *29*, 119–130. [CrossRef]
14. Grimaccia, F.; Leva, S.; Niccolai, A.; Cantoro, G. Assessment of PV plant monitoring system by means of unmanned aerial vehicles. In Proceedings of the 2018 IEEE International Conference on Environment and Electrical Engineering and 2018 IEEE Industrial and Commercial Power Systems Europe (EEEIC/I&CPS Europe), Palermo, Italy, 12–15 June 2018; pp. 1–6.
15. Guerrero-Liquet, G.C.; Oviedo-Casado, S.; Sánchez-Lozano, J.; García-Cascales, M.S.; Prior, J.; Urbina, A. Determination of the Optimal Size of Photovoltaic Systems by Using Multi-Criteria Decision-Making Methods. *Sustainability* **2018**, *10*, 4594. [CrossRef]

16. Grimaccia, F.; Aghaei, M.; Mussetta, M.; Leva, S.; Quater, P.B. Planning for PV plant performance monitoring by means of unmanned aerial systems (UAS). *Int. J. Energy Environ. Eng.* **2015**, *6*, 47–54. [CrossRef]
17. Shen, K.; Qiu, Q.; Wu, Q.; Lin, Z.; Wu, Y. Research on the Development Status of Photovoltaic Panel Cleaning Equipment Based on Patent Analysis. In Proceedings of the 2019 3rd International Conference on Robotics and Automation Sciences (ICRAS), Wuhan, China, 1–3 June2019; pp. 20–27.
18. Azaiz, R. Flying Robot for Processing and Cleaning Smooth, Curved and Modular Surfaces. U.S. Patent App. 15/118,849, 2 March 2017.
19. Libra, M.; Daneček, M.; Lešetickỳ, J.; Poulek, V.; Sedláček, J.; Beránek, V. Monitoring of Defects of a Photovoltaic Power Plant Using a Drone. *Energies* **2019**, *12*, 795. [CrossRef]
20. Li, X.; Li, W.; Yang, Q.; Yan, W.; Zomaya, A.Y. An unmanned inspection system for multiple defects detection in photovoltaic plants. *IEEE J. Photovoltaics* **2019**, *10*, 568–576. [CrossRef]
21. Alsafasfeh, M.; Abdel-Qader, I.; Bazuin, B.; Alsafasfeh, Q.; Su, W. Unsupervised fault detection and analysis for large photovoltaic systems using drones and machine vision. *Energies* **2018**, *11*, 2252. [CrossRef]
22. Gallardo-Saavedra, S.; Franco-Mejia, E.; Hernández-Callejo, L.; Duque-Pérez, Ó.; Loaiza-Correa, H.; Alfaro-Mejia, E. Aerial thermographic inspection of photovoltaic plants: Analysis and selection of the equipment. In Proceedings of the 2017 Proceedings ISES Solar World Congress, IEA SHC, Abu Dhabi, UAE, 29 October–2 November 2017.
23. Leva, S.; Aghaei, M.; Grimaccia, F. PV power plant inspection by UAS: Correlation between altitude and detection of defects on PV modules. In Proceedings of the 2015 IEEE 15th International Conference on Environment and Electrical Engineering (EEEIC), Rome, Italy, 10–13 June 2015; pp. 1921–1926.
24. Aghaei, M.; Dolara, A.; Leva, S.; Grimaccia, F. Image resolution and defects detection in PV inspection by unmanned technologies. In Proceedings of the 2016 IEEE Power and Energy Society General Meeting (PESGM), Boston, MA, USA, 17–21 July 2016; pp. 1–5.
25. Quater, P.B.; Grimaccia, F.; Leva, S.; Mussetta, M.; Aghaei, M. Light unmanned aerial vehicles (UAVs) for cooperative inspection of PV plants. *IEEE J. Photovoltaics* **2014**, *4*, 1107–1113. [CrossRef]
26. Oliveira, A.K.V.; Aghaei, M.; Madukanya, U.E.; Rüther, R. Fault inspection by aerial infrared thermography in a pv plant after a meteorological tsunami. *Rev. Bras. Energ. Sol.* **2019**, *10*, 17–25.
27. Tsanakas, J.A.; Chrysostomou, D.; Botsaris, P.N.; Gasteratos, A. Fault diagnosis of photovoltaic modules through image processing and Canny edge detection on field thermographic measurements. *Int. J. Sustain. Energy* **2015**, *34*, 351–372. [CrossRef]
28. Niccolai, A.; Grimaccia, F.; Leva, S. Advanced Asset Management Tools in Photovoltaic Plant Monitoring: UAV-Based Digital Mapping. *Energies* **2019**, *12*, 4736. [CrossRef]
29. Thrower, N.J.; Jensen, J.R. The orthophoto and orthophotomap: Characteristics, development and application. *Am. Cartogr.* **1976**, *3*, 39–56. [CrossRef]
30. Yao, Y.Y.; Hu, Y.T. Recognition and location of solar panels based on machine vision. In Proceedings of the 2017 2nd Asia-Pacific Conference on Intelligent Robot Systems (ACIRS), Wuhan, China, 16–18 June 2017; pp. 7–12.
31. Leva, S.; Aghaei, M. Power Engineering: Advances and Challenges Part B: Electrical Power. In *Chapter 3: Failures and Defects in PV Systems Review and Methods of Analysis*; Taylor & Francis Group, CRC Press: Boca Raton, FL, USA, 2018.
32. Kim, D.; Youn, J.; Kim, C. Automatic photovoltaic panel area extraction from uav thermal infrared images. *J. Korean Surv. Soc.* **2016**, *34*, 559–568. [CrossRef]
33. Li, X.; Yang, Q.; Lou, Z.; Yan, W. deep learning Based Module Defect Analysis for Large-Scale Photovoltaic Farms. *IEEE Trans. Energy Convers.* **2019**, *34*, 520–529. [CrossRef]
34. Ding, Y.; Cao, R.; Liang, S.; Qi, F.; Yang, Q.; Yan, W. Density-Based Optimal UAV Path Planning for Photovoltaic Farm Inspection in Complex Topography. In Proceedings of the 2020 Chinese Control And Decision Conference (CCDC), Hefei, China, 22–24 August 2020; pp. 3931–3936.
35. Luo, X.; Li, X.; Yang, Q.; Wu, F.; Zhang, D.; Yan, W.; Xi, Z. Optimal path planning for UAV based inspection system of large-scale photovoltaic farm. In Proceedings of the 2017 Chinese Automation Congress (CAC), Jinan, China, 20–22 October 2017; pp. 4495–4500. [CrossRef]
36. Salahat, E.; Asselineau, C.A.; Coventry, J.; Mahony, R. Waypoint Planning for Autonomous Aerial Inspection of Large-Scale Solar Farms. In Proceedings of the IECON 2019-45th Annual Conference of the IEEE Industrial Electronics Society, Lisbon, Portugal, 14–17 October 2019; Volume 1, pp. 763–769. [CrossRef]
37. Sizkouhi, A.M.M.; Esmailifar, S.M.; Aghaei, M.; De Oliveira, A.K.V.; Rüther, R. Autonomous path planning by unmanned aerial vehicle (UAV) for precise monitoring of large-scale PV plants. In Proceedings of the 2019 IEEE 46th Photovoltaic Specialists Conference (PVSC), Chicago, IL, USA, 16–21 June 2019; pp. 1398–1402.
38. Choset, H.M.; Lynch, K.M.; Hutchinson, S.; Kantor, G.; Burgard, W.; Kavraki, L.; Thrun, S.; Arkin, R.C. *Principles of Robot Motion: Theory, Algorithms, and Implementation*; MIT Press: Cambridge, MA, USA, 2005.
39. Cabreira, T.M.; Brisolara, L.B.; Ferreira Jr, P.R. Survey on coverage path planning with unmanned aerial vehicles. *Drones* **2019**, *3*, 4. [CrossRef]
40. Sebbane, Y.B. *Intelligent Autonomy of UAVs: Advanced Missions and Future Use*; CRC Press: Boca Raton, FL, USA, 2018.
41. Choset, H. Coverage for robotics–a survey of recent results. *Ann. Math. Artif. Intell.* **2001**, *31*, 113–126. [CrossRef]
42. Veerajagadheswar, P.; Ping-Cheng, K.; Elara, M.R.; Le, A.V.; Iwase, M. Motion planner for a Tetris-inspired reconfigurable floor cleaning robot. *Int. J. Adv. Robot. Syst.* **2020**, *17*, 1729881420914441. [CrossRef]

43. Coombes, M.; Fletcher, T.; Chen, W.H.; Liu, C. Optimal Polygon Decomposition for UAV Survey coverage path planning in Wind. *Sensors* **2018**, *18*, 2132. [CrossRef] [PubMed]
44. Islam, N.; Rashid, M.M.; Pasandideh, F.; Ray, B.; Moore, S.; Kadel, R. A Review of Applications and Communication Technologies for Internet of Things (IoT) and unmanned aerial vehicle (UAV) Based Sustainable Smart Farming. *Sustainability* **2021**, *13*, 1821. [CrossRef]
45. Pham, H.X.; La, H.M.; Feil-Seifer, D.; Deans, M. A distributed control framework for a team of unmanned aerial vehicles for dynamic wildfire tracking. In Proceedings of the 2017 IEEE/RSJ International Conference on Intelligent Robots and Systems (IROS), Vancouver, BC, Canada, 24–28 September 2017; pp. 6648–6653.
46. Kaljaca, D.; Vroegindeweij, B.; van Henten, E. Coverage trajectory planning for a bush trimming robot arm. *J. Field Robot.* **2020**, *37*, 283–308. [CrossRef]
47. Chang, W.; Yang, G.; Yu, J.; Liang, Z.; Cheng, L.; Zhou, C. Development of a power line inspection robot with hybrid operation modes. In Proceedings of the 2017 IEEE/RSJ International Conference on Intelligent Robots and Systems (IROS), Vancouver, BC, Canada, 24–28 September 2017; pp. 973–978.
48. Mansouri, S.S.; Kanellakis, C.; Fresk, E.; Kominiak, D.; Nikolakopoulos, G. Cooperative coverage path planning for visual inspection. *Control Eng. Pract.* **2018**, *74*, 118–131. [CrossRef]
49. Galceran, E.; Carreras, M. A survey on coverage path planning for robotics. *Robot. Auton. Syst.* **2013**, *61*, 1258–1276. [CrossRef]
50. Bormann, R.; Jordan, F.; Hampp, J.; Hägele, M. Indoor coverage path planning: Survey, implementation, analysis. In Proceedings of the 2018 IEEE International Conference on Robotics and Automation (ICRA), Brisbane, QLD, Australia, 21–25 May 2018; pp. 1718–1725.
51. Nam, L.; Huang, L.; Li, X.; Xu, J. An approach for coverage path planning for UAVs. In Proceedings of the 2016 IEEE 14th International Workshop on Advanced Motion Control, AMC 2016, Auckland, New Zealand, 22–24 April 2016; pp. 411–416. [CrossRef]
52. Dai, R.; Fotedar, S.; Radmanesh, M.; Kumar, M. Quality-aware UAV coverage and path planning in geometrically complex environments. *Ad Hoc Netw.* **2018**, *73*, 95–105. [CrossRef]
53. Yao, P.; Cai, Y.; Zhu, Q. Time-optimal trajectory generation for aerial coverage of urban building. *Aerosp. Sci. Technol.* **2019**, *84*, 387–398. [CrossRef]
54. Majeed, A.; Lee, S. A new coverage flight path planning algorithm based on footprint sweep fitting for unmanned aerial vehicle navigation in urban environments. *Appl. Sci.* **2019**, *9*, 1470. [CrossRef]
55. Elfes, A.; Campos, M.; Bergerman, M.; Bueno, S.; Podnar, G. A robotic unmanned aerial vehicle for environmental research and monitoring. In Proceedings of the First Scientific Conference on the Large Scale Biosphere-Atmosphere Experiment in Amazonia (LBA), LBA Central Office, CPTEC/INPE, Belém, Pará, Brazil, 25–30 June 2000; pp. 12630–12645.
56. Saeed, A.; Abdelkader, A.; Khan, M.; Neishaboori, A.; Harras, K.A.; Mohamed, A. On realistic target coverage by autonomous drones. *ACM Trans. Sens. Netw. (TOSN)* **2019**, *15*, 1–33. [CrossRef]
57. Lingelbach, F. Path planning using probabilistic cell decomposition. In Proceedings of the IEEE International Conference on Robotics and Automation, 2004. Proceedings, ICRA'04, New Orleans, LA, USA, 26 April–1 May 2004; Volume 1, pp. 467–472.
58. Khanam, Z.; Saha, S.; Ehsan, S.; Stolkin, R.; Mcdonald-Maier, K. coverage path planning Techniques for Inspection of Disjoint Regions With Precedence Provision. *IEEE Access* **2021**, *9*, 5412–5427. [CrossRef]
59. Khiati, W.; Moumen, Y.; Habchi, A.E.; Zerrouk, I.; Berrich, J.; Bouchentouf, T. Grid Based approach (GBA): A new approach based on the grid-clustering algorithm to solve a CPP type problem for air surveillance using UAVs. In Proceedings of the 2020 Fourth International Conference On Intelligent Computing in Data Sciences (ICDS), Fez, Morocco, 21–23 October 2020; pp. 1–5. [CrossRef]
60. Juliá, M.; Gil, A.; Reinoso, O. A comparison of path planning strategies for autonomous exploration and mapping of unknown environments. *Auton. Robot.* **2012**, *33*, 427–444. [CrossRef]
61. Rodriguez-Esparza, E.; Zanella-Calzada, L.A.; Oliva, D.; Heidari, A.A.; Zaldivar, D.; Pérez-Cisneros, M.; Foong, L.K. An efficient Harris hawks-inspired image segmentation method. *Expert Syst. Appl.* **2020**, *155*, 113428. [CrossRef]
62. Garcia-Garcia, A.; Orts-Escolano, S.; Oprea, S.; Villena-Martinez, V.; Martinez-Gonzalez, P.; Garcia-Rodriguez, J. A survey on deep learning techniques for image and video semantic segmentation. *Appl. Soft Comput.* **2018**, *70*, 41–65. [CrossRef]
63. Long, J.; Shelhamer, E.; Darrell, T. Fully convolutional networks for semantic segmentation. In Proceedings of the IEEE conference on computer vision and pattern recognition, Boston, MA, USA, 7–12 June 2015; pp. 3431–3440.
64. Ronneberger, O.; Fischer, P.; Brox, T. U-net: Convolutional networks for biomedical image segmentation. In *International Conference on Medical Image Computing and Computer-Assisted Intervention*; Springer: Munich, Germany, 2015; pp. 234–241.
65. Pérez-González, A.; Jaramillo-Duque, A.; Cano-Quintero, J.B. Automatic Boundary Extraction for Photovoltaic Plants Using the Deep Learning UNet Model. *Appl. Sci.* **2021**, *11*, 6524. [CrossRef]
66. Sizkouhi, A.M.M.; Aghaei, M.; Esmailifar, S.M.; Mohammadi, M.R.; Grimaccia, F. Automatic boundary extraction of large-scale photovoltaic plants using a fully convolutional network on aerial imagery. *IEEE J. Photovoltaics* **2020**, *10*, 1061–1067. [CrossRef]
67. Di Franco, C.; Buttazzo, G. coverage path planning for UAVs photogrammetry with energy and resolution constraints. *J. Intell. Robot. Syst.* **2016**, *83*, 445–462. [CrossRef]
68. Zefri, Y.; ElKettani, A.; Sebari, I.; Ait Lamallam, S. Thermal infrared and visual inspection of photovoltaic installations by UAV photogrammetry application case: Morocco. *Drones* **2018**, *2*, 41. [CrossRef]

69. Coopmans, C.; Podhradský, M.; Hoffer, N.V. Software-and hardware-in-the-loop verification of flight dynamics model and flight control simulation of a fixed-wing unmanned aerial vehicle. In Proceedings of the 2015 Workshop on Research, Education and Development of Unmanned Aerial Systems (RED-UAS), Cancun, Mexico, 23–25 November 2015; pp. 115–122.

70. Roggi, G.; Niccolai, A.; Grimaccia, F.; Lovera, M. A Computer Vision Line-Tracking Algorithm for Automatic UAV Photovoltaic Plants Monitoring Applications. *Energies* **2020**, *13*, 838. [CrossRef]

71. Coates, E.M.; Fossen, T.I. Geometric Reduced-Attitude Control of Fixed-Wing UAVs. *Appl. Sci.* **2021**, *11*, 3147. [CrossRef]

72. Tullu, A.; Endale, B.; Wondosen, A.; Hwang, H.Y. Machine Learning Approach to Real-Time 3D Path Planning for Autonomous Navigation of unmanned aerial vehicle. *Appl. Sci.* **2021**, *11*, 4706. [CrossRef]

73. Pitonakova, L.; Giuliani, M.; Pipe, A.; Winfield, A. Feature and performance comparison of the V-REP, Gazebo and ARGoS robot simulators. In *Annual Conference Towards Autonomous Robotic Systems*; Springer: Bristol, UK, 25–27 July 2018; pp. 357–368.

74. Akcakoca, M.; Atici, B.M.; Gever, B.; Oguz, S.; Demirezen, U.; Demir, M.; Saldiran, E.; Yuksek, B.; Koyuncu, E.; Yeniceri, R.; et al. A simulation-based development and verification architecture for micro uav teams and swarms. In *AIAA Scitech 2019 Forum*; AIAA: San Diego, CA, USA, 2019; p. 1979.

75. Unoeste Terá Maior Usina Solar de Geração Distribuída de SP-Unoeste. Available online: http://www.unoeste.br/noticias/2019/3/unoeste-tera-maior-usina-solar-de-geracao-distribuida-de-sp (accessed on 11 October 2021).

76. ABA Newsletter. Available online: http://www.csus.edu/aba2/newsletters/fall2012/abagreennews.html (accessed on 11 October 2021).

77. Abadi, M.; Barham, P.; Chen, J.; Chen, Z.; Davis, A.; Dean, J.; Devin, M.; Ghemawat, S.; Irving, G.; Isard, M.; et al. TensorFlow: A System for Large-Scale Machine Learning. In *12th USENIX Symposium on Operating Systems Design and Implementation (OSDI 16)*; USENIX Association: Savannah, GA, USA, 2016; pp. 265–283.

78. Flask. 2021. Available online: https://flask.palletsprojects.com/en/2.0.x/ (accessed on 31 October 2021).

79. Mavlink/Mavlink. 2021. Available online: https://github.com/mavlink/mavlink (accessed on 24 October 2021).

80. Li, J.; Zhou, Y.; Lamont, L. Communication architectures and protocols for networking unmanned aerial vehicles. In Proceedings of the 2013 IEEE Globecom Workshops (GC Wkshps), Atlanta, GA, USA, 9–13 December 2013; pp. 1415–1420. [CrossRef]

81. Sarker, I.H. Machine Learning: Algorithms, Real-World Applications and Research Directions. *SN Comput. Sci.* **2021**, *2*, 160. [CrossRef] [PubMed]

82. Bradski, G. The openCV library. *Dobb'S J. Softw. Tools Prof. Program.* **2000**, *25*, 120–123.

83. OpenCV: OpenCV-Python Tutorials. Available online: https://docs.opencv.org/master/d6/d00/tutorial_py_root.html (accessed on 18 October 2021).

84. Choset, H.; Pignon, P. coverage path planning: The boustrophedon cellular decomposition. In *Field and Service Robotics*; Springer: London, UK, 1998; pp. 203–209.

85. Gabriely, Y.; Rimon, E. Spanning-tree based coverage of continuous areas by a mobile robot. *Ann. Math. Artif. Intell.* **2001**, *31*, 77–98. [CrossRef]

86. Zelinsky, A.; Jarvis, R.A.; Byrne, J.; Yuta, S. Planning paths of complete coverage of an unstructured environment by a mobile robot. In Proceedings of the 1993 IEEE/Tsukuba International Workshop on Advanced Robotics, Tsukuba, Japan, 8–9 November 1993; Volume 13, pp. 533–538.

87. Ceballos, N.M.; Valencia, J.; Giraldo, A.A. Simulation and assessment educational framework for mobile robot algorithms. *J. Autom. Mob. Robot. Intell. Syst.* **2014**, *8*, 53–59. [CrossRef]

88. Jaeyoung, L. PX4 Gazebo Plugin Suite for MAVLink SITL and HITL. 2021. Available online: https://github.com/PX4/PX4-SITL_gazebo/blob/ffb87ef4a312564cf91791bd5a9d683aacd085a6/models/iris/iris.sdf.jinja (accessed on 7 December 2021).

89. Torres, M.; Pelta, D.A.; Verdegay, J.L.; Torres, J.C. coverage path planning with unmanned aerial vehicles for 3D terrain reconstruction. *Expert Syst. Appl.* **2016**, *55*, 441–451. [CrossRef]

90. Araújo, J.; Sujit, P.; Sousa, J. Multiple UAV area decomposition and coverage. In Proceedings of the 2013 IEEE Symposium on Computational Intelligence for Security and Defense Applications (CISDA), Singapore, 16–19 April 2013; pp. 30–37. [CrossRef]

91. PX4 Drone Autopilot. 2021. Available online: https://github.com/PX4/PX4-Autopilot (accessed on 23 September 2021).

92. Koenig, N.; Howard, A. Design and use paradigms for Gazebo, an open-source multi-robot simulator. In Proceedings of the 2004 IEEE/RSJ International Conference on Intelligent Robots and Systems (IROS) (IEEE Cat. No.04CH37566), Sendai, Japan, 28 September–2 October 2004; Volume 3, pp. 2149–2154. [CrossRef]

93. Quigley, M.; Conley, K.; Gerkey, B.; Faust, J.; Foote, T.; Leibs, J.; Wheeler, R.; Ng, A.Y. ROS: An open-source Robot Operating System. In Proceedings of the ICRA Workshop on Open Source Software, Kobe, Japan, 12–17 May 2009; p. 5.

94. Silano, G.; Iannelli, L. CrazyS: A Software-in-the-Loop Simulation Platform for the Crazyflie 2.0 Nano-Quadcopter. In *Robot Operating System (ROS): The Complete Reference (Volume 4)*; Koubaa, A., Ed.; Springer International Publishing: Cham, Swotzerland, 2020; pp. 81–115. [CrossRef]

95. Salamh, F.E.; Karabiyik, U.; Rogers, M.K. RPAS Forensic Validation Analysis Towards a Technical Investigation Process: A Case Study of Yuneec Typhoon H. *Sensors* **2019**, *19*, 3246. [CrossRef]

96. QGroundControl Ground Control Station. 2021. Available online: https://github.com/mavlink/qgroundcontrol (accessed on 31 October 2021).

97. Yuneec Typhoon H-Yuneec Futurhobby. Available online: https://yuneec-futurhobby.com/yuneec-typhoon-h-pro-realsense (accessed on 28 October 2021).

98. 3DR Iris-The Ready to fly UAV Quadcopter. Available online: http://www.arducopter.co.uk/iris-quadcopter-uav.html (accessed on 28 October 2021).
99. Tan, S.H.; Md Ali, J. Quartic and quintic polynomial interpolation. In Proceedings of the 20th National Symposium on Mathematical Sciences: Research in Mathematical Sciences: A Catalyst for Creativity and Innovation, Putrajaya, Malaysia, 18–20 December 2012; Volume 1522, pp. 664–675
100. Peréz-González, A. Andresperez86/CPP_GUI. 2021. Available online: https://github.com/andresperez86/CPP_GUI (accessed on 29 October 2021).
101. Jordan, S.; Moore, J.; Hovet, S.; Box, J.; Perry, J.; Kirsche, K.; Lewis, D.; Tse, Z.T.H. State-of-the-art technologies for UAV inspections. *IET Radar Sonar Navig.* **2018**, *12*, 151–164. [CrossRef]
102. Ghaddar, A.; Merei, A.; Natalizio, E. PPS: Energy-Aware Grid-Based coverage path planning for UAVs Using Area Partitioning in the Presence of NFZs. *Sensors* **2020**, *20*, 3742. [CrossRef] [PubMed]
103. Öst, G. Search Path Generation with UAV Applications Using Approximate Convex Decomposition. 2012; p. 52. Available online: http://urn.kb.se/resolve?urn=urn:nbn:se:liu:diva-77353 (accessed on 29 October 2021).

Article

Short-Term Forecasting of Photovoltaic Solar Power Production Using Variational Auto-Encoder Driven Deep Learning Approach

Abdelkader Dairi [1,†], **Fouzi Harrou** [2,*,†], **Ying Sun** [2,†] and **Sofiane Khadraoui** [3,†]

1   Computer Science Department Signal, Image and Speech Laboratory (SIMPA) Laboratory, University of Science and Technology of Oran-Mohamed Boudiaf (USTO-MB), El Mnaouar, BP 1505, Bir El Djir 31000, Algeria; abdelkader.dairi@univ-usto.dz
2   Computer, Electrical and Mathematical Sciences and Engineering (CEMSE) Division, King Abdullah University of Science and Technology (KAUST), Thuwal 23955-6900, Saudi Arabia; ying.sun@kaust.edu.sa
3   Department of Electrical Engineering, University of Sharjah, Sharjah 27272, UAE; skhadraoui@sharjah.ac.ae
*   Correspondence: fouzi.harrou@kaust.edu.sa
†   These authors contributed equally to this work.

Received: 7 October 2020; Accepted: 19 November 2020; Published: 25 November 2020

Abstract: The accurate modeling and forecasting of the power output of photovoltaic (PV) systems are critical to efficiently managing their integration in smart grids, delivery, and storage. This paper intends to provide efficient short-term forecasting of solar power production using Variational AutoEncoder (VAE) model. Adopting the VAE-driven deep learning model is expected to improve forecasting accuracy because of its suitable performance in time-series modeling and flexible nonlinear approximation. Both single- and multi-step-ahead forecasts are investigated in this work. Data from two grid-connected plants (a 243 kW parking lot canopy array in the US and a 9 MW PV system in Algeria) are employed to show the investigated deep learning models' performance. Specifically, the forecasting outputs of the proposed VAE-based forecasting method have been compared with seven deep learning methods, namely recurrent neural network, Long short-term memory (LSTM), Bidirectional LSTM, Convolutional LSTM network, Gated recurrent units, stacked autoencoder, and restricted Boltzmann machine, and two commonly used machine learning methods, namely logistic regression and support vector regression. The results of this investigation demonstrate the satisfying performance of deep learning techniques to forecast solar power and point out that the VAE consistently performed better than the other methods. Also, results confirmed the superior performance of deep learning models compared to the two considered baseline machine learning models.

Keywords: photovoltaic power forecasting; data-driven; deep learning; variational autoencoders; RNN

1. Introduction

The accurate modeling and forecasting of solar power output in photovoltaic (PV) systems are certainly essential to improve their management and enable their integration in smart grids [1,2]. Namely, the output power of a PV system is highly correlated with the solar irradiation and the weather conditions that explain the intermittent nature of PV system power generation. Particularly, the characteristic of fluctuation and intermittent of the temperature and solar irradiance could impact solar power production [3]. In practice, a decrease of larger than 20% of power output can be recorded in PV plants [4]. Hence, the connected PV systems to the public power grid can impact the stability and the expected operation of the power plant [5]. Given reliable real-time solar power forecasting, the integration of PV systems into the power grid can be assured. Also, power forecasting becomes an indispensable component of smart grids to efficiently manage power grid generation, storage, delivery, and energy market [6,7].

Long-and short-term forecasting methods are valuable tools for efficient power grid operations [8,9]. The success of integrating PV systems in smart grids depends largely on the accuracy of the implemented forecasting methods. Numerous models have been developed to enhance the accuracy of solar power forecasting, including autoregressive integrated moving average (ARIMA), and Holt-Winters methods. In Reference [10], a short term PV power forecasting based on the Holt-Winters algorithm (also called triple exponential smoothing method) has been introduced. This model is simple to construct and convenient to use. In Reference [11], different time series models including Moving average models, exponential smoothing, double exponential smoothing (DES), and triple exponential smoothing (TES) have been applied for short-term solar power forecasting. In Reference [12], a coupled strategy integrating discrete wavelet transform (DWT), random vector functional link neural network hybrid model (RVFL), and SARIMA has been proposed to a short-term forecast of solar PV power. This study showed that the use of the DWT negatively affects the accuracy of solar PV power forecasting under a clear sky. While the quality of the forecast model is improved when using DWT in cloudy and rainy sky weather. In addition, the coupled model showed superior forecasting performance in comparison to individuals models (i.e., SARIMA or RVFL). However, switching between two forecast models is not an easy task, particularly for real-time forecasting. In Reference [13], a hybrid model merging seasonal decomposition and least-square support vector regression was developed for forecasting monthly solar power output. Improved results have been obtained with this hybrid model compared to those obtained with ARIMA, SARIMA, and generalized regression neural network.

In recent years, shallow machine learning (ML) as non-parametric models, which are more flexible, have been widely exploited in improving solar PV forecasting. These models possess desirable characteristics and can model the complicated relationship between process variables and do not need an explicit model formulation to be specified, as is generally required. In Reference [14], a hybrid approach combining support vector regression (SVR) and improved adaptive genetic algorithm (IAGA) is developed for an hourly electricity demand forecasting. It has been shown that this hybrid approach outperformed the traditional feed-forward neural networks, the extreme learning machine (ELM) model, and the SVR model. In Reference [15], an approach for forecasting PV and wind-generated power using the higher-order multivariate Markov Chain. This approach considers the time-adaptive stochastic correlation between the wind and PV output power to achieve the 15-min ahead forecasting. The observation interval of the last measured samples are included to follow the pattern of PV/wind power fluctuations. In Reference [16], a univariate method is developed for multiple steps ahead of solar power forecasting by integrating a data re-sampling approach with machine learning procedures. Specifically, machine learning algorithms including Neural Networks (NNs), Support Vector Regression (SVR), Random Forest (RF), and Multiple Linear Regression (MLR) are applied to re-sampled time-series for computing multiple steps ahead predictions. However, this approach is designed only for univariate time series data. In Reference [17], a forecasting strategy combining the gradient boosting trees algorithm with feature engineering techniques is proposed to uncover information from a grid of numerical weather predictions (NWP) using both solar and wind data. Results indicate that appropriate features extraction from the raw NWP could improve the forecasting. In Reference [18], a modified ensemble approach based on an adaptive residual compensation (ARC) algorithm is introduced for solar power forecasting. In Reference [19], an analog method for day-ahead regional photovoltaic power forecasting is introduced based on meteorological data, and solar time and earth declination angle. This method exhibited better day-ahead regional power forecasting compared to the persistence model, System advisor model, and SVM model.

Over the last few years, deep learning has emerged as a promising research area both in academia and industry [20–24]. The deep learning technology has realized advancement in different areas, such as computer vision [25], natural language processing [26], speech recognition [27], renewable energy forecasting [4,28], anomaly detection [29–31], and reinforcement learning [32]. Owing to its data-driven approaches, deep learning has brought a paradigm shift in the way relevant information in time series data are extracted and analyzed. By concatenating multiple layers into the neural network structures, deep learning-driven methods enable flexible and efficient modeling of implicit interactions between process variables and automatic extraction of relevant information from a voluminous dataset with limited human instruction. Various deep techniques have been employed in the literature for improving solar power forecasting. For instance, in Reference [33], Recurrent Neural Networks (RNNs) is adopted for PV power forecasting. However, simple RNN is not suited to learn long-term evolution due to the vanishing gradient and exploding gradient. To bypass this limitation, several variants of RNN have been developed including Long Short-Term Memory Networks (LSTM) and gated recurrent unit (GRU) networks. Essentially, compared to a simple RNN model, LSTM and GRU models possess the superior capacity in modeling time-dependent data within a longer time span. In Reference [4], the LSTM model, which is a powerful tool in modeling time-dependent data, is applied to forecast solar power time series data. In Reference [34], a GRU network, which is an extended version of the LSTM model, has been applied to forecast short-term PV power. In Reference [35], at first, an LSTM recurrent neural network (LSTM-RNN) is applied for independent day-ahead PV power forecasting. Then, the forecasting results have been refined using a modification approach that takes into consideration the correlation of diverse PV power patterns. Results showed that the forecasting quality is improved by considering time correlation modification. In Reference [36], by using the LSTM model, a forecasting framework is introduced for residential load forecasting to address volatility problems, such as variability of resident's activities and individual residential loads. Results show that the forecasting accuracy could be enhanced by incorporating appliance measurements in the training data. In Reference [37], a hybrid forecasting approach is introduced by combining a convolutional neural network (CNN) and a salp swarm algorithm (SSA) for PV power output forecasting. After classifying the PV power data and associated weather information in five weather classes: rainy, heavy cloudy, cloudy, light cloudy, and sunny, the CNN is applied to predict the next day's weather type. To this end, five CNN models are constructed and SSA is applied to optimize each model. However, using several CNN models makes this hybrid approach not suitable for real-time forecasting. In Reference [38], a method combining deep convolutional neural network and wavelet transform technique is proposed for deterministic PV power forecasting. Then, the PV power uncertainty is quantified using quantile regression. Results demonstrated the deterministic model possesses reasonable forecasting stability and robustness. Of course, deep learning models possess the capacity to efficiently learn nonlinear features and pertinent information in time-series data that should be exploited in a wide range of applications.

This study offers a threefold contribution. Firstly, to the best of our knowledge, this the first study introducing a variational autoencoder (VAE) and Restricted Boltzmann Machine (RBM) methods to forecast PV power. Secondly, this study provides a comparison of forecasting outputs of eight deep learning models, including simple RNN, LSTM, ConvLSTM, Bidirectional LSTM (BiLSTM), GRUs, stacked autoencoders, VAE, and RBM, which takes into account temporal dependencies inherently and nonlinear characteristics. The eight deep learning methods and two commonly used machine learning methods, namely logistic regression (LR) and support vector regression (SVR), were applied to forecast PV power time-series data. Finally, for the guidance of short- and long-term operational strategies for PV systems, both single- and multi-step-ahead forecasting are examined and compared in this paper. Data sets from two grid-connected plants are adopted to assess the outputs of the deep learning-driven forecasting methods. Section 2 introduces the eight used deep learning methods. Section 3 describes the deep learning-based PV power forecasting strategy. Section 4 assesses the forecasting methods and compares their performance using two actual datasets. Finally, Section 5 concludes this study and sheds light on potential future research lines.

2. Methodologies

Deep learning techniques, which possess good capabilities in automatically learning pertinent features embed in data, are examined in this study to forecast PV power output. Table 1, summarizes the pros and cons of the seven considered benchmark deep learning architectures: RNN [39], LSTM [40], GRU [41], Bi-LSTM [42], ConvLSTM [43], SAE [44], RBM [45,46], and VAE [20,47].

Table 1. The considered benchmark deep learning methods.

Model	Description	Key Points
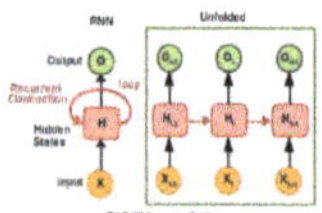 RNN model	• RNNs are able to include historical information in the forecasting process via their recurrent structure and memory units • Simple RNN do not have gates [39]. • The RNNs are entirely trained in a supervised way.	+ Modeling time dependencies - Simple RNNs fail to catch the long-term evolution due to the vanishing gradient and exploding gradient [33].
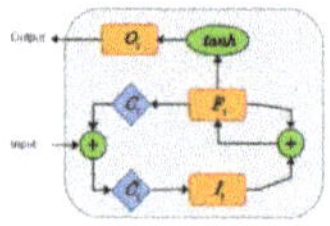 LSTM model	• LSTM consists of three gates regulating the information flow called input, forge, and output gates [40]. • Gate mechanism is used to store and memorize historical data features. • GRU use two gates, while LSTM is based on three gates.	+ LSTM showed good performance for learning long-term dependencies more easily than the conventional RNN - Its training is relatively longer than that of other RNN algorithms - The architecture of typical LSTM is very complex
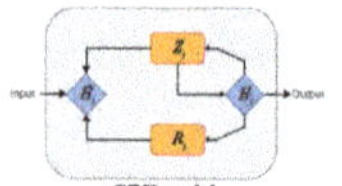 GRU model	• The major demarcation of GRU from LSTM is that only one unit is used to control both the forgetting factor and the decision to update the state unit [41]. • GRU contains only two gates, the update, and the reset gates. • The GRU has been widely used in time-series, data sequence (e.g., speech and text processing), temporal features extraction, prediction, and forecasting.	+ The attractive features of the GRU model are the shorter training time compared to the LSTM and the fewer parameters that the GRU model possesses compared to the LSTM [41]. - GRU models have problems such as slow convergence rate and low learning efficiency, resulting in too long training time and even under-fitting.
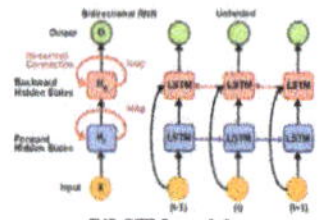 BiLSTM model	• Compared to the LSTM model that passes the input data through the network in one direction from past to future (forward), the BiLSTM processes the input also in the backward direction from the future to the past [42]. • This architecture improves the learning of complex temporal dependencies through double processing.	+ Modeling time dependencies + Improved accuracy in state reconstruction is achieved by BiLSTM that merges the desirable features of both bidirectional RNN and LSTM [42]. - Complex architecture
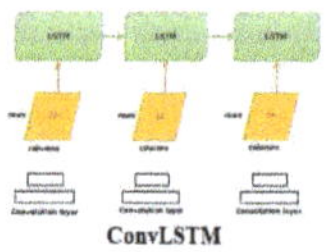 ConvLSTM	• The ConvLSTM is a special variant of the traditional LSTM, in which the fully-connected layer operators are replaced with convolutional operators [43]. • LSTM with recurrent connection to deal with data sequences. • The convolutional layer can deal with 2D inputs like a sequence of images.	+ The ConvLSTM can process 2D input through convolutional transformations to learn the spatial features and then feed the LSTM module. + It has been used in modeling time dependencies, feature extraction, and spatiotemporal modeling - Complex architecture

Table 1. *Cont.*

Model	Description	Key Points
Stacked Autoencoder (SAE)	• Autoencoders are neural networks that aim to create a compact representation of a given input x like images or any type of data [44]. • The network learns how to compress the input features by keeping the most important information by minimizing the reconstruction error between the compressed input and the original input x [44]. • Autoencoders are usually stacked to build a deep-stacked autoencoder.	+ Powerful compression capabilities + The SAEs are trained in an unsupervised way + They are applied for features extraction, data generation, dimensionality reduction, classification, prediction, and forecasting. - Suffers from the error vanishing and the overfitting
Restricted Boltzmann machines	• RBMs are stochastic and generative neural networks [45] consisting of visible units and hidden units. There are no connections between visible-to-visible and hidden-to-hidden; however, visible and hidden units are fully connected. • Usually, RBMs are trained based on the contrastive divergence learning method. • Contrastive divergence uses Gibbs sampling to compute the intractable negative phase.	+ Simple architecture with two layers + Generative model, + Strong data distribution approximation. + Can be stacked to build a deep learning model like DBN or DBM. - Slow training due to Contrastive Divergence approach.

Variational Autoencoders Model

VAEs are an essential class of generative-based techniques that are efficient to automatically extract information from data in an unsupervised manner [20,47]. One desirable characteristic of VAEs is their ability for reducing the input dimensionality enabling them to compress large dimensional data into a compressed representation. Moreover, they are very effective for approximating complex data distributions using stochastic gradient descent [47]. There are two major advantages of VAEs compared the conventional autoencoders, one is they are efficient to solve the overfitting problem in the conventional autoencoders by using a regulation mechanism in the training phase, and the second advantage is that they have proved effective when handling various kinds of complex data in different applications, including handwritten digits, and urban networks modeling [48]. Here, VAE is adopted for solar PV production forecasting. Figure 1 shows a schematic diagram of the construction of a VAE.

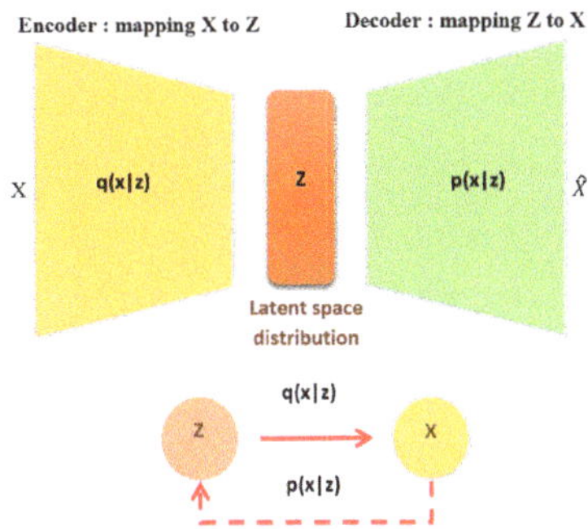

Figure 1. Basic schematic illustration of a variational autoencoder (VAE).

Basically, the VAE, as a variant of autoencoders, contains two neural networks an encoder and a decoder, where the encoder mission is to encode a given observed set, $\mathbf{X}$ into a latent space Z as distribution, $q(\mathbf{z}|\mathbf{x})$. The latent (termed hidden) space dimension is decreased in comparison to the dimension of the observed set. Indeed, the encoder is built to compress the observed set toward this reduced dimensional space efficiently. Then, a sample is generated via, $z \sim q(\mathbf{z}|\mathbf{x})$, using the learned probability distribution. On the other hand, the key purpose of the decoder, $\mathbf{p}(\mathbf{x}|\mathbf{z})$, consists in generating the observation x based on the input z. It should be emphasized that the reconstruction of data using the decoder results in some deviation of reconstruction, which is calculated and backpropagated through the network. This error is minimized in the training phase of the VAE model by the minimization of the deviation between the observed set and the encoded-decoded set.

To summarize, the VAE encoder is gotten via an approximate posterior $\mathbf{q}_\theta(\mathbf{z}|\mathbf{x})$, and the decoder is obtained by a likelihood $\mathbf{p}_\phi(\mathbf{x}|\mathbf{z})$, where θ and ϕ refers respectively to the parameters of encoder and decoder. Here a neural network is constructed for learning θ and ϕ. Essentially, the VAE encoder's role is learning latent variable $\mathbf{z}$ based on gathered sensor data, and the decoder employs the learned latent variable $\mathbf{z}$ for recovering the input data. The deviation between the reconstructed data and the input data should be close to zero as possible. Notably, the learned latent variable $\mathbf{z}$ from the encoder is used for feature extraction based on the input data. Usually, the dimension of the output of the encoder is smaller than that of the original data, which leads to the dimensionality reduction of input data. Note that the encoder is trained by training the entire VAE comprising encoder and decoder.

It is worth pointing out that the loss function has an essential effect on feature extraction for training VAE. Assume that $\mathbf{X}_t = [x_{11}, x_{2t}, \ldots, x_{Nt}]$ is the input data points of VAE at time point t, and $\mathbf{X}'_t$ is the reconstructed data using the VAE model. Furthermore, it is assumed maximizing the marginal likelihood learning of parameters, expressed as [49]:

$$logp_\phi(\mathbf{x}') = D_{KL}\big[q_\theta(\mathbf{z}|\mathbf{x})\|p_\phi(\mathbf{x})\big] + \mathcal{L}(\theta, \phi; \mathbf{x}), \tag{1}$$

where $D_{KL}[.]$ denotes the Kulback-Leibler divergence, and $\mathcal{L}$ refers to the likelihood of the parameters of encoder and decoder (i.e., θ and ϕ). Hence, the loss can be expressed as

$$\mathcal{L}(\theta, \phi) = \underbrace{\mathbb{E}_{\mathbf{z} \sim q_\theta(\mathbf{z}|\mathbf{x})}\big(log\mathrm{p}(\mathbf{x}'|\mathbf{z})\big)}_{\text{Reconstruction term}} - \underbrace{D_{KL}\big(q_\theta(\mathbf{z}|\mathbf{x})\|\mathrm{p}_\phi(\mathbf{z})\big)}_{\text{Regularization term}}. \tag{2}$$

The VAE's loss function is composed of two parts: the reconstruction loss and a regularizer. Reconstruction loss tries to get an efficient encoding-decoding procedure. In contrast, a regularizer part permits the regularization of the latent space construction to approximate the distributions out of the encoder as near as feasible to a prefixed distribution (e.g., Normal distribution). Figure 2 schematically summarizes the procedure for computing the loss function.

The term (2) permits reinforcing the decoder capacity to learn data reconstruction. Higher values of the reconstruction loss mean that the performed reconstruction is not suitable , while lower values mean that the model is converging. The regularizer is reported using the Kulback-Leibler (KL) divergence separating the distribution of the encoder function ($q_\theta(\mathbf{z}|\mathbf{x})$) and of the latent variable prior $(\mathbf{z}, |\mathrm{p}_\phi(\mathbf{z}))$. Indeed, KL is employed to compute the distance that separates two given probability distributions. The gradient descent method is used to minimize the loss function with respect to the encoder's parameters and decoder in the training phase. Overall, we minimize the loss function to ensure getting a regular latent space,z, and adequate sampling of new observation using $\mathbf{z} \sim \mathrm{p}_\phi(\mathbf{z})$ [50].

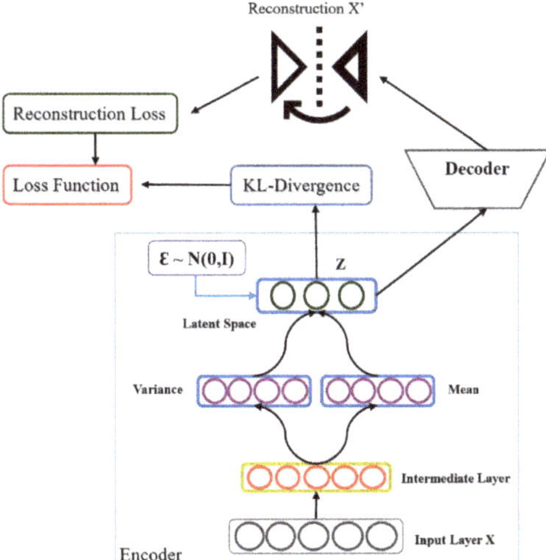

Figure 2. Reconstruction loss and Kulback-Leibler (KL) divergence to train VAE.

Let assume that $p_\phi(\mathbf{z}) = \mathcal{N}(z; 0, I)$, we can write $q_\theta(\mathbf{z}|\mathbf{x})$ in the following form:

$$log q_\theta(\mathbf{z}|\mathbf{x}) = log \mathcal{N}(\mathbf{z}; \mu, \sigma^2 I). \tag{3}$$

The mean and standard deviation of the approximate posterior are denoted by (μ, σ), respectively. Note here that a layer is dedicated to both of them. Moreover, the latent space z is constructed using a deterministic function g parameterized by ϕ and an auxiliary noise variable $\varepsilon \sim p(\varepsilon)$ or more specifically $\varepsilon \sim \mathcal{N}(0, I)$.

$$z = g_\phi(x, \varepsilon) = \mu + \sigma \odot \varepsilon. \tag{4}$$

The reconstruction error term can be expressed in the following form:

$$\mathcal{L}(\theta, \phi, \mathbf{x}) = \frac{1}{2} \sum_i \left(1 + log((\sigma_i)^2) - (\mu_i)^2 - (\sigma_i)^2\right) + \frac{1}{L} \sum_{l=1}^{L} log(p_\phi(\mathbf{x}|\mathbf{z}^{(l)})), \tag{5}$$

where the $\odot$ denotes the element-wise product.

Overall, the encoder and decoder's parameters are obtained by minimizing the loss function, $\mathcal{L}(\theta, \phi)$, using the training observations. The VAE is trained using the procedure tabulated Algorithm 1.

Algorithm 1: VAE training algorithm.

Input: : Training dataset $X = \{x^1, \ldots, x^k\}$
Output: : $\{\theta, \phi\}$
θ : Encoder parameters;
ϕ : Decoder parameters;
M : number of mini-batch (drawn from full dataset) ;
$\{\theta, \phi\} \longleftarrow$ Initialize model parameters randomly ;
repeat
$\quad$ $X_m \longleftarrow RandomMinibatch(X, M)$;
$\quad$ Draw L samples from $\epsilon \sim \mathcal{N}(0,1)$;
$\quad$ $z = g_\phi(X_m, \varepsilon)$;
$\quad$ $\mathcal{G} = \sum_j KL \left(q_j \left(z|x^{(j)} \right) || p\left(z\right) \right) + \frac{1}{L}\Sigma_{l=1}^{L} log(p(x^{(l)}|z^{(i,l)}))$;
$\quad$ $\{\theta, \phi\} \longleftarrow OptimizerUpdate(\mathcal{G}, \theta, \phi)$;
until *parameters convergence* : $\{\theta, \phi\}$;

---

3. Deep Learning-Based PV Power Forecasting

The input data consists of PV power output that variates between 0 and the rated output power. Thus, when handling some large-value data with the RNN model, a gradient explosion can be occurred and negatively affects the performance of the RNN. Furthermore, the learning effectiveness of RNN will be reduced. To remedy this issue, the input data is normalized via min-max normalization within the interval $[0, 1]$, and then used for constructing the deep learning models. The normalization of the original measurements, **y** is defined as:

$$\widetilde{y} = \frac{(y - y_{min})}{(y_{max} - y_{min})}, \tag{6}$$

where y_{min} and y_{max} refer to the minimum and maximum of the raw PV power data, respectively. After getting forecasting outputs, we applied a reverse operation to ensure that the forecasted data match to the original PV power time-series data.

$$y = \widetilde{y} * (y_{max} - y_{min}) + y_{min}. \tag{7}$$

As discussed above, the generated PV power shows a high level of variability and volatility because of its high correlation with the weather conditions. Hence, for mitigating the influence of uncertainty on the accuracy of the PV power forecasting this work presents a deep-learning framework to forecast PV power output time-series. Essentially, deep learning models are an efficient tool to learn relevant features and process nonlinearity from complex datasets. In this study, a set of eight deep learning models have been investigated and compared for one-step and multiple steps ahead forecasting of solar PV power. The overall structure of the proposed forecasting procedures is depicted in Figure 3. As shown in Figure 3, solar PV power forecasting is accomplished in two phases: training and testing. The original PV power data is split into a training sub-data and a testing sub-data. At first, the raw data is normalized to build deep learning models. Adam optimizer is used to select the values of parameters of each model by minimizing the loss function based on training data. Once the models are constructed, they are exploited for PV power output forecasting. The quality of models are quantified using several statistical indexes including the Coefficient of determination (R^2), explained variance (EV), mean absolute error (MAE), Root Mean Square Error (RMSE), and normalized RMSE (NRMSE).

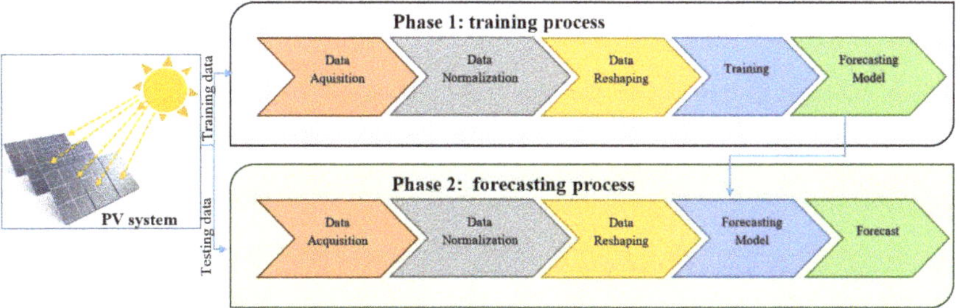

Figure 3. Schematic presentation of deep learning-based photovoltaic (PV) power forecasting.

Essentially, the deep learning-driven forecasting methods learn the temporal correlation hidden on the PV power output data and expected to uncover and captures the sequential features in the PV power time series. The main objective of this study is to investigate the capability of learning models namely RNN, LSTM, BiLSTM, ConvLSTM, GRU, RBM, SAE, and VAE for one-step and multiple-steps ahead solar PV power forecasting.

3.1. Training Procedure

The eight models investigated in this study can be categorized into two classes: autoencoders and recurrent neural networks. The autoencoders represented include RBM, VAE, and SAEs while the RNN-based models contain RNN, LSTM, GRU, BiLSTM, and ConvLSTM. The dataset used for training and testing are normalized first, and more data preprocessing is needed for the autoencoder models. For instance, data reshaping is needed to transform the univariate PV power time-series data to a two-dimension matrix to be used as input for the autoencoders including the SAE, VAE, and RBM. The main difference between the two classes in the training phase is the learning way, the RNNs are entirely supervised trained while the auto-encoders are first pre-trained in an unsupervised manner and then the training is completed based on supervised learning. Specifically, RNNs models are trained in a supervised way by using a subset of training as input sequence ($\mathcal{X} = x_1, \ldots, x_k$) and an output variable $\mathcal{Y} = x_{k+1}$. The sequence length l, called the lag, is a crucial parameter used in the data preparation phase. The mapping sequence to the next value is constructed using a window sliding algorithm. The value of l is determined using the Grid Search approach [51]. Here, the value of l is chosen 6, which is the lowest value that maximizes the overall performance of the proposed approach.

RNN—based models are trained to learn the mapping function from the input to the output. After that, these trained models are used to forecast new data that complete the sequence. On the other hand, the greedy layer-wise unsupervised plus fine-tuning were applied to RBM, VAE, and SAES. It should be noted that PV power output forecasting based on autoencoder is accomplished as a dimensionality reduction. That is these models do not have the possibility to discover time dependencies or model time series data. Hinton [44] shows that a greedy layerwise unsupervised learning for each layer followed by a fine-tuning improves the features extraction and learning process of the neural networks dedicated to prediction problems or for dimensionality reduction like autoencoders. The VAE-driven forecasting procedure including the pretreatment step is illustrated in Figure 4.

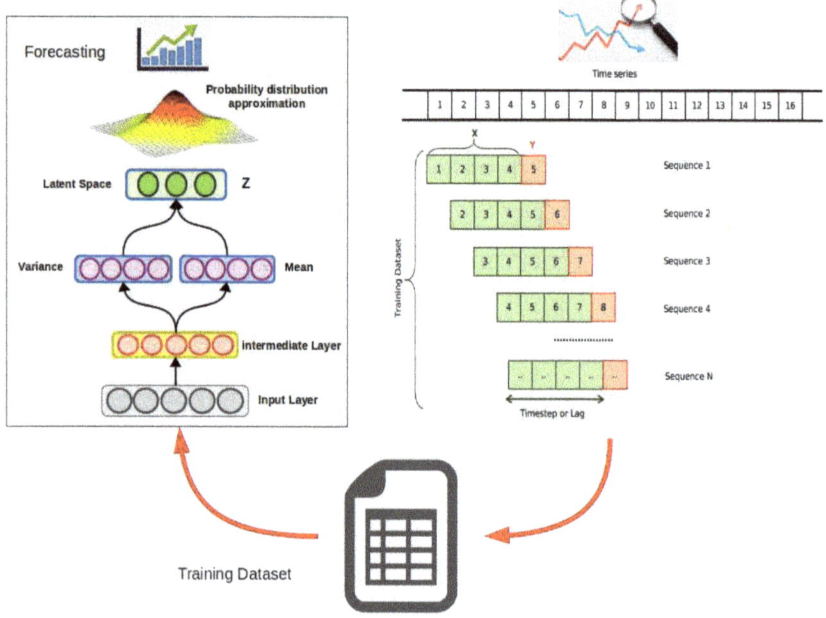

Figure 4. VAE-driven procedure.

3.2. Measurements of Effectiveness

The deep learning-driven forecasting methods will be evaluated using the following metrics: R^2, RMSE, MAE, EV, and NRMSE.

$$R^2 = \frac{\sum_{i=1}^{n}[(y_{i,} - \bar{y}) \cdot (\hat{y}_i - \bar{\hat{y}})]^2}{\sqrt{\sum_{i=1}^{n}(y_i - \bar{y})^2} \cdot \sqrt{\sum_{i=1}^{n}(\hat{y}_i - \bar{\hat{y}})^2}}, \tag{8}$$

$$RMSE = \sqrt{\frac{1}{n}\sum_{t=1}^{n}(y_t - \hat{y}_t)^2}, \tag{9}$$

$$MAE = \frac{\sum_{t=1}^{n}|y_t - \hat{y}_t|}{n}, \tag{10}$$

$$EV = 1 - \frac{\mathrm{Var}(\hat{\mathbf{y}} - \mathbf{y})}{\mathrm{Var}(\mathbf{y})}, \tag{11}$$

$$NRMSE = \left(1 - \sqrt{\frac{\sum_{i=1}^{N}(y_i - \hat{y})^2}{\sum_{i=1}^{N}(y_i - \bar{y})^2}}\right).100\%, \tag{12}$$

where y_t are the actual values, $\hat{y}_t$ are the corresponding estimated values, $\bar{y}$ is the mean of measured power data points, and n is the number of measurements. Instead of using RMSE that relies on the range of the measured values, the benefit of using NRMSE as the statistical indicator is that it does not rely on the range of the measured values. NRMSE metric indicates how well the forecasted model response matches the measurement data. A value of 100% for NRMSE denotes perfect forecasting and lower values characterize the poor forecasting performance. Lower RMSE and MAE values and EV and R2 closer to 1 are an indicator of accurate forecasting.

4. Results and Discussion

4.1. Data Description

In this study, solar PV power data from two PV systems are adopted to verify the performance of the eight deep learning-driven forecasting methods.

- **Data Set 1:** The first historical solar-PV power dataset used are collected from a parking lot canopy array monitored by the National Institute of Standard and Technology (NIST) [52]. The PV system contains eight canopies tilted 5 degrees down from horizontal, four canopies tilt to the west, and the other four canopies tilt to the east. The modules are installed with their longer dimension running east-west. Each shed contains 129 modules laid out in a 3 (E − W) × 43 (N − S) grid. This power system has a rated DC power output of 243 kW. The first dataset is collected from January 2015 to December 2017 with a one-minute temporal resolution. The distribution of the Parking Lot Canopy Array dataset collected from January 2015 to December 2015 are shown in Figure 5a.
- **Data Set 2:** The second solar-PV power dataset is collected from a grid-connected plant in Algeria with a peak power of 9 MWp from January 2018 to December 2018 with 15 min temporal resolution. This PV plant consists of nine identical mini-PV plants of one mega each. Indeed, a set of 93 PV array provides one MWp of DC power, two central inverters with 500 kVA each, allow to connect the 93 PV array to one transformer of 1250 kVA. The hourly distribution of the first dataset are shown in Figure 5b.

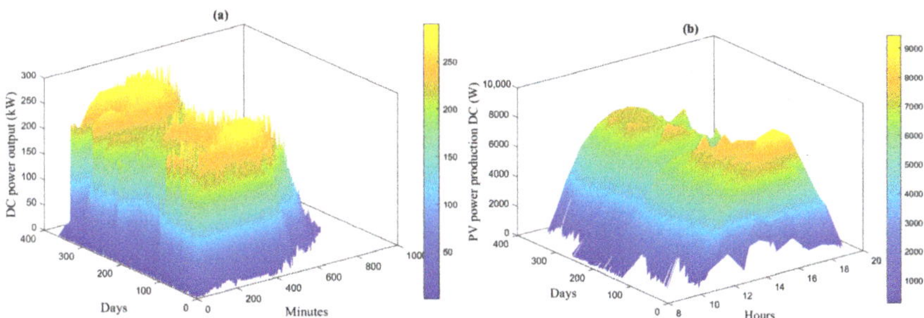

Figure 5. (**a**) distribution of solar PV power output from Parking Lot Canopy Array dataset. (**b**) Hourly distribution of solar PV power output from January to December 2018.

Figure 6 depicts the boxplots of DC power output (Data Set 1 and Data Set 2) in Figure 5 to show the distribution of DC power in the daytime. The maximum power is generated around mid-day.

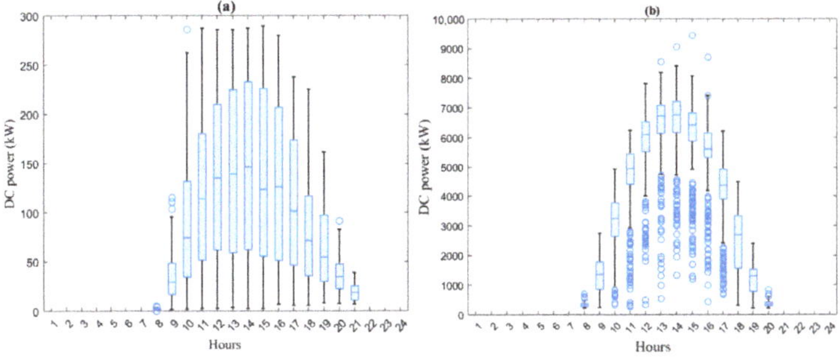

Figure 6. Boxplots of PV power output during daytime hours: (**a**) Data Set 1 and (**b**) Data Set 2.

4.2. Forecasting Results

Accurate short-term forecasting of PV power output gives pertinent information for maintaining the desired power grid production delivery and storage [7,53]. This section assesses the eight models (i.e., RNN, GRU, LSTM, BiLSTM, ConvLSTM, RBM, AE, and VAE) and compares their forecasting performance using PV power output collected from two different PV systems. Towards this ends, we first build each to capture the maximum variance in training data and then use them to forecast the future trend of PV power output. The training data in Data Set 1 consists of one-minute power data collected from 1 January 2017 to 29 June 2017. The training data in Data Set 2 is collected from 1 January 2018 to 19 October 2018. The hyper-parameters of the built deep learning methods based on training datasets are tabulated in Table 2. For all models, we used the cross-entropy as loss function and Rmsprop as an optimizer in training.

Table 2. Tuned parameters in the considered methods.

Methods	Parameter	Value
RBM	learning rate	0.0005
	Gibbs sampling (k)	5
	Training epochs	500
	Layers	01
SEAS	Learning rate	0.0005
	Training epochs	500
	Layers	04
VAE	Learning rate	0.0005
	Training epochs	500
	Layers	05
RNN	Learning rate	0.0005
	Training epochs	500
GRU	Learning rate	0.0005
	Training epochs	200
LSTM	Learning rate	0.0005
	Training epochs	200
BiLSTM	Learning rate	0.0005
	Training epochs	200
ConvLSTM	Learning rate	0.0005
	Training epochs	200

4.2.1. Forecasting Results Based on Data Set 1: Parking Lot Canopy Array Datasets

The principal feature of the PV power output is its intermittency. This unpredicted fluctuation in solar PV power could lead to many challenges including power generation control and storage management. Essentially, it is crucial to appropriately forecast PV power output to guarantee reliable operation and economic integration in the power grid. In the first case study, the above-trained models will be evaluated using the testing solar PV power output starting from 30 June to 6 July 2017 collected from Parking lot canopy array. Forecasting outputs using the eight deep learning models using test measurements are displayed in Figure 7. These results illustrate the goodness of deep learning models for PV power forecasting.

Also, to show clearly the accordance of the measured and the forecast outputs from the investigated deep learning models, the scatter plots are presented in Figure 8. Figure 8 shows that the forecasted data from RBM and SAE models are moderately correlated with the actual PV power output. The forecasting with ConvLSTM is relatively weakly correlated with the measured power data. On the other hand, the forecasted power with RNN-based models and the VAE model are strongly correlated with the measured PV power.

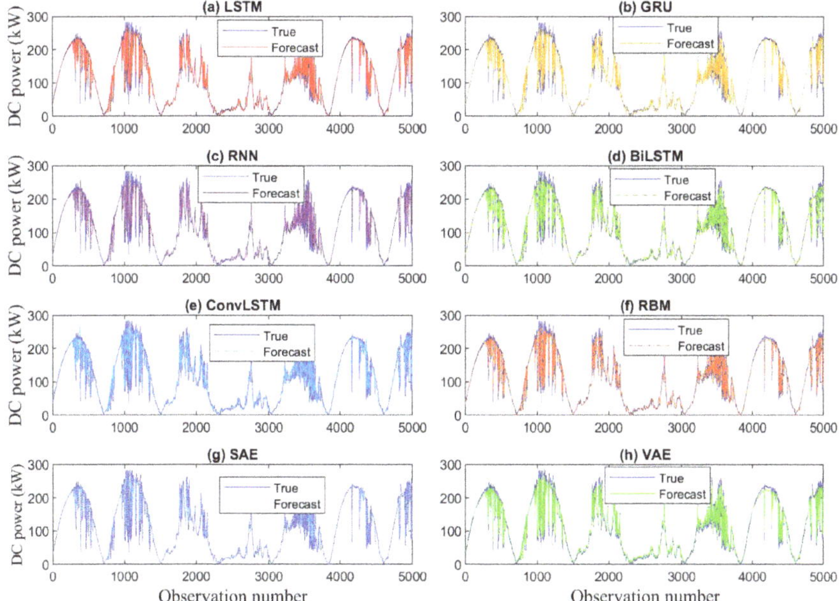

Figure 7. Forecasting results from the eight models using the testing datasets: (**a**) Long Short-Term Memory Networks (LSTM), (**b**) gated recurrent unit (GRU), (**c**) recurrent neural network (RNN), (**d**) Bidirectional LSTM (BiLSTM), (**e**) Convolutional LSTM (ConvLSTM), (**f**) Restricted Boltzmann Machine (RBM), (**g**) stacked autoencoder (SAE) and (**h**) VAE.

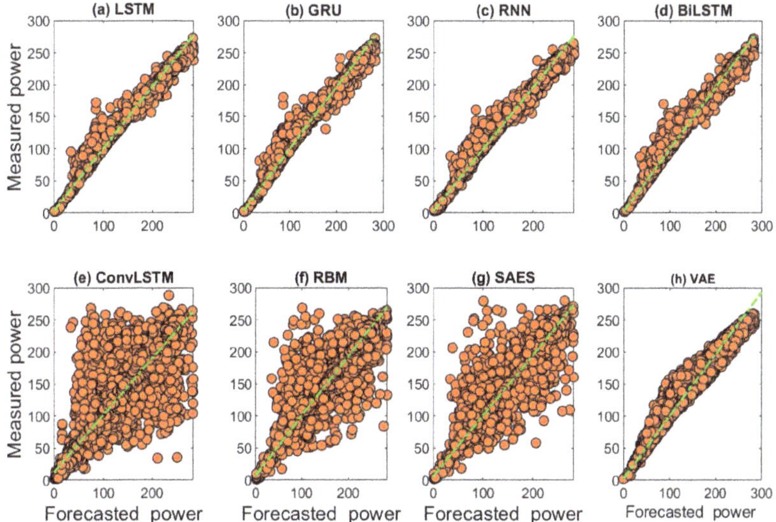

Figure 8. Scatter graphs of PV power forecasting and measurements using the eight models: (**a**) LSTM, (**b**) GRU, (**c**) RNN, (**d**) BiLSTM, (**e**) ConvLSTM, (**f**) RBM, (**g**) SAE, and (**h**) VAE.

Now, to quantitatively evaluate the forecasting accuracy of the eight considered models based on the testing data, five statistical indexes are computed and listed in Table 3. Also, we compared the eight the forecasting results of the ten deep learning models with two baseline machine learning models:

LR and SVR (Table 3). For this application, ConvLSTM performs poorly in terms of the forecasting accuracy compared to the other models and cannot track well the variability of PV power and does not describe the most variance in the data (i.e., EV = 0.832). Moderate forecasting performance are obtained using RBM and SAE by explaining respectively 0.929 and 0.932 of the total variance in the testing PV power data. The results of this investigation exhibit also that the VAE model provides accurate forecast in comparison to the other models by achieving PV power forecast with lower RMSE, MAE, and higher NRMSE (%) as well as the highest R2, EV values closer to 1 that means that most of the variance in the data is captured by the VAE model. Specifically, the VAE model achieved the highest R2 of 0.992 and the lower RMSE (6.891) and MAE (5.595). We highlight that this is the first time that the VAE model is used for solar PV power output forecasting. This application showed that the VAE method for PV power forecasting has superior performance. Also, it is noticed that RNN and its extended variants LSTM, BiLSTM, and GRU achieve slightly comparable performance to the VAE in terms of the statistical indexes (RMSE, RMSE, MAE, EV, and NRMSE). Table 3 indicates that deep learning models exhibited improved forecasting performance compared to the baseline methods (i.e., LR and SVR).

Table 3. Forecasting performance of the eight models based on testing data of the first dataset.

Method	R2	RMSE	MAE	EV
LSTM	0.990	7.672	3.911	0.990
GRU	0.990	7.768	4.022	0.990
RNN	0.991	7.539	4.072	0.991
BiLSTM	0.990	7.722	3.910	0.990
ConvLSTM	0.832	31.842	16.338	0.832
RBM	0.929	20.672	8.910	0.929
SAE	0.932	20.301	8.300	0.932
SVM	0.971	14.174	12.889	0.972
LG	0.965	15.45	9.942	0.966
VAE	0.995	5.471	3.232	0.995

4.2.2. Forecasting Results Based on Data Set 2: Algerian PV Array Datasets

Now, the effectiveness of the eight methods will be tested based on power output data collected from the 9 MWp PV plant in Algeria (Data Set 2). In this experiment, the above-trained models will be evaluated using the testing solar PV power output collected from 20 October to 31 December 2018. The measured test set together with model forecasts are charted in Figure 9. Similar conclusions are also valid for these datasets. One major reason is that RNN-based models have a strong capability to describe time dependents data and can better model the complicated relationship between historical and future PV power output data than other methods. The RNN-based models and the VAE model again confirm the superior forecasting performance of PV power output as shown by the scatter plots in Figure 9. The ConvLSTM model shows poor forecasting performance results (Figure 9).

And then, the statistical indicators are computed to compare the forecasting performance between the eight models, and baseline machine learning models: LR and SVR based on testing datasets (Table 4). It is worth noting that the RNN-based models (i.e., RNN, LSTM, BiLSTM, ConvLSTM, and GRU) and the VAE model show the improved forecasting performance compared to the RBM, and SAE.

Results in Figure 9 and Table 4 indicate that using RNN-based models and VAE method has led to improved forecasting performance. Furthermore, the error analysis highlights that the forecasting accuracy obtained by these models can satisfy practical needs and can be useful for PV power management. It should be noted that the VAE model is trained in an unsupervised manner in order to forecast solar PV power. This means that the forecast is based only on the information from past data. However, the other models are trained in a supervised way by using a subset of training as input sequence $(x_1, \ldots, x_k)$ and an output variable x_k and we train RNN-based models to learn the mapping function from the input to the output. After that, these trained models are used to forecast new data. Even if the VAE model is trained in an unsupervised way, it can provide comparable forecasting performance to those obtained by the supervised RNN-based models. Accordingly, the VAE-based forecasting approach is a more flexible and powerful tool to be used in real-time PV power forecasting.

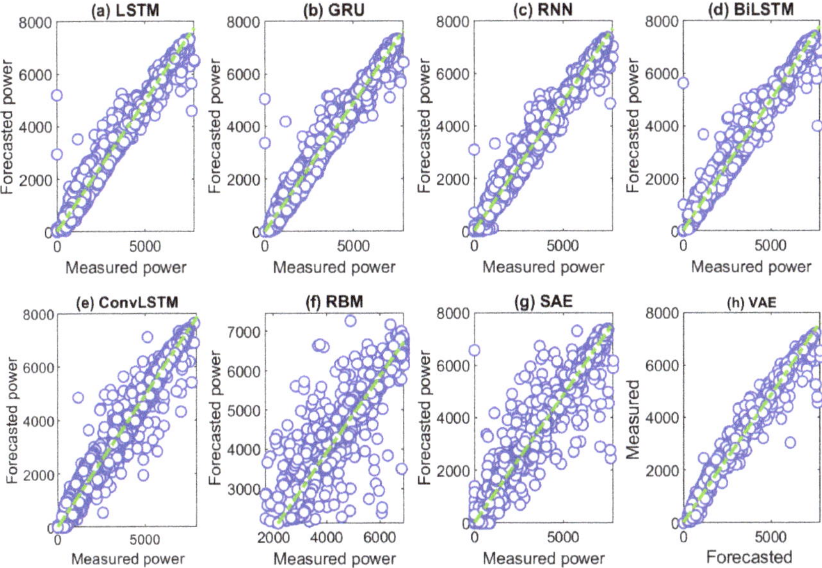

Figure 9. Scatter graphs of PV power forecasting and measurements using the eight models: (**a**) LSTM, (**b**) GRU, (**c**) RNN, (**d**) BiLSTM, (**e**) ConvLSTM, (**f**) RBM, (**g**) SAE, and (**h**) VAE.

Table 4. Forecasting performance of the eight methods using the test set of the second dataset.

Method	R2	RMSE	MAE	EV
LSTM	0.992	246.781	107.799	0.992
GRU	0.991	250.004	117.622	0.992
RNN	0.992	241.170	125.960	0.992
BiLSTM	0.991	259.884	117.579	0.991
ConvLSTM	0.692	1500.070	846.750	0.692
RBM	0.977	407.238	170.568	0.977
SAE	0.983	349.066	124.772	0.983
SVM	0.924	770.996	699.837	0.954
LG	0.899	886.665	837.721	0.926
VAE	0.995	199.645	99.838	0.995

Overall, the NRMSE (%) quantifies the quality of power forecasting between the measured and forecasted PV power output time-series data, where the larger value indicates a better prediction performance. A visual display of the NRMSE (%) derived with the eight considered deep learning methods based on the testing datasets from the two PV systems is displayed in Figure 10. The first dataset is with a one-minute resolution and the second dataset is with fifteen minutes resolution. The VAE model achieves better PV power flow forecast performance compared to the RBM and SAE models and RNN-based models. Furthermore, the results show that VAE models are efficient in capturing the linear and nonlinear features in PV power data with different time resolutions.

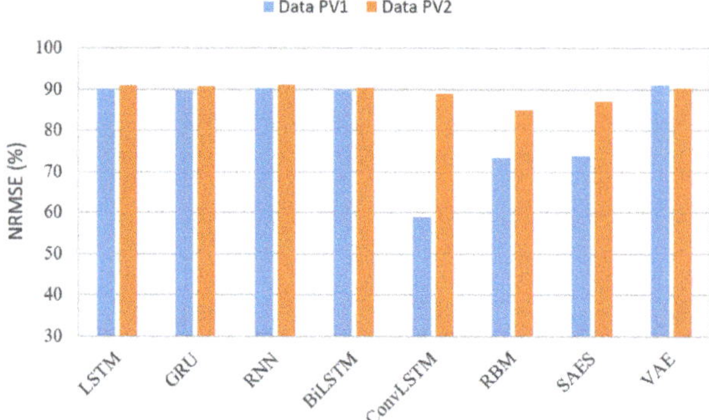

Figure 10. NRMSE by method based on the testing datasets from the two considered PV systems.

4.3. Multi-Step Ahead PV Power Forecasts

Precise multi-step forecasts are essential to managing the operation of PV systems appropriately. Now, we assess the capability of the eight methods for multi-step ahead forecasting of PV power output using data from Data Set 1 (a 243 kW parking lot canopy array in the US) and Data Set 2 (a 9 MW PV system in Algeria). Based on the past measurements, $x = [x_1, x_2, \ldots, x_l]$, the computed single-, two-, and multistep-ahead forecast are respectively x_{l+1}, x_{l+2}, and x_{l+n}. The 5, 10, 15 steps-ahead forecastings of PV power data based on the testing data of the Parking lot canopy array dataset and the Adrar PV system are tabulated in Table 5.

Table 5. Validation metrics for multistep-step-ahead forecasts.

Model	Steps	Minutes	Algerian RMSE	PV MAE	System R2	EV	Parking RMSE	Lot MAE	Canopy R2	Array EV
BiLSTM	2	30	1591.639	1303.28	0.675	0.876	18.309	10.011	0.951	0.953
CNN	2	30	1007.54	613.344	0.87	0.888	24.105	16.534	0.915	0.916
ConvLSTM2D	2	30	359.128	262.526	0.983	0.991	17.31	7.786	0.956	0.956
GRU	2	30	910.589	770.907	0.894	0.936	17.454	7.897	0.955	0.955
LSTM	2	30	376.359	285.324	0.982	0.982	17.47	7.914	0.955	0.955
RBM	2	30	390.581	213.174	0.98	0.98	17.564	7.847	0.955	0.955
RNN	2	30	1448.063	1170.279	0.731	0.868	18.101	8.474	0.952	0.953
SAE	2	30	477.986	255.739	0.971	0.974	17.724	8.783	0.954	0.955
VAE	2	30	303.091	117.701	0.988	0.988	17.31	7.566	0.956	0.956
BiLSTM	3	45	374.303	186.576	0.982	0.982	21.855	12.028	0.93	0.93
CNN	3	45	1166.323	749.863	0.826	0.84	26.135	18.617	0.9	0.9
ConvLSTM2D	3	45	370.848	260.922	0.982	0.986	21.315	10.397	0.933	0.934
GRU	3	45	395.163	225.154	0.98	0.98	21.78	11.596	0.93	0.932
LSTM	3	45	367.048	194.691	0.983	0.983	21.485	9.707	0.932	0.932
RBM	3	45	579.104	341.151	0.957	0.958	21.76	10.434	0.931	0.931
RNN	3	45	428.615	284.6	0.976	0.977	22.458	11.99	0.926	0.93
SAE	3	45	521.723	255.686	0.965	0.969	21.823	10.335	0.93	0.93
VAE	3	45	344.468	164.635	0.985	0.985	21.461	9.878	0.932	0.932
BiLSTM	4	60	432.448	229.34	0.976	0.977	23.687	13.131	0.918	0.92
CNN	4	60	949.999	567.358	0.884	0.897	27.022	17.502	0.893	0.9
ConvLSTM2D	4	60	462.93	360.618	0.972	0.978	23.174	12.113	0.921	0.921
GRU	4	60	538.857	381.686	0.963	0.965	23.252	11.681	0.921	0.921
LSTM	4	60	533.916	355.042	0.963	0.966	23.298	11.912	0.92	0.921
RBM	4	60	614.722	379.406	0.951	0.951	23.321	11.605	0.92	0.92
RNN	4	60	476.67	297.196	0.971	0.971	24.152	13.208	0.914	0.918
SAE	4	60	554.917	274.536	0.96	0.961	23.604	11.768	0.918	0.918
VAE	4	60	420.029	193.157	0.977	0.978	23.134	11.664	0.921	0.921

We can easily observe that, for all data sets, except BiLSTM and ConvLSTM, the other models performed consistently reasonable forecasting results five-, ten-, fifteen-step-ahead forecasting.

For instance, the VAE model achieved R2 values of 0.902,0.873, 0.856 for five-, ten-, fifteen-step-ahead forecasting when using the first for Data Set 1, R2 values of 0.951,0.877, 0.818 for Data Set 2. The RNN, GRU, RBM, SAE, and VAE models performed about equally in terms of R2, MAPE, and RMSE in all cases.

For Data Set 1, the five-step-ahead forecasting R2's for all models except ConvLSTM is around 0.90 (Table 5). Results in Table 5 show that for five-steps ahead forecasting based Data Set 2 almost all models provide relatively good forecasting accuracy in terms of $R2$ which is around 0.94. It is worthwhile noticing that for a ten-step -ahead forecast, the accuracy of all models starts to decrease and achieve R_2 values around 0.86. In the fifteen-step -ahead forecasting, we observed that LSTM, BiLSTM, and ConvLSTM achieved poor forecasting performance. The other models are still providing acceptable forecasting accuracy. We notice that the SAE model outperforms slightly the other models with higher R_2 and lowest forecasting errors. The overall forecasting performance of the RNN, GRU, RBM, SAE, and VAE model was satisfying, and they can maintain a reasonable forecasting performance to forecast solar PV power output as the number of steps increases. The error for the second dataset is large compared to the first one. The first dataset is 15 min time resolution recorded for one year, while the second data is of one-minute time resolution recorded for three years. Moreover, we used 90% of data for both datasets for training and 10% for testing. The one-minute data is very dynamic, which explains the large error compared to the first dataset.

It is challenging to tell which models were absolutely superior on the basis of the R2, MAPE, and RMSE values. The results of this study show that RNN, GRU, and VAE performs slightly better on average than the other models in most cases for one- and multi-step-ahead forecasting. The obtained results demonstrate that both RNNs with supervised learning and VAE with unsupervised learning can perform a one-step and multi-step forecasting accurately. Overall, the VAE deep learning model gives an effective way to model and forecast PV power output, and it has emerged as a serious competitor to the RNN-driven models (i.e., RNN, GRU, and LSTM).

5. Conclusions

PV power output possesses high volatility and intermittency because of its great dependency on environmental factors. Hence, a reliable forecast of solar PV power output is indispensable for efficient operations of energy management systems. This paper compares eight deep learning-driven forecasting methods for solar PV power output modeling and forecasting. The considered models can be categorized into two categories: supervised deep learning methods, including RNN, LSTM, BiLSTM, GRU, and ConvLSTM, and unsupervised methods, including AE, VAE, and RBM. We also compared the performance of the deep learning methods with two baseline machine learning models (i.e., LR and SVR). It is worth highlighting that this study introduced the VAE and RBM methods to forecast PV power. For efficiently managing the PV system, both single- and multi-step-ahead forecasts are considered. The forecasting accuracy of the ten models has been evaluated using two real-world datasets collected from two different PV systems. Results show the domination of the VAE-based forecasting methods due to its ability to learn higher-level features that permit good forecasting accuracy.

To further enhance the forecasting performance, in future study, we plan to consider multivariate forecasting by incorporating weather data. Also, these deep learning models can be applied and compared using data from other renewable energy systems, such as forecasting the power generated by wind turbines. Further, it will be interesting to conduct comparative studies to investigate the impacts of data from different technologies, such as monocrystalline, and polycrystalline.

Author Contributions: A.D.: Conceptualization, Formal analysis, Investigation, Methodology, Software, Writing-original draft, Writing-review & editing F.H.: Conceptualization, Formal analysis, Investigation, Methodology, Software, Supervision, Writing-original draft, Writing-review & editing Y.S.: Investigation, Conceptualization, Formal analysis, Methodology, Writing-review & editing, Funding acquisition, Supervision. S.K.: Investigation, Conceptualization, Formal analysis, Methodology, Writing- original draft. All authors have read and agreed to the published version of the manuscript.

Funding: This work was supported by funding from King Abdullah University of Science and Technology (KAUST), Office of Sponsored Research (OSR) under Award No: OSR-2019-CRG7-3800.

Conflicts of Interest: The authors declare no conflict of interest.

References

1. Masa-Bote, D.; Castillo-Cagigal, M.; Matallanas, E.; Caamaño-Martín, E.; Gutiérrez, A.; Monasterio-Huelín, F.; Jiménez-Leube, J. Improving photovoltaics grid integration through short time forecasting and self-consumption. *Appl. Energy* **2014**, *125*, 103–113. [CrossRef]
2. Behera, M.K.; Majumder, I.; Nayak, N. Solar photovoltaic power forecasting using optimized modified extreme learning machine technique. *Eng. Sci. Technol. Int. J.* **2018**, *21*, 428–438. [CrossRef]
3. Wang, K.; Qi, X.; Liu, H. A comparison of day-ahead photovoltaic power forecasting models based on deep learning neural network. *Appl. Energy* **2019**, *251*, 113315. [CrossRef]
4. Harrou, F.; Kadri, F.; Sun, Y. Forecasting of Photovoltaic Solar Power Production Using LSTM Approach. In *Advanced Statistical Modeling, Forecasting, and Fault Detection in Renewable Energy Systems*; IntechOpen: London, UK, 2020.
5. Fu, M.P.; Ma, H.W.; Mao, J.R. Short-term photovoltaic power forecasting based on similar days and least square support vector machine. *Power Syst. Prot. Control* **2012**, *40*, 65–69.
6. Sun, M.; Feng, C.; Zhang, J. Probabilistic solar power forecasting based on weather scenario generation. *Appl. Energy* **2020**, *266*, 114823. [CrossRef]
7. Qing, X.; Niu, Y. Hourly day-ahead solar irradiance prediction using weather forecasts by LSTM. *Energy* **2018**, *148*, 461–468. [CrossRef]
8. Chitalia, G.; Pipattanasomporn, M.; Garg, V.; Rahman, S. Robust short-term electrical load forecasting framework for commercial buildings using deep recurrent neural networks. *Appl. Energy* **2020**, *278*, 115410. [CrossRef]
9. Li, P.; Zhou, K.; Lu, X.; Yang, S. A hybrid deep learning model for short-term PV power forecasting. *Appl. Energy* **2020**, *259*, 114216. [CrossRef]
10. Kanchana, W.; Sirisukprasert, S. PV Power Forecasting with Holt-Winters Method. In Proceedings of the 2020 8th International Electrical Engineering Congress (iEECON), Chiang Mai, Thailand, 4–6 March 2020; pp. 1–4.
11. Prema, V.; Rao, K.U. Development of statistical time series models for solar power prediction. *Renew. Energy* **2015**, *83*, 100–109. [CrossRef]
12. Kushwaha, V.; Pindoriya, N.M. A SARIMA-RVFL hybrid model assisted by wavelet decomposition for very short-term solar PV power generation forecast. *Renew. Energy* **2019**, *140*, 124–139. [CrossRef]
13. Lin, K.P.; Pai, P.F. Solar power output forecasting using evolutionary seasonal decomposition least-square support vector regression. *J. Clean. Prod.* **2016**, *134*, 456–462. [CrossRef]
14. Zhang, G.; Guo, J. A Novel Method for Hourly Electricity Demand Forecasting. *IEEE Trans. Power Syst.* **2020**, *35*, 1351–1363. [CrossRef]
15. Sanjari, M.J.; Gooi, H.B.; Nair, N.C. Power Generation Forecast of Hybrid PV—Wind System. *IEEE Trans. Sustain. Energy* **2020**, *11*, 703–712. [CrossRef]
16. Rana, M.; Rahman, A. Multiple steps ahead solar photovoltaic power forecasting based on univariate machine learning models and data re-sampling. *Sustain. Energy Grids Netw.* **2020**, *21*, 100286. [CrossRef]
17. Andrade, J.R.; Bessa, R.J. Improving Renewable Energy Forecasting With a Grid of Numerical Weather Predictions. *IEEE Trans. Sustain. Energy* **2017**, *8*, 1571–1580. [CrossRef]
18. Su, H.; Liu, T.; Hong, H. Adaptive Residual Compensation Ensemble Models for Improving Solar Energy Generation Forecasting. *IEEE Trans. Sustain. Energy* **2020**, *11*, 1103–1105. [CrossRef]
19. Zhang, X.; Li, Y.; Lu, S.; Hamann, H.F.; Hodge, B.; Lehman, B. A Solar Time Based Analog Ensemble Method for Regional Solar Power Forecasting. *IEEE Trans. Sustain. Energy* **2019**, *10*, 268–279. [CrossRef]

20. Harrou, F.; Sun, Y.; Hering, A.S.; Madakyaru, M.; Dairi, A. *Statistical Process Monitoring Using Advanced Data-Driven and Deep Learning Approaches: Theory and Practical Applications*; Elsevier: Amsterdam, The Netherlands, 2020.

21. Dairi, A.; Cheng, T.; Harrou, F.; Sun, Y.; Leiknes, T. Deep learning approach for sustainable WWTP operation: A case study on data-driven influent conditions monitoring. *Sustain. Cities Soc.* **2019**, *50*, 101670. [CrossRef]

22. Harrou, F.; Dairi, A.; Sun, Y.; Kadri, F. Detecting abnormal ozone measurements with a deep learning-based strategy. *IEEE Sens. J.* **2018**, *18*, 7222–7232. [CrossRef]

23. Harrou, F.; Dairi, A.; Sun, Y.; Senouci, M. Statistical monitoring of a wastewater treatment plant: A case study. *J. Environ. Manag.* **2018**, *223*, 807–814. [CrossRef]

24. Dairi, A.; Harrou, F.; Sun, Y.; Senouci, M. Obstacle detection for intelligent transportation systems using deep stacked autoencoder and k-nearest neighbor scheme. *IEEE Sens. J.* **2018**, *18*, 5122–5132. [CrossRef]

25. Krizhevsky, A.; Sutskever, I.; Hinton, G.E. ImageNet Classification with Deep Convolutional Neural Networks. In *Advances in Neural Information Processing Systems 25*; Pereira, F., Burges, C.J.C., Bottou, L., Weinberger, K.Q., Eds.; Curran Associates, Inc.: New York, NY, USA, 2012; pp. 1097–1105.

26. Young, T.; Hazarika, D.; Poria, S.; Cambria, E. Recent trends in deep learning based natural language processing. *IEEE Comput. Intell. Mag.* **2018**, *13*, 55–75. [CrossRef]

27. Graves, A.; Rahman Mohamed, A.; Hinton, G.E. Speech Recognition with Deep Recurrent Neural Networks. In Proceedings of the 2013 IEEE International Conference on Acoustics, Speech and Signal Processing, Vancouver, BC, Canada, 26–31 May 2013.

28. Zeroual, A.; Harrou, F.; Dairi, A.; Sun, Y. Deep learning methods for forecasting COVID-19 time-Series data: A Comparative study. *Chaos Solitons Fractals* **2020**, *140*, 110121. [CrossRef] [PubMed]

29. Harrou, F.; Hittawe, M.M.; Sun, Y.; Beya, O. Malicious attacks detection in crowded areas using deep learning-based approach. *IEEE Instrum. Meas. Mag.* **2020**, *23*, 57–62. [CrossRef]

30. Hittawe, M.M.; Afzal, S.; Jamil, T.; Snoussi, H.; Hoteit, I.; Knio, O. Abnormal events detection using deep neural networks: Application to extreme sea surface temperature detection in the Red Sea. *J. Electron. Imaging* **2019**, *28*, 021012. [CrossRef]

31. Wang, W.; Lee, J.; Harrou, F.; Sun, Y. Early Detection of Parkinson's Disease Using Deep Learning and Machine Learning. *IEEE Access* **2020**, *8*, 147635–147646. [CrossRef]

32. Silver, D.; Huang, A.; Maddison, C.J.; Guez, A.; Sifre, L.; van den Driessche, G.; Schrittwieser, J.; Antonoglou, I.; Panneershelvam, V.; Lanctot, M.; et al. Mastering the game of Go with deep neural networks and tree search. *Nature* **2016**, *529*, 484–503. [CrossRef]

33. Li, G.; Wang, H.; Zhang, S.; Xin, J.; Liu, H. Recurrent neural networks based photovoltaic power forecasting approach. *Energies* **2019**, *12*, 2538. [CrossRef]

34. Wang, Y.; Liao, W.; Chang, Y. Gated recurrent unit network-based short-term photovoltaic forecasting. *Energies* **2018**, *11*, 2163. [CrossRef]

35. Wang, F.; Xuan, Z.; Zhen, Z.; Li, K.; Wang, T.; Shi, M. A day-ahead PV power forecasting method based on LSTM-RNN model and time correlation modification under partial daily pattern prediction framework. *Energy Convers. Manag.* **2020**, *212*, 112766. [CrossRef]

36. Kong, W.; Dong, Z.Y.; Hill, D.J.; Luo, F.; Xu, Y. Short-Term Residential Load Forecasting Based on Resident Behaviour Learning. *IEEE Trans. Power Syst.* **2018**, *33*, 1087–1088. [CrossRef]

37. Aprillia, H.; Yang, H.T.; Huang, C.M. Short-Term Photovoltaic Power Forecasting Using a Convolutional Neural Network-Salp Swarm Algorithm. *Energies* **2020**, *13*, 1879. [CrossRef]

38. Wang, H.; Yi, H.; Peng, J.; Wang, G.; Liu, Y.; Jiang, H.; Liu, W. Deterministic and probabilistic forecasting of photovoltaic power based on deep convolutional neural network. *Energy Convers. Manag.* **2017**, *153*, 409–422. [CrossRef]

39. Dorffner, G. Neural networks for time series processing. *Neural Netw. World* **1996**, *6*, 447–468.

40. Hochreiter, S.; Schmidhuber, J. Long short-term memory. *Neural Comput.* **1997**, *9*, 1735–1780. [CrossRef] [PubMed]

41. Cho, K.; van Merrienboer, B.; Gulcehre, C.; Bahdanau, D.; Bougares, F.; Schwenk, H.; Bengio, Y. Learning Phrase Representations using RNN Encoder-Decoder for Statistical Machine Translation. *arXiv* **2014**, arXiv:1406.1078.

42. Schuster, M.; Paliwal, K.K. Bidirectional recurrent neural networks. *IEEE Trans. Signal Process.* **1997**, *45*, 2673–2681. [CrossRef]

43. Xingjian, S.; Chen, Z.; Wang, H.; Yeung, D.Y.; Wong, W.K.; Woo, W.C. Convolutional LSTM network: A machine learning approach for precipitation nowcasting. *Adv. Neural Inf. Process. Syst.* **2015**, *28*, 802–810.

44. Hinton, G.E.; Osindero, S.; Teh, Y.W. A fast learning algorithm for deep belief nets. *Neural Comput.* **2006**, *18*, 1527–1554. [CrossRef]

45. Smolensky, P. *Information Processing in Dynamical Systems: Foundations of Harmony Theory*; CU-CS-321-86; University of Colorado: Boulder, CO, USA, 1986.

46. Bengio, Y. *Learning Deep Architectures for AI*; Now Publishers Inc.: Delft, The Netherlands, 2009.

47. Kingma, D.P.; Welling, M. Auto-encoding variational bayes. *arXiv* **2013**, arXiv:1312.6114.

48. Kempinska, K.; Murcio, R. Modelling urban networks using Variational Autoencoders. *Appl. Netw. Sci.* **2019**, *4*, 1–11. [CrossRef]

49. Kingma, D.; Salimans, T.; Josefowicz, R.; Chen, X.; Sutskever, I.; Welling, M. Improving variational autoencoders with inverse autoregressive flow. In Proceedings of the 30th Annual Conference on Neural Information Processing Systems 2016, Barcelona, Spain, 5–10 December 2016.

50. Doersch, C. Tutorial on variational autoencoders. *arXiv* **2016**, arXiv:1606.05908.

51. Bergstra, J.; Bengio, Y. Random Search for Hyper-Parameter Optimization. *J. Mach. Learn. Res.* **2012**, *13*, 281–305.

52. Boyd, M. Performance Data from the NIST Photovoltaic Arrays and Weather Station. *J. Res. Natl. Inst. Stand. Technol.* **2017**, *122*, 40. [CrossRef]

53. Harrou, F.; Sun, Y. *Advanced Statistical Modeling, Forecasting, and Fault Detection in Renewable Energy Systems*; BoD—Books on Demand: Norderstedt, Germany, 2020.

Article

Solar Power System Assessments Using ANN and Hybrid Boost Converter Based MPPT Algorithm

Imran Haseeb [1], Ammar Armghan [2], Wakeel Khan [3], Fayadh Alenezi [2], Norah Alnaim [4], Farman Ali [1,*], Fazal Muhammad [5], Fahad R. Albogamy [6,*] and Nasim Ullah [7]

1. Department of Electrical Engineering, Qurtuba University of Science and IT, Dera Ismail Khan 29050, Pakistan; imranhaseeb@qurtuba.edu.pk
2. Department of Electrical Engineering, College of Engineering, Jouf University, Sakaka 72388, Saudi Arabia; aarmghan@ju.edu.sa (A.A.); fshenezi@ju.edu.sa (F.A.)
3. Department of Electrical Engineering, Foundation University Islamabad, Islamabad 44000, Pakistan; wakeel.khan@fui.edu.pk
4. Department of Computer Science, College of Sciences and Humanities in Jubail, Imam Abdulrahman bin Faisal University, Dammam 31441, Saudi Arabia; Nmalnaim@iau.edu.sa
5. Department of Electrical Engineering, University of Engineering Technology, Mardan 23200, Pakistan; fazal.muhammad@uetmardan.edu.pk
6. Computer Sciences Program, Turabah University College, Taif University, P.O. Box 11099, Taif 21944, Saudi Arabia
7. Department of Electrical Engineering, College of Engineering TAIF University KSA, Taif 21944, Saudi Arabia; nasimullah@tu.edu.sa
* Correspondence: drfarmanali.optics@qurtuba.edu.pk (F.A.); f.alhammdani@tu.edu.sa (F.R.A.)

Citation: Haseeb, I.; Armghan, A.; Khan, W.; Alenezi, F.; Alnaim, N.; Ali, F.; Muhammad, F.; Albogamy, F.R.; Ullah, N. Solar Power System Assessments Using ANN and Hybrid Boost Converter Based MPPT Algorithm. *Appl. Sci.* **2021**, *11*, 11332. https://doi.org/10.3390/app112311332

Academic Editors: Luis Hernández-Callejo, Maria del Carmen Alonso García and Sara Gallardo Saavedra

Received: 5 October 2021
Accepted: 26 November 2021
Published: 30 November 2021

Publisher's Note: MDPI stays neutral with regard to jurisdictional claims in published maps and institutional affiliations.

Abstract: The load pressure on electrical power system is increased during last decade. The installation of new power generators (PGs) take huge time and cost. Therefore, to manage current power demands, the solar plants are considered a fruitful solution. However, critical caring and balance output power in solar plants are the highlighted issues. Which needs a proper procedure in order to minimize balance output power and caring issues in solar plants. This paper investigates artificial neural network (ANN) and hybrid boost converter (HBC) based MPPT for improving the output power of solar plants. The proposed model is analyzed in two steps, the offline step and the online step. Where the offline status is used for training various terms of ANNs in terms of structure and algorithm while in the online step, the online procedure is applied with optimum ANN for maximum power point tracking (MPPT) using traditional converter and hybrid converter in solar plants. Moreover, a detail analytical framework is studied for both proposed steps. The mathematical and simulation approaches show that the presented model efficiently regulate the output of solar plants. This technique is applicable for current installed solar plants which reduces the cost per generation.

Keywords: artificial neural network based MPPT; hybrid boost converter; renewable energies; solar power system

---

1. Introduction

The electrical energy plays a key role in the economic development of a country. From last decade, rapid increase in consumption of electricity and demand is recorded [1,2]. In order to fulfill power demands, installation of new power plants are time consumption and high expensive procedures [3]. The renewable energies, such as wind and solar, appear to be appropriate solutions to cover energy demand while reducing environmental pollution and toxic materials [4]. Furthermore, the renewable energy resources are followed by many countries, which comes from geothermal heat, tides, rain, wind, and sunlight. Major remote sites in each country of the globe utilizing renewable energy [5].

The renewable energy generated power depends on site selection and climate condition. Furthermore, there is a need of storage system power induced by renewable energy to insure the continuity of available power [6]. Various types of resources are existed in the entire globe, however, among of all these resources solar energy is considered economical, inexhaustible, and sustainable energy. Photovoltaic (PV) cells mostly use for bulk and small power and gradually increased day by day [7]. The output of PV depends on different factor such as temperature, weather conditions, module materials, and is used in different applications, i.e., light sources, battery charging, water pumping, space, satellite power system, remote islanded power system, etc. This work explores the applications of artificial neural network (ANN) and hybrid boost converter (HBC) for enhancing the efficiency for MPPT to increase the smoothness of solar power system. The deep learning based ANN mechanism has key features to optimize the input data and produce the purified desire outcomes. Furthermore, currently the ANN approach is widely used in several fields for the purpose of optimizing structure complexity and predict the expected outputs. For example, in [8] the ANN technique is utilized for predicting droughts, in [9] the ANN method is studied for accurately measure the wind speed and predict its results, in [10] the particular matter is discussed for the Ankara city through ANN and similarly, in the field of solar system the ANN is applied in [11] for forecasting the generation of solar system. Thus, in this proposed work the ANN mechanism is estimated for optimize the output of MPPT and HBC.

1.1. Major Contribution

The use of MPPT has increased the performance of solar power system and now the demand of solar plant installation is enhanced. The efficiency of the MPPT based solar system can be further improved applying ANN and HBC. Thus, this paper studies the contributions of ANN and HBC based MPPT in the solar system. The major contributions of this work are discussed as follows.

1. The outcomes of solar system are enhanced in this model using ANN and HBC based MPPT algorithm;
2. The mathematical model is elaborated in detail to show how the ANN and HBC based MPPT algorithm improve the quality of service of a solar power system;
3. The model is designed for simulation analysis and is compared among various procedures like Elman neural network (ENN), with install HBC and ANN mechanisms and without install HBC and ANN procedures in order to evaluate proposed model executions;
4. The proposed ANN model presents reliable outcomes in view of simple structure, fast training and robust performance. This effectiveness is further modified by applying HBC in the proposed model.

1.2. Organization of Paper

The rest of the paper is organized as follows: Section 2 explains the proposed framework; Section 3 investigates the analytical approach; Section 4 defines the result and discussions; similarly, Section 5 presents the conclusion.

1.3. Related Work

As it has been shown that solar energy is a cost effective solution for managing current electric load demands, a number of research work have been carried out so far, which are analyzed as follows:

In [12,13], the authors have discussed the two artificial based mechanisms in term of variable and fixed step load. The theoretical and simulation analysis are investigation for MPPT controller, and simulation. In [14], authors have proposed the MPPT technique which is based on enhanced neural network (ENN). The proposed ENN based control have the ability to adjust the step size to track MPPT automatically and it can improve the dynamic and steady performance of PV panels. The proposed ENN technique can easily be implemented because of less constructed data. In [15], authors have proposed a MPPT converter using online learning neural network (OLNN) and perturbation and observation technique. The proposed work can improve the fast tracking of the solar panels when the radiation is changing constantly. The simulation results show improved output when the radiation is changing rapidly. In [16], authors have investigated an optimal power operating point (OPOP) using photovoltaic and thermal neural network (PV/TNN). The NN has been used to calculate the OPOP. The OPOP computes the optimum flow rate of PV/T for maximum electrical and thermal power. The results show that this technique can hold fast and accurate PV/T flow rate control. In [17], authors have presented the super twisting sliding mode MPPT controller (ST-SMC) with ANN. The robust sliding mode controller is used against disturbance and parametric variations while the ANN controls the peak power voltage for the efficiency of MPPT. The simulation results show the improved performance in terms of dynamic response and robustness. In [18], authors have suggested the two fast and digital MPPT techniques for fast changing environment. They approximate the MPP locus by using piecewise line segment or cubic equation. The MPP locus parameters are found with the help of NN based program. The results show that the proposed research required less computation requirement, high static, and dynamic tracking and fast speed. In [19], authors have proposed the different intelligence technique, i.e., neural network (NN), fuzzy logic (FL), genetic algorithm (GA), and neuro-fuzzy and their possible implementation into a field programmable gate array (FPGA). The authors have developed intelligent MPP controller using simulink/MATLAB and then different step to design and implement the controller into FPGA. The best controller among all these is tested in real time using FPGA Virtex 5. The comparative study of all controllers describe that the effectiveness of developed intelligence technique in term of accuracy, quick response, power consumption, flexibility, and simplicity. In [20], authors have presented the two techniques ANN and Fuzzy logic controller (FLC) for MPPT of PV cell. They investigate the proposed work using MATLAB/Simulink. The results show that the efficiency and response under variable irradiation conditions are satisfactory. In [21], authors have investigated the new combined method that is established by a three point comparing method and an ANN based PV model method. During the fast variations of solar irradiance, the exact MPP is searched by three point comparison and ANN methods. The results show that the proposed method search the exact and fast MPP under different irradiance condition with feedback voltage and current. In [22], authors have proposed the ANN method for the non-linear and time-varying output of PV panels. ANN is the suitable solution for non-linear outputs. The results also denote that the ANN algorithm have better MPPT characteristics as compared with the traditional perturbation and observation technique. In [23], authors have suggested the intelligent MPPT method for a standalone PV system using ANN and fuzzy logic controller (FLC). The ANN is used for different solar irradiance and temperature to find MPP voltage. FLC uses MPP voltage as a reference voltage to generate a control signal for the DC-DC converter. The results explain good performance of ANN-FLC as compared to traditional incremental conductance (IC). In [24], authors have investigated the method to achieve acceptable tracking time and less power oscillation by adjusting the changing step size of Flyback inverter. Solar irradiance is adopted as an input of ANN which is used to appropriate modulation step size. The simulation results show that for any solar irradiance ANN based Flyback inverter can find appropriate step size. In [25], authors have presented the MPPT using ANN, and the hysteresis current controlled inverter with fixed band and variation of load value is determined with output current total harmonic distraction (THD) is lower than 5%. The results confirm that the efficiency of controller and flexibility of inverter is satisfactory. In [26], authors have proposed the MPPT of a grid connected 20 kW neuro-fuzzy network based PV system. The neuro-fuzzy system consists of fuzzy based classifier and three multilayered feed forwarded ANN. The inputs of ANN are irradiance and temperature, and, after the estimation process, output is the reference voltage. The reference voltage guarantees the optimal power transfer between PV generator and the main utility grid. The simulation results proved the best efficiency as compared to conventional single ANN and perturb and observe (P&O) algorithm. The study in [27] reveals the performance of ANN based different algorithms, i.e., Levenberg–Marquardt (LM), Bayesian regularization (BR), and scaled conjugate gradient (SCG) algorithms are used in MPPT energy harvesting in solar photovoltaic system. The simulated results show that SCG algorithm reveal superior performance compared to LM and BR algorithms. However, the LM algorithm performs better in the correlation between input–output, dataset training, and error. A fuzzy logic, particle swarm optimization (PSO), and imperialist competitive algorithm are proposed in [28], where it is declared that the fuzzy logic is less complicated, faster, more accurate, and more stability then the other three algorithms. The authors have explored a hybrid shuffled frog leaping and pattern search (HSFL–PS) algorithm in [29] for optimizing ANN-based MPPT in a grid-tied PV system. The simulation results show that the performance of MPPT is improved in comparison with the conventional MPP methods. A novel ANN-ACO MPPT controller is developed by authors in [30] based on an ant colony optimization (ACO) algorithm which is useful to train the developed ANN. The ANN-ACO technique has improved the MPPT and reduced the drawbacks in conventional MPPT. The authors have also improved the power quality and distortion free signal to the grid. The simulation and experimental results show that ANN-ACO controller can track the MPP rapidly and robustness with a minimum steady-state oscillation than the conventional INC method. ANN based controller is used in [31] to control the converter fed by an autonomous photovoltaic generator (PVG) instead of classical MPPT algorithms such as perturb and observe (P&O). The results present that the ANN based PVG provide low oscillation and better performance.

2. Proposed ANN Based MPPT and Hybrid Boost Converter Model

Figure 1 explains the HBC and ANN based MPPT presented model for minimizing the solar power system critical caring and balancing output power issues. The output of PV is attained by ANN based MPPT in order to improve the performance of solar power system. The neural network technique consist of major amount of interconnected processors called neurons. Each neuron includes a huge number of weighted links for transforming signals. Thus, it has potential to manage the difficult task of data processing and interpretation. In this proposed model, the feed forward back propagation ANN is installed which contains logsig purelin and purelin activation functions based hidden layers, as depicted in Figure 2. In case of offline step the ANN training is performed in terms of activation function structure and training algorithm. On the other side for online step the trained ANN based MPPT is applied to track the MPP. The output of PV array voltage derivation (dv) and power derivation (dp) is given to ANN based MPPT which depends on solar radiation and temperature conditions. Table 1 summarizes the basic principle of ANN based MPPT which output is corresponding normalized increasing or decreasing duty cycle (+1, 0, −1).

Table 1 explains the basic mechanism of ANN based MPPT controller.

Table 1. ANN based MPPT controller basic principle.

dv	dp	dp/dv	Duty Cycle
−1	−1	−1	D(n) = D(n − 1)
+1	−1	+1	D(n) = D(n − 1)
−1	+1	+1	D(n) = D(n − 1)
+1	+1	−1	D(n) = D(n − 1)

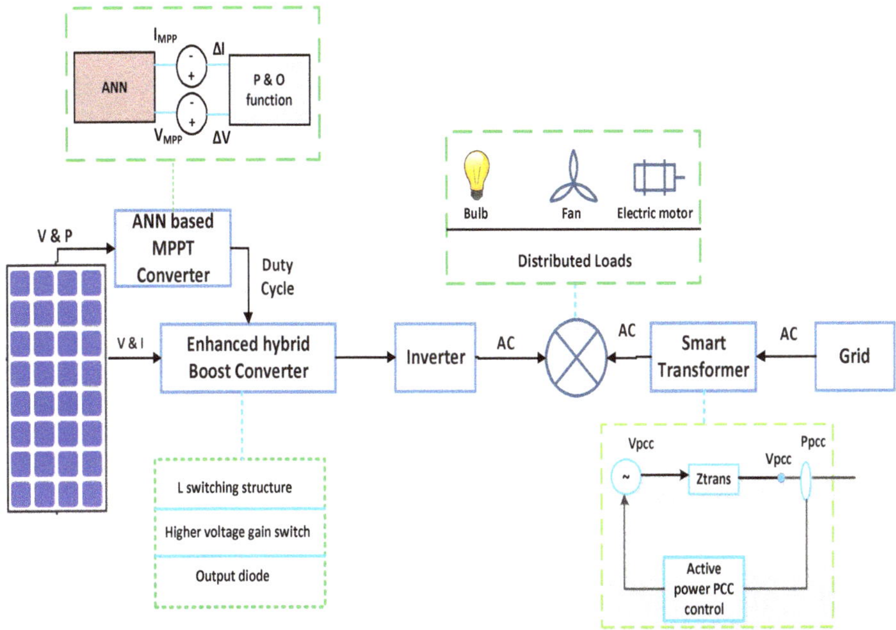

Figure 1. Proposed framework using ANN MPPT and enhanced hybrid boost converter.

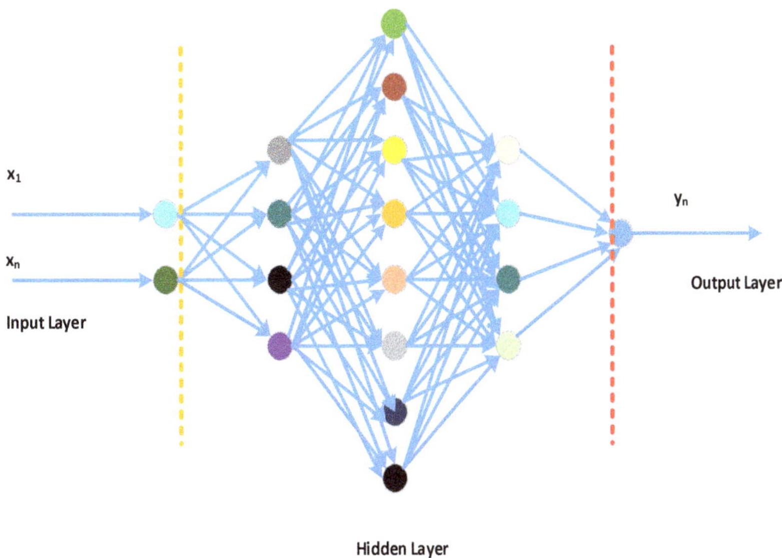

Figure 2. Internal structure of ANN based MPPT .

3. Analytical Approach

3.1. Analytical Model of PV Module

The main background of the presented approach is discussed in Section 2. This section elaborates the mathematical mechanism for the modern MPPT based ANN system, considering the array of PVs at input side and utility grid at the receiving end. The description of used symbols are declared in Table 2. The circuitry of the PV cell contain diode for paralleling current as shown in Figure 3, which can be mathematically [32–34] defined as

$$i_{total} = i_{sc} - i_{diode}, \tag{1}$$

where i_{diode} is the diode current, i_{sc} is the current through the parallel resistor. The i_{diode} is further explained as

$$i_{diode} = i_o(e^{\frac{ev_d}{\eta K T_c}} - 1) \tag{2}$$

So, the Equation (1) can be modified with Equation (2) the total current from PV cell is measured as

$$i_{total} = i_{sc} - i_o(e^{\frac{ev_d}{\eta K T_c}} - 1) \tag{3}$$

Parallel resistance R_p and series resistance R_s is also considered for the dynamic behavior of PV cell, however, for the proposed PV cell simulation there is computational limitation if taking both series and parallel resistance. Therefore, parallel resistance is usually neglected in PV systems bibliography. For the proposed PV cell simulation series resistance is considered and the PV cell total current is mathematically written as [35,36]

$$i_{total} = i_{sc} - i_o\{exp(\frac{e(v + i_{total} R_s)}{\eta K T_c})\} - 1 \tag{4}$$

The PV system consist of N numbers of branches in parallel and each parallel branch consist of N number of PV cells in series. So, the total current from PV module under constant weather condition is

$$i_{sc}^N = i_{sc}^N[1 - exp(\frac{V^m - V_{oc}^m + R_s^N \cdot i^N}{N_s \cdot V_T^c})] \tag{5}$$

The module current depends on some different parameters of the cell. Each variable depends on different parameters of each cell. The module short circuit current depends on number of branches and short circuit of the cell [37,38].

$$i_{sc}^M = N_p \cdot i_{sc}^c \tag{6}$$

The open circuit voltage of the module depends on the number of cells in series and the open circuit voltage of the cell.

$$V_{oc}^m = N_s \times V_{oc}^c \tag{7}$$

The thermal voltage of the semiconductor in the module depends on Boltzmann's constant, cell temperature and the charge of electron, i.e., 1.602×10^{-19} [39].

$$V_T^c = \frac{mK T_c}{e} \tag{8}$$

The equivalent series resistant of the module depends on series resistance of the cell, number of parallel branches in the module and the number of cell in series in each branch.

$$R_s^M = \frac{R_s^c \cdot N_p}{N_s} \tag{9}$$

Assume that all the cell in PV system are same, so the power, voltage, and current under standard condition are

$$P_{max}^c = \frac{P_{max}^M}{N_s \cdot N_p} \tag{10}$$

$$V_{oco}^c = \frac{V_{oco}^M}{N_s} \tag{11}$$

$$i_{sco}^c = \frac{i_{sco}^M}{N_p} \tag{12}$$

Now the instantaneous series resistance of the PV cell can be expressed as

$$R_s^c = \left(1 - \frac{FF}{\frac{P_{mxo}^c}{V_{oco}^c i_{xo}^c}}\right) \cdot \frac{V_{oco}^c}{I_{sco}^c} \tag{13}$$

Fill factor is denoted with FF and is given as

$$FF = \frac{\frac{V_{oco}^c}{V_{To}^c} - L_n\left(\frac{V_{oco}^c}{V_{To}^c} + 0.72\right)}{\frac{V_{oco}^c}{V_{To}^c} + 1} \tag{14}$$

The short circuit current of the cell itself depends on the irradiance and can be expressed as

$$i_{sc}^c = \frac{i_{sco}^c G_a}{G_{ao}} \tag{15}$$

The open circuit voltage of the cell depends on the nominal open circuit and the actual weather condition, i.e., ambient temperature, temperature of cell, and irradiance. The open circuit voltage is expressed [40,41] as

$$V_{oc}^c = V_{oco}^c + 0.03(T_a + 0.03 \cdot G_a - T_{co}) \tag{16}$$

Now the current generated from the PV module in term of all parameters, i.e., irradiance, temperature, voltage, etc., is

$$i^M = N_p \cdot i_{sc}^c \left[1 - exp\left(\frac{V^m - N_s V_{oc}^c + i^m R_s^c \frac{N_s}{N_p}}{N_s V_T^c}\right)\right] \tag{17}$$

Now assume the case that PV array consist of B number of parallel branches and each branch contain M number of module so, the array current will be equal to $I = \sum_{i=1}^{N_B}$. If we consider all the panels are same and under same temperature and irradiance then the output current will be

$$i_{total} = N_B \cdot i^M \tag{18}$$

Table 2. List of symbols used for analytical model.

Name of Parameter	Description
i_{diode}	Diode Current
i_{sc}	Short Circuit Current
i_o	Reserve Saturation Current
i_{sc}^M	Short circuit current under standard condition
V_{oc}^m	open voltage under standard condition
V_T	Thermal voltage in the semiconductor
P_{max}^c	Power under standard condition
N_s	Cell in series in module
N_p	parallel branches in module
N_m	module in series in an array
N_B	parallel branches in array
T_a	Ambient temperature
T_c	Cell temperature
G_a	Irradiance
K	Boltzmann Constant
FF	Fill Factor
D	Duty Cycle
T_{ON}	On time of switch
V_{pu}^*	Reference Voltage
i_L^*	Reference inductor current
m^*	Optimal duty cycle
$L_{1,2}$	Inductor 1 and 2
C_{IN}	Capacitance
I_{LAV}	Inductor Average Current
V_{OUT}	Output Voltage

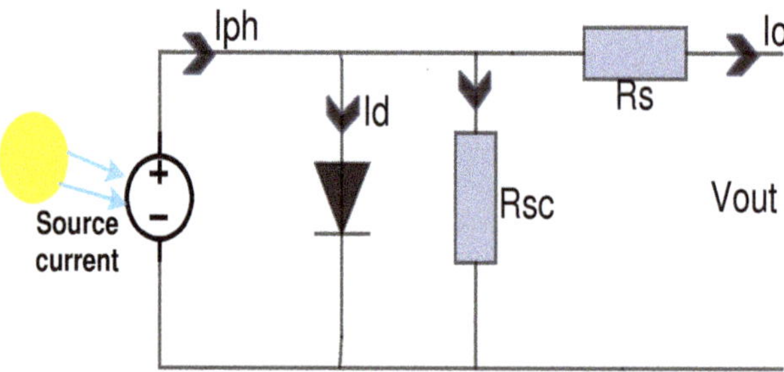

Figure 3. Equivalent circuit of photovoltaic cell.

3.2. Analytical Modeling of a Traditional Boost Converter

The electrical MPPT can be achieved through a DC to DC converter inserted between the photovoltaic module and the load. The DC to DC converter ensures the matching of load resistance with the varying source resistance. The converter is basically used to transfer maximum energy from source to load.

The framework of traditional boost converter is represented in Figure 4 which transfers maximum energy by adjusting the PV module output voltage to the reference voltage. The traditional boost converter has two controllable variables: Voltage V_{pu} and the inductor current i_L. The mathematical model that describe the voltage and inductor current relation is

$$\begin{cases} V_m \\ i_L \end{cases} = (1 - D) \begin{cases} V_m \\ i_L \end{cases} \tag{19}$$

where D is the duty cycle of the boost converter, which is expressed as

$$D = \frac{T_{on}}{T_{total}} = T_{on} f_s \tag{20}$$

where f_s is the frequency of switching of converter's switch and T_{total} is the total time and T_{ON} is the on time of the switch.

$$\frac{dI_L}{T} = \frac{1}{L_{pu}}(V_m - V_{pu}) - \frac{P_{pu}}{L_{pu}} \tag{21}$$

where V_{pu} is the reference voltage generated by the ANN based MPPT.

$$\frac{dV_{pu}}{dt} = \frac{(I_L - I_{pu})}{Pu} \tag{22}$$

By using Equation (21) we obtain the reference inductor current

$$i_L^* = (V_{pu}^* - V_{pu}) + I_{pu} \tag{23}$$

The optimal switching voltage can be expressed by using the Equations (22) and (23)

$$V_m^* = PI(V_{pu}^* - V_{pu}) + Vpu + \frac{R_{pu}}{L_{pu}} I_L \tag{24}$$

while the boost converter command is obtained by the conversion of Equation (19)

$$D^* = 1 - \frac{V_m^*}{V_{DC}} \tag{25}$$

The DC link capacitor at the output play a key role to ensure energy balance between the PV module and the power injected into the system. The DC link capacitor charge and discharge that oscillate between two levels depending on actual weather condition and power injection.

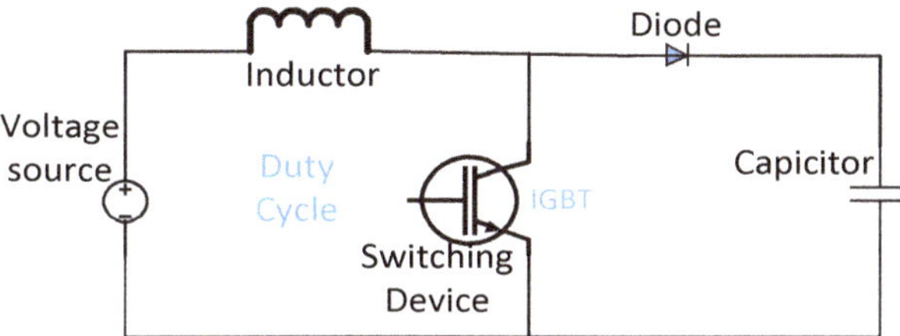

Figure 4. Traditional boost converter.

3.3. Analytical Modelling of Enhanced Single Phase Hybrid Boost Converter

Figure 4 shows the traditional boost converter that theoretically, voltage gain is very high at very high duty ratio near 100% but practically, traditional boost converter cannot work efficiently at 100% duty cycle because of diodes, equivalent resistance of inductor and capacitor, and the saturation of both capacitor and inductor [40,41]. In this paper, enhanced single phase HBC is introduced which has replaced the inductor in traditional boost converter into two inductors and 3 diodes, this structure provides high gain and efficiency. The gain of proposed converter is also increased than the traditional boost converter by a factor of (d + 1). Figure 5 shows the enhanced single phase HBC, for the mathematical modeling, we assumed that all the components are lossless and the outputs and the inputs are pure DC signals. For these consideration voltage across the inductor depends on input and output voltages. The voltage across the inductor during ON and OFF time are estimated as

$$T_{ON} : V_L = V_{IN} \tag{26}$$

$$T_{OFF} : V_L = -\frac{V_{OUT} - V_{IN}}{2} \tag{27}$$

The above formula indicates that during the switch is at ON state the voltage across the inductor is positive and at OFF state voltage will be negative. The positive and negative area of the voltage of the inductor will be same. With the above condition the output voltage is calculated as

$$V_{OUT} = \frac{1+d}{1-d} . V_{IN} \; with \; \; d = \frac{T_{ON}}{T_P}, \; 1 - d = \frac{T_{OFF}}{T_P} \tag{28}$$

The average inductor current is the function of input current from the solar module and the duty cycle of the switch

$$I_{LAV} = \frac{I_{IN}}{1+d} \tag{29}$$

To find the appropriate inductor and capacitor value in the circuit we need to know the output and input voltage of the converter and the rated power which should be transferred to the output. With these known parameters input current and average inductor current can be calculated and assumed that the average inductor current is the max current in the circuit. To find the inductor value in the circuit the input and output voltage is considered and the duty cycle is measured as

$$L_1 = L_2 = \frac{V_{IN} \cdot T_{ON}}{\triangle I_L} \tag{30}$$

$$L_1 = L_2 = \frac{V_{OUT} \cdot T_P}{\triangle I_L} \cdot \frac{d.(1-d)}{1+d} \tag{31}$$

The AC voltage is produced in the capacitor which overlaps with DC voltage so, because of this reason voltage variation is the key factor in capacity dimension of capacitor.

$$C_{IN} = \frac{I_c \cdot T_{ON}}{\triangle \cdot V_c} \tag{32}$$

$$C_{IN} = \frac{I_{LAV} \cdot T_P}{\triangle \cdot V_{INmax}} \cdot d \cdot (1-d) \tag{33}$$

To design the capacitor for boost converter the RMS current of the load is the important factor. Duty cycle of the average inductor current and the maximum current variation are used for the capacitor current, which are defined as

$$I_{INC} = I_{LAV}\sqrt{d \cdot (1-d) + \left(\frac{\triangle \cdot I_{Lmax}}{I_{LAV}}\right)^2 \cdot \frac{d2 \cdot (1-d)^2 \cdot (1+3 \cdot d)}{12 \cdot (1+d)^2 (3-2 \cdot \sqrt{2})}} \tag{34}$$

The maximum voltage variation that is acceptable for the capacitor is important for the selection of output capacitor. The current time area of capacitor is the function of duty cycle and output current, which are calculated as

$$C_{OUT} = \frac{I_{OUT} \cdot T_P \cdot d}{\triangle V_{OUTmax}} \tag{35}$$

$$C_{OUT} = \frac{I_{LAV} \cdot T_P}{\triangle V_{OUTmax}} \cdot d \cdot (1-d) \tag{36}$$

Practically, acceptable voltage variation for capacitor is less than 1% of the output voltage so the output capacitor dimension is calculated is given as

$$C_{OUT} = \frac{I_{LAV\,R} \cdot T_P}{4 \triangle V_{OUTmax}} \quad with \quad \triangle V_{OUTmax} \leq 0.01 \cdot V_{out\,R} \tag{37}$$

The output current across the capacitor is the function of duty cycle for an average inductor current and maximum inductive current variation. So the output current will be

$$I_{INC} = I_{LAV}\sqrt{d \cdot (1-d) + \left(\frac{\triangle \cdot I_{Lmax}}{I_{LAV}}\right)^2 \cdot \frac{d2 \cdot (1-d)^3}{12 \cdot (1+d)^2 (3-2 \cdot \sqrt{2})}} \tag{38}$$

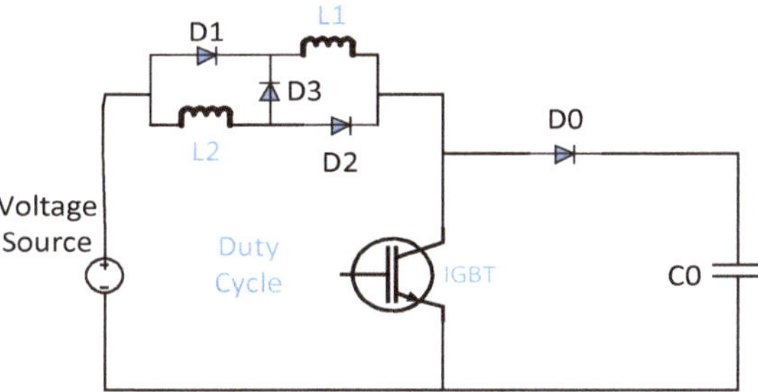

Figure 5. Enhanced single phase hybrid boost converter.

4. Results and Discussion

In the presented model, the ANN and HBC based MPPT is used for enhancing the productivity and performance. The results of the proposed model are compared among ANN based traditional boost converter and ANN based proposed HBC. The results of the proposed model present that the ANN based HBC MPPT give smooth output powers as compare to traditional boost converter. The proposed model is analyzed using MATLAB simulation software. The data are collected in terms of short circuit current, maximum current, open circuit voltage, maximum voltage, current temperature and voltage temperature, standard irradiance, standard temperature, maximum current of the given irradiance and temperature, maximum voltage of the given irradiance and temperature, and maximum power of the given irradiance and temperature. The temperature and irradiance are applied as input variables while the MPP voltage is used as a output of the ANN model. The current and power of the MPPT are estimated using V-! characteristics and multiplication of obtained current and voltage, respectively. The collected data are divided among training and testing for ANN approach. Table 3 explains the list of elements used for calculating the simulation results. Whereas the flow chart of applied algorithm is depicted in Figure 6. Figure 7 presents the correlation among tradition boost converter and proposed HBC, which clarifies that HBC gain is several folds higher than traditional boost converter. Hence, with applying the proposed HBC the performance of ANN based MPPT is enhanced than without installed HBC MPPT system. Figure 8 shows the analysis of ANN based MPPT controller and advance hybrid converter in terms of power and time. Which declares that the irradiance level is instantly changed at 0 s, 0.04 s, 0.07 s, 0.11 s, and 0.15 s, and temperature is constant at 25 °C. At 0.04 s, the range of power is constant till 0.5 s. At 0.5 s the power jumped from 140 W to 220 where fluctuation is recorded. Similarly, this slight variations are continued in whole process as mentioned in Figure 8. Furthermore, it is depicted from Figure 8, that the ANN based MPPT has increased the smoothness of outcomes. The recorded instability in Figure 8 is highlighted in Figure 9, which shows that variation is increased from 0.0 to 0.2 s. The constant outcomes are gained at 0.04 s. Thus, Figure 9 defines that, in view of ANN and advance hybrid converter, the variation in generated power is optimized. The smoothness of the proposed induced power compared with conventional boost converter, which is illustrated in Figure 10 for power against time. Figure 10 clarifies that without the use of the proposed advance hybrid converter, the maximum variation in the produced power occurs. The signals achieve constant position after maximum delay. The alteration of the induced power mentioned in Figure 10 is zoomed out in Figure 11. Which explains that the fluctuations in generated power is larger than the proposed advance HBC and ANN based MPPT model. Another important term is the ripples in the output power during the constant irradiance, shown in Figure 12, which is 0.02 W in the proposed model. Figures 13 and 14 declare the relation among train, test, validation, and best data from the ANN approach in terms of means square error and epochs. Similarly, Figure 15 depicts the output results of the utilized AN as a function of time. The parameters like training target, training outputs, validation targets, validation outputs, test targets, test outputs, errors, and responses are evaluated in Figure 15. Figure 16a,b mention the outcomes of solar system excluding ANN and HBC based MPPT and including ANN and HBC based MPPT, respectively. In the last, the fruitful outcomes of the proposed ANN and advance hybrid boost converter based MPPT for enhancing the fidelity of the solar power system is compared with current work, as shown in Table 4.

Table 3. Parameter description used for evaluation the proposed model.

Name of Parameter	Description
Short circuit current	10.5 A
Open circuit voltage	22.1 V
Irradiance	700–1000 W/m^2
No of cells in module	60
No of cells in series module	8
No of cells in parallel	6
Temperature	25 °C
Maximum power	120 W
Ideal factor	1.9
Maximum voltage	620 v

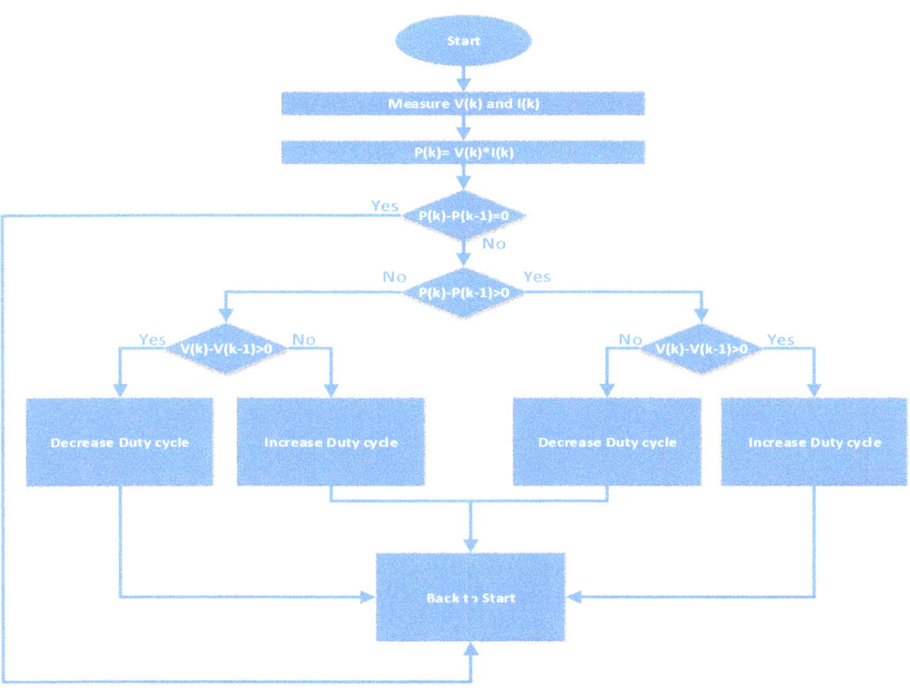

Figure 6. Flow chart description for proposed model.

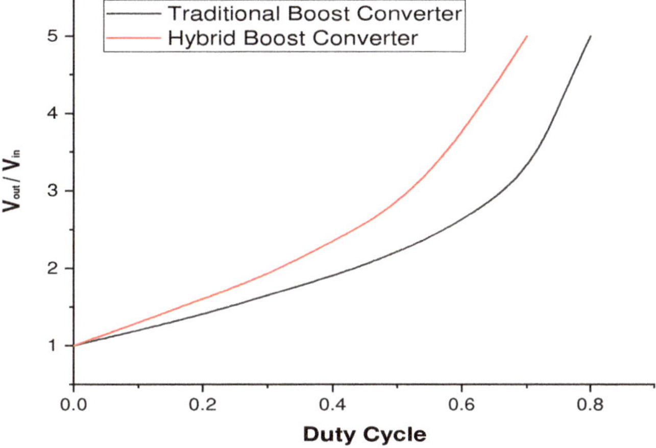

Figure 7. Comparison of traditional and hybrid boost converters in terms of voltage gain and duty cycle.

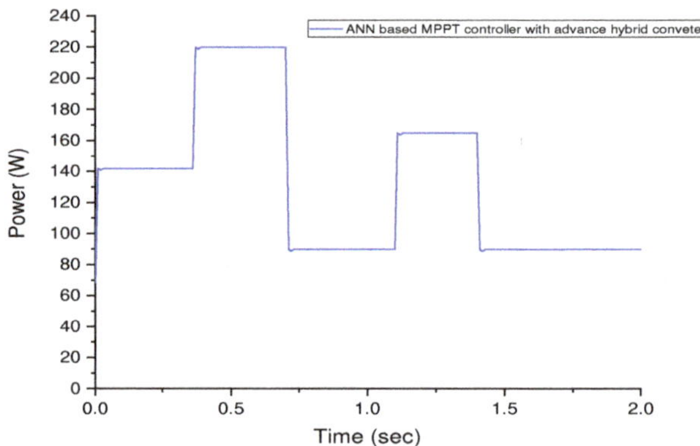

Figure 8. ANN based MPPT controller with hybrid boost converter.

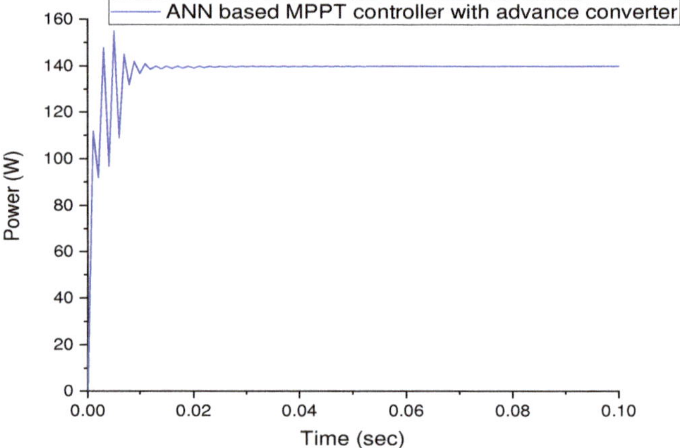

Figure 9. Zoom out portion of the ANN based MPPT controller with hybrid boost converter.

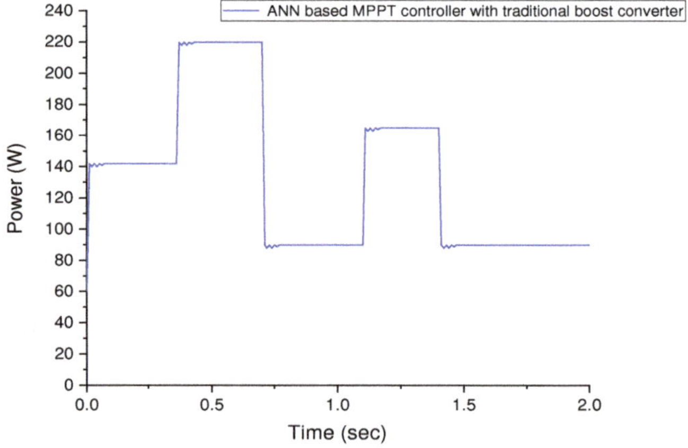

Figure 10. ANN based MPPT controller with traditional boost converter.

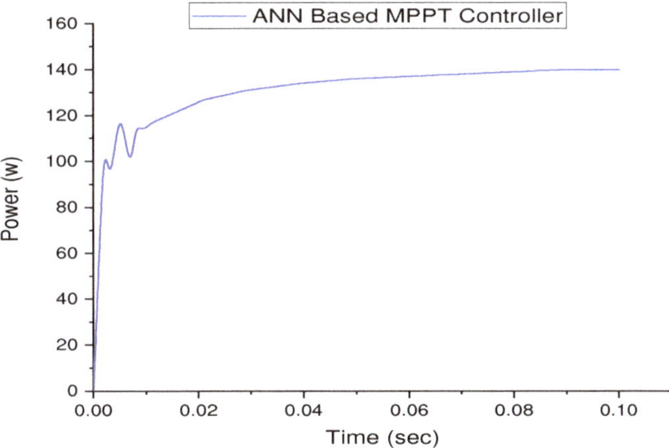

Figure 11. Analysis of varied area of the ANN based MPPT controller with traditional boost converter.

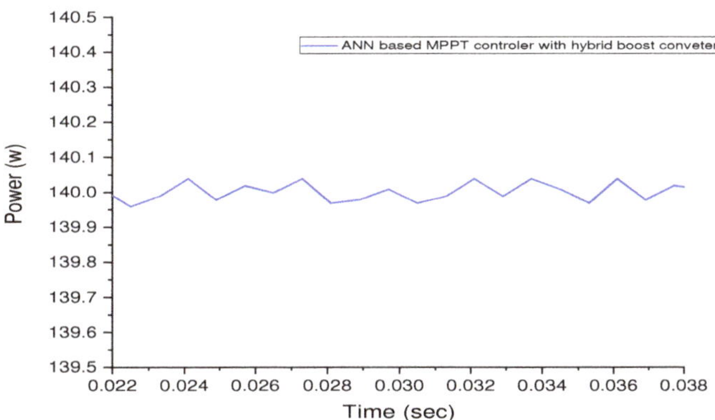

Figure 12. Power ripples of the ANN and hybrid boost converter based MPPT converter.

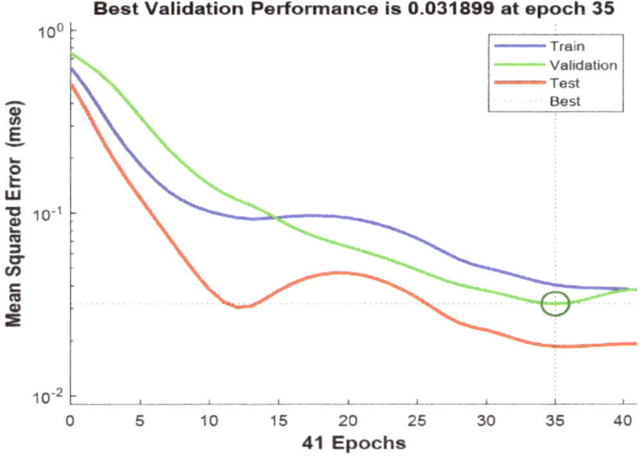

Figure 13. Maximum square error against epoch for evaluation the performance (41 Epochs).

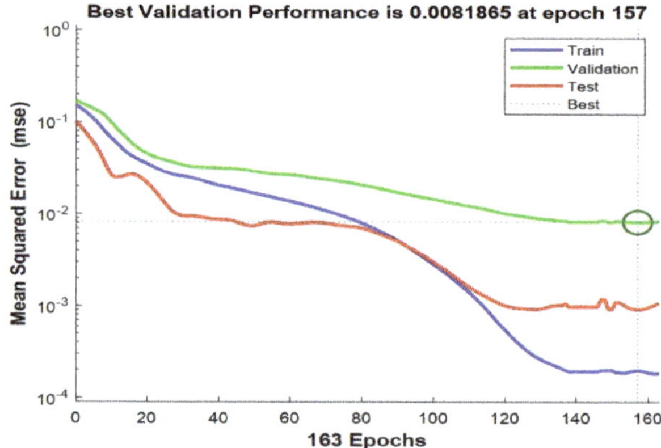

Figure 14. Maximum square error against epoch for evaluation the performance (163 Epochs).

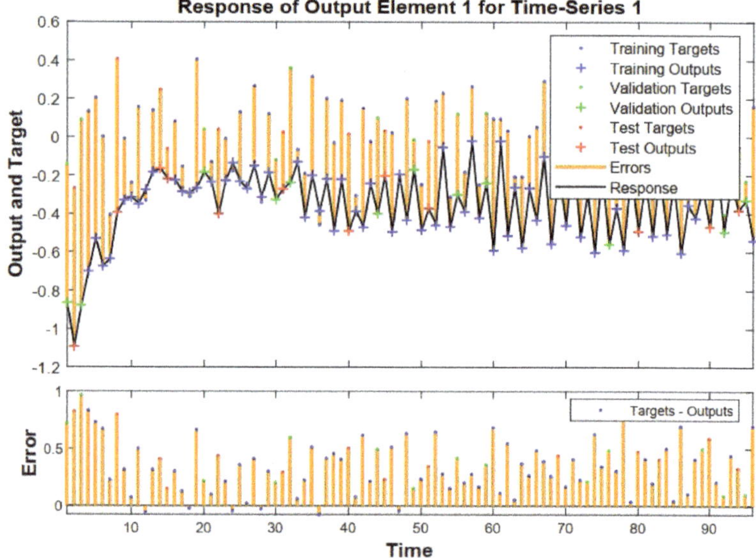

Figure 15. Measuring the output response in terms of error and output targets.

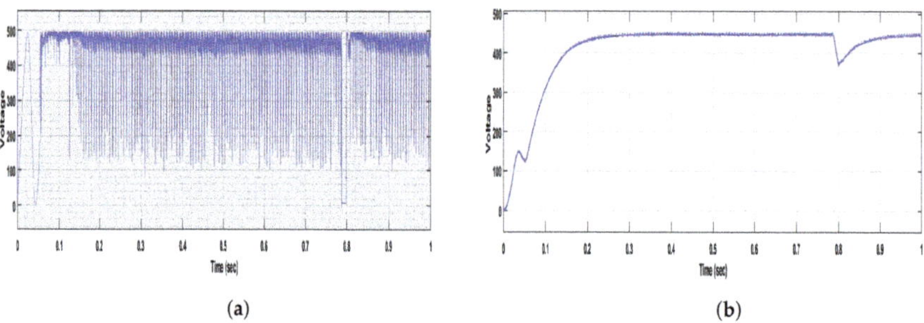

(a)

(b)

Figure 16. (**a**): Output results of the proposed model excluding ANN and HBC enabled MPPT, (**b**): Proposed model outcomes with ANN and HBC MPPT.

Table 4. Performance comparison with current published system.

Name of Parameter	[42]	[43]	Presented Model
Power extraction efficiency	68%	80%	85%
Maximum error of optimum	29%	23%	0.56%
MPPT accuracy	2.6%	2.6%	0.4%
Transient response	0.7	4	0.3

5. Conclusions

The increase in electric load is recorded during last decade. Solar plants are the fruitful solution to minimize pressure on the main power grids. In this paper, load management is discussed using an ANN based MPPT and hybrid boost converter. Furthermore, to maintain output power smart transformer is installed before load distribution. The ANN based MPPT controller with hybrid converter have fast tracking and less fluctuations. The mechanism of ANN based is explained with mathematical background. The traditional and proposed HBC are compared theoretically and the features of HBC are highlighted. The results of proposed ANN and HBC based MPPT are evaluated in terms of power and time for using traditional boost converter with ANN and proposed HBC with ANN. The results show the superiority of solar power system with hybrid converter to improve the efficiency of solar power system. In this work, the performance and productivity of the solar power system are strengthened by utilizing the ANN and HBC based MPPT. However, in future the power generation can be further improved by using advance algorithms and machine learning methodologies, such as the flower pollination algorithm (PFA), optical swarm optimization (PSO), convolution neural network (CNN), and coyote optimization algorithm (COA). In addition, for forecasting the solar power system condition the hybrid based ANN and adaptive neural-fuzzy inference system (ANFIS) using hybrid whale optimization and pattern search (HWO-PS) algorithm can be evaluated.

Author Contributions: Conceptualization, I.H., F.A. (Farman Ali), and A.A.; methodology, F.A. (Farman Ali), and I.H.; software, F.A. (Farman Ali), and A.M., F.A. (Fayadh Alenezi); validation, F.M., F.A. (Farman Ali) and N.A.; formal analysis, F.M., F.A. (Farman Ali), A.M., F.A. (Fayadh Alemezi); investigation, A.M., F.A. (Fayadh Alenezi), A.A. and N.A. data analysis, F.A. (Farman Ali), I.H.; writing—original draft preparation, I.H.; writing—review and editing, W.K., F.A. (Fayadh Alenezi); and A.A.; visualization, F.M., F.A. (Fayadh Alenezi); supervision, F.A. (Farman Ali), A.A.; project administration, F.A. (Farman Ali); funding acquisition, W.K., N.U., F.R.A. All authors have read and agreed to the published version of the manuscript.

Funding: The authors would like to acknowledge the support from Taif University Researchers Supporting Project Number(TURSP-2020/331), Taif University, Taif, Saudi Arabia.

Institutional Review Board Statement: Not applicable

Informed Consent Statement: Not applicable

Data Availability Statement: Data will be available as per request.

Conflicts of Interest: The authors declare no conflict of interest.

References

1. Markvart, T. *Solar Electricity*; John Wiley & Sons: Chichester, UK, 1994.
2. Rekioua, D.; Matagne, E. *Optimization of Photovoltaic Power Systems: Modelization, Simulation and Control*; Springer: London, UK, 2012.
3. De Soto, W.; Klein, S.A.; Beckman, W.A. Improvement and validation of a model for photovoltaic array performance. *Sol. Energy* **2006**, *80*, 78–88. [CrossRef]
4. Koutroulis, E.; Kalaitzakis, K.; Voulgaris, N.C. Development of a microcontroller-based, photovoltaic maximum power point tracking control system. *IEEE Trans. Power Electron.* **2001**, *16*, 46–54. [CrossRef]

5. Sera, D.; Teodorescu, R.; Hantschel, J.; Knoll, M. Optimized maximum power point tracker for fast-changing environmental conditions. *IEEE Trans. Ind. Electron.* **2008**, *55*, 2629–2637. [CrossRef]
6. Inthamoussou, F.A.; Battista, H.D.; Mantz, R.J. New concept in maximum power tracking for the control of a photovoltaic/hydrogen system. *Int. J. Hydrog. Energy* **2012**, *37*, 14951–14958. [CrossRef]
7. Femia, N.; Petrone, G.; Spagnuolo, G.; Vitelli, M. A technique for improving P & O MPPT performances of double-stage grid-connected photovoltaic systems. *IEEE Trans. Ind. Electron.* **2009**, *56*, 4473–4482.
8. Munir, H.K.; Nur, S.M.; Ahmed, E.-S. Wavelet based hybrid ANN-ARIMA models for meteorological drought forecasting. *J. Hydrol.* **2020**, *590*, 125380.
9. Ruiz-Aguilar, J.J.; Turias, I.; González-Enrique, J. A permutation entropy-based EMD–ANN forecasting ensemble approach for wind speed prediction. *Neural Comput. Appl.* **2021**, *33*, 2369–2391. [CrossRef]
10. Akbal, Y.; Ünlü, K.D. A deep learning approach to model daily particular matter of Ankara: Key features and forecasting. *Int. J. Environ. Sci. Technol.* **2021**. [CrossRef]
11. Abdul, R.P.; Damhuji, R.; Kharudin, A.; Muhammad, Z.M.; Ahmed, N.A.; Moneer, A.F. Solar irradiance measurement instrumentation and power solar generation forecasting based on Artificial Neural Networks (ANN): A review of five years research trend. *Sci. Total Environ.* **2020**, *715*, 136848.
12. Esram, T.; Chapman, P.L. Comparison of photovoltaic array maximum power point tracking techniques. *IEEE Trans. Energy Convers.* **2007**, *22*, 439–449. [CrossRef]
13. Femia, G.N.; Petrone, G.; Spagnuolo, G.; Vitelli, M. Optimization of perturb and observe maximum power point tracking method. *IEEE Trans. Power Electron.* **2005**, *20*, 963–973. [CrossRef]
14. Li, G.; Wang, H.A. Novel stand-alone PV generation system based on variable step size INC MPPT and SVPWM control. In Proceedings of the IEEE 6th International Power Electronics and Motion Control Conference, IEEE-IPEMC'09, Wuhan, China, 17–20 May 2009; p. 2155e60.
15. Safari, A.; Mekhilef, S. Simulation and hardware implementation of incremental conductance MPPT with direct control method using cuk converter. *IEEE Trans. Ind. Electron.* **2011**, *58*, 1154–11561. [CrossRef]
16. Reisi, A.R.; Moradi, M.H.; Jamasb, S. Classification and comparison of maximum power point tracking techniques for photovoltaic system: A review. *Renew. Sustain. Energy Rev.* **2013**, *19*, 433–443. [CrossRef]
17. Kamarzaman, N.A.; Tan, C.W. A comprehensive review of maximum power point tracking algorithms for photovoltaic systems. *Renew. Sustain. Energy Rev.* **2014**, *37*, 585–598. [CrossRef]
18. Xiao, W.; Dunford, W.G. A modified adaptive hill climbing MPPT method for photovoltaic power systems. In Proceedings of the 35th Annual IEEE Power Electronics Specialists Conference, Aachen, Germany, 20–25 June 2004; pp. 1957–1963.
19. Liu, F.; Kang, Y.; Zhang, Y.; Duan, S. Comparison of P & O and hill climbing MPPT methods for grid-connected PV converter. In Proceedings of the 3rd IEEE Conference on Industrial Electronics and Applications, Singapore, 3–5 June 2008; pp. 804–807.
20. Mutoh, N.; Matuo, T.; Okada, K.; Sakai, M. Prediction-databased maximum-power-pointtracking method for photovoltaic power generation systems. In Proceedings of the IEEE 33rd Annu. Power Electronics Specialists Conference, Cairns, QLD, Australia, 23–27 June 2002; pp. 1489–1494.
21. Chao, K.H.; Li, C.J.; Wang, M.H. *A Maximum Power Point Tracking Method Based on Extension Neural Network for PV Systems [Part I, LNCS 5551]*; Springer: Wuhan, China, 2009; pp. 745–755.
22. Yasushi, K.; Koichiro, Y.; Masahito, K. Quick Maximum Power Point Tracking of Photovoltaic Using Online Learning Neural Network. In Proceedings of the International Conference on Neural Information Processing ICONIP 2009: Neural Information Processing, Bangkok, Thailand, 1–5 December 2009; pp. 606–613.
23. Majed, B.A.; Maher, C.; Zied, C. Artificial Neural Network based control for PV/T panel to track optimum thermal and electrical power. *Energy Convers. Manag.* **2013**, *65*, 372–380.
24. Shahzad, A.; Hafiz, M.; Muhammad, A. Iftikhar, A.; Muhammad, K.A.; Zil, H.; Safdar, A.K. Supertwisting Sliding Mode Algorithm Based Nonlinear MPPT Control for a Solar PV System with Artificial Neural Networks Based Reference Generation. *Energies* **2020**, *13*, 3695. [CrossRef]
25. Liu, Y.H.; Liu, C.L.; Huang, J.W.; Chen, J.H. Neural-network-based maximum power point tracking methods for photovoltaic systems operating under fast changing environments. *Sol. Energy* **2013**, *89*, 42–53. [CrossRef]
26. Chekired, A.; Mellit, S.A.; Kalogirou, C.L. Intelligent maximum power point trackers for photovoltaic applications using FPGA chip: A comparative study. *Sol. Energy* **2014**, *101*, 83–99. [CrossRef]
27. Rajib, B.R.; Rokonuzzaman, M.; Amin, N.; Mishu, M.K.; Rahman, S.; Mithulananthan, N.; Rahman, K.S.; Shakeri, M.; Pasupuleti, J. A Comparative Performance Analysis of ANN Algorithms for MPPT Energy Harvesting in Solar PV System. *IEEE Access* **2021**, *9*, 102137–102152.
28. Fathi, M.; Parian, J.A. Intelligent MPPT for photovoltaic panels using a novel fuzzy logic and artificial neural networks based on evolutionary algorithms. *Energy Rep.* **2021**, *7*, 1338–1348. [CrossRef]
29. Jiang, M.; Ghahremani, M.; Dadfar, S.; Chi, H.; Abdallah, Y.N.; Furukawa, N. A novel combinatorial hybrid SFL–PS algorithm based neural network with perturb and observe for the MPPT controller of a hybrid PV-storage system. *Control Eng. Pract.* **2021**, *114*, 104880. [CrossRef]
30. Badreddine, B.; Amar, B.; Noureddine, H. A novel nature-inspired maximum power point tracking (MPPT) controller based on ACO-ANN algorithm for photovoltaic (PV) system fed arc welding machines. *Neural Comput. Appl.* **2021**. [CrossRef]

31. Zerglaine, A.; Mohammedi, A.; Bentata, K.; Rekioua, D.; Oubelaid, A.; Mebarki, N.E. Enhancement of Extracted Photovoltaic Power Using Artificial Neural Networks MPPT Controller. *Adv. Green Energies Mater. Technol.* **2021**, 265–272._35. [CrossRef]
32. Faisal, S.; Muhammad, H.Y.; Haider, A.T.; Muhammad, R.A.; Zeeshan, A.A.; Muhammad, H.K. Performance Benchmark of Multi-Layer Neural Network Based Solar MPPT for PV Applications. In Proceedings of the 2021 International Conference on Emerging Power Technologies (ICEPT), Topi, Pakistan, 10–11 April 2021.
33. Ankit, G.; Pawan, K.; Rupendra, K.P.; Yogesh, K.C. Performance Analysis of Neural Network and Fuzzy Logic Based MPPT Techniques for Solar PV Systems. In Proceedings of the 2014 6th IEEE Power India International Conference (PIICON), Delhi, India, 5–7 December 2014.
34. Zhang, H.; Cheng, S. A new MPPT algorithm based on ANN in solar PV systems. Advances in Computer. In *Communication, Control and Automation*; Springer: Berlin/Heidelberg, Germany, 2011; pp. 77–84.
35. Yong, Z.; Hong, L.; Liqun, L.; Xiao, F.G. The MPPT control method by using BP neural networks in PV generating system. In Proceedings of the 2012 International Conference on Industrial Control and Electronics Engineering, Hangzhou, China, 23–25 March 2012.
36. Bendib, B.; Krim, F.; Belmili, H.; Almi1, M.F.; Bolouma, S. An Intelligent MPPT Approach based on Neural- Network Voltage Estimator and Fuzzy Controller, Applied to a Stand-alone PV System. In Proceedings of the 2014 IEEE 23rd International Symposium on Industrial Electronics (ISIE), Istanbul, Turkey, 1–4 June 2014.
37. Poom, K.; Somyot, K. Maximum Power Point Tracking Using Neural Network in Flyback MPPT inverter for PV systems. In Proceedings of the SCIS-ISIS 2012, Kobe, Japan, 20–24 November 2012.
38. Rahul, D. Neural Network MPPT Control Scheme With Hysteresis Current Controlled Inverter For Photovoltaic System. In Proceedings of the 2014 RAECS UIET Panjab University, Chandigarh, India, 6–8 March 2014.
39. Aymen, C.; Rashad, M.K.; Ken, N. A novel multi-model neuro-fuzzy-based MPPT for three-phase grid-connected photovoltaic system. *Sol. Energy* **2010**, *84*, 2219–2229.
40. Wai, R.J.; Lin, C.Y.; Duan, R.Y.; Chang, Y.R. Highefficiency dc-dc converter with high voltage gain and reduced switch stress. *IEEE Trans. Ind. Electron.* **2007**, *54*, 354–364. [CrossRef]
41. Wu, T.F.; Lai, Y.S.; Hung, J.C.; Chen, Y.M. Boost converter with coupled inductors and buck-boost type of active clamp. *IEEE Trans. Ind. Electron.* **2008**, *55*, 154–162. [CrossRef]
42. Vahedi, H.; Sharifzadeh, M.; Al-Haddad, K. Modified Seven-Level Pack U-Cell Inverter for Photovoltaic Applications. *IEEE J. Emerg. Sel. Top. Power Electron.* **2018**, *6*, 1508–1516. [CrossRef]
43. Babaie, M.; Sharifzadeh, M.; Mehrasa, M.; Chouinard, G.; AlHaddad, K. PV Panels Maximum Power Point Tracking based on ANN in Three-Phase Packed E-Cell Inverter. In Proceedings of the 2020 IEEE International Conference on Industrial Technology (ICIT), Buenos Aires, Argentina, 26–28 February 2020; pp. 854–859. [CrossRef]

Article

Online Distributed Measurement of Dark I-V Curves in Photovoltaic Plants

José Ignacio Morales-Aragonés [1], María del Carmen Alonso-García [2], Sara Gallardo-Saavedra [1], Víctor Alonso-Gómez [1], José Lorenzo Balenzategui [2], Alberto Redondo-Plaza [1] and Luis Hernández-Callejo [1,*]

[1] Campus Universitario Duques de Soria, Universidad de Valladolid, 42004 Soria, Spain; ziguratt@coit.es (J.I.M.-A.); sara.gallardo@uva.es (S.G.-S.); victor.alonso.gomez@uva.es (V.A.-G.); alberredon@gmail.com (A.R.-P.)

[2] Centro de Investigaciones Energéticas, Medioambientales y Tecnológicas (CIEMAT), Photovoltaic Solar Energy Unit, Energy Department, 28040 Madrid, Spain; carmen.alonso@ciemat.es (M.d.C.A.-G.); jl.balenzategui@ciemat.es (J.L.B.)

* Correspondence: luis.hernandez.callejo@uva.es; Tel.: +34-975129418

Abstract: The inspection techniques for defects in photovoltaic modules are diverse. Among them, the inspection with measurements using current–voltage (I-V) curves is one of the most outstanding. I-V curves, which can be carried under illumination or in dark conditions, are widely used to detect certain defects in photovoltaic modules. In a traditional way, these measurements are carried out by disconnecting the photovoltaic module from the string inside the photovoltaic plant. In this work, the researchers propose a methodology to perform online dark I-V curves of modules in photovoltaic plants without the need of disconnecting them from the string. For this, a combination of electronic boards in the photovoltaic modules and a bidirectional inverter are employed. The results are highly promising, and this methodology could be widely used in upcoming photovoltaic plants.

Keywords: dark I-V curves; bidirectional power inverter; online distributed measurement of dark I-V curves

Citation: Morales-Aragonés, J.I.; Alonso-García, M.d.C.; Gallardo-Saavedra, S.; Alonso-Gómez, V.; Balenzategui, J.L.; Redondo-Plaza, A.; Hernández-Callejo, L. Online Distributed Measurement of Dark I-V Curves in Photovoltaic Plants. *Appl. Sci.* **2021**, *11*, 1924. https://doi.org/10.3390/app11041924

Academic Editor: Francesco Calise

Received: 20 January 2021
Accepted: 19 February 2021
Published: 22 February 2021

Publisher's Note: MDPI stays neutral with regard to jurisdictional claims in published maps and institutional affiliations.

1. Introduction

Solar photovoltaic (PV) energy is a reliable renewable energy source that has increased its cumulative installed capacity up to 633.7 GW by the end of 2019 [1,2]. This means an increase of 23% from the 516.8 GW achieved in 2018 and represents a progression by almost 400 times the installed capacity at the beginning of the century. In terms of annual installed capacity, the 116.9 GW installed in 2019 reflects a growth of 13% with respect to 2018, and it positions solar photovoltaic as the champion power generation source installed in 2019, with 48% of annual share [1]. The COVID-19 pandemic and its associated crisis slowed down the progression of PV solar energy in 2020. A recent study of the International Energy Agency analyzes different consequences of the pandemic and accordingly proposes several scenarios of possible energy futures. In all scenarios analyzed, renewable energies will grow rapidly, with solar in the lead of this new era of electricity generation [3].

Therefore, solar PV energy is a key factor in the diversification of energy sources that promotes a gradual decarbonation and encourages the large-scale implementation of energy models free of greenhouse gas emissions. The reason for the spectacular growth and prospect of this energy source lays in the constant development of the technology and its high reliability. This has made possible a drastic reduction of costs, which has favored its large-scale implementation. Moreover, PV applications have a variety of configurations and power levels, ranging from a few watts for consumer products or small home applications to large utility-scale power plants. These power plants have represented the largest share in the solar market in preceding years, with great expectations to keep this trend in next years [3]. In these high-power applications, the long-term reliability of the plant results in the recovery of the investment and the profit of the plant. In order to ensure this long-term reliability and a return on investment, the components of a PV plant, and the plant itself, are subjected to careful control procedures before, during, and after their construction. There exists a large collection of quality control mechanisms and international standards to asseverate the guaranty of the PV systems, being one of most widely used the ones issued by Technical Committee 82 of the International Electrotechnical Commission [4].

Ensuring energy production is a key factor in warranting plant profitability, and this has forced the design of increasingly intelligent and advanced operation and maintenance (O&M) strategies [5–8]. The maintenance operation included in most O&M PV power plant contracts can be divided into Preventive, Corrective, and Predictive. While corrective maintenance is performed after some failure has been detected, preventive and predictive seek to anticipate the fault. Both need specialized personnel to perform the tasks and to analyze the data, and the second one requires "intelligent" equipment for the monitoring that can supply information about the state of the plant, allowing the evaluation of subtle trends that could be unnoticed in regular inspections.

Among the advanced operation and maintenance techniques that provide information about the PV modules, infrared thermography (IRT), electroluminescence (EL), and string current–voltage (I-V) curve measurement can be highlighted.

IRT has been widely used to detect failures in PV modules and plants [9–15]. It has the advantage of being non-intrusive, and it can be done while the plant is in operation. EL, on the other hand, is an excellent technique for the detection of failures such as cracks, interconnection defects, broken cell fingers, or other cell and string issues [16,17]. It can be used at the laboratory level in PV modules [18,19] or in PV plants where bigger areas can be analyzed with ground-mounted or aerial equipment [20]. To perform EL analysis, it is necessary to forward bias the module or array. This implies disconnecting the module or array and the use of an electronic load or an external power supply to bias the module or string.

Finally, I-V curve measurement gives more comprehensive information about the state of a PV module or string. The main characteristic parameters of the device are obtained from the I-V curve. Some of them are inferred directly from the measurement and give information about the performance: short circuit current (I_{sc}), open circuit voltage (V_{oc}), and maximum power point (P_{max}) with its associate current (I_{mpp}) and voltage (V_{mpp}). Others give information about the physical properties of the device under study, and they are obtained from the I-V curve by applying models. The most widely used are the double or single diode models [21–23], which supply the parameters: photogenerated current I_L, series resistance R_s, shunt resistance R_{sh}, diode saturation current I_0, and ideality factor m. By analyzing the magnitude and evolution of these parameters, the degradation of a PV device can be quantified. There are several techniques and devices used to measure the I-V curve of PV devices in the field. An excellent review can be found elsewhere [24]. In addition, some authors developed a mobile laboratory to perform I-V curve measurements of PV modules in the plant at standard test conditions according to IEC standard [25,26]. This has the advantage of ensuring uniform environmental conditions during the tests and it allows a precise quality control of selected PV modules throughout their shelf life.

The measurement of the I-V curve of PV modules or strings in a PV plant under illumination implies in most of the cases the disconnection of the module/array in order to sweep the operating point from I_{sc} to V_{oc}. This has the drawback, on the one hand, of requiring the disconnection of the plant and thus production, and on the other, the high voltages and/or currents when strings are measured, which forces taking rigorous personal security measures. In recent times, electronic boards to be installed in photovoltaic modules and inverters have been developed, from which it is possible to make I-V traces without disconnecting the photovoltaic modules. For example, in [27], the authors present two I-V tracing strategies at the photovoltaic module level without the need to disconnect them. One strategy uses electronic boards at the photovoltaic module level, while the second strategy combines electronic boards at the photovoltaic module level together with an electronic string card (common for all photovoltaic modules). In both cases, the measures are of high quality, and the strategies have low costs.

The measurement at dark conditions has the clear advantage that it is performed when the plant is not in operation. While the normal operating mode of a plant or PV modules is under illumination, the dark curve gives very important information about the characteristics of the device, and it allows analyzing various failure modes. It has been used for many years to determine solar cell parameters [21,28,29] and as a diagnosis to detect certain defects such as mismatch, handling, soldering, or lamination problems in module manufacturing [30]. More recent works propose the use of a dark I-V curve alone or in combination with a light I-V curve for the identification of several failure modes. In [31], complementary analysis of dark and light current voltage characteristic is used to characterize failure modes such as degradation of the electrical circuit of the PV module, mechanical damage to the solar cells, and potential-induced degradation (PID). In [32], the analysis of a dark I-V curve to in situ monitor the degradation of PV modules undergoing thermal cycling and mechanical loading stress testing is proposed. The same group proposes an extension of their method to analyze in situ the temperature dependence of power loss estimations in PID experiments [33].

In this paper, we go a step further and propose the use of the developed technology to measure online distributed dark I-V curves of the PV modules in the plant. The online measurement refers to the acquisition without the need to disconnect the PV module from the string to which it is connected. It is a distributed strategy, since all the electronics required for the I-V measurement are located in the PV modules. Therefore, the proposed strategy implies the automatic acquisition of dark I-V curves by the developed electronic device, which is included in the PV modules, without the need for the onsite disconnection of any component in the plant. This is especially interesting for PV plant operators, who can obtain the dark I-V curves of the modules, which give relevant information of the individual modules, as has been outlined (characteristics of the device, to analyze various failure modes, to determine solar cell parameters, for monitoring the degradation of modules, to diagnosis or characterize certain defects and failure modes) with the advantage of online measures, significantly reducing the onsite workforce, acquisition time, and costs. The paper has been structured as follows: it starts with an introduction to the research in Section 1, followed by the materials and method presented in Section 2, the results and their discussion in Section 3, and the main conclusions in Section 4.

2. Materials and Method

This section presents the materials and method used in the research performed. It has been divided in five subsections. The first one revises the dark I-V curves, highlighting their importance in PV inspections; the second and the third subsections explain the bidirectional inverter and the electronic board used in the research, respectively. Both have been developed by this research group. In the fourth subsection, the main characteristics of the PV modules tested are detailed. Finally, subsection five describes the methodology followed in the research proposed in this article.

2.1. Dark I-V Curves

Solar PV cells convert sunlight into electricity. All potential combinations of current and voltage pairs of points of the PV device under certain conditions of irradiance and temperature are represented in an I-V curve. There are different models that describe the electrical behavior of the I-V curve of a PV device. The one exponential model is widely used and it includes the series and shunt resistances, as detailed in the following Equation (1) [10,21]. Figure 1 shows the equivalent circuit of the one-diode model of a PV cell under illumination conditions (a) and dark conditions (b).

$$I = I_L - I_0 \left[\exp\left(\frac{V + I\,R_s}{n\,v_t} \right) - 1 \right] - \left(\frac{V + I\,R_s}{R_{sh}} \right) \tag{1}$$

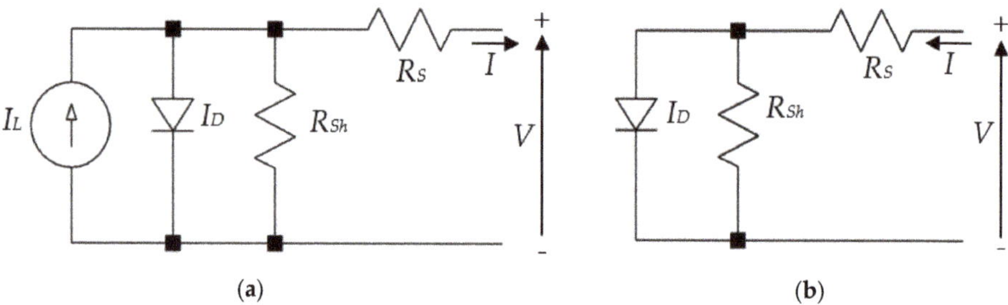

(a)   (b)

Figure 1. One-diode model of a photovoltaic (PV) cell under illumination conditions (**a**) and dark conditions (**b**), including the series and shunt resistances.

In Equation (1), I_L is the photogenerated current, I_0 is the saturation current of the diode, n is the diode ideality factor, R_s is the series resistance, R_{sh} is the shunt resistance, and v_t is the thermal voltage ($K\,T/e$) with k as the Boltzmann constant, e as the electron charge, and T as the temperature in Kelvin. In the dark conditions model circuit, as the photogenerated current I_L due to illumination is zero, the current generator is missing. In Figure 2, there is a common I-V curve under illumination represented in the fourth quadrant, which overlaps the dark diode current with the photogenerated current due to illumination and the dark I-V curve in the first quadrant. The main characteristic points of the curve in illumination are marked in this figure. The criteria that have been considered in this research are that the current is negative when the PV device is generating power (fourth quadrant).

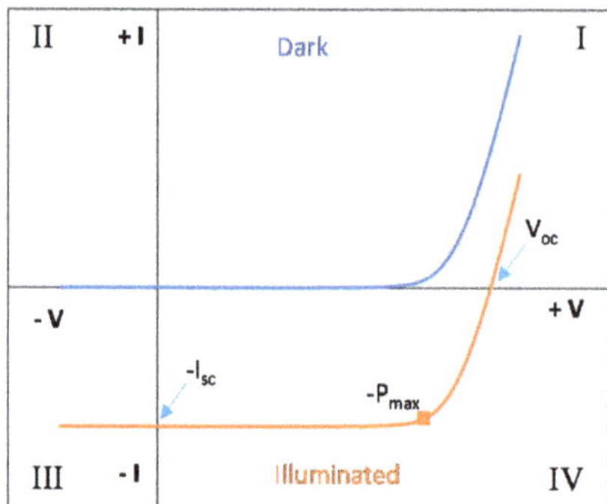

Figure 2. Illuminated and dark current–voltage (I-V) curves with their characteristic points.

As it has been remarked in the introduction, the measurement at dark conditions has the clear advantage that it is performed when the plant is not in operation, giving very important information about the characteristics of the device. It allows analyzing various failure modes, determining solar cell parameters, and monitoring the degradation of modules and it is used as a diagnosis to detect or characterize certain defects and failure modes. The application of the model of Equation (1) with the calculation of the parameters for the dark I-V curve can give information about the changes of a PV module performance, indicating that some failure or degradation has occurred. To illustrate how the variation of the parameters affect the dark I-V curve, Figure 3a,b show the changes in the dark I-V curve of a standard PV cell in which each one of the model parameters of Equation (1) has been changed one at a time. Parameters are expressed in relation to the area for being more comparable, so current is presented in mA/cm^2, and J_0 represents the diode saturation current density. The reference curve corresponds to a standard cell with the following parameters: $J_0 = 1.5 * 10^{-9}$ A/cm^2, $n = 1.5$, $R_s = 0.6$ Ω cm^2, and $R_{sh} = 60.000$ Ω cm^2. From these initial values, parameters have been changed to simulate worse device behavior by increasing R_s, J_0, and n, and decreasing R_{sh}. The graph has been split into two parts to better appreciate the modifications that the change of each parameter introduces in the I-V curve. For the case of variations in "n" and R_{sh}, the graph is presented in logarithm scale to better distinguish its influence and the voltage area in which these changes produce their effect.

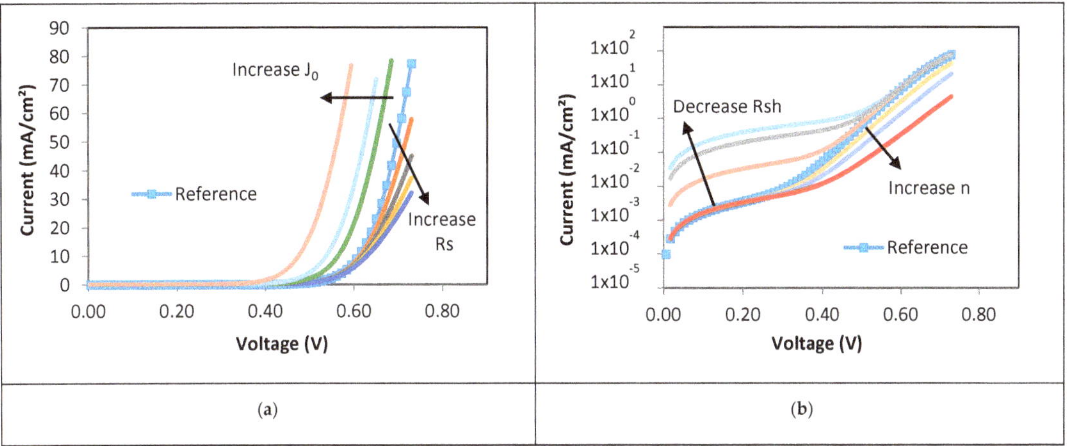

Figure 3. Simulated dark I-V curves by introducing changes in one of the exponential parameters. Reference curve: $J_0 = 1.5 \times 10^{-9}$ A/cm^2, $n = 1.5$, $R_s = 0.6$ Ω cm^2, and $R_{sh} = 60.000$ Ω cm^2, (**a**) increase in R_s from 0.6 to 2.5 Ω cm^2 and J_0 from 1.5×10^{-9} to 5×10^{-8} A/cm^2; (**b**) increase in n from 1.5 to 1.9 and decrease in R_{sh} from 60,000 to 500 Ω cm^2. Graph (**b**) is presented in logarithm scale.

The variations of the curves presented in Figure 3 can be indicative of certain degradation modes. For example, the degradation of the electrical circuit of a PV module usually contributes to the increasing of series resistance [31], which is appreciated in the high-voltage area of the I-V curve. Mechanical damage would contribute also to the increase of series resistance, but it will also increase recombination and shunt losses [31], as appreciated by the changes in J_0 and R_{sh}. Shunting losses are visible in the low-voltage region. These types of analysis through the dark I-V curve have been shown to be especially interesting to detect degradation provoked by PID [31,34,35].

An additional way to evaluate the performance of a PV device through the dark I-V curve is by obtaining the parameters of the dark curve equivalent to the illuminated one by transposing the dark I-V curve to the short circuit current of the illuminated one and calculating the *fill factor dark (FF$_D$)* [31,34,35]. A decrease in this *FF$_D$* implies degradation in the performance of the PV device.

In a common arrangement, it is necessary to forward bias the module with an external power supply in order to perform dark I-V curves. As a novel approach, a power inverter with bidirectional power flow capability has been used in this work for biasing the modules instead of the external power supply by applying the procedure explained in the following subsections [36].

2.2. Power Inverter with Bidirectional Power Flow Capability

Nowadays, common solar inverters used in PV plants only have the capability to invert the electricity produced in the solar panels to lately be injected to the grid. However, the use of a power inverter with bidirectional power flow capability is proposed in recent research [36] as a novel system that extremely facilitates the operation and maintenance of PV plants, allowing the on-site outdoor EL and IRT inspections [10]. In this scenario, this device is suitable for the acquisition of dark I-V curves, as permits the current injection to the modules (I quadrant).

The 3 kW bidirectional inverter pilot presented in [36] has been used in this research. In the DC input side, the PV string voltage can be set between 330 and 550 V. The bidirectional inverter allows fixing ten different levels of current injection to the string of modules, from 10% of I_{sc} to 100% of I_{sc} in steps of 10% of I_{sc}. For the dark I-V curves tracing presented in this paper, the bidirectional inverter has been set to 100% of I_{sc}, and the electronic board is in charge of tracing the curve, as it is explained in the following section. Although the leakage current is dependent on the voltage, the temperature, and the humidity conditions, at low temperatures (< 25°C), the influence of the voltage in the leakage current is minor [37]. Therefore, achieving slightly higher values of voltage in defective modules during outdoor dark I-V curve measurement, in which the temperature is low and the acquisition time is rapid (40 ms) should not affect the leakage current. As well, the authors in [38] proved how injecting current for long periods does not degrade the panel. Hence, the degradation of the panels should be insignificant during a 40 ms measurement. That is why the maximum limit of 100% of I_{sc} has been selected, obtaining complete dark I-V curves. However, the maximum current limit can be set in the bidirectional inverter if desired in any specific case.

This bidirectional inverter allows EL to be done at night in individual modules or in the string with this system, without the need for an external source. The current set point, sent from the computer to the inverter, is controlled by means of the PV side voltage. Figure 4 shows the power inverter with bidirectional power flow capability used in on-site measurements.

Figure 4. Bidirectional power inverter of the Campus Duques de Soria of the University of Valladolid.

2.3. Electronic Board Integrated in Photovoltaic Modules

For the local measurements over each PV module, an electronic board has been designed to be installed within the module connecting box. In order to obtain a full I-V curve of the PV module under dark conditions, it is necessary to sweep the forward current injected from the bidirectional inverter explained above between zero and its maximum value (100% of I_{sc}). This way, a set of module current values and their corresponding voltage values can be measured and stored as points of the I-V curve.

It is possible to perform this sweep controlling the current supplied by the bidirectional inverter, but if a good resolution in the current values is needed, it will require a precise regulation of string voltages of some hundreds of volts, increasing the complexity of the bidirectional inverter, and in addition, a communications system between the inverter and the module electronics will be required to synchronize the sweep with the measurements. For these reasons, we have opted for a local sweeping electronics integrated in the PV module.

Figure 5 shows the schematic of the PV module electronic board. An 8-bit microcontroller is the heart of the circuit as it stores and executes the firmware for generating all the necessary signals that perform the I-V tracing process. For the measurement of current and voltage values, two of the microcontroller pins (labelled as V and I in Figure 5) can be configured as analog inputs for an internal analog-to-digital converter (ADC) of 10-bits. The module voltage analog input (V) is fed from the positive module terminal through a voltage divider. The resistive values of the voltage divider are much greater than the PV module R_{sh}; therefore, they do not alter the module parallel resistance. This divider reduces the module voltage for a full-scale matching with the positive reference of the ADC (5 V). Consequently, the resolution of the 10-bits ADC is 5 mV (5 V/1024), which corresponds with the full scale of the I-V curve measurement, which results in a resolution of 50 mV (50V/1024) in voltage measurements.

The module current is measured with an AMR (Anisotropic Magnetoresistance) sensor (I_s), which outputs a voltage proportional to the current with a 5 volts swing for a current value from zero to 5 Amps, so it can be connected directly to the 10 bit ADC input (I). This setup allows for a current resolution of 5 mA (5 Amps/1024). The MOSFET transistor M1 will be responsible for the local current sweeping process. The M1 gate is driven directly by the microcontroller pin (Cm), which is configured as an analog output from an internal digital-to-analog converter (DAC). For a smooth and as linear as possible sweep, a closed control loop has been implemented in firmware, where the current and voltage values sampled by the ADC are used as a feedback for adjusting the gate voltage of M1. This strategy allows for the tuning of the control loop by simply changing some parameters in the firmware (as delay or gain).

The power supply for the entire system is provided by the voltage module established when the forward current from the bidirectional inverter flows along the string, via a DC/DC converter that outputs the 5 volts needed by the electronics. Since during the I-V tracing, the current sweep forces the module to different operating points, including the zero voltage, a power supply holding circuit has been implemented in order to keep the supply voltage at the input of the DC/DC converter high enough during the time that the I-V tracing is performed. This circuit is composed by the capacitor (C) and the diode (D) in such a way that capacitor (C) is charged up to the module voltage when the circuit is idle. When during an I-V tracing the module voltage drops, the diode (D) is reverse biased, avoiding the capacitor discharge over the module, and this capacitor can then supply the energy to the circuit until the I-V tracing is finished.

For the experiments, the communications with the microcontroller for tracing demand or data download have been implemented using the serial port integrated in it, which is connected to an external computer that send the commands for starting the measurements and receives the numerical data of the I-V curves.

The process applied for I-V curve tracing is based on these consecutive steps:

- First, in dark conditions, the I_{sc} current is driven from the bidirectional inverter towards the modules string in a forward biased way (opposite of the current flow in production while in daylight). This establishes a voltage in the module terminals, and the circuit is powered up.
- When the I-V tracing is demanded from the external computer via a serial port, the MOSFET M1 is closed, subtracting all the current from the module, thus driving the module to zero voltage.
- The current sweep and the current and voltage measurements are started simultaneously. The MOSFET is pushed from its initial closed state to an open state; then, the

module voltage returns to its maximum, and the I-V tracing is finished. The maximum current value is the one fixed initially in the bidirectional inverter (I_{sc}).

- The microcontroller sends the data corresponding to the I-V curve to the external computer, and the circuit returns to the idle mode.

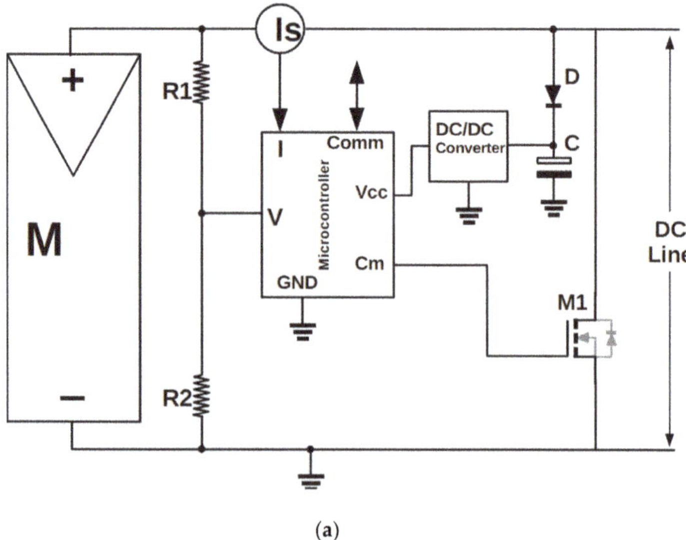

(a)

(b)

Figure 5. Schematic of the module electronic board (**a**) and a real picture of the card (**b**).

2.4. Tested Modules

On-site outdoor tests have been performed in the PV installation in the School of Forestry, Agronomic, and Bioenergy Industry Engineering (EIFAB) of the University of Valladolid, in Soria, Spain. It is shown in the upper row of PV modules in Figure 6. It is composed by eleven mono-crystalline PV modules with different kinds of defects. Their main characteristics are presented in Table 1. The area of each cell is 156.25 cm^2.

Figure 6. Photovoltaic (PV) installation in Campus Duques de Soria.

Table 1. Main characteristics of the PV modules tested.

Module Number	Module Model	Cells	Power (P) [W]	V_{oc} [V]	V_{mpp} [V]	I_{sc} [A]	I_{mpp} [A]
1, 2, 3, 4, 5, 7, 8 and 9	EOPLLY		175	44.35	36.26	5.45	4.83
6	EOPLLY	72 cells	165	43.92	35.64	5.23	4.63
10	EGNG	(125 mm × 125 mm)	180	44.4	35.4	5.35	5.08
11	SKY GLOBAL		175	42.6	35.5	5.52	4.93

2.5. Method

As it has been introduced, the objective of the research presented in this paper is the use of the developed electronic board integrated in photovoltaic modules presented in Section 2.3 to measure the online distributed dark I-V curves of the PV modules in the plant. To obtain and validate the dark I-V curves, the followed methodology has been used: firstly, the dark I-V curves of all the modules presented are measured with the developed device, and secondly, the main parameters of the curves are extracted to draw some conclusions on the state of the analyzed PV modules.

I-V curve measurements are carried out after sunset. It is not necessary to cover the panels, since the influence of the moon is completely negligible, 3.916 * 10^{-3} W/m^2 (perigee, perihelion) for a full moon [37]. For the I-V curves acquisition, 100% of the I_{sc} setpoint is selected in a computer connected to the bidirectional inverter, which gives the order. The eleven PV modules under analysis are all connected in series to the bidirectional inverter (which is unique for the entire string). To make the dark I-V curve measurements, each module has one electronic board integrated (the developed card presented). In order, one of the cards makes a short circuit in the module, so that the rest of the modules of the string remain connected in series to the inverter while the card makes the dark I-V curve of that first module. When the acquisition of the curve ends, the card re-integrates the first module measured in the string. Successively, the measurements of all the modules are made. The time it takes to acquire the I-V curve of each module is 40 ms. A better definition, reliability, and low disturbance in the low-current region of the I-V curve can be obtained, if required, by the acquisition of several I-V curves and averaging [39], but at the expense of a longer time of exploration. For installations with a large number of PV modules, a balance between total time of measurement and curve definition should be considered.

Ambient and module temperatures in the I-V curve acquisition moment have been measured and will be presented within the results. The temperatures were measured using an external PT1000 temperature probe, as this electronic board prototype does not have any temperature nor irradiance sensor incorporated. The temperature probe has been attached to a healthy cell in the middle of the PV module. A schema of the system is presented in Figure 7, showing the global operating diagram, where it is possible to see the bidirectional inverter, as well as the electronic boards installed in each photovoltaic module. The injection of current in dark conditions will allow the realization of the dark I-V curve.

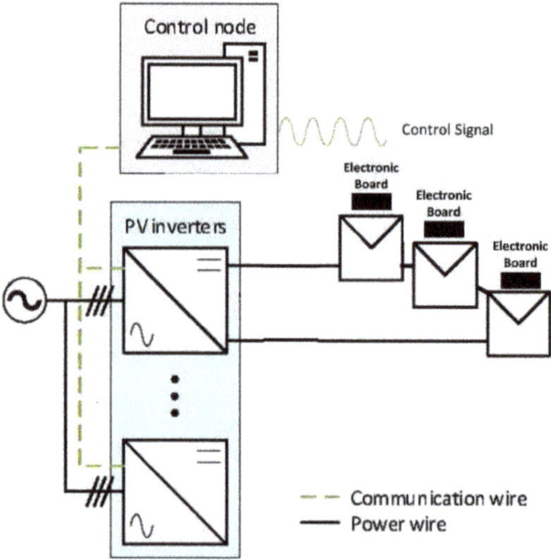

Figure 7. Global operating diagram of the system.

Once the dark I-V curves are captured, the main parameters of the one exponential model (single-diode model) of each module are obtained using the 2/3 Diode Fit software [40] with the objective of drawing some conclusions on the state of the analyzed PV modules. The free software 2/3 Diode Fit permits calculating model parameters for PV devices using a 1-diode model, 2-diode model, and other more complex options. The software provides initial guesses of the parameters, calculating the slopes and applying conditions in the areas in which each parameter has more influence. From these starting parameters, optimized ones are obtained through an iterative procedure, in which it is possible to select the precision and stopping criteria. The algorithm to calculate the I-V curve is described in detail in [41], and all program functionalities and a comprehensive explanation of the equations and options available can be found in the program manual available at [40]. As it has been detailed, these parameters are extremely useful for analyzing the appearance of different failure modes, monitoring the degradation of modules, and detecting or characterizing certain defects, among others.

3. Results and Discussion

3.1. Dark I-V Curves Onsite Measurements

The following Table 2 shows the ambient temperature and the temperature of the photovoltaic cell for the measurements performed.

Table 2. Ambient and cell temperatures of each module at the I-V curve measurement.

Module Number	Ambient Temperature [°C]	Cell Temperature [°C]
1	14.5	14.7
2	14.4	14.7
3	14.4	14.5
4	14.4	14.7
5	14.6	15.4
6	15.3	16.2
7	14.6	16.7
8	14.8	16.1
8	14.7	17.0
9	14.8	16.5
10	15.0	16.9
11	15.0	21.5

Figure 8 shows the I-V curves in darkness for all photovoltaic modules. These measurements have been taken at the temperature values indicated in Table 2. Each I-V curve has been presented at the cell temperature at which it has been captured. Conversion to 25 °C cell temperature has not been performed, as PV modules with different kinds of defects have been used, and the nominal conversion parameters (alpha, beta, and gamma) could have changed over the years associated with the degradation of the modules.

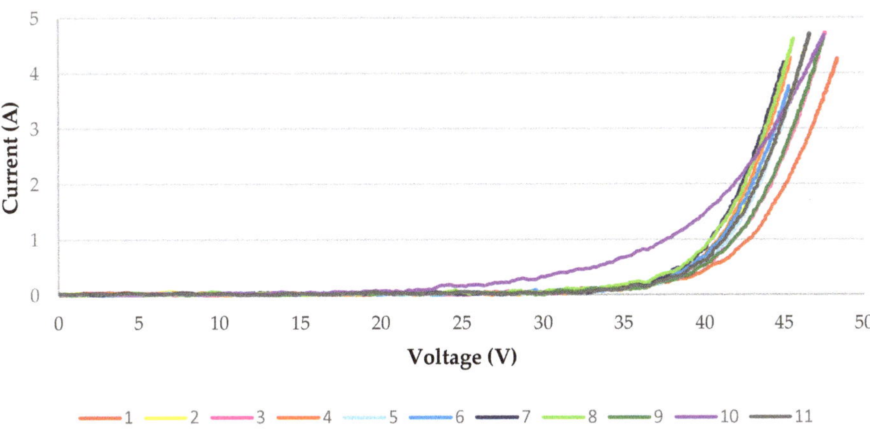

Figure 8. Dark I-V curves of the eleven modules measured with the developed device at the PV installation in Campus Duques de Soria.

3.2. Extraction of Solar Module Parameters from the Measured Dark I-V Curves

Once the dark I-V curves onsite measurements have been presented, the main parameters of the one exponential model (single-diode model) of each module are obtained using the 2/3 Diode Fit software. These results are presented in Table 3. In order to have comparative results between the different modules, parameters are presented in relation to the area. With the objective of drawing some conclusions on the state of the analyzed PV modules and for comparison purposes, all dark I-V curves have been presented in the same graph in Figure 8 and are analyzed together with their main parameters in this subsection.

Table 3. Absolute and referred to unit area main parameters of the one exponential model (single-diode model) of each module obtained using the 2/3 Diode Fit software. The values referred to unit area correspond to the cell current density J_0 [A/cm2] and cell shunt R_{sh} [kΩ·cm^2] and series R_s [Ω cm^2] resistances. The absolute values of current and resistance included in the table correspond to the module parameters, and they have been obtained from the cell values referred to unit area.

Module Number	J_0 [A/cm^2]	I_0 [A]	n	R_{sh} [kΩ cm^2]	R_{sh} [kΩ]	R_s [Ω·cm^2]	R_s [Ω]
1	7.87×10^{-9}	1.23×10^{-6}	1.73	24.79	11.42	1.17	0.54
2	4.22×10^{-9}	6.60×10^{-7}	1.61	3.75	1.73	0.51	0.24
3	3.64×10^{-10}	5.69×10^{-8}	1.38	12.33	5.68	1.29	0.59
4	2.83×10^{-11}	4.42×10^{-9}	1.17	9.56	4.41	0.99	0.46
5	2.15×10^{-9}	3.36×10^{-7}	1.55	7.72	3.56	0.49	0.23
6	7.57×10^{-10}	1.18×10^{-7}	1.41	4.15	1.91	1.00	0.46
7	3.98×10^{-12}	6.22×10^{-10}	1.04	2.64	1.21	1.45	0.67
8	3.68×10^{-9}	5.75×10^{-7}	1.56	2.02	0.93	0.34	0.16
9	4.00×10^{-9}	6.25×10^{-7}	1.63	89.47	41.23	0.54	0.25
10	7.49×10^{-6}	1.17×10^{-3}	3.05	2.15	0.99	0.86	0.40
11	3.91×10^{-9}	6.11×10^{-7}	1.57	60.57	27.91	0.47	0.22

First, it can be observed from Figure 8 is that there is one curve—number 10 from module 10—that has different characteristics from the rest. This is reflected in the extremely high value of the saturation current that, combined with the high value of the diode factor, dominates the rest of the model parameters.

For the curves that belong to the same module type (1, 2, 3, 4, 5, 7, 8, and 9), some differences can also be appreciated. If we fix a current equivalent to I_{sc}, we find that different voltages are found for the I-V characteristics. Given that all the curves are measured at approximately the same temperature, these variations would indicate different degradation in these modules. Furthermore, changes in the slope in the high-voltage area indicate changes in series resistance (higher slope, lower series resistance). It has to be taken into account that these modules have a lot of damages, some of them appreciated by the naked eye (see the upper row of modules in Figure 6), and various degradation modes are mixed, influencing the measured dark I-V curves. Higher series resistance modules indicate degradation in the module electrical circuit (modules 1, 3, and 7). High series resistance, combined with lower shunt resistance and diode saturation current could inform about mechanical damage [31] (for example, module 3).

It is also observed that some modules, especially those numbered 7, 8, and 2 present a very low value of shunt resistance (see Table 3). This low value of shunt resistance is also found in the other EOPLLY module (165W type), and in the number 10 module, which presents high degradation in all parameters. These low values of shunt resistance would imply shunting loses in the module. The combination of a lower R_{sh} and higher J_0 in modules 2 and 8 would indicate that the recombination losses are also enhanced in these modules.

With reference to the I-V curves presented, other interesting ideas can be pointed out. Differences between curves (in the 35–50 V range) are a consequence of the combination of parameters that can have opposite effects (for example, J_0 and n), so it is very difficult to grasp differences among the degradation states of several modules by solely comparing their I-V curves in a certain moment (except for cases as that of module 10), one to each other. In this case, the rightmost curve corresponds to module 1 ($J_0 = 7.87 \times 10^{-9}$ A/cm^2 and $n = 1.73$), while the leftmost is of module 7 ($J_0 = 3.98 \times 10^{-12}$ A/cm^2 and $n = 1.04$). The important point is also to compare the I-V curves of the same module in different periods along its life, because a check of possible shifts or changes in the curves or in the extracted parameters along time could reveal degradation processes being started or affecting the performance of the module.

4. Conclusions

This work demonstrates the possibility of tracing dark I-V curves in the modules of a PV plant without disconnecting them from the string. This is a breakthrough, as the dark I-V curve gives important information about the PV device and certain defects. These measures have been possible thanks to the combination of a bidirectional inverter and the developed electronic boards installed at the PV module level.

The results of the measurements have been satisfactory, and this allows obtaining the values of R_s and R_{sh}, among others, of the different PV modules of a plant online, without disconnection. The values obtained for R_s and R_{sh}, adjusted with the 2/3 Diode Fit software perfectly fit to the PV module installed.

With regard to future work, the authors will research the combination of dark and light I-V curve measurements. All measurements will be carried out without disconnection in the PV plant when combined with the bidirectional inverter. In addition, these measures will be managed and controlled through a low-cost communications system based on power line communications. In addition, with the measurements carried out, this research group will work with models based on artificial intelligence to locate defects in modules and cells, with the aim of keeping the performance of the photovoltaic plant high. Advances in O&M are key for this research team. Models based on artificial intelligence will work with images (electroluminescence and thermography) and I-V curves (light and dark).

Author Contributions: Conceptualization, L.H-C. and J.I.M.-A.; methodology, L.H-C., J.I.M.-A., S.G-S., M.d.C.A.-G., V.A-G., and J.L.B.; software, J.I.M.-A. and A.R-P.; validation, L.H-C., J.I.M.-A., S.G-S., M.d.C.A.-G., V.A-G., and J.L.B.; formal analysis, L.H-C. and J.I.M.-A.; investigation, L.H-C., J.I.M.-A., S.G-S., M.d.C.A.-G., and V.A-G.; resources, L.H-C. and V.A-G.; data curation, J.I.M.-A., S.G-S., M.d.C.A.-G., and A.R-P.; writing—original draft preparation, L.H-C., J.I.M.-A., S.G-S., M.d.C.A.-G., and J.L.B.; writing—review and editing, L.H-C., J.I.M.-A., S.G-S., M.d.C.A.-G., V.A-G., J.L.B., and A.R-P.; visualization, L.H-C., J.I.M.-A., S.G-S., M.d.C.A.-G., and V.A-G.; supervision, L.H-C., M.d.C.A.-G., and V.A-G.; project administration, L.H-C.; funding acquisition, L.H-C. and V.A-G. All authors have read and agreed to the published version of the manuscript.

Funding: This research was funded by the "Ministerio de Industria, Economía y Competitividad" grant number "RTC-2017-6712-3" with name "Desarrollo de herramientas Optimizadas de operaCión y manTenimientO pRedictivo de Plantas fotovoltaicas—DOCTOR-PV".

Data Availability Statement: To request the data, please, contact the corresponding author

Conflicts of Interest: The authors declare no conflict of interest.

References

1. Solar Power Europe Global Market Outlook For Solar Power/2020–2024. Available online: www.solarpowereurope.org (accessed on 15 July 2020).
2. International Energy Agency. Photovoltaic Power Systems. In *Snapshot of Global PV Markets 2020*; International Energy Agency: Paris, France, 2020.
3. International Energy Agency. *World Energy Outlook 2020. Executive Summary*; International Energy Agency: Paris, France, 2020.
4. International Electrotechnical Comission. *International Electrotechnical Comission. Working Group 2: Solar Photovoltaic Energy Systems*; International Electrotechnical Comission: Geneva, Switzerland, 2010.
5. Klise, G.T.; Balfour, J.R.; Keating, T.J. *Solar PV O&M Standards and Best Practices—Existing Gaps and Improvement Efforts*; Sandia National Lab. (SNL-NM): Albuquerque, NM, USA, 2014.
6. Hill, R.R.; Klise, G.; Balfour, J.R. *Precursor Report of Data Needs and Recommended Practices for PV Plant Availability, Operations and Maintenance Reporting*; Sandia National Laboratories: Albuquerque, NM, USA, 2015.
7. National Renewable Energy Laboratory (NREL). *Best Practices for Operation and Maintenance of Photovoltaic and Energy Storage Systems*, 3rd ed.; National Renewable Energy Laboratory (NREL): Golden, CO, USA, 2018.
8. Solar Power Europe. *Operation & Maintenance Best Practice Guidelines/Version 4.0*; SolarPower Europe: Brussels, Belgium, 2019.
9. Tsanakas, J.A.; Ha, L.; Buerhop, C. Faults and infrared thermographic diagnosis in operating c-Si photovoltaic modules: A review of research and future challenges. *Renew. Sustain. Energy Rev.* **2016**, *62*, 695–709. [CrossRef]
10. Gallardo-Saavedra, S.; Hernández-Callejo, L.; Alonso-García, M.D.C.; Muñoz-Cruzado-Alba, J.; Ballestín-Fuertes, J. Infrared thermography for the detection and characterization of photovoltaic defects: Comparison between illumination and dark conditions. *Sensors* **2020**, *20*, 4395. [CrossRef] [PubMed]

11. Roumpakias, E.; Bouroutzikas, F.; Stamatelos, A. On-site Inspection of PV Panels, Aided by Infrared Thermography. *Adv. Appl. Sci.* **2016**, *1*, 53–62. [CrossRef]

12. Teubner, J.; Buerhop, C.; Pickel, T.; Hauch, J.; Camus, C.; Brabec, C.J. Quantitative assessment of the power loss of silicon PV modules by IR thermography and its dependence on data-filtering criteria. *Prog. Photovolt. Res. Appl.* **2019**, *27*, 856–868. [CrossRef]

13. Dunderdale, C.; Brettenny, W.; Clohessy, C.; van Dyk, E.E. Photovoltaic defect classification through thermal infrared imaging using a machine learning approach. *Prog. Photovolt. Res. Appl.* **2020**, *28*, 177–188. [CrossRef]

14. Jahn, U.; Herz, M.; Rheinland, T. *Review on Infrared (IR) and Electroluminescence (EL) Imaging for Photovoltaic Field Applications*; International Energy Agency: Paris, France, 2018.

15. Muttillo, M.; Nardi, I.; Stornelli, V.; de Rubeis, T.; Pasqualoni, G.; Ambrosini, D. On field infrared thermography sensing for PV system efficiency assessment: Results and comparison with electrical models. *Sensors* **2020**, *20*, 1055. [CrossRef] [PubMed]

16. Köntges, M.; Kurtz, S.; Packard, C.; Jahn, U.; Berger, K.A.; Kato, K.; Friesen, T.; Liu, H.; Van Iseghem, M. *Review of Failures of Photovoltaic Modules*; International Energy Agency: Paris, France, 2014.

17. Berardone, I.; Lopez Garcia, J.; Paggi, M. Quantitative analysis of electroluminescence and infrared thermal images for aged monocrystalline silicon photovoltaic modules. In Proceedings of the 2017 IEEE 44th Photovoltaic Specialist Conference (PVSC), Washington, DC, USA, 25–30 June 2017; IEEE: Piscataway, NJ, USA, 2018; pp. 402–407.

18. Köntges, M.; Siebert, M.; Hinken, D. Quantitative analysis of PV-modules by electroluminescence images for quality control. In Proceedings of the 24th European Photovoltaic Solar Energy Conference, Hamburg, Germany, 21–25 September 2009; pp. 3226–3231.

19. Santos, J.D.; Valverde, A.; Alonso-García, M.C. Quantitative Analysis of Electroluminescence Imaging of a pv module with different mismatch levels. In Proceedings of the 36th European Photovoltaic Solar Energy Conference and Exhibition, Marseille, France, 9–13 September 2019; pp. 1127–1133.

20. Koch, S.; Weber, T.; Sobottka, C.; Fladung, A.; Clemens, P.; Berghold, J.; Koch, S.; Weber, T.; Sobottka, C.; Fladung, A.; et al. Outdoor Electroluminescence Imaging of Crystalline Photovoltaic Modules: Comparative Study Between Manual Ground-Level Inspections and Drone-Based Aerial Surveys. In Proceedings of the 32nd European Photovoltaic Solar Energy Conference and Exhibition, Munich, Germany, 20–24 June 2016; pp. 1736–1740.

21. Wolf, M.; Rauschenbach, H. Series resistance effects on solar cell measurements. *Adv. Energy Convers.* **1963**, *3*, 455–479. [CrossRef]

22. Celik, A.N.; Acikgoz, N. Modelling and experimental verification of the operating current of mono-crystalline photovoltaic modules using four- and five-parameter models. *Appl. Energy* **2007**, *84*, 1–15. [CrossRef]

23. Ciulla, G.; Lo Brano, V.; Di Dio, V.; Cipriani, G. A comparison of different one-diode models for the representation of I-V characteristic of a PV cell. *Renew. Sustain. Energy Rev.* **2014**, *32*, 684–696. [CrossRef]

24. Zhu, Y.; Xiao, W. A comprehensive review of topologies for photovoltaic I–V curve tracer. *Sol. Energy* **2020**, *196*, 346–357. [CrossRef]

25. Pérez, L.; Coello, J.; Domínguez, F.; Navarrete, M. Development and implementation of a mobile laboratory—PV mobile Lab quality assurance of photovoltaic modules "on site". In Proceedings of the 28th European Photovoltaic Solar Energy Conference and Exhibition, Paris, France, 30 September–4 October 2013; pp. 3500–3504.

26. Coello, J.; Pérez, L.; Domínguez, F.; Navarrete, M. On-site quality control of photovoltaic modules with the PV MOBILE LAB. *Energy Procedia* **2014**, *57*, 89–98. [CrossRef]

27. Morales-Aragonés, J.I.; Hernández-Callejo, L.; Gallardo-Saavedra, S.; Alonso-Gómez, V.; Sánchez-Pacheco, F.J.; González-Rebollo, M.A.; Martínez-Sacristán, O.; Muñoz-García, M.A.; Alonso-García, M.C. Low cost electronics for online I-V tracing at photovoltaic module level: Development of two strategies and comparison between them. *Electronics* **2021**. under review.

28. Bouzidi, K.; Chegaar, M.; Aillerie, M. Solar cells parameters evaluation from dark I-V characteristics. *Energy Procedia* **2012**, *18*, 1601–1610. [CrossRef]

29. Martil, I.; Gonzalez Diaz, G. Determination of the dark and illuminated characteristic parameters of a solar cell from I-V characteristics. *Eur. J. Phys.* **1992**, *13*, 193–197. [CrossRef]

30. King, D.L.; Hansen, B.R.; Kratochvil, J.A.; Quintana, M.A. Dark current-voltage measurements on photovoltaic modules as a diagnostic or manufacturing tool. In Proceedings of the Conference Record of the IEEE Photovoltaic Specialists Conference, Anaheim, CA, USA, 29 September–3 October 1997; pp. 1125–1128.

31. Spataru, S.V.; Sera, D.; Hacke, P.; Kerekes, T.; Teodorescu, R. Fault identification in crystalline silicon PV modules by complementary analysis of the light and dark current–voltage characteristics. In Proceedings of the Progress in Photovoltaics: Research and Applications, 29th EU PVSEC, Amsterdam, The Netherlands, 2014; Volume 24, pp. 517–532.

32. Spataru, S.; Hacke, P.; Sera, D. *In-Situ Measurement of Power Loss for Crystalline Silicon Modules Undergoing Thermal Cycling and Mechanical Loading Stress Testing*; National Renewable Energy Laboratory (NREL): Golden, CO, USA, 2015.

33. Spataru, S.; Hacke, P.; Sera, D.; Packard, C.; Kerekes, T.; Teodorescu, R. Temperature-dependency analysis and correction methods of in situ power-loss estimation for crystalline silicon modules undergoing potential-induced degradation stress testing. *Prog. Photovolt. Res. Appl.* **2015**, *23*, 1536–1549. [CrossRef]

34. Alonso-Garcia, M.C.; Hacke, P.; Glynn, S.; Muzzillo, C.P.; Mansfield, L.M. Analysis of Potential-Induced Degradation in Cu (In, Ga) Se2 Samples. *IEEE J. Photovolt.* **2019**, *9*, 331–338. [CrossRef]

35. Oh, W.; Bae, S.; Kim, D.; Park, N. Initial detection of potential-induced degradation using dark I–V characteristics of crystalline silicon photovoltaic modules in the outdoors. *Microelectron. Reliab.* **2018**, *88–90*, 998–1002. [CrossRef]
36. Ballestín-Fuertes, J.; Muñoz-Cruzado-Alba, J.; Sanz-Osorio, J.F.; Hernández-Callejo, L.; Alonso-Gómez, V.; Morales Aragonés, I.; Gallardo-Saavedra, S.; Martínez-Sacristan, O.; Moretón-Fernández, Á. Novel Utility-Scale Photovoltaic Plant Electroluminescence Maintenance Technique by Means of Bidirectional Power Inverter Controller. *Appl. Sci.* **2020**, *10*, 3084. [CrossRef]
37. Miller, S.D.; Turner, R.E. A dynamic lunar spectral irradiance data set for NPOESS/VIIRS day/night band night time environmental applications. *IEEE Trans. Geosci. Remote Sens.* **2009**, *47*, 2316–2329. [CrossRef]
38. Moretón, A.; Gallardo, S.; Jiménez, M.M.; Alonso, V.; Hernández, L.; Morales, J.I.; Martínez, O.; González, M.A.; Jiménez, J. Influence of large periods of dc current injection in c-si photovoltaic panels. In Proceedings of the 36th European Photovoltaic Solar Energy Conference and Exhibition (PVSEC2019), Marseille, France, 9–13 September 2019; pp. 107–108.
39. Balenzategui, J.L.; Cuenca, J.; Rodríguez-Outón, I.; Chenlo, F. Intercomparison and Validation of Solar Cell I-V Characteristic Measurement Procedures. In Proceedings of the 27th European Photovoltaic Solar Energy Conference and Exhibition, Frankfurt, Germany, 24–28 September 2012; pp. 1471–1476.
40. Suckow, S. 2/3-Diode Fit. Available online: http://nanohub.org/resources/14300 (accessed on 15 August 2020).
41. Suckow, S.; Pletzer, T.M.; Kurz, H. Fast and reliable calculation of the two-diode model without simplification. *Prog. Photovolt. Res. Appl.* **2014**, *22*, 494–501. [CrossRef]

35. Oh, W.; Bae, S.; Kim, D.; Park, N. Initial detection of potential-induced degradation using dark I–V characteristics of crystalline silicon photovoltaic modules in the outdoors. *Microelectron. Reliab.* **2018**, *88–90*, 998–1002. [CrossRef]
36. Ballestín-Fuertes, J.; Muñoz-Cruzado-Alba, J.; Sanz-Osorio, J.F.; Hernández-Callejo, L.; Alonso-Gómez, V.; Morales-Aragones, I.; Gallardo-Saavedra, S.; Martínez-Sacristan, O.; Moretón-Fernández, Á. Novel Utility-Scale Photovoltaic Plant Electroluminescence Maintenance Technique by Means of Bidirectional Power Inverter Controller. *Appl. Sci.* **2020**, *10*, 3084. [CrossRef]
37. Miller, S.D.; Turner, R.E. A dynamic lunar spectral irradiance data set for NPOESS/VIIRS day/night band night time environmental applications. *IEEE Trans. Geosci. Remote Sens.* **2009**, *47*, 2316–2329. [CrossRef]
38. Moretón, A.; Gallardo, S.; Jiménez, M.M.; Alonso, V.; Hernández, L.; Morales, J.I.; Martínez, O.; González, M.A.; Jiménez, J. Influence of large periods of dc current injection in c-si photovoltaic panels. In Proceedings of the 36th European Photovoltaic Solar Energy Conference and Exhibition (PVSEC2019), Marseille, France, 9–13 September 2019; pp. 107–108.
39. Balenzategui, J.L.; Cuenca, J.; Rodríguez-Outón, I.; Chenlo, F. Intercomparison and Validation of Solar Cell I-V Characteristic Measurement Procedures. In Proceedings of the 27th European Photovoltaic Solar Energy Conference and Exhibition, Frankfurt, Germany, 24–28 September 2012; pp. 1471–1476.
40. Suckow, S. 2/3-Diode Fit. Available online: http://nanohub.org/resources/14300 (accessed on 15 August 2020).
41. Suckow, S.; Pletzer, T.M.; Kurz, H. Fast and reliable calculation of the two-diode model without simplification. *Prog. Photovolt. Res. Appl.* **2014**, *22*, 494–501. [CrossRef]

Article

A Study on the Improvement of Efficiency by Detection Solar Module Faults in Deteriorated Photovoltaic Power Plants

Myeong-Hwan Hwang [1,2], Young-Gon Kim [1], Hae-Sol Lee [1,3], Young-Dae Kim [4] and Hyun-Rok Cha [1,3,*]

[1] Smart Mobility Material & Component R&D Group, Korea Institute of Industrial Technology, 6 Cheomdan-gwagiro 208 beon-gil, Buk-gu, Gwangju 61012, Korea; han9215@kitech.re.kr (M.-H.H.); ygkim1@kitech.re.kr (Y.-G.K.); eddylee12@kitech.re.kr (H.-S.L.)

[2] Department of Electrical Engineering, Chonnam National University, 77 Youngbong-ro, Buk-gu, Gwangju 61186, Korea

[3] Robotics and Virtual Engineering, Korea University of Science and Technology, Daejeon 34113, Korea

[4] Department of R&D Group, TOPINFRA, Ltd., 69 Chuam-ro, Buk-gu, Gwangju 61009, Korea; ydkim@topinfra.co.kr

* Correspondence: hrcha@kitech.re.kr; Tel.: +82-62-600-6210

Abstract: In recent years, photovoltaic (PV) power generation has attracted considerable attention as a new eco-friendly and renewable energy generation technology. With the recent development of semiconductor manufacturing technologies, PV power generation is gradually increasing. In this paper, we analyze the types of defects that form in PV power generation panels and propose a method for enhancing the productivity and efficiency of PV power stations by determining the defects of aging PV modules based on their temperature, power output, and panel images. The method proposed in the paper allows the replacement of individual panels that are experiencing a malfunction, thereby reducing the output loss of solar power generation plants. The aim is to develop a method that enables users to immediately check the type of failures among the six failure types that frequently occur in aging PV panels—namely, hotspot, panel breakage, connector breakage, busbar breakage, panel cell overheating, and diode failure—based on thermal images by using the failure detection system. By comparing the data acquired in the study with the thermal images of a PV power station, efficiency is increased by detecting solar module faults in deteriorated photovoltaic power plants.

Keywords: photovoltaic module; defect detection; power plant; efficiency; thermal image; photovoltaic aging

Citation: Hwang, M.-H.; Kim, Y.-G.; Lee, H.-S.; Kim, Y.-D.; Cha, H.-R. A Study on the Improvement of Efficiency by Detection Solar Module Faults in Deteriorated Photovoltaic Power Plants. *Appl. Sci.* **2021**, *11*, 727. https://doi.org/10.3390/app11020727

Received: 21 December 2020
Accepted: 7 January 2021
Published: 13 January 2021

Publisher's Note: MDPI stays neutral with regard to jurisdictional claims in published maps and institutional affiliations.

1. Introduction

In recent years, photovoltaic (PV) power has been receiving considerable attention as an alternative renewable energy source. Numerous studies on PV systems have attempted to improve the electrical performance of PV panels and have proposed maximum power estimation techniques and advanced power conversion techniques [1,2].

Owing to their environmental friendliness and good productivity, PV modules have been installed in various areas, including building rooftops, forests, and open fields. According to the "Korea Renewable Energy 2030 Plan", Korea should increase the proportion of renewable energy production from the current value of 4% to 20% by 2030. However, the amount of waste from spent PV modules is expected to exceed 1900 tons in 2030 (the target year of the 2030 Plan) and 85,000 tons in 2040. Therefore, it is necessary to develop a method for the disposal of solar equipment in preparation for the end-of-life management of PV modules, understand the environmental and social issues of PV modules, and reduce the cost of PV power generation [3–5].

PV modules have a lifespan of approximately 20 years when exposed to the outside environment, and they can be used as semipermanent systems with low maintenance costs [6]. Further, it was reported that the electrical performance of systems operated for a long time in more technologically advanced countries degraded significantly because of the degradation of the electrical characteristics and discoloration of the buffer materials, such as ethylene-vinyl acetate, in addition to physical damage [7].

In general, defect diagnosis of an aging PV panel is performed by comparing its system status and output with that of a normally operating PV system. The defects of the system can be evaluated by measuring the voltage and current of the PV panel terminals. In a previous study, ZigBee communication models were used to identify the types of PV panel defects, and it is expected that these models can be used to obtain thermal imaging data of the PV panels that are determined to be defective through their voltage drops [8].

For the accurate localization of the defects and determination of the causes of the efficiency decrease in PV panels, a system that thoroughly reflects the natural environment should be developed to compare the PV panel output under various control conditions and to obtain experimental data, including thermal images for each failure type. When implementing an actual system with traditional models or electrical performance-based failure diagnosis techniques, errors and inaccuracies in the identification of the defects may be experienced since the system status, and the output contains random signal components. To reduce such errors, in this paper, a system has been proposed that detects the failures and malfunctions of aging PV panels based on image data [9–12].

Furthermore, it is necessary to develop a method for the detection of PV panel failures at a minimum cost and avoid replacing all the panels in a PV power station. To accelerate the defect detection in PV panels, in this study, we first designed and fabricated an environmental chamber for accelerated life tests to simulate the actual natural environment by controlling key variables, such as temperature, lighting irradiation (illumination intensity), humidity, and vibration. Further, the fabricated chamber was used to conduct artificial tests and analyze the key factors affecting the normal PV panel output through the Taguchi method. Subsequently, an aging PV panel failure detection system was developed by analyzing images of frequently occurring failure types.

Ultimately, the aim was to develop a method that enables users to immediately check the type of failures among the six failure types that frequently occur in aging PV panels—namely, hotspot, panel breakage, connector breakage, busbar breakage, panel cell overheating, and diode failure—based on thermal images by using the failure detection system. To this end, we first checked the output of a normally operating PV panel and then induced various failures on the panels to obtain image data by failure type through output comparison and thermal image capturing process. Subsequently, we checked whether the same image is generated when images are captured after establishing a small-scale power generation facility using the normal panels and faulty panels.

It is expected that the proposed method will enable efficient operation of PV power generation facilities by standardizing the thermal images that are used indiscriminately and allowing easy detection of failure types. Additionally, the method will also contribute to reducing the maintenance cost as it will enable users to determine which panel is producing the lowest output in the power generation facility. Further, we propose a detection system that enables easy detection of failure types by standardizing thermal images for each failure type after capturing PV panel images with a thermal imaging camera. In addition, a method that can increase the output in the PV power generation facility by detecting whether the output is the highest or lowest in the failure types and replacing the faulty panels in a timely manner is presented.

2. Structure of PV Panel and Causes of Power Loss

2.1. Structure of a PV Panel

Once the photons from sunlight enter the semiconductor composed of a P–N junction, the electrons separated by the internal electric field accumulate in the N-type silicon, generating a charge; a path is formed to enable current to flow in the direction opposite to electron flow [13–15].

In a typical PV power generation system, several solar cells are wired in series to form a panel-type module. Then, several modules are wired in series and parallel. Depending on the type of light-absorbing material, solar cells can be classified into three types: silicon-based, compound semiconductor-based, and organic semiconductor-based. Further, the order of commercialization is as follows: first-generation cells (crystalline silicon), second-generation cells (amorphous silicon, copper indium gallium selenide (CIGS), and cadmium telluride (CdTe)), third-generation cells (dye-sensitized and organic cells), and next-generation cells (quantum dots and plasmons).

Among these materials, crystalline silicon solar cells were the first to be commercialized, and they currently account for more than 90% of the solar cells on the world market. Further, thin-film solar cells (CdTe, CIGS, etc.) account for approximately 8% of the world market. Dye-sensitized solar cells and organic solar cells are attracting interest owing to their potential implementation in a variety of applications, such as building-integrated PV and mobile devices. In recent years, perovskite solar cells have been widely studied as potential alternatives to silicon solar cells [16].

2.2. Power-Loss Factors of PV Panels

Excluding manufacturing errors and deterioration, the factors that affect the power loss in PV panels can be categorized into three types: mismatch effect due to shading or dirt that occurs when panels are connected in series and parallel to increase the output, changes in solar altitude, and temperature increases. These three types of factors that affect the power loss in PV panels are described in detail in the following subsections.

2.2.1. Module Mismatch Losses and Application of Inverters

The required output of PV panels is generated by arranging them into strings, i.e., connecting several panels in series or parallel (arrays), rather than using the panels individually. During operation, shading effects due to obstruction factors such as clouds or fallen leaves may cause mismatch losses, which could lead to a decrease in the power output. Several methods to decrease the module mismatch losses have been developed, including the use of built-in module inverters, string inverters, and module built-in DC–DC converters [17–21]. However, these methods require a separate circuit for compensating for the voltage loss due to shadowed strings. For the implementation of these methods, the cost of the additional modules to be installed, and the power losses due to the presence of these modules should be analyzed.

2.2.2. Power Loss Caused by Solar Altitude Change and Tracking System

The altitude of the sun changes from sunrise to sunset and over different seasons. PV panels are most efficient when they are positioned perpendicular to the position of the sun; thus, it is efficient to use a solar tracking system. The methods for positioning panels at the optimal angle include tracking the sun through a software program, utilizing an optical sensor, or employing both methods. In a previous study, the use of a tracking system for one month led to an increase in the output power of a 100 W panel of approximately 18% compared to a case where no tracking system was used. Nevertheless, the implementation of the corresponding systems required additional structures and power systems and incurred maintenance costs and additional power consumption.

2.3. Power-Loss Factors of PV Panels

Excluding manufacturing errors and deterioration, the factors that affect the power loss in PV panels can be categorized into three types: mismatch effect due to shading or dirt that occurs when panels are connected in series and parallel to increase the output, changes in solar altitude, and temperature increases. These three types of factors that affect the power loss in PV panels are described in detail in the following subsections.

2.3.1. Module Mismatch Losses and Application of Inverters

The required output of PV panels is generated by arranging them into strings, i.e., connecting several panels in series or parallel (arrays), rather than using the panels individually. During operation, shading effects due to obstruction factors such as clouds or fallen leaves may cause mismatch losses, which could lead to a decrease in the power output. Several methods to decrease the module mismatch losses have been developed, including the use of built-in module inverters, string inverters, and module built-in DC–DC converters [16–20]. However, these methods require a separate circuit for compensating for the voltage loss due to shadowed strings. For the implementation of these methods, the cost of the additional modules to be installed, and the power losses due to the presence of these modules should be analyzed.

2.3.2. Power Loss Caused by Solar Altitude Change and Tracking System

The altitude of the sun changes from sunrise to sunset and over different seasons. PV panels are most efficient when they are positioned perpendicular to the position of the sun; thus, it is efficient to use a solar tracking system. The methods for positioning panels at the optimal angle include tracking the sun through a software program, utilizing an optical sensor, or employing both methods. In a previous study, the use of a tracking system for one month led to an increase in the output power of a 100 W panel of approximately 18% compared to a case where no tracking system was used [21]. Nevertheless, the implementation of the corresponding systems required additional structures and power systems and incurred maintenance costs and additional power consumption.

3. Methods and Materials

3.1. Analysis of Aging PV Panel Failure Types

There are two main types of PV panel failure modes: corrosion and solar cell or connection problems. However, corrosion problems are usually caused by reactions such as those occurring when a metal is exposed to oxygen, solutions such as water, and other microscopic organisms. Thus, corrosion can be neglected because it is rare and insignificant in most cases; thus, only solar cell and connection problems are analyzed. Table 1 summarizes the various solar cell failure mode types analyzed in this study. Among the failure modes presented in Table 1, the most critical is the hotspot type.

Table 1. Solar cell failure types.

No.	Failure Type	Failure Cause
1	Hotspot	Dust, bird droppings, shade, snow
2	Panel breakage	Hail
3	Breakages of diode and connector	Connection problem
4	Overheating of solar cells	Connection problem
5	Busbar	Connection problem

3.2. Specifications of Experimental PV Panels

Three PV panel models with different outputs and sizes were selected for the PV panel output comparison test; their parameters are presented in Table 2. The panel shown in the fourth column of the table was selected to analyze whether the failure types can be found in the panel.

Table 2. Photovoltaic (PV) panel output and specifications.

Panel Size	530 × 370 (mm)	1640 × 980 (mm)	1950 × 980 (mm)
Panel power	35 W (18 V, 2.1 A)	220 W (25 V, 8.8 A)	364 W (39 V, 9.3 A)

3.3. Configuration of the Environmental Chamber

Figure 1 shows photographs of the interior and exterior of the experimental chamber constructed in this study. The chamber had an external size of 3000 (W) × 4500 (D) × 2800 (H) mm, rated power of 4800 W, a correlated color temperature of 5500 K, and lamp luminous flux of 180,000 Lm. (PV panel experimental chamber, Korea) The interior of the experimental chamber was sealed and designed to allow the control of the temperature and humidity using a thermo-hygrostat and induce vertical vibrations through a vibrator installed on the support structure area. Additionally, xenon lamps with a wavelength and light irradiation similar to those of the sun were installed on top of the chamber to irradiate the PV panel. Specifically, a total of six xenon lamps with a power of 1 kW (total: 6 kW) were installed inside the chamber, along with three 220 V power supplies. (UXL Xenon Short-Arc Lamps, UXL-16SB, Germany).

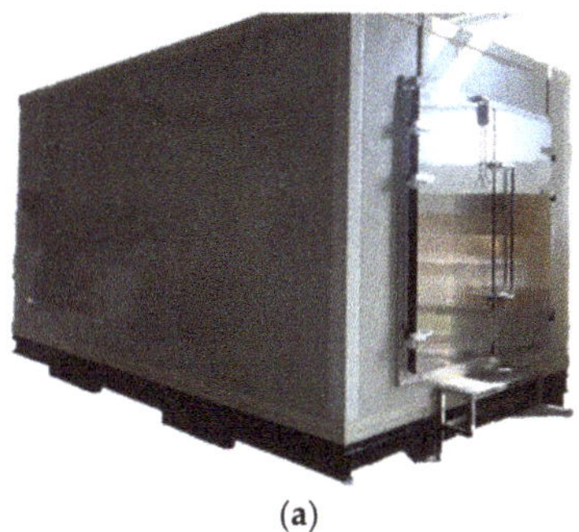

(**a**)

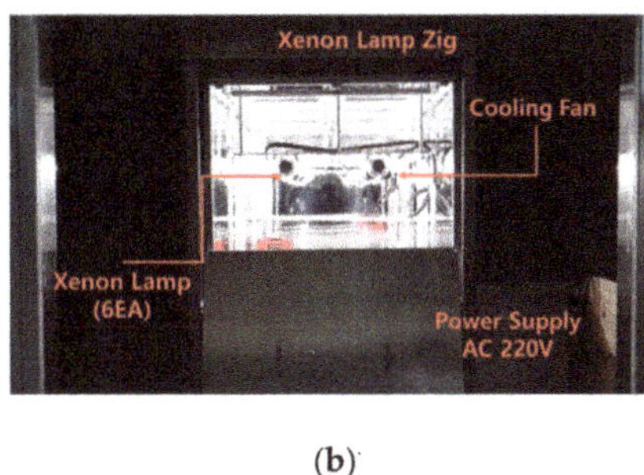

(**b**)

Figure 1. Photographs of the (**a**) exterior and (**b**) the interior of the PV panel experimental chamber.

3.4. PV Panel Performance Test Method

The cells constituting the PV module generate heat during electricity generation. The cells display 100% of their normal starting efficiency when the temperature is maintained at 25 °C, and for every temperature increase of 1 °C, the efficiency decreases by 0.5%. Thus, it can be inferred that PV panels are sensitive to the temperature and that higher cell temperatures cause output degradation. Accordingly, the experiment was conducted by acquiring thermal images of the PV panel failures and considering the output decrease when the temperature increases in certain areas of the PV panels.

Assuming that the panel output degrades when a specific area of the panel heats up, the types of failure can be determined by taking a wide range of images of the power station with a drone equipped with a thermal imaging camera.

Table 3 summarizes the image acquisition methods for the most frequently occurring failure types. The thermal images used to determine the failure types can serve as a basis for reducing the maintenance cost of power generation stations by allowing only partial replacement of power generation equipment.

Figure 2 shows photographs of typical PV modules with different failure types. In figure, a total of six images are secured on failures by panel breakage, diode failure, connector degradation, hotspot, busbar breakage, and panel cell overheating to obtain thermal images that can immediately differentiate the type of failure in an aging PV panel.

Table 3. Panel image acquisition method for different panel failure types.

No.	Failure Type	Panel Image Acquisition Method
1	Panel breakage	Applying external physical forces on the panel to induce damages of various sizes
2	Diode failure	Diode breakage and reverse polarity connection
3	Connector breakage	Connecting the connectors with reversed poles
4	Hotspot	Generating shades with pollutants, boxes, fallen leaves, etc.
5	Busbar breakage	Connecting busbars, used for connecting cells to each PV panel, to different cells
6	Panel cell overheating	Connecting additional cells through wiring to each PV panel cell

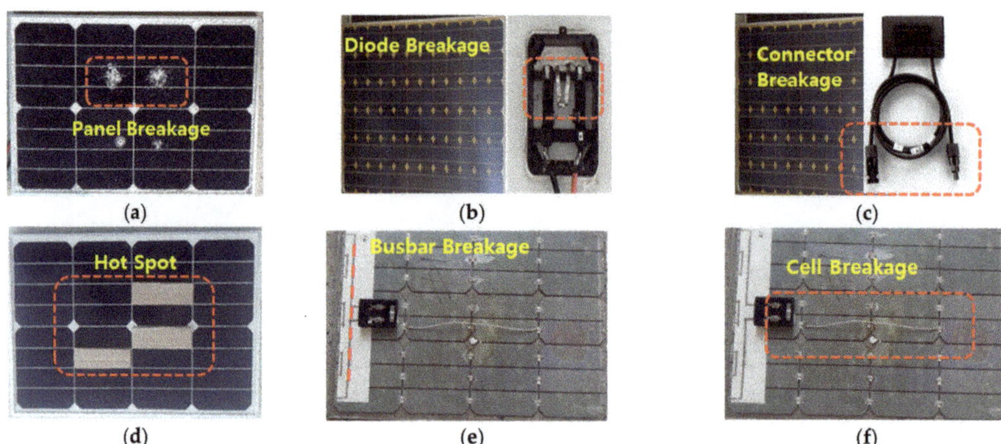

Figure 2. Photographs of PV panels with different failure types: (**a**) panel breakage, (**b**) diode failure, (**c**) connector breakage, (**d**) hotspot, (**e**) busbar, and (**f**) overheating of panel cells.

3.5. Method for Securing PV Panel Failure Images by Constructing a Small-Scale Power Generation Facility

We established a small-scale power generation facility composed of normal and faulty panels using the failure types of Figure 2 and detected the images by failure type. The facility was equipped with a total of 32 panels, comprising two parallel modules, each composed of 16 panels connected in series. As each module generated 1.7 kW output, the facility was designed to generate a total of 3.4 kW when the two modules were connected. Additionally, the small-scale power generation facility was arranged so that the images of normal and faulty panel types could be analyzed. In this study, the PV panel images were analyzed after capturing them using a drone equipped with a thermal imaging camera. Table 4 shows Normal and faulty panel placement layout in the small-scale power generation facility.

Table 4. Normal and faulty panel placement layout in the small-scale power generation facility.

Normal Panel	Panel Breakage	Normal Panel	Diode Failure	Hot Spot	Busbar	Connector Breakage	Panel Cell Overheating
Normal panel	Panel cell overheating	Normal panel	Panel breakage	Diode failure	Busbar	Connector breakage	Hotspot
Normal panel	Connector breakage	Normal panel	Busbar	Panel breakage	Hotspot	Panel cell overheating	Diode failure
Normal panel	Hotspot	Normal panel	Diode failure	Busbar	Panel breakage	Connector breakage	Panel cell overheating

4. Results

4.1. Key Factors of PV Module Output

Figure 3 shows the key factors affecting the PV module output obtained using the Taguchi experimental design method. In addition to the four main influential factors, the operator and maintenance time factors were analyzed. The operator factor was categorized as either professional or nonprofessional, denoted as U0 and U1, respectively; the maintenance time factor was set to 1 or 3 h, denoted as V0 and V1, respectively.

PU	30	50	200	58.17	58.31	56.57	58.61
PU	45	65	0	37.90	35.32	38.14	34.54
PUA	15	65	200	123.34	124.07	131.24	130.82
PUA	30	80	0	104.93	99.94	93.56	95.08
PUA	45	50	100	45.42	39.89	24.36	24.00

Response table for signal-to-noise ratio
Closer to goal level is better (10*Log(Ybar**2/s**2)

Level	Lighting	Temperature	Humidity	Vibration
1	44.96	47.60	36.54	34.24
2	45.30	43.73	37.85	42.47
3	22.05	20.98	37.91	35.60
Delta	23.25	26.62	1.37	0.23
Ranking	2	1	4	3

Response table for Means

Level	Lighting	Temperature	Humidity	Vibration
1	44.96	47.60	36.54	34.24
2	45.30	43.73	37.85	42.47
3	22.05	20.98	37.91	35.60
Delta	23.25	26.62	1.37	0.23
Ranking	2	1	4	3

Figure 3. Key factors affecting PV module output obtained using the Taguchi experimental design method.

From Figure 4, it can be seen that the level of influence of the factors was ranked based on signal-to-noise ratio and mean. Temperature and lighting have a high ranking in terms of both the signal-to-noise ratio and mean. It can be seen that the factors with higher response ranks were more likely to be recognized as key influential factors.

Figure 4 shows the analysis results of the main effects of the signal-to-noise ratio. Similar to the response tables, the temperature and lighting factors show a greater influence (steeper slope) than the other factors.

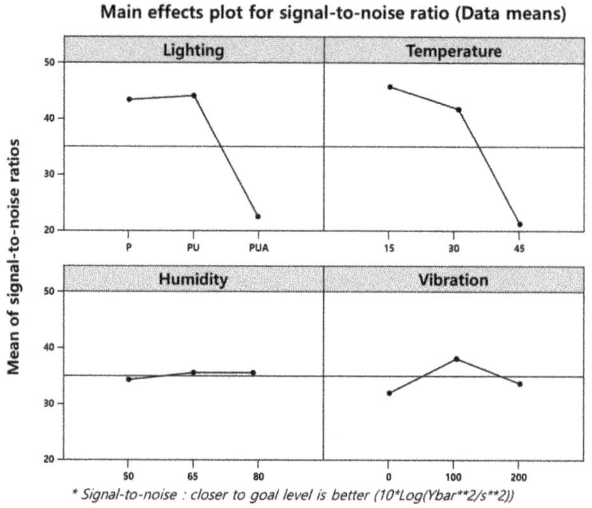

Figure 4. Analysis results of the main effects of the signal-to-noise ratio.

4.2. Image Analysis According to Failure Type

We imaged the failures of twenty 35 W panels, twenty 220 W panels, and ten 365 W panels, for a total of 50 panels. Because the 35 W panels did not contain diodes, a total of five images were acquired for the failures of panel breakage, connector degradation, hotspots, busbars, and cell overheating. Furthermore, because the 220 and 365 W panels contain diodes, a total of six images, including the diode failure, were acquired for each panel type.

4.3. Analysis of the 365 W PV Panel

Table 5 presents the power output of the 365 W PV panel measured before and after different failure types. From the table, it can be seen that before failure, the output of the 365 W panel is 355.79 W, and after failure, it decreased by at least 10% in all cases.

Table 5. Power output of the 365 W PV panel after different failure types.

Failure Type	Output before Failure (W)	Output after Failure (W)	Voltage after Failure (V)	Current after Failure (A)
Hot Spot	355.76	316.96	39	8.135
Diode	355.79	199.95	25	7.99
Connector breakage	355.79	0	0	0
Panel breakage	355.79	110	39	2.846
Busbar	355.79	214.37	25	8.575
Cell overheating	355.79	239.47	25	9.589

Figure 5 shows photographs and thermal images of the 365 W PV panel after different types of failure. From the figure, it can be seen that overheating occurs in a certain part of the panel in the case of the hotspot problem. Furthermore, sporadic heat generation is observed in the case of diode malfunction. In the case of connector failure, heat generation is observed at the edges of the panel. In the case of cell overheating, heat generation is observed throughout the damaged areas. Lastly, in the case of busbar failure, overheating is observed in the area where the busbar is located.

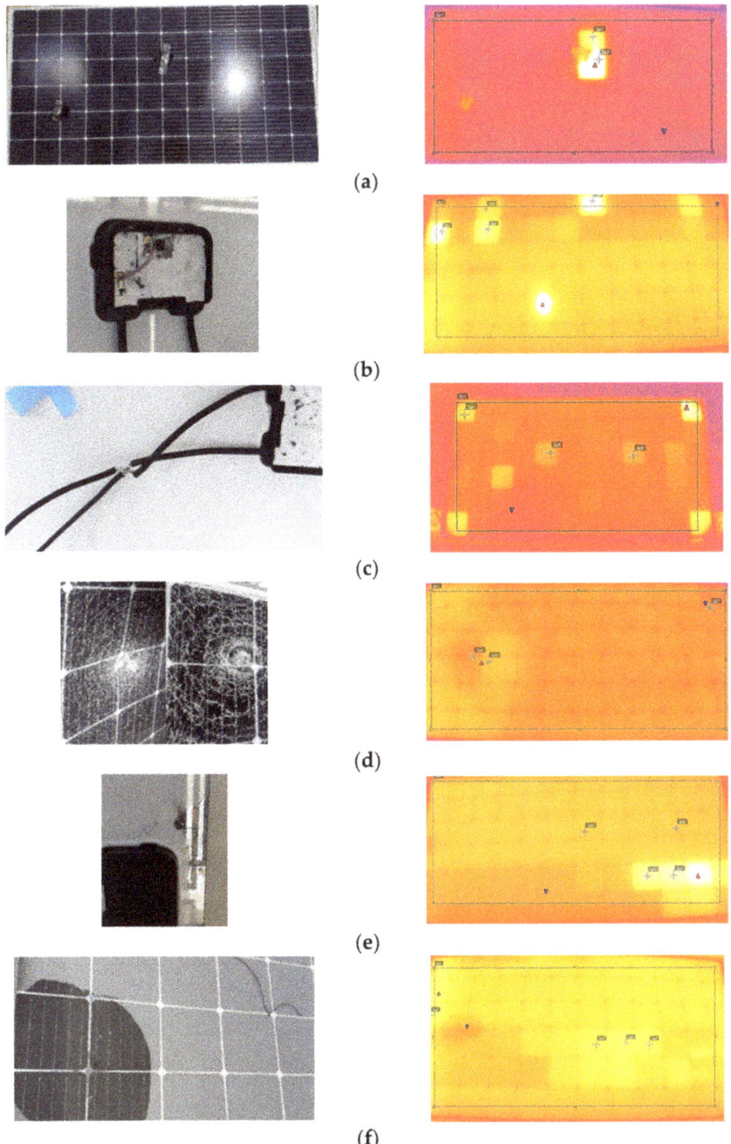

Figure 5. Photographs (left) and thermal images (right) of the 365 W panel with different failure types: (**a**) hotspot, (**b**) diode failure, (**c**) connector breakage, (**d**) panel breakage, (**e**) busbar, and (**f**) overheating of panel cells.

4.4. Analysis of the 220 W PV Panel

Table 6 presents the power output of the 220 W PV panel measured before and after different failure types. From the table, it can be seen that before failure, the output of the panel is 175.10 W. After diode failure, connector breakage, and cell overheating, the panel output is 0 W.

Table 6. Power output of the 220 W PV panel after different failure types.

Failure Type	Output before Failure (W)	Output after Failure (W)	Voltage after Failure (V)	Current after Failure (A)
Hot Spot	175.12	107.16	25	4.298
Diode	175.12	0	0	0
Connector breakage	175.12	0	0	0
Panel breakage	175.12	40.68	25	1.618
Busbar	175.12	78.83	20	3.934
Cell overheating	175.12	0	0	0

Figure 6 shows photographs and thermal images of the 220 W PV panel after different types of failures. It can be seen that the thermal images of the 220 W PV panels after different types of failures are similar to those of the 365 W PV panels.

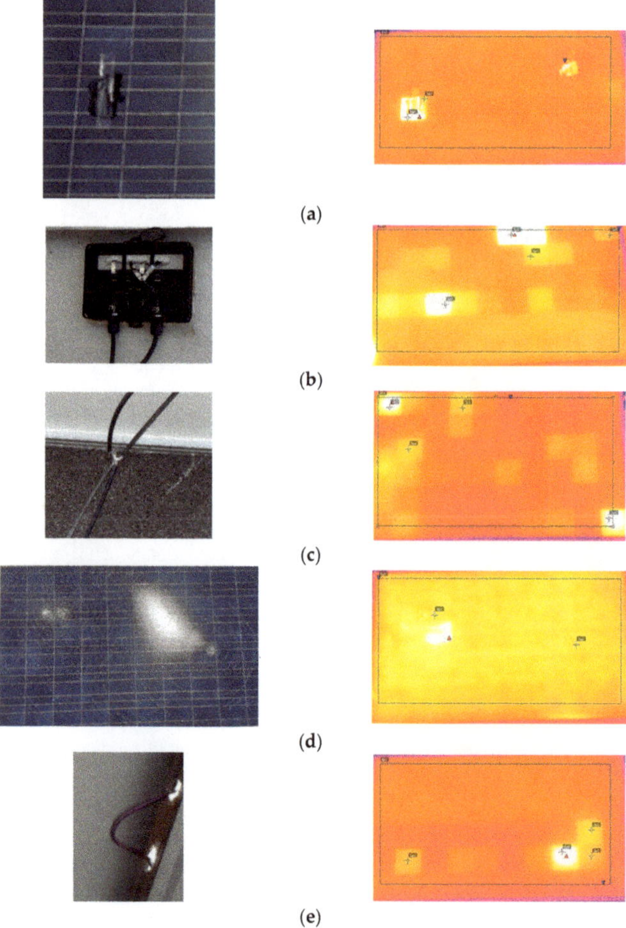

(a)

(b)

(c)

(d)

(e)

Figure 6. *Cont.*

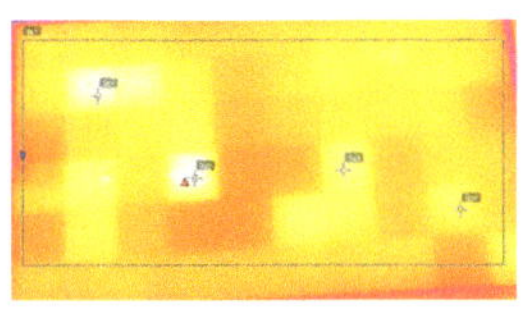

(f)

Figure 6. Photographs (left) and thermal images (right) of the 220 W panel with different failure types: (**a**) hotspot, (**b**) diode failure, (**c**) connector breakage, (**d**) panel breakage, (**e**) busbar, and (**f**) overheating of panel cells.

4.5. Analysis of the 35 W PV Panel

Table 7 presents the power output of the 35 W PV panel measured before and after different failure types (excluding diode failure because low-output power panels do not contain diodes). After connector breakage and cell overheating, the panel output is 0 W.

Table 7. Power output of the 35 W PV panel after different failure types.

Failure Type	Output before Failure (W)	Output after Failure (W)	Voltage after Failure (V)	Current after Failure (A)
Hotspot	31.4	8.16	18	0.454
Connector breakage	31.4	0	0	0
Panel breakage	31.4	0	0	0
Busbar	31.4	28.89	18	1.598
Cell overheating	31.4	14.84	10	1.482

Figure 7 shows photographs and thermal images of the 35 W PV panel after different types of failures. It can be seen that the thermal images of the 35 W PV panels after different types of failures are similar to those of the 365 and 220 W PV panels.

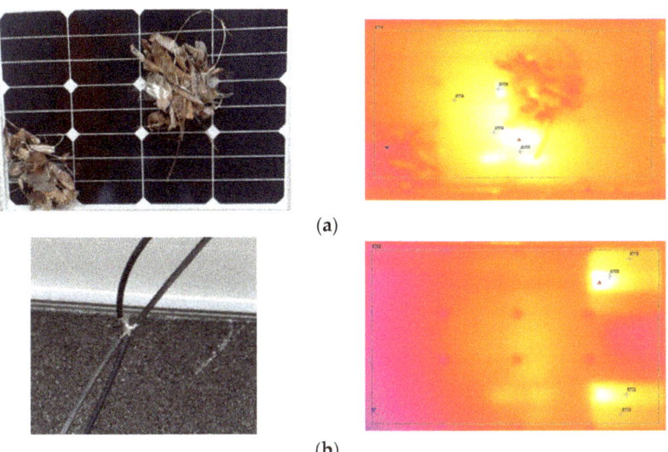

(a)

(b)

Figure 7. *Cont.*

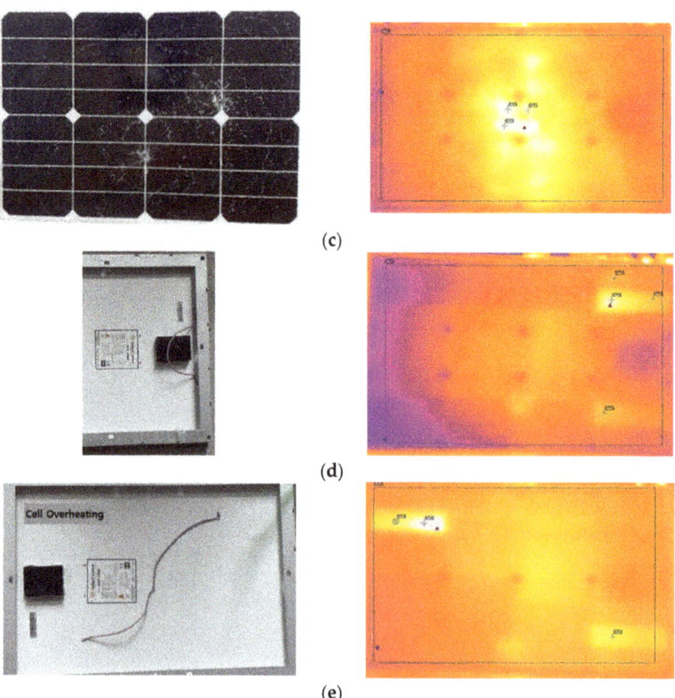

Figure 7. Photographs (left) and thermal images (right) of the 35 W panel with different failure types: (**a**) hotspot, (**b**) connector breakage, (**c**) panel breakage, (**d**) busbar, and (**e**) overheating of panel cells.

4.6. PV Panel Thermal Image Analysis Results

Figure 8 shows thermal images of the investigated panels with different failure types. From figure, it can be seen that most of the failure images are similar. It can be concluded that by photographing a PV power station with a drone equipped with a thermal imaging camera, malfunctions in the panels can be discovered immediately. As a result, faulty panels can be replaced, enhancing the power output and efficiency of the station.

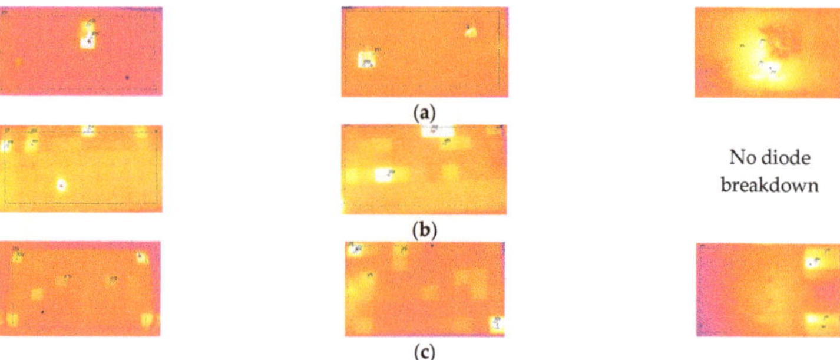

Figure 8. *Cont.*

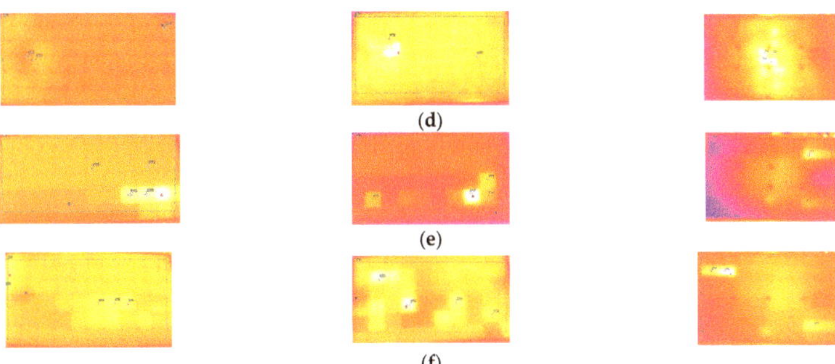

Figure 8. Thermal images of the 365 W (left), 220 W (middle), and 35 W (right) panels after different failure types: (**a**) hotspot, (**b**) diode failure, (**c**) connector breakage, (**d**) panel breakage, (**e**) busbar, and (**f**) overheating of panel cells.

4.7. Output Data Comparison Analysis by Failure Type in the 365 W, 220 W, and 35 W Panels

As the output of the PV panels varies by size, failure types were generated on a total of 50 panels composed of 365 W, 220 W, and 35 W PV panels to compare the output data. Figure 5 illustrates failure images of a 365 W panel by failure type, while Figures 6 and 7 illustrate failure images of a 220 W panel and a 35 W panel, respectively. Figure 8 shows combined thermal images for each output. The important factor here is to analyze the changes in the output in each failure type and propose a method for increasing the efficiency of the power generation facility by prioritizing replacement of the PV panels having the lowest output among the failure types.

Figure 9 compares the output values of each PV panel by failure type. In the figure, (a) to (f) indicate the following failure types: (a) hotspot, (b) diode failure, (c) connector breakage, (d) panel breakage, (e) busbar, and (f) overheating of panel cells. The failure type yielding the lowest panel output was analyzed as the (c) connector breakage, followed by (d) panel breakage, (b) diode failure, (e) busbar, (f) overheating of panel cells, and (a) hotspot. When the connector breakage was detected through PV panel imaging, the corresponding panel was analyzed as the panel requiring a priority in replacement. It is inferred that the panel that requires immediate replacement can be derived based on the images obtained using a thermal imaging camera and the panel output data.

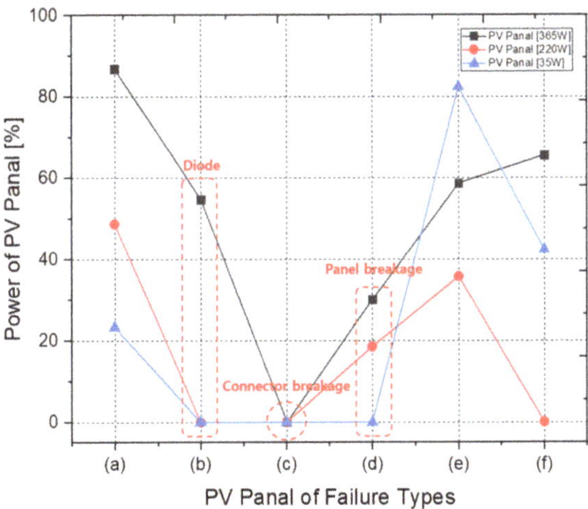

Figure 9. Output comparison of each PV panel by failure type.

4.8. Thermal Image Results of Small-Scale Power Generation Facility

Based on the panel placement layout shown in Table 4, the PV panels were arranged using normal and faulty panels. The panels were arranged in parallel modules, with each module comprising16 panels in series. Further, a thermal imaging camera was attached to a drone to obtain the images of the normal and faulty PV panels. The images were captured during clear weather, and the drone was set to fly at the height of about 20 m from the ground.

Figure 10. Faulty panel images acquired using a thermal imaging camera. While panels #1, 3, 9, 11, 17, 19, 25, and 27 were normal panels without any failure shown in the thermal images, it can be seen that the rest of the panels other than the normal ones showed overheating of cells in certain areas of the panel as a result of taking images with a thermal imaging camera for each failure type, thereby generating various images. As for the connector breakage, it was possible to detect the failure with only the image analysis as severe overheating is generated throughout the PV panel. In the connector breakage, which yields the lowest output, the phenomenon in which the cells encounter overheating throughout the panels is observed. Through this image analysis, it was possible to check the results of image data for panels requiring immediate replacement by acquiring images of connector breakage, which displays the lowest panel output among the failure types.

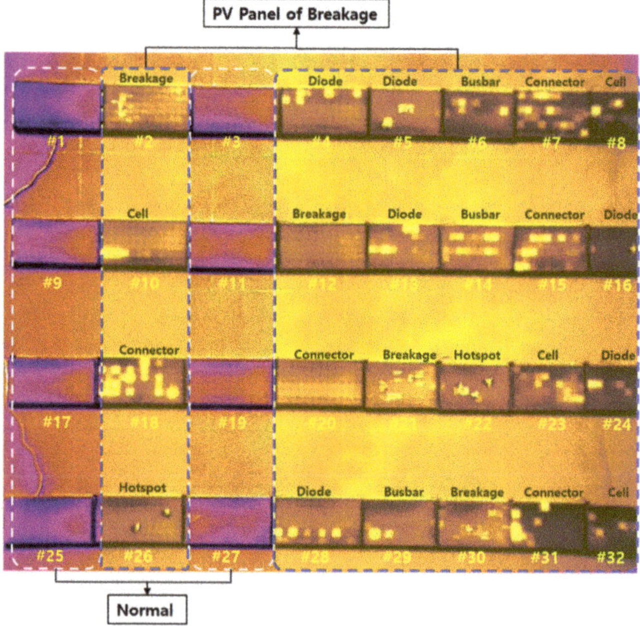

Figure 10. Faulty panel images acquired using a thermal imaging camera.

5. Conclusions

In this study, we analyzed the power output of PV panels after different types of failure. To this end, we designed and fabricated an experimental chamber and placed PV panels inside the chamber. Xenon lamps were used to simulate sunlight irradiation. After generating different types of damage to PV panels with power outputs of 365, 220 and 35 W and irradiating these in the experimental chamber, failure images were acquired using a thermal imaging camera to analyze the damaged PV panels. Many similarities were found in the thermal images of the PV panels with different failures.

By comparing the data acquired in this study with the thermal images of a PV power station, PV panels experiencing malfunction could be identified, and the power output loss due to the defects could be calculated.

In the future, more failure images of panels in PV power stations should be captured using drones. By analyzing these images in real time, panel failures can be discovered, improving the efficiency of PV power stations.

Based on the experiment results of this study, it was possible to obtain images of PV panels showing the lowest output and overheating throughout the panels by conducting analysis on the thermal images of a PV power generation facility captured using a drone. While the output can be increased by replacing the entire panel when failure types occur, if the replacement priority is given on the connector breakage failure type that yields the lowest output among all failure types in consideration of the cost aspect, the output of the power generation facility can be improved, and the maintenance cost can be minimized as proposed in this study.

Author Contributions: Conceptualization, M.-H.H. and H.-S.L.; data curation, H.-S.L.; formal analysis, Y.-G.K.; methodology, Y.-D.K.; supervision, H.-R.C.; validation, H.-R.C. and M.-H.H.; visualization, H.-S.L.; writing—original draft, M.-H.H.; writing—review and editing, Y.-G.K. All authors have read and agreed to the published version of the manuscript.

Funding: This study was conducted with the support of the Korea Institute of Energy Technology Evaluation and Planning as "Developing image big data based real time detection system for detecting defective module applied to solar power plant (KETEP 20183010014230)."

Conflicts of Interest: The authors declare no conflict of interest.

References

1. Larbes, C.; Cheikh, S.A.; Obeidi, T.; Zerguerras, A. Genetic algorithms optimized fuzzy logic control for the maximum power point tracking in photovoltaic system. *Renew. Energy* **2009**, *34*, 2093–2100. [CrossRef]
2. Ram, J.P.; Babu, T.S.; Rajasekar, N. A comprehensive review on solar PV maximum power point tracking techniques. *Renew. Sustain. Energy Rev.* **2017**, *67*, 826–847. [CrossRef]
3. Hong, Y.Y.; Buay, P.M.P. Robust design of type-2 fuzzy logic-based maximum power point tracking for photovoltaics. *Sustain. Energy Technol. Assess.* **2020**, *38*, 100669. [CrossRef]
4. Kim, K.J. Development of hybrid type photovoltaic energy storage system for improvement of durability and electric power efficiency. *J. Korean Soc. Mech. Technol.* **2018**, *20*, 306–311.
5. Madeti, S.R.; Singh, S.N. A comprehensive study on different types of faults and detection techniques for solar photovoltaic system. *Sol. Energy* **2017**, *158*, 161–185. [CrossRef]
6. Qais, M.H.; Hasanien, H.M.; Alghuwainem, S.; Nouh, A.S. Coyote optimization algorithm for parameters extraction of three-diode photovoltaic models of photovoltaic modules. *Energy* **2019**, *187*, 116001. [CrossRef]
7. Ilse, K.K.; Figgis, B.W.; Naumann, V.; Hagendorf, C.; Bagdahn, J. Fundamentals of soiling processes on photovoltaic modules. *Renew. Sustain. Energy Rev.* **2018**, *98*, 239–254. [CrossRef]
8. Cho, H.C.; Jung, Y.J.; Lee, G.H. A study on fault detection for photovoltaic power modules using statistical comparison scheme. *J. Korean Sol. Energy Soc.* **2013**, *33*, 153–156. [CrossRef]
9. Ko, S.W.; So, J.H.; Hwang, H.M.; Ju, Y.C.; Song, H.J.; Shin, W.G.; Kang, G.H.; Choi, J.R.; Kang, I.C. The monitoring system with PV module-level fault diagnosis algorithm. *J. Korean Sol. Energy Soc.* **2018**, *38*, 21–28.
10. So, J.H.; Ko, S.W.; Ju, Y.C. Losses comparison and analysis for fault modes of grid-connected photovoltaic system. *J. Korean Sol. Energy Soc.* **2017**, *37*, 23–32. [CrossRef]
11. Kim, Y.H.; Shim, K.S. Efficiency computation and failure detection of solar power generation panels. *Comput. Syst. Theory* **2013**, *40*, 1–7.
12. Jeong, J.Y.; Choi, S.S.; Choi, H.Y.; Ryu, S.W.; Lee, I.C.; Rho, D.S. A study on the optimal configuration algorithm for modeling and improving the performance of PV module. *J. Korean Acad. Industr. Coop. Soc.* **2016**, *17*, 723–730.
13. Bollinger, B.; Gillingham, K. Peer effects in the diffusion of solar photovoltaic panels. *Mark. Sci.* **2012**, *31*, 900–912. [CrossRef]
14. Sohani, A.; Sayyaadi, H. Employing genetic programming to find the best correlation to predict temperature of solar photovoltaic panels. *Energy Convers. Manag.* **2020**, *224*, 113291. [CrossRef]
15. Ezzaeri, K.; Fatnassi, H.; Bouharroud, R.; Guardo, L.; Bazgaou, A.; Wifaya, A.; Demarti, H.; Bekkaoui, A.; Aharoune, A.; Poncet, C.; et al. The effect of photovoltaic panels on the microclimate and on the tomato production under photovoltaic canarian greenhouses. *Sol. Energy* **2018**, *173*, 1126–1134. [CrossRef]
16. Bayrak, F.; Oztop, H.F.; Selimefendigil, F. Effects of different fin parameters on temperature and efficiency for cooling of photovoltaic panels under natural convection. *Sol. Energy* **2019**, *188*, 484–494. [CrossRef]

17. Picault, D.; Raison, B.; Bacha, S.; Aguilera, J.; Casa, J. Changing photovoltaic array interconnections to reduce mismatch losses: A case study. In Proceedings of the 2010 9th International Conference on Environment and Electrical Engineering, Prague, Czech Republic, 16–19 May 2010; pp. 37–40.
18. Aoun, N.; Bailek, N. Evaluation of mathematical methods to characterize the electrical parameters of photovoltaic modules. *Energy Convers. Manag.* **2019**, *193*, 25–38. [CrossRef]
19. Ma, T.; Gu, W.; Shen, L.; Li, M. An improved and comprehensive mathematical model for solar photovoltaic modules under real operating conditions. *Sol. Energy* **2019**, *184*, 292–304. [CrossRef]
20. Ko, J.S.; Chung, D.H. Reconfiguration of PV module considering the shadow influence of photovoltaic system. *J. Korean Inst. Illum. Electr. Install. Eng.* **2013**, *27*, 36–44.
21. Kish, G.J.; Lee, J.J.; Lehn, P.W. Modelling and control of photovoltaic panels utilising the incremental conductance method for maximum power point tracking. *IET Renew. Power Gener.* **2012**, *6*, 259–266. [CrossRef]

Article

Automatic Boundary Extraction for Photovoltaic Plants Using the Deep Learning U-Net Model

Andrés Pérez-González *, Álvaro Jaramillo-Duque and Juan Bernardo Cano-Quintero

Research Group in Efficient Energy Management (GIMEL), Electrical Engineering Department, Universidad de Antioquia, Calle 67 No. 53-108, Medellín 050010, Colombia; alvaro.jaramillod@udea.edu.co (Á.J.-D.); bernardo.cano@udea.edu.co (J.B.C.-Q.)
* Correspondence: afernando.perez@udea.edu.co; Tel.: +57-322-639-7758

Abstract: Nowadays, the world is in a transition towards renewable energy solar being one of the most promising sources used today. However, Solar Photovoltaic (PV) systems present great challenges for their proper performance such as dirt and environmental conditions that may reduce the output energy of the PV plants. For this reason, inspection and periodic maintenance are essential to extend useful life. The use of unmanned aerial vehicles (UAV) for inspection and maintenance of PV plants favor a timely diagnosis. UAV path planning algorithm over a PV facility is required to better perform this task. Therefore, it is necessary to explore how to extract the boundary of PV facilities with some techniques. This research work focuses on an automatic boundary extraction method of PV plants from imagery using a deep neural network model with a U-net structure. The results obtained were evaluated by comparing them with other reported works. Additionally, to achieve the boundary extraction processes, the standard metrics Intersection over Union (IoU) and the Dice Coefficient (DC) were considered to make a better conclusion among all methods. The experimental results evaluated on the Amir dataset show that the proposed approach can significantly improve the boundary and segmentation performance in the test stage up to 90.42% and 91.42% as calculated by IoU and DC metrics, respectively. Furthermore, the training period was faster. Consequently, it is envisaged that the proposed U-Net model will be an advantage in remote sensing image segmentation.

Keywords: deep learning (DL); unmanned aerial vehicle (UAV); photovoltaic (PV) systems; image-processing; image segmentation; semantic segmentation

Citation: Pérez-González, A.; Jaramillo-Duque, Á.; Cano-Quintero, J.B. Automatic Boundary Extraction for Photovoltaic Plants Using the Deep Learning U-Net Model. *Appl. Sci.* **2021**, *11*, 6524. https://doi.org/10.3390/app11146524

Academic Editors: Luis Hernández-Callejo, Maria del Carmen Alonso García and Sara Gallardo Saavedra

Received: 2 June 2021
Accepted: 13 July 2021
Published: 15 July 2021

Publisher's Note: MDPI stays neutral with regard to jurisdictional claims in published maps and institutional affiliations.

1. Introduction

In the last decade, the world began the transition towards renewable energy the harvesting of solar energy one of the most promising sources used today. Photovoltaic (PV) energy production is a fast-growing market: The Compound Annual Growth Rate (CAGR) of cumulative PV plants was 35% from year 2010 to 2019. The main reasons for this accelerated growth are: production cost of PV panels have decreased, return on investment ranging from 0.7 to 1.5 years. Some countries offer economic benefits for new facilities and the performance ratio (which informs how energy-efficient and reliable PV plants are against its theoretical production) is better nowadays. Before 2000 it was 70%, today performance ranges from 80% to 90% [1,2].

Nonetheless, PV plants present some challenges for maintaining proper performance with failures and defects being the most common ones. In general, failures on PV systems are more concentrated in the inverters and PV modules. In the PV modules, because of dirty equipment, environmental conditions, or manufacturing problems the PV plant energy output can be reduced by 31% [3–5]. To detect these problems, it is necessary to consider that the PV systems are commonly located on roofs, rooftops, and farms. Therefore the access, maintenance, and detection of possible problems in the panels should be carried out by trained and qualified personnel working at heights to detect these problems. These procedures can put the integrity of people, equipment, and PV Plants at risk [6]. Manual inspection can take up to 8 h/MW, depending on the number of test modules. This period can be more than double for rooftop systems, depending on the characteristics of the installation [7].

As an alternative to use trained personnel for maintenance, the use of an Unmanned Aerial Vehicle (UAV) has many advantages: it reduces the risks in maintenance labours, increases reliability, and increases effectiveness of PV plants. As a result, research teams are currently working on developing equipment that can automatically inspect and clean PV systems, as shown in [8,9].

Compared to traditional methods, UAVs could perform an automatic inspection and monitoring with lower costs, cover larger areas, and achieve faster detection. The cameras installed on UAVs take photos [10], and through image processing, the area of the PV systems can be identified in a process called boundary extraction [11]. Once the area is identified, the ground control station calculates the Coverage Path Planning (CPP) that guides the UAV in the automatic plant inspection. Any faults are detected with the inspection, the required maintenance is scheduled.

This work focused on the boundary extraction of PV systems which is a key aspect for UAVs to conduct autonomous inspections and enhance Operation and Maintenance (O&M) [11].

Several inspections and defect detection methods have been proposed in the literature. Lately, UAVs have been used for the inspection of different PV plants, to identify the correlation between altitude and the PV panel defects detection as: shape, size, location, color, among others [12–16]. Many attempts have been committed to developing a reliable and cost-effective aerial robot with optimum efficiency over PV plant inspection [10,17–19]. For autonomous inspection, large volumes of information or big data are required from PV systems. These datasets improve the inspection by means of automatic learning algorithms during the O&M process [7]. The O&M process of photovoltaic plants is an important aspect for the profitability of investors. Autonomous inspection of PV systems is a technology with great potential, mainly for large PV plants, roofs, facades and where manual techniques have notable restrictions in terms of human risk, performance, time and cost.

Traditional Image Processing (TIP) has been used extensively by other authors. In this study [13,20–24], the authors used TIP to defect recognition in the inspection of photovoltaic plants. Furthermore, using HSV transformation, color filtering and segmentation, techniques have been implemented in many projects, especially for defect detection [25], to enumerate photovoltaic modules [20,26] and identification of limits [27]. This technique has a restriction for unsupervised procedures; the user should assist in the image processing by adjusting the filter to the particular color of each target the technique aims to find. Therefore, TIP is not a proper method for autonomous aerial inspection of photovoltaic plants.

The boundary extraction is referred to as an image segmentation technique. This technique divides an image into a set of regions, and it is performed by dividing the image histogram into optimal threshold values [28,29]. The aim is to substitute the representation of an image into something easily analyzable to obtain detailed information on the region of interest in an image and aid to annotate the scene of the object [30]. Image segmentation is necessary to identify the content of the photo. Accordingly, edge detection is an essential tool for image segmentation [31] and can be achieved by means of traditional image processing techniques [27,32] or through artificial vision techniques [33].

The image segmentation techniques with TIP were developed to identify objects such as the area of PV Plants out of an orthophoto [10,34,35]. Later, the Machine Learning (ML) and Deep Learning (DL) image segmentation techniques, also known as semantic segmentation, were proposed [36,37]. In semantic segmentation each pixel is labeled with the class of its enclosing object or region [33]. Convolutional Neural Networks have been used for semantic segmentation , such as the Fully Convolutional Network (FCN) model [33], and U-Net network model [38], which drastically enhances the segmentation certainty compared with TIP method results, and ML technique results [36,37].

The convolutional neural networks are used for extracting dense semantic representations from input images and to predict labels at the pixel level. To perform this task, it is necessary to obtain or create a dataset, perform a pre-processing of the data, select an appropriate model and train it based on metrics, and then evaluate the results as shown in [11]. This is a fundamental challenge in computer vision with wide applications in scene interpretation, medical imaging, robot vision, etc. [39]. Once the segmentation is done, the next step is to obtain the automatic Coverage Path Planning (CPP).

Although advances in GPS systems have improved and accuracy is around 10 cm in low-cost Real Time Kinematics (RTK) GPS systems [40]. Most of the projects use software tools that provide companies like Mission Planner [41] or development groups as Qground-Control [42]. These tools are based on simple polygonal coverage areas and a coverage pattern of zigzag path. They require time when the area is of complex geometry, or when the plant is in continuous expansion. Additionally, the programmer preloads waypoints without optimal coverage. As a consequence, to develop a real-time path-planning algorithm for an autonomous monitoring system, it is a hard task in this platform. Therefore, it is first necessary to determine the boundary of the PV plant. By taking out the boundaries of PV plants, aerial photogrammetry and mapping can be faster, effective, economical and customizable [27], they motivate to make this work.

The key contributions of this work are as follows:

- In the revised literature, there is no report of U-net model to extract the boundaries of PV Plants; this work proposes such a model.
- The *IoU* and *DC* metrics were not used in previous related research works. For the trained and tested of U-net and FCN model, this work uses these metrics and finds a better solution.

This paper is structured as follows. In Section 2, the necessary definitions and techniques to obtain the results are described. In Section 3, the three techniques implemented for boundary extraction are compared to show the best method. Finally, in Section 4 some conclusions are shown.

2. Materials and Methods

2.1. Samples Collection

Before the segmentation, training samples were collected, based on the orthoimage and PV plant on-farm, rooftop, and roof photos. The samples collected to cover the spectral variability of each class of PV panel and consider the lighting variation in the scene, also in different parts of the world. For CNN, the samples were converted in a tagged image file format (.jpg) file and mask image file format (.png) with a shape of 240×320. The total of this dataset was found in the Amir dataset [43].

2.2. Boundary Extraction Procedure

UAVs must have a precise set of coordinates to define the coverage path planning correctly and thus fly over the total area of PV Plants in the inspection mission. To achieve this task automatically, it is necessary to explore how to extract the boundary of photovoltaic facilities with some techniques. There is a process called semantic segmentation, where each pixel is labeled with the class of its enclosing object or region, which can extract the PV Plants as a particular object in an image [11], but with the constraints that this work addresses. Two techniques have been implemented so far: Traditional Image Processing (TIP) [10] and Deep Learning (DL) [11]. Figure 1 shows the steps followed to reach that result by TIP and DL-based techniques.

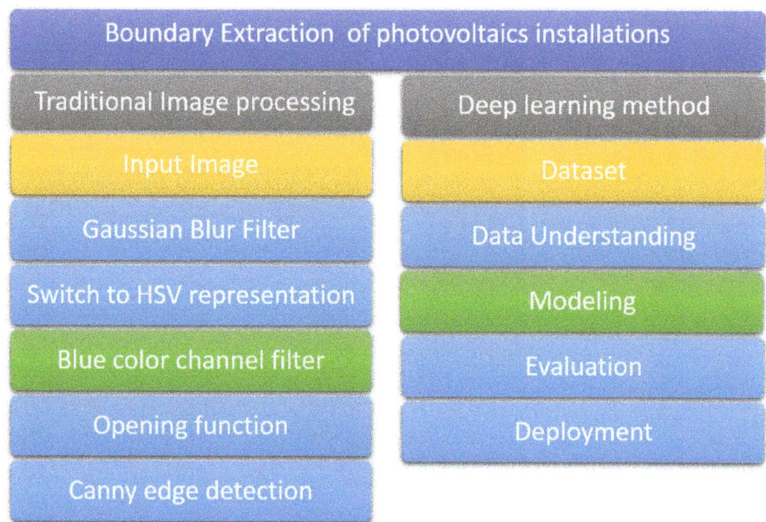

Figure 1. Steps of boundary extraction by image analysis with two techniques.

2.3. Traditional Image Processing (TIP)

The process to obtain the boundary pixels of a target can be achieved by means of traditional image processing techniques with functions that extract, increase, filter, and detect the features of an image and obtain its segmentation [27,32]. The main stages were used to remove the borders of PV plants out of an image, as shown in Figure 1 [10]. In the first stage, the original image was filtered using "filter2D" function from OpenCV, that is a convolution filter with 5×5 averaging filter kernel, as shown at Algorithm 1. This filter is compound with various Low-Pass Filters (LPF) and High-Pass Filters (HPF). LPF helps in removing noise, blurring images. HPF filters help in finding edges in images.

In the second stage, the filtered image is transformed into the HSV (hue, saturation, and value) representation. The transformation lessens reflection caused by environmental light during aerial image collection. Furthermore, this transformation helps in the color-based segmentation required in the next stages.

In the third stage, each channel was processed separately to extract the area of the PV plants. This was achieved by applying thresholding operations on the HSV image. To extract the PV blue color out of the image, the HSV range limits for thresholding where determined: from $(50, 0, 0)$ to $(110, 255, 255)$. Thresholding was implementing using the inRange function of OpenCV.

At the fourth stage, two morphological operators were applied: the "erode" and "dilate" functions. Together these operations helps to reduce noise and to better define the boundaries of the PV devices, the application of erosion followed by dilation is also known as opening operation. Erosion and dilation requires an structuring element (also known as kernel) to be applied to the images. In this case, a rectangular kernel of 2×2 pixels (MORPH_RECT,$(2, 2)$) was used for both operations. Lines 13, 14 and 15 from Algorithm 1 show the creation of the structuring element and the successive use of the erode and dilate functions.

Then, the "findCountours" function was used to help in extracting the contours from the image. The contour can be defined as a curve joining all the continuous points in the boundary of the PV installation. The input parameters for this function are: the image (dilated image from previous stage), the type of contour to be extracted (in this case only the external contours, RETR_EXTERNAL) and the contour approximation method (in this case not approximation, CHAIN_APPROX_NONE). Finally the area was recognized using a multi-stage algorithm to detect a wide range of edges in images, known as the Canny edge detection "Canny" [44].

The pseudo-code of the Traditional Image Processing is shown in Algorithm 1, and was implemented in Python 3 using OpenCV library.

Algorithm 1: TIP algorithms

---

1. **input** : A image Im of size $w \times l$
2. **output:** A Boundary of the image

3. initialization
4. import cv 2, np
5. **for** i in range(n):
6.   I $\leftarrow$ cv 2.imread(Im)
7.   Irgb $\leftarrow$ cv 2.cvtColor$(I[w,l],COLOR_BRG2RGB)$
8.   Kernel $\leftarrow$ np.ones$([5,5])$
   nIldst $\leftarrow$ cv 2.filter 2D(Irgb$[w,l],-1$,Kernel)
9.   Ihsv $\leftarrow$ cv 2.cvtColor(Idst$[w,l],COLOR_BRG2hsv$)/* Transformation from RGB to HSV   */
10.   LowerBlue $\leftarrow$ np.array$([55,0,0])$
11.   UpperBlue $\leftarrow$ np.array$([110,255,255])$
12.   mask $\leftarrow$ cv 2inRange(Ihsv$[w,l]$,LowerBlue,UpperBlue)
13.   K $\leftarrow$ cv 2.getStructuringElement$(MORPH_RECT,(2,2))$
14.   Ierosion $\leftarrow$ cv 2.erode(mask$[w,l],K,iteration = 3$)
15.   Idilation $\leftarrow$ cv 2.dilate(Ierosion$[w,l],K,iteration = 8$)
16.   cnt $\leftarrow$ cv 2.findCountours(Idilation$[w,l],RETRE_XTERNAL,CHAIN_APPROX_NONE$)
17.   mask2 $\leftarrow$ cv2.zeros$([w,l])$
18.   cc $\leftarrow$ cv 2.drawContours(mask2,cnt,-1,255,-1)
19.   edge $\leftarrow$ cv2.Canny(cc,100,105)

2.4. Deep Learning

Another approach to ascertain the boundaries of PV plants uses a DL-based technique which consists of several steps:

2.4.1. Data Specifications

The first step is to select the data for training the Neural Networks. The parameters to take into account are: PV Plants in orthophotos and aerial images with the respective masks for each image [11].

2.4.2. Data Understanding

The data preparation phase can be subdivided, into at least four steps. The first step is data selection inside the dataset. The second step involves correcting the individual data, which are assumed to be noisy, apparently incorrect, or absent. The third step involves resizing the data as needed. Finally, most of the available implementations assume that the data are given in a single table, so if the data are in several tables, they must be parsed together in a single one [45].

2.4.3. Modeling

In the literature, the semantic segmentation task has many existing models that can be selected for the desired task. In this work, two methods based on deep learning have been selected, taking into account the following criteria: the most competent for the type of task, the amount of data to be processed, the execution time, and the ease of implementation to predict each label for each pixel. The methods were selected according to [11,46–49]. The FCN model was the first one selected, which was proposed by [33] and used by [11]. The network architecture is delineated in Figure 2. The second one is the U-Net model, first proposed by [38] and selected for this project. The network architecture is illustrated in Figure 3.

Fully Convolutional Network

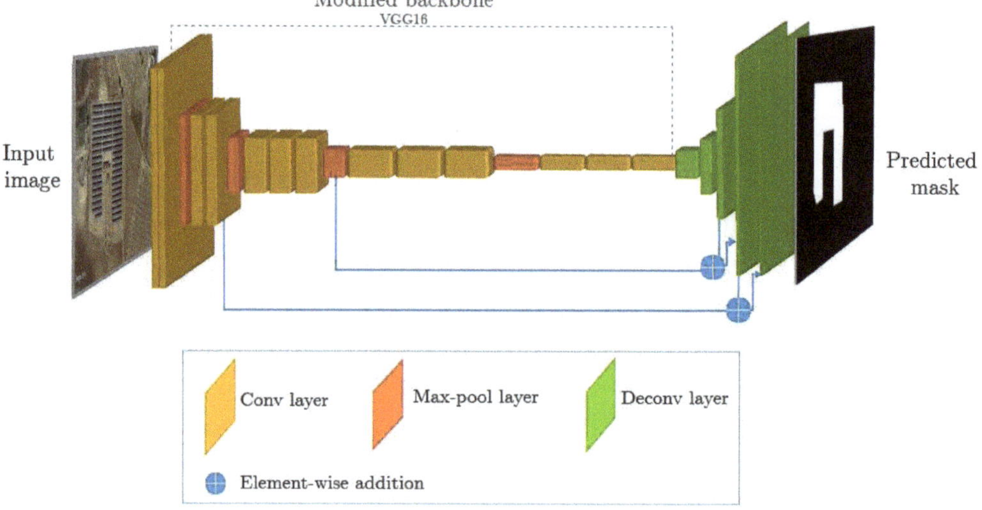

Figure 2. FCN model.

U-net

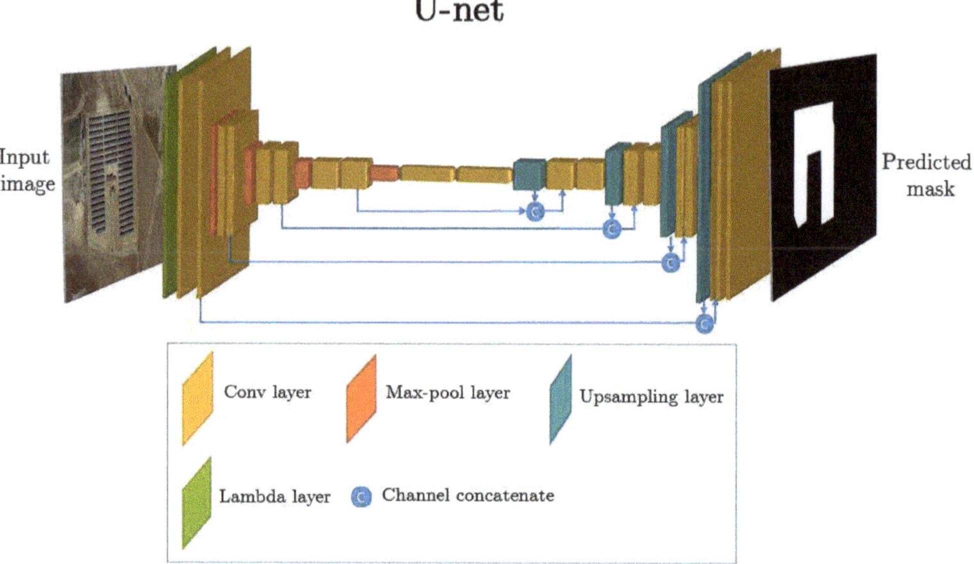

Figure 3. U-net Model.

(a). Fully Convolutional Network (FCN) model: This model has two blocks. The first block is a series of 13 layers in order to create a modified version of a VGG16 backbone Figure 2, which was introduced for the first time by [50]. The VGG16 backbone has 16 convolutional layers and its creators belong to the team "Visual Geometry Group", hence its name VGG16. The backbone is the network that takes the image as input and extracts the feature map upon which the rest of the network is based. The second block consists of a series of deconvolutional layers that simply reverses the forward and backward passes of convolution. The last layer uses a softmax function to predict the probability of the category as shown in Figure 2. As a result, the input of FCN model is an RGB image, and the output is the predicted mask of the PV plants. For more details, read [33]. The parameters for the training process were depicted in Table 1.

(b). The U-net network model: This model has two blocks: a decreasing path and an increasing path, respectively, which gives it the u-shaped architecture or horizontal hourglass shape [51]. The decreasing path is a typical convolutional network that consists of repeated application of convolutions, each followed by a rectified linear unit (ReLU) and a max-pooling operation. During the decrease, the spatial information is reduced whereas feature information is increased. The increasing pathway combines the feature and spatial information through a sequence of upsampling layers followed by two layers of transposed convolution for each step [38,52], as illustrated in Figure 3. The parameters for the training process were depicted in Table 1. Its architecture is shown in Table 2. The platform used for FCN and Unet models by this work was Tensorflow with Keras backend [53]. The U-net model had never been used for this kind of application so far.

Table 1. Summary of the FCN and U-net model parameters for the training process.

Activation (Last Layers)	Activation (Inner Layers)	Optimizer	Loss Function	Metrics	Epoch	Batch Size
Sigmoid	Relu	RMS	Binary_cross	N/A	150	1
Sigmoid	Elu	Adam	Binary_cross	IoU,F1 score	15	8

The FCN and U-net models additionally have a binary cross-entropy function (H_p) to calculate the loss in the process of training the neuronal network [54]. As the problem at hand is a semantic segmentation task, Equation (1) is used. This function examines each pixel and compares the binary-predicted values vector with the binary-encoded target vector.

$$H_p(q) = -\frac{1}{N} \sum_{i=1}^{N} y_i \cdot \log(p(y_i)) + (1 - y_i) \cdot \log(1 - p(y_i)) \tag{1}$$

where y is the label of each pixel, it takes the value of 1 for the PV plants area and 0 to indicate other areas or elements, and $p(y)$ is the probability of the pixel belonging to the PV plants area for all N points. The Adam optimization function is used to optimize the models [55]. Because semantic segmentation is the task at hand, it is essential to implement metrics to ensure the model performs well.

Table 2. Architecture of the U-net.

Layer (Type)	Output Shape	Parameters
Input Layer	$(None, 240, 320, 3)$	0
Lambda	$(None, 240, 320, 3)$	0
Conv2D	$(None, 240, 320, 16)$	448
Dropout	$(None, 240, 320, 16)$	0
Conv2D	$(None, 240, 320, 16)$	2320
MaxPooling2D	$(None, 120, 160, 16)$	0
Conv2D	$(None, 120, 160, 32)$	4640
Dropout	$(None, 120, 160, 32)$	0
Conv2D	$(None, 120, 160, 32)$	9248

Table 2. *Cont.*

Layer (Type)	Output Shape	Parameters
MaxPooling2D	$(None, 60, 80, 32)$	0
Conv2D	$(None, 60, 80, 64)$	18,496
Dropout	$(None, 60, 80, 64)$	0
Conv2D	$(None, 60, 80, 64)$	36,928
MaxPooling2D	$(None, 30, 40, 64)$	0
Conv2D	$(None, 30, 40, 128)$	73,856
Dropout	$(None, 30, 40, 128)$	0
Conv2D	$(None, 30, 40, 128)$	147,584
MaxPooling2D	$(None, 15, 20, 128)$	0
Conv2D	$(None, 15, 20, 256)$	295,168
Dropout	$(None, 15, 20, 256)$	0
Conv2D	$(None, 15, 20, 256)$	590,080
Conv2D_Transpose	$(None, 30, 40, 128)$	131,200
Concatenate	$(None, 30, 40, 128)$	73,856
Conv2D	$(None, 30, 40, 128)$	295,040
Dropout	$(None, 30, 40, 128)$	0
Conv2D	$(None, 30, 40, 128)$	147,584
Conv2D_Transpose	$(None, 60, 80, 64)$	32,832
Concatenate	$(None, 60, 80, 128)$	0
Conv2D	$(None, 60, 80, 64)$	73,792
Dropout	$(None, 60, 80, 64)$	0
Conv2D	$(None, 60, 80, 64)$	36,928
Conv2D_Transpose	$(None, 120, 160, 32$	8224
Concatenate	$(None, 120, 160, 64)$	0
Conv2D	$(None, 120, 160, 32)$	18,464
Dropout	$(None, 120, 160, 32)$	0
Conv2D	$(None, 120, 160, 32)$	9248
Conv2D_Transpose	$(None, 240, 320, 16)$	2064
Concatenate	$(None, 240, 320, 32)$	0
Conv2D	$(None, 240, 320, 16)$	4624
Dropout	$(None, 240, 320, 16)$	0
Conv2D	$(None, 240, 320, 16)$	2320
Conv2D	$(None, 240, 320, 1)$	17

2.4.4. Metrics

The metrics evaluate the similarities between the predicted mask (N) and the original mask (S). Such similarity assessment can be performed by considering spatial overlapping information, that is, by computing the true positives (TP), false positives (FP) and false negatives (FN) given by $TP = |N \cap S|$, $FP = |N \setminus S|$, and $FN = |S \setminus N|$, respectively.

There are three standard metrics commonly employed to evaluate the effectiveness of the proposed semantic segmentation technique [29,48,49,56]. The three metrics, namely, pixel accuracy (Acc), region Intersection over Union (*IoU*), and Dice Coefficient (*DC*).

Pixel accuracy is the ratio of correctly classified PV plants pixels to the total number of PV plants pixels in the original mask image [57], which can be mathematically represented as Equation (2).

$$Accuracy = \frac{TP}{TP + FN} \tag{2}$$

The *IoU* metric (the Jaccard index) is defined by Equation (3). This equation is a ratio between the intersection of the predicted mask N, and the original mask S and the union of both. More details can be found in [58].

$$IoU(N,S) = \frac{|N \cap S|}{|N \cup S|} = \frac{TP}{TP + FP + FN} \tag{3}$$

The *DC* metric [56,58,59] is expressed as Equation (4). This equation divides the intersection of the predicted mask N, and the original mask S times two by the sum of N and S.

$$DC(N,S) = \frac{2.|N \cap S|}{|N| + |S|} = \frac{2.TP}{2.TP + FP + FN} \tag{4}$$

To validate the results of the techniques described above, the FCN and U-net models were trained and their performance was evaluated by validating and testing samples of the Amir dataset [43]. The next section describes such results and compares the models in detail.

3. Results and Discussion

3.1. Database Specification

For this work, the DeepSolar [60], Google Sun-Roof [61], OpenPV [62], and Amir's databases were accessed [43]. Only the last database met the established parameters. It contained PV plants in orthophotos and aerial images with their respective masks. Furthermore, the PV plants images were from different countries around the world. Therefore, the "Amir" dataset was selected.

3.2. Results with TIP Technique

The results obtained in this work were compared with the results obtained in previous investigations where the TIP and the deep learning techniques were used alongside the FCN model [11].

The stages to obtain the results are shown in Figure 4. In the First Stage, a 2D filter was applied, depicted in Figure 4a. In the second stage, the filtered image is transformed into the HSV representation, Figure 4b. In the third stage, the blue color was filtered out, Figure 4c. At the fourth stage, the opening function was used, as seen in Figure 4d. Finally, the area was recognized using the canny method illustrated in Figure 4e. The results were satisfactory and can be modified depending on the environment.

The results are shown in Table 3. The TIP technique was obtained by randomly selecting images out of the test dataset, then applying the functions described in the methodology section (Section 2), and finally comparing the mask obtained with the original mask. The *IoU* metric obtained was 71.62% and the *DC* was 71.62%.

Original	(a) Filter2D	(b) HSV filter	(c) Blue filter	(d) Opening	(e) Edge
	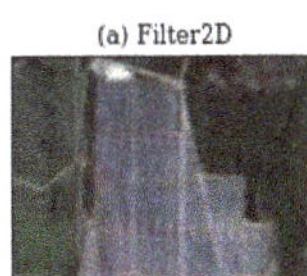	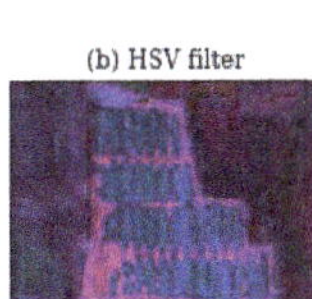	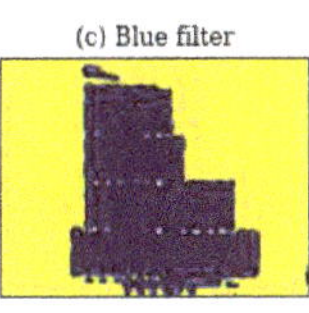	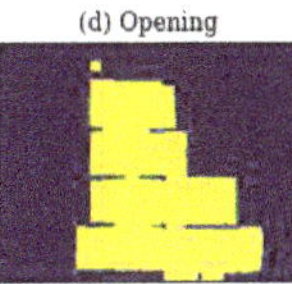	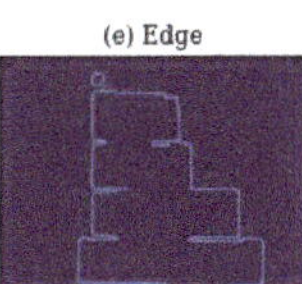

Figure 4. Steps of boundary extraction by TIP.

3.3. Results with DL-Based Techniques

The training data consisted of 2864 aerial images selected at random: 90% of the training dataset in the Amir database. The validation data were the remaining 10% of the same training dataset. Figure 5a shows the loss function and IoU metric of the FCN model during the training and validation process. The general trend of the two curves is consistent, showing that the network converges rapidly and is stable at iteration 30, and the loss value tends to 0.04%. Figure 5b shows the DC metric of the model during the training and validation stage. The general trend of the two curves is consistent at iteration 30.

On the other hand, using the same metrics, the U-net model proposed in this work shows a better performance. Figure 6a shows the loss function and IoU metric of the model during the training and validation stage. The common trend of the two curves shows the network converges quickly and is stable at iteration 16, and the loss value tends to 0.03%. Figure 6b shows the DC metric of the model all along the training and validation phase. The prevalent trend of the two curves is consistent and in iteration 16.

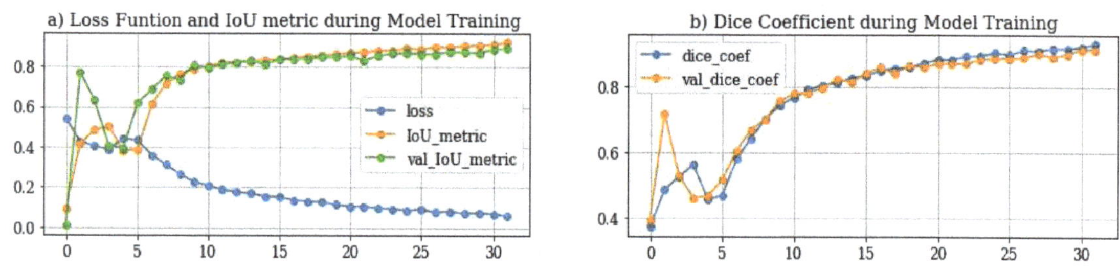

Figure 5. Performance and metrics of the FCN model using the training and validation sets.

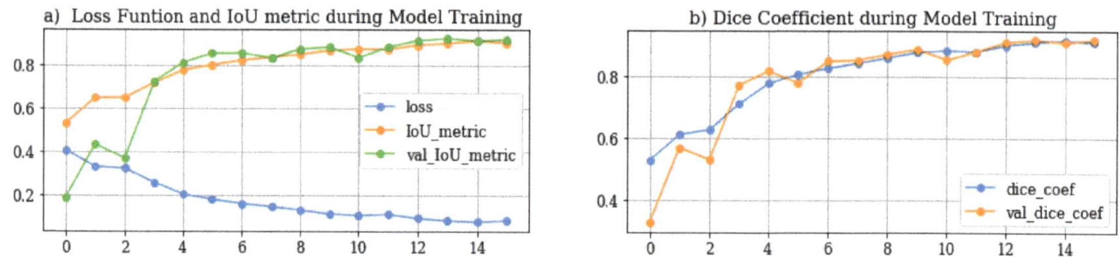

Figure 6. Performance and metrics of the U-net model obtained using the training and validation sets.

In the evaluation stage, 716 images were used along with the trained FCN model for PV plant detection. Some relevant results are shown in Figure 7. In this figure, the columns correspond to different PV plants. The first row contains the original images; the second row, the original masks; and, the third one, the predicted masks. The images used were taken in deserted regions and vegetation zones. The FCN model detects the PV plants in vegetation zones with some false positives. As an example, the second and third predictions of Figure 7 identify a lake and vegetation as part of the PV plants. In deserted regions, PV plants are detected more precisely. Although these images have very high precision, their predicted shape does not fully correspond to the original mask. Hence, it was necessary to review the performance metrics of the algorithm [63].

The segmentation results in the evaluation stage, using the same 716 images and the trained U-Net model, are shown in Figure 8. The arrangement is the same as in the previous Figure 7. It is noteworthy that this model correctly segments the photovoltaic plant while the other model does not achieve this result, as can be seen in the second and third predictions in Figure 8.

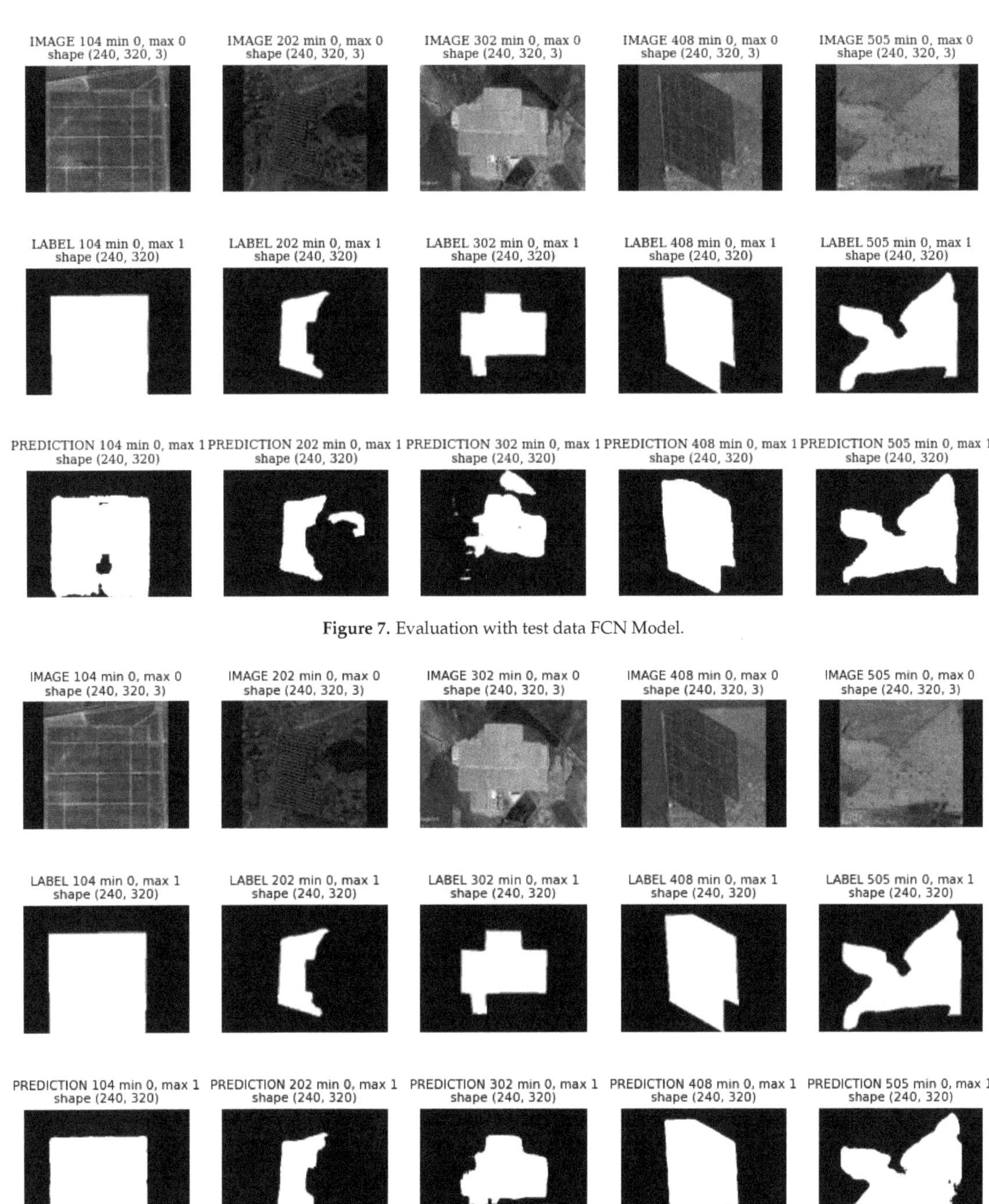

Figure 7. Evaluation with test data FCN Model.

Figure 8. Evaluation with test data U-net Model.

Afterwards, the trained model tested 716 samples. Table 3 shows the results and comparison among the TIP technique, the U-net proposed model and the FCN model used by [11], which was replicated in this study. The FCN and the proposed U-net models were compared. The accuracies obtained for the FCN model in the stages of training and testing were 97.99% and 94.16% respectively [11]. For U-Net proposed, the accuracy obtained in the stages of training and testing were 97.07% and 95.44%, respectively. Both results can be seen in Table 3.

To compare the FCN model proposed by Amir [11], and the U-net model proposed in this work, the two most used metrics in semantic segmentation problems were used. The FCN model was implemented with the standard IoU metric, whose result for the training stage was 94.13%,and the validation stage was 90.91% and for test stage was 87.47%. The DC metric of the validation 92.96% and test 89.61% which deviates a little from the training 95.10%. However, using the same metrics the U-net model proposed in this work shows a better performance. The IoU metric obtained was 93.57% in the training stage, 93.51% in the validation stage, and 91.42% in the test stage. The DC metric of the validation 94.44% was almost the same as that of the training 94.03% which deviates a little from the test 91.42%. Table 3 shows these results. Due to this, a difference was found between the FCN and U-net model for the first metric of 2.95% and for the second metric used of 1.81% difference was calculated. All files and logs from the experiments are available at GitHub in [64].

Table 3. Comparison between three techniques.

Parameter	TIP Method	FCN Model Amir [11]	U-Net Proposed
Metrics	N/A	N/A	IoU, F1 scor
Acc train	N/A	97.99%	97.07%
Acc test	N/A	94.16%	95.44%
IoU metric	N/A	94.13%	93.57%
Dice coef metric	N/A	95.10%	94.03%
val IoU metric	N/A	90.91%	93.51%
val Dice coef	N/A	92.96%	94.44%
test IoU metric	71.62%	87.47%	90.42%
test Dice coef metric	71.62%	89.61%	91.42%

3.4. Discussion

The U-net model proposed reconstructs the segmented image and protects the original image shape characteristics by storing the grouping indices of the max-pooling layer, a process that is not done in the FCN model.

The training and testing accuracy is the percentage of pixels in the image that are classified correctly and cannot be taken as indicators of how similar the predicted PV plants and the original mask are [65]. For the purpose of comparing the similarity in the results, the IoU metric was used. This metric varies from 0 to 1 (0–100%) with 0 meaning no similarity and 1 meaning total similarity between original and predicted masks [63].

The U-net model proposed in this work aimed to obtain a value closer to 1 in the IoU metric. The iteration times show the model used is faster and therefore reliable for the training and processing stages obtaining results virtually in real time [66]. The DC is the other metric used in this work. This metric also ranges from 0 to 1, with 1 signifying the greatest similarity between the predicted and original masks [63]. Both metrics were used to determine if the U-net model was better than the FCN model in the validation and test stages. The values of the IoU and Dice metrics in Table 3 showed the U-net model had a better performance when compared to the FCN model. This work was implemented with VGG16 as an encoder because it was the encoder used by Amir [11], which is a comparison work, but in future work, it is possible to use other encoders like ResNet, AlexNet, etc. [37].

Finally, the results obtained with the TIP and FCN model agree with the results obtained by other authors [11,13]. The authors mentioned they did not use the standard metrics for these kinds of problems and the bias in the results were expected. On the contrary, this work did take these metrics into account and found satisfactory results. The U-net network increased the processing speed, veracity in the segmentation process, and the overall performance of the model.

4. Conclusions

This work used three techniques, namely, the TIP technique, the DL-based FCN and U-net models. This work applied the U-net model to PV plants. All the models were used for the extraction of the PV plants boundaries out of an image. As a consequence, the TIP technique can be very precise but requires constant adjustment depending on the image, whereas the FCN and U-net network models are more useful when it comes to unknown PV plants.

The U-net network model is novel for this kind of problem. It allows greater processing speeds and performance when predicting the area of PV plants, also better features. The results obtained open the door for further investigation of this model in this problem.

The U-net technique turned out to be satisfactory compared to the TIP technique and the FCN model used in previous studies. The values obtained in the implemented metrics guarantee that the areas predicted for the PV plants are similar to the real ones. The results also help to predict possible false positives, such as lakes in the vicinity of photovoltaic plants. The relevant features of an object can be obtained using this technique while using the FCN technique is not possible.

Author Contributions: Conceptualization, A.P.-G., Á.J.-D. and J.B.C.-Q.; methodology, A.P.-G.; software, A.P.-G.; validation, A.P.-G.; formal analysis, A.P.-G.; investigation, A.P.-G.; resources, A.P.-G., Á.J.-D. and J.B.C.-Q.; data curation, A.P.-G.; writing—original draft preparation, A.P.-G.; writing—review and editing, A.P.-G., Á.J.-D. and J.B.C.-Q.; visualization, A.P.-G.; supervision, Á.J.-D. and J.B.C.-Q.; project administration, Á.J.-D. and J.B.C.-Q.; funding acquisition, A.P.-G., Á.J.-D. and J.B.C.-Q. All authors have read and agreed to the published version of the manuscript.

Funding: This research was funded by the Colombia Scientific Program within the framework of the so-called Ecosistema Científico (Contract No. FP44842-218-2018).

Institutional Review Board Statement: Not applicable.

Informed Consent Statement: Not applicable.

Data Availability Statement: The models used in the computational experiment are available at GitHub in [64].

Acknowledgments: The authors gratefully acknowledge the support from the Colombia Scientific Program within the framework of the call Ecosistema Científico (Contract No. FP44842-218-2018). The authors also want to acknowledge Universidad de Antioquia for its support through the project "estrategia de sostenibilidad".

Conflicts of Interest: The authors declare no conflict of interest.

List of Symbols

H_p	binary cross-entropy
$IoU(N,S)$	IoU metric, the mask predicted N and ground-trouth S original mask
$DC(N,S)$	The DC, the mask predicted N and ground-trouth S original mask
FP	false positives
TP	true positives
FN	false negatives
FP	false positives

Abbreviations

The following abbreviations are used in this manuscript:

DL	Deep Learning
ML	Machine Learning
UAV	Unmanned Aerial Vehicle
PV	Photovoltaic
TIP	Traditional Image Processing
O&M	Operation and Maintenance
FCN	Fully Convolutional Network
CAGR	Compound Annual Growth Rate
MV	Machine Vision

References

1. Donovan, C.W. *Renewable Energy Finance: Funding the Future of Energy*; World Scientific Publishing Co. Pte. Ltd.: London, UK, 2020.
2. Philipps, S.; Warmuth, W. Photovoltaics Report Fraunhofer Institute for Solar Energy Systems. In *ISE with Support of PSE GmbH November 14th*; Fraunhofer ISE: Freiburg, Germany, 2019.
3. Jamil, W.J.; Rahman, H.A.; Shaari, S.; Salam, Z. Performance degradation of photovoltaic power system: Review on mitigation methods. *Renew. Sustain. Energy Rev.* **2017**, *67*, 876–891. [CrossRef]
4. Kaplani, E. PV cell and module degradation, detection and diagnostics. In *Renewable Energy in the Service of Mankind Vol II*; Springer: Cham, Switzerland, 2016; pp. 393–402.
5. Di Lorenzo, G.; Araneo, R.; Mitolo, M.; Niccolai, A.; Grimaccia, F. Review of O&M Practices in PV Plants: Failures, Solutions, Remote Control, and Monitoring Tools. *IEEE J. Photovolt.* **2020**, *10*, 914–926.
6. Guerrero-Liquet, G.C.; Oviedo-Casado, S.; Sánchez-Lozano, J.; García-Cascales, M.S.; Prior, J.; Urbina, A. Determination of the Optimal Size of Photovoltaic Systems by Using Multi-Criteria Decision-Making Methods. *Sustainability* **2018**, *10*, 4594. [CrossRef]
7. Grimaccia, F.; Leva, S.; Niccolai, A.; Cantoro, G. Assessment of PV plant monitoring system by means of unmanned aerial vehicles. In Proceedings of the 2018 IEEE International Conference on Environment and Electrical Engineering and 2018 IEEE Industrial and Commercial Power Systems Europe (EEEIC/I&CPS Europe), Palermo, Italy, 12–15 June 2018; pp. 1–6.
8. Shen, K.; Qiu, Q.; Wu, Q.; Lin, Z.; Wu, Y. Research on the Development Status of Photovoltaic Panel Cleaning Equipment Based on Patent Analysis. In Proceedings of the 2019 3rd International Conference on Robotics and Automation Sciences (ICRAS), Wuhan, China, 1–3 June 2019; pp. 20–27.
9. Azaiz, R. Flying Robot for Processing and Cleaning Smooth, Curved and Modular Surfaces. U.S. Patent 15/118,849, 2 March 2017.
10. Grimaccia, F.; Aghaei, M.; Mussetta, M.; Leva, S.; Quater, P.B. Planning for PV plant performance monitoring by means of unmanned aerial systems (UAS). *Int. J. Energy Environ. Eng.* **2015**, *6*, 47–54. [CrossRef]
11. Sizkouhi, A.M.M.; Aghaei, M.; Esmailifar, S.M.; Mohammadi, M.R.; Grimaccia, F. Automatic boundary extraction of large-scale photovoltaic plants using a fully convolutional network on aerial imagery. *IEEE J. Photovolt.* **2020**, *10*, 1061–1067. [CrossRef]
12. Leva, S.; Aghaei, M.; Grimaccia, F. PV power plant inspection by UAS: Correlation between altitude and detection of defects on PV modules. In Proceedings of the 2015 IEEE 15th International Conference on Environment and Electrical Engineering (EEEIC), Rome, Italy, 10–13 June 2015; pp. 1921–1926.
13. Aghaei, M.; Dolara, A.; Leva, S.; Grimaccia, F. Image resolution and defects detection in PV inspection by unmanned technologies. In Proceedings of the 2016 IEEE Power and Energy Society General Meeting (PESGM), Boston, MA, USA, 17–21 July 2016; pp. 1–5.
14. Grimaccia, F.; Leva, S.; Dolara, A.; Aghaei, M. Survey on PV modules' common faults after an O&M flight extensive campaign over different plants in Italy. *IEEE J. Photovolt.* **2017**, *7*, 810–816.
15. Quater, P.B.; Grimaccia, F.; Leva, S.; Mussetta, M.; Aghaei, M. Light Unmanned Aerial Vehicles (UAVs) for cooperative inspection of PV plants. *IEEE J. Photovolt.* **2014**, *4*, 1107–1113. [CrossRef]
16. Oliveira, A.K.V.; Aghaei, M.; Madukanya, U.E.; Rüther, R. Fault inspection by aerial infrared thermography in a pv plant after a meteorological tsunami. *Rev. Bras. Energ. Sol.* **2019**, *10*, 17–25.
17. De Oliveira, A.K.V.; Amstad, D.; Madukanya, U.E.; Do Nascimento, L.R.; Aghaei, M.; Rüther, R. Aerial infrared thermography of a CdTe utility-scale PV power plant. In Proceedings of the 2019 IEEE 46th Photovoltaic Specialists Conference (PVSC), Chicago, IL, USA, 16–21 June 2019; pp. 1335–1340.
18. AGHAEI, M. Novel Methods in Control and Monitoring of Photovoltaic Systems. Ph.D. Thesis, Politecnico di Milano, Milan, Italy, 2016.
19. Li, X.; Li, W.; Yang, Q.; Yan, W.; Zomaya, A.Y. An Unmanned Inspection System for Multiple Defects Detection in Photovoltaic Plants. *IEEE J. Photovolt.* **2019**, *10*, 568–576. [CrossRef]
20. Aghaei, M.; Leva, S.; Grimaccia, F. PV power plant inspection by image mosaicing techniques for IR real-time images. In Proceedings of the 2016 IEEE 43rd Photovoltaic Specialists Conference (PVSC), Portland, OR, USA, 5–10 June 2016; pp. 3100–3105.

21. Aghaei, M.; Gandelli, A.; Grimaccia, F.; Leva, S.; Zich, R.E. IR real-time analyses for PV system monitoring by digital image processing techniques. In Proceedings of the 2015 International Conference on Event-Based Control, Communication, and Signal Processing (EBCCSP), Krakow, Poland, 17–19 June 2015; pp. 1–6.

22. Menéndez, O.; Guamán, R.; Pérez, M.; Auat Cheein, F. Photovoltaic modules diagnosis using artificial vision techniques for artifact minimization. *Energies* **2018**, *11*, 1688. [CrossRef]

23. López-Fernández, L.; Lagüela, S.; Fernández, J.; González-Aguilera, D. Automatic evaluation of photovoltaic power stations from high-density RGB-T 3D point clouds. *Remote Sens.* **2017**, *9*, 631. [CrossRef]

24. Niccolai, A.; Grimaccia, F.; Leva, S. Advanced asset management tools in photovoltaic plant monitoring: UAV-based digital mapping. *Energies* **2019**, *12*, 4736. [CrossRef]

25. Tsanakas, J.A.; Chrysostomou, D.; Botsaris, P.N.; Gasteratos, A. Fault diagnosis of photovoltaic modules through image processing and Canny edge detection on field thermographic measurements. *Int. J. Sustain. Energy* **2015**, *34*, 351–372. [CrossRef]

26. Yao, Y.Y.; Hu, Y.T. Recognition and location of solar panels based on machine vision. In Proceedings of the 2017 2nd Asia-Pacific Conference on Intelligent Robot Systems (ACIRS), Wuhan, China, 16–18 June 2017; pp. 7–12.

27. Sizkouhi, A.M.M.; Esmailifar, S.M.; Aghaei, M.; De Oliveira, A.K.V.; Rüther, R. Autonomous path planning by unmanned aerial vehicle (UAV) for precise monitoring of large-scale PV plants. In Proceedings of the 2019 IEEE 46th Photovoltaic Specialists Conference (PVSC), Chicago, IL, USA, 16–21 June 2019; pp. 1398–1402.

28. Rodriguez-Esparza, E.; Zanella-Calzada, L.A.; Oliva, D.; Heidari, A.A.; Zaldivar, D.; Pérez-Cisneros, M.; Foong, L.K. An efficient Harris hawks-inspired image segmentation method. *Expert Syst. Appl.* **2020**, *155*, 113428. [CrossRef]

29. Garcia-Garcia, A.; Orts-Escolano, S.; Oprea, S.; Villena-Martinez, V.; Martinez-Gonzalez, P.; Garcia-Rodriguez, J. A survey on deep learning techniques for image and video semantic segmentation. *Appl. Soft Comput.* **2018**, *70*, 41–65. [CrossRef]

30. Pal, N.R.; Pal, S.K. A review on image segmentation techniques. *Pattern Recognit.* **1993**, *26*, 1277–1294. [CrossRef]

31. Hoeser, T.; Kuenzer, C. Object detection and image segmentation with deep learning on Earth observation data: A review-part I: Evolution and recent trends. *Remote Sens.* **2020**, *12*, 1667. [CrossRef]

32. Henry, C.; Poudel, S.; Lee, S.W.; Jeong, H. Automatic detection system of deteriorated PV modules using drone with thermal camera. *Appl. Sci.* **2020**, *10*, 3802. [CrossRef]

33. Long, J.; Shelhamer, E.; Darrell, T. Fully convolutional networks for semantic segmentation. In Proceedings of the IEEE Conference on Computer Vision and Pattern Recognition, Boston, MA, USA, 7–12 June 2015; pp. 3431–3440.

34. Puttemans, S.; Van Ranst, W.; Goedemé, T. Detection of photovoltaic installations in RGB aerial imaging: A comparative study. In Proceedings of the GEOBIA 2016 Proceedings, Enschede, The Netherlands, 14–16 September 2016.

35. Karoui, M.S.; Benhalouche, F.Z.; Deville, Y.; Djerriri, K.; Briottet, X.; Houet, T.; Weber, C. Partial linear NMF-based unmixing methods for detection and area estimation of photovoltaic panels in urban hyperspectral remote sensing data. *Remote Sens.* **2019**, *11*, 2164. [CrossRef]

36. Bhatnagar, S.; Gill, L.; Ghosh, B. Drone Image Segmentation Using Machine and Deep Learning for Mapping Raised Bog Vegetation Communities. *Remote Sens.* **2020**, *12*, 2602. [CrossRef]

37. Sothe, C.; Almeida, C.M.D.; Schimalski, M.B.; Liesenberg, V.; Rosa, L.E.C.L.; Castro, J.D.B.; Feitosa, R.Q. A comparison of machine and deep-learning algorithms applied to multisource data for a subtropical forest area classification. *Int. J. Remote Sens.* **2020**, *41*, 1943–1969. [CrossRef]

38. Ronneberger, O.; Fischer, P.; Brox, T. U-net: Convolutional networks for biomedical image segmentation. In Proceedings of the International Conference on Medical Image Computing and Computer-Assisted Intervention, Munich, Germany, 5–9 October 2015; Springer: Cham, Switzerland, 2015; pp. 234–241.

39. Ren, M.; Zemel, R.S. End-to-end instance segmentation with recurrent attention. In Proceedings of the IEEE Conference on Computer Vision and Pattern Recognition, Honolulu, HI, USA, 21–26 July 2017; pp. 6656–6664.

40. McCollum, B.T. *Analyzing GPS Accuracy through the Implementation of Low-Cost Cots Real-Time Kinematic GPS Receivers in Unmanned Aerial Systems*; Technical Report; Air Force Institute of Technology Wright-Patterson AFB OH Wright-Patterson: Fort Belvoir, VA, USA, 2017.

41. Mission Planner Home–Mission Planner Documentation. Available online: https://ardupilot.org/planner/ (accessed on 2 May 2021).

42. QGroundControl, Intuitive and Powerful Ground Control Station for the MAVLink Protocol. Available online: http://qgroundcontrol.com/ (accessed on 15 April 2021).

43. Sizkouhi, M.; Aghaei, M.; Esmailifar, S.M. Aerial Imagery of PV Plants for Boundary Detection. 2020. Available online: https://ieee-dataport.org/documents/aerial-imagery-pv-plants-boundary-detection (accessed on 14 December 2020).

44. Canny, J. A Computational Approach to Edge Detection. *IEEE Trans. Pattern Anal. Mach. Intell.* **1986**, *PAMI-8*, 679–698. [CrossRef]

45. Berthold, M.R.; Borgelt, C.; Höppner, F.; Klawonn, F.; Silipo, R. Data preparation. In *Guide to Intelligent Data Science*; Springer: Cham, Switzerland, 2020; pp. 127–156.

46. Abdollahi, A.; Pradhan, B.; Alamri, A.M. An ensemble architecture of deep convolutional Segnet and Unet networks for building semantic segmentation from high-resolution aerial images. *Geocarto Int.* **2020**, 1–16. [CrossRef]

47. Minaee, S.; Boykov, Y.Y.; Porikli, F.; Plaza, A.J.; Kehtarnavaz, N.; Terzopoulos, D. Image segmentation using deep learning: A survey. *IEEE Trans. Pattern Anal. Mach. Intell.* **2021**. [CrossRef]

48. Hao, S.; Zhou, Y.; Guo, Y. A brief survey on semantic segmentation with deep learning. *Neurocomputing* **2020**, *406*, 302–321. [CrossRef]

49. Lateef, F.; Ruichek, Y. Survey on semantic segmentation using deep learning techniques. *Neurocomputing* **2019**, *338*, 321–348. [CrossRef]

50. Simonyan, K.; Zisserman, A. Very deep convolutional networks for large-scale image recognition. *arXiv* **2014**, arXiv:1409.1556.

51. Cao, M.; Zou, Y.; Yang, D.; Liu, C. GISCA: Gradient-inductive segmentation network with contextual attention for scene text detection. *IEEE Access* **2019**, *7*, 62805–62816. [CrossRef]

52. Shibuya, N. Up-Sampling with Transposed Convolution. Towards Data Science. 2017. Available online: https://naokishibuya.medium.com/up-sampling-with-transposed-convolution-9ae4f2df52d0 (accessed on 24 February 2021).

53. Abadi, M.; Agarwal, A.; Barham, P.; Brevdo, E.; Chen, Z.; Citro, C.; Corrado, G.S.; Davis, A.; Dean, J.; Devin, M.; et al. TensorFlow: Large-Scale Machine Learning on Heterogeneous Systems. 2015. Available online: https://www.tensorflow.org/ (accessed on 1 May 2021).

54. Alain, G.; Bengio, Y. What regularized auto-encoders learn from the data-generating distribution. *J. Mach. Learn. Res.* **2014**, *15*, 3563–3593.

55. Yi, D.; Ahn, J.; Ji, S. An Effective Optimization Method for Machine Learning Based on ADAM. *Appl. Sci.* **2020**, *10*, 1073. [CrossRef]

56. Rizzi, M.; Guaragnella, C. Skin Lesion Segmentation Using Image Bit-Plane Multilayer Approach. *Appl. Sci.* **2020**, *10*, 3045. [CrossRef]

57. Talal, M.; Panthakkan, A.; Mukhtar, H.; Mansoor, W.; Almansoori, S.; Al Ahmad, H. Detection of water-bodies using semantic segmentation. In Proceedings of the 2018 International Conference on Signal Processing and Information Security (ICSPIS), Dubai, United Arab Emirates, 7–8 November 2018; pp. 1–4.

58. Powers, D.M. Evaluation: from precision, recall and F-measure to ROC, informedness, markedness and correlation. *arXiv* **2020**, arXiv:2010.16061.

59. Alalwan, N.; Abozeid, A.; ElHabshy, A.A.; Alzahrani, A. Efficient 3D Deep Learning Model for Medical Image Semantic Segmentation. *Alex. Eng. J.* **2021**, *60*, 1231–1239. [CrossRef]

60. Yu, J.; Wang, Z.; Majumdar, A.; Rajagopal, R. DeepSolar: A machine learning framework to efficiently construct a solar deployment database in the United States. *Joule* **2018**, *2*, 2605–2617. [CrossRef]

61. Elkin, C. Sun Roof Project. 2015. Available online: http://google.com/get/sunroof (accessed on 14 December 2020).

62. NREL. *Open PV Project*; U.S. Department of Energy's Solar Energy Technologies Office: Denver, CO, USA, 2018. Available online: https://www.nrel.gov/pv/open-pv-project.html (accessed on 14 December 2020).

63. Costa, M.G.F.; Campos, J.P.M.; e Aquino, G.D.A.; de Albuquerque Pereira, W.C.; Costa Filho, C.F.F. Evaluating the performance of convolutional neural networks with direct acyclic graph architectures in automatic segmentation of breast lesion in US images. *BMC Med. Imaging* **2019**, *19*, 85. [CrossRef] [PubMed]

64. Perez, A. 2021. Available online: https://github.com/andresperez86/BoundaryExtractionPhotovoltaicPlants (accessed on 26 May 2021).

65. Qiongyan, L.; Cai, J.; Berger, B.; Okamoto, M.; Miklavcic, S.J. Detecting spikes of wheat plants using neural networks with Laws texture energy. *Plant Methods* **2017**, *13*, 83. [CrossRef]

66. Ling, Z.; Zhang, D.; Qiu, R.C.; Jin, Z.; Zhang, Y.; He, X.; Liu, H. An accurate and real-time method of self-blast glass insulator location based on faster R-CNN and U-net with aerial images. *CSEE J. Power Energy Syst.* **2019**, *5*, 474–482.

applied
sciences

Article

Generation of Data-Driven Expected Energy Models for Photovoltaic Systems

Michael W. Hopwood [1,2] and **Thushara Gunda** [1,*]

1   Sandia National Laboratories, Albuquerque, NM 87123, USA; mwhopwo@sandia.gov
2   Department of Statistics and Data Science, University of Central Florida, Orlando, FL 32826, USA
*   Correspondence: tgunda@sandia.gov

Abstract: Although unique expected energy models can be generated for a given photovoltaic (PV) site, a standardized model is also needed to facilitate performance comparisons across fleets. Current standardized expected energy models for PV work well with sparse data, but they have demonstrated significant over-estimations, which impacts accurate diagnoses of field operations and maintenance issues. This research addresses this issue by using machine learning to develop a data-driven expected energy model that can more accurately generate inferences for energy production of PV systems. Irradiance and system capacity information was used from 172 sites across the United States to train a series of models using Lasso linear regression. The trained models generally perform better than the commonly used expected energy model from international standard (IEC 61724-1), with the two highest performing models ranging in model complexity from a third-order polynomial with 10 parameters ($R^2_{adj} = 0.994$) to a simpler, second-order polynomial with 4 parameters ($R^2_{adj} = 0.993$), the latter of which is subject to further evaluation. Subsequently, the trained models provide a more robust basis for identifying potential energy anomalies for operations and maintenance activities as well as informing planning-related financial assessments. We conclude with directions for future research, such as using splines to improve model continuity and better capture systems with low ($\leq$1000 kW DC) capacity.

Keywords: photovoltaic systems; expected energy models; fleet-scale; lasso regression; performance modeling; machine learning

Citation: Hopwood, M.W.; Gunda, T. Generation of Data-Driven Expected Energy Models for Photovoltaic Systems. *Appl. Sci.* **2022**, *12*, 1872. https://doi.org/10.3390/app12041872

Academic Editor: Giovanni Petrone

Received: 23 December 2021
Accepted: 31 January 2022
Published: 11 February 2022

Publisher's Note: MDPI stays neutral with regard to jurisdictional claims in published maps and institutional affiliations.

1. Introduction

The increasing penetration of photovoltaic (PV) systems within the energy markets has established the need for evaluating and ensuring high system reliability. In particular, a large emphasis has been placed on monitoring algorithms that can contextualize observed energy generation at a site with information about how the system would have performed in a nominal state [1]. The latter are commonly estimated through expected energy models. Expected energy models are incorporated into many PV performance monitoring tasks, including anomaly detection [2–5], financial planning [6], fleet-level (site vs. site) comparisons [7], degradation analysis [7], and the evaluation of extreme weather effects [8]. The comparison of observed energy values to those derived from expected energy models serves as the basis for informing both tactical (i.e., short-term tasks such as field repair) and strategic (i.e., long-term activities such as site planning) operations and maintenance (O&M) activities.

Expected energy models can vary from asset-level to site-level estimates [9]. Asset-level models typically focus on using parameters provided by the manufacturer (e.g., maximum power) [9,10]. However, such approaches do not always work well for in-field performance since the parameters were developed in standardized test conditions and thus do not reflect operational conditions [11]. In response to these limitations, empirical methods that use field observations and regression methods have emerged to derive parameters across non-standardized test conditions (e.g., [12,13]). At the site-level, most expected energy models leverage the correlation between power production and meteorological covariates [14,15]. For example, the standard expected energy model from the International Electrotechnical Commission (IEC) uses irradiance and site capacity information to develop an expected energy estimate [15]. Similarly, the PVUSA model trains a regression model for a given site by estimating power production using local irradiance, temperature, and wind speed conditions [16]. Industry research shows that most expected energy estimates tend to be overestimate production by a median of 3% but could be up to 20% [17]. Although the mismatch between observed and expected generation are well-recognized [18], limited attention has been given to date for improving the accuracy of expected energy models at the site-level, especially suited for fleet-level (i.e., site vs. site) comparisons.

This work aims to address this knowledge gap by generating a standardized, interpretable data-driven expected energy model that can be used for fleet-level comparisons. Although gradient-boosted and neural network-based methods have demonstrated significant successes for output performance [19–21], they often lack in model interpretability. In particular, models with high complexity can hide prediction biases or other vulnerabilities [22]. Thus for this work, we opted for more interpretable, regression-based models to increase the transparency of the implemented methods. In addition to identifying a more robust alternate for expected energy modeling, the associated publication of code used for training models (in the open source software pvOps) enables the extension of these methods to develop site-specific expected energy models for PV systems anywhere in the world or to other renewable energy systems. Such advancements in expected energy model estimates are needed to continue supporting better planning and field O&M activities, both of which ultimately influence the sustainability of PV sites. The following sections describes the data processing and model construction activities (Section 2), the performance of trained models (Section 3), and summarize primary findings (Section 4).

2. Methodology

The data-driven expected energy model training activities were supported by Sandia National Laboratories' PV Reliability, Operations, and Maintenance (PVROM) database [23]. Information about the PVROM database, as well as data processing, model training, and model evaluations, are described in the following Sections.

2.1. Data

The PVROM database contains 1.3 million data points of hourly production data across 176 sites in the United States [23], spanning multiple states (Figure 1) and generally ranging between 2017 and 2020. The database contains hourly measurements of expected energy in kiloWatt-hours (kWh), irradiance (Watts per square meter; $\frac{W}{m^2}$), ambient temperature, and module temperature; site-level direct current (DC) capacity is provided by the industry partners. The DC capacity (C_{DC}) for the sites within the database span from 37.8 kilowatts (kW) to 130,000 kW; a majority of the sites (140) are under 10,000 kW, with 67 of those sites under 1000 kW. A subset of the sites (100) contain industry-partner-provided expected energy estimates generated from proprietary models; these values serve as a basis for model validation activities (see Section 2.5).

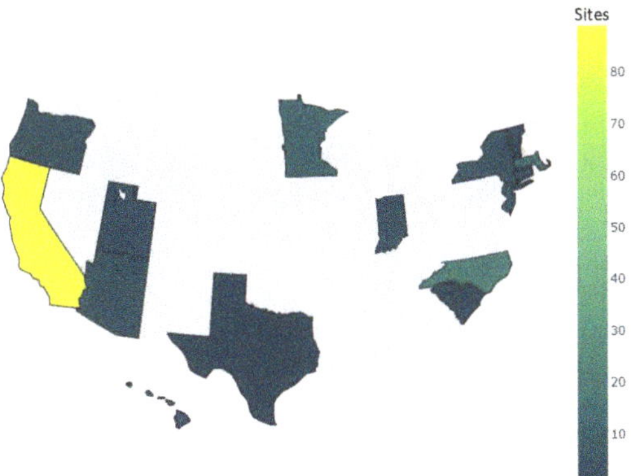

Figure 1. Geographical coverage of sites within the PVROM database. A majority of the sites are located in California.

2.2. Preprocessing

Data quality issues stemming from measurement errors and system anomalous conditions (reflecting local field failures, such as communication loss) could introduce signal variations in field data that would hinder model performance. Problematic data convolute the relationships between features, making it more difficult to measure the true parameter estimates; these potential irreducible errors are decreased through numerous data quality filters (Figure 2). Missing values (i.e., NaN or None values) were removed prior to applying data quality filters. An evaluation of these missing values revealed that a majority of them (~88%) occurred during nighttime hours (~7 p.m. to 8 a.m.), indicating that some sites captured night-time entries as null (Figure 2). After removing these missing values, ~900 K data points remained, which were then subject to a series of data quality filtering steps.

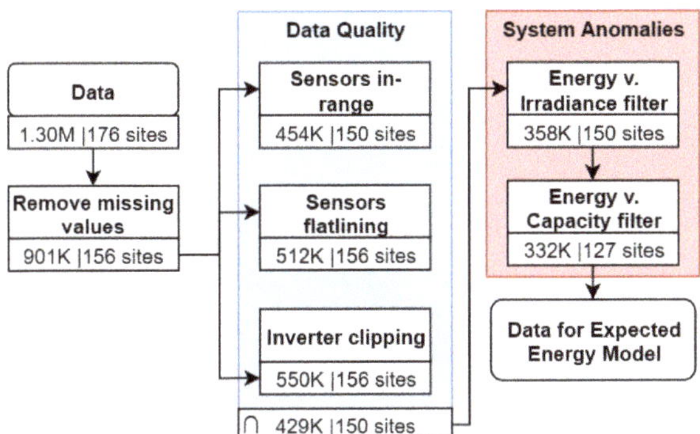

Figure 2. Data preprocessing activities included both data quality- and anomaly-related filters. Data quality filters were conducted independently; only data points that passed all quality-based filters were subject to the anomaly-based filters.

Data were filtered to ensure they are within nominal sensor ranges, using thresholds following [24] and the IEC 61724-1 standard [15]. Specifically, we retained data that met the following criteria:

- $20 \frac{W}{m^2} \leq$ Irradiance $(I) \leq 1500 \frac{W}{m^2}$;
- Energy $(E) > 0$ kWh;
- Ambient temperature $(T_{amb}) \leq 50\,°C$ and module temperature $(T_{mod}) \leq 90\,°C$.

Wind speed was not consistently available from partners and thus was excluded from analysis. Although available temperature data were used in the preprocessing steps, they are not used as a predictor variable in the regression models, since they ate not included in current standard models [15].

Flatlining values—determined by periods where consecutive data changed by less than a threshold—were flagged for removal using the pecos package [25], which follows the IEC 61724-3 standard [26]. Specifically, four consecutive hours with either $\Delta E < 0.01\%$ of the site's capacity or $\Delta I < 10 \frac{W}{m^2}$ were filtered. Lastly, inverter clipping, which occurs when the DC energy surpasses an inverter's DC energy rating, was addressed by mathematically observing plateaus in the energy signal using the pvanalytics package [27]. Dropping energy measurements during inverter clipping, which manifest as a static value across high irradiance levels, would create a better linear fit. After data quality checks, 429 K data points across 150 sites remained (Figure 2).

Data points that passed all quality checks were also assessed for system-level anomalies. These anomalies likely reflect abnormal operating conditions (i.e., local failures) and thus require removal to ensure the trained baseline energy models reflect nominal system performance. Anomalous entries were detected using a comparison of observed energy to irradiance and site capacities (Figure 3). The comparison of observed energy and irradiance filter focuses on removing data where the E–I ratio (λ) is outside its nominal distribution by 3 standard deviations $\{\lambda : \lambda < \mu_\lambda - 3\sigma_\lambda \cup \lambda > \mu_\lambda + 3\sigma_\lambda\}$, where μ_λ and σ_λ are the mean and standard deviation of the E–I ratio, respectively [28]. This filter was implemented for each site separately to capture site-specific variations (including system capacity) and resulted in the removal of 70 K data points (Figure 2). The second system anomaly filter focused on removing sites with mismatches between observed energy and site capacity. Namely, if a site's maximum recorded energy was over $1.2 \times C_{DC}$ or under $0.7 \times C_{DC}$, then all data points for that site were excluded from subsequent analysis. This method filtered 23 sites; 50%+ of these sites were under 1000 kW, and only 1 was over 10,000 kW. Approximately 26 K data points were removed with this filter, resulting in a final dataset that contained 332 K data points across 127 sites for model training and testing activities (Figure 2). The age of the sites within the final dataset ranged from newly installed sites up to 10 years, with a majority being less than 5 years in age (Figure A2).

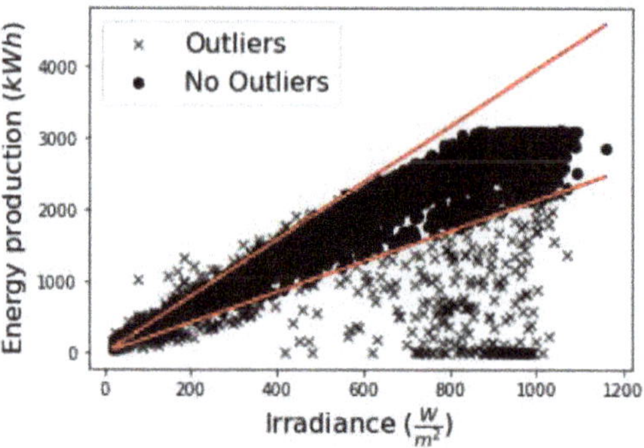

Figure 3. An example of anomaly-based filter (energy production vs. irradiance) for a particular site. Anomalous data points (visualized as Xs) are often lower than non-anomalous values within the distribution-derived bands (red lines).

2.3. Variable Standardization

The specific inputs used for model training mimic commonly available parameters used in current expected energy models (e.g., [15]), such as irradiance and site capacity. However, with covariates at different scales (e.g., $\{0\frac{W}{m^2} < I < 1.2 \times 10^3 \frac{W}{m^2}\}$ while $\{1 \times 10^2 \text{ kW} < C_{DC} < 1.3 \times 10^5 \text{ kW}\}$), variable standardization is required to reduce model sensitivity to parameter scales. In particular, without standardization, weights generated for each parameter are more likely to reflect scalar nuances rather than the relative importance of the parameter to the outcome of interest. Variable standardization centers the data by subtracting data points in a feature from its associated mean value (μ) and then scales the data by dividing by the associated standard deviation (σ)—i.e., $Z = \frac{\mu - \hat{\mu}}{\sigma_X}$. The resulting standardized variables have a mean of zero and a standard deviation of one. This process makes parameters easier to rank in terms of influence; the variable with the larger parameter holds a more important effect on the output response. Thus, variable standardization also aids in the interpretability of the derived parameters, especially when variable interactions (e.g., $I \times C_{DC}$) are introduced. The mean and standard deviation parameters used to standardize irradiance, capacity, and energy values are captured in Table 1.

Table 1. Mean and standard deviation (StDev) parameters used to standardize variables prior to training regression models.

Parameter	>1000 kW Systems		<1000 kW Systems	
	Mean	**StDev**	**Mean**	**StDev**
Irradiance	571.459	324.199	413.533	286.110
Capacity	14,916.234	20,030.000	375.919	234.151
Energy	7449.152	12,054.525	119.008	119.829

2.4. Model Design and Training

Similar to other machine learning models, regression techniques leverage input data to learn relationships and use those relationships to predict unseen quantities. These relationships are generally contained in model parameters ($\hat{\beta}$), which map predictors, as summarized in a design matrix $\mathbf{X}$, to an output $\hat{Y} = \mathbf{X}\hat{\beta} + \epsilon$ with residual model error ϵ. Many different regression techniques exist; these techniques typically vary in the structure of the cost function, which quantifies the error between predicted and expected values. This cost function (C) is usually captured as a summation of loss functions (calculated on each data point) across the training set. The set $\hat{\beta}$, which renders the smallest cost, is defined as the learned parameters, mathematically notated as:

$$\hat{\beta} = \underset{\theta}{arg\min}\, C. \tag{1}$$

A popular regression model is the ordinary least squares (OLS), which defines its best model ($\hat{\beta}_{OLS} = \underset{\theta}{arg\min}\, SSE$) with an objective function equal to the sum of squared errors (SSE):

$$SSE = \sum_{i=1}^{n}(y_i - \hat{y}_i)^2 = \sum_{i=1}^{n}(y - \sum_{j=0}^{p} \hat{\beta}_j x_{ij})^2 \tag{2}$$

where n is the number of samples, p is the number of predictors, and x_{ij} is the ith value for the jth explanatory variable. As shown in the equation, the SSE sums the squared difference between each sample (y) and its associated model estimate ($\hat{y}$). High emphasis is naturally placed on reducing high-error samples. Therefore, outliers can have a large effect on the learned parameters, so data preprocessing steps are required for robust model development. Additionally, OLS renders non-zero coefficients on all $\hat{\beta}$, which can create small, insubstantial parameters which are likely components of the training dataset and therefore contribute to model overfitting and thus should be removed from the model.

Alternate approaches to OLS include the Theil–Sen regressor [29], which is robust against outliers since it chooses the median of the slopes of all lines between pairs of points, as well as techniques such as Lasso regression [30] that explicitly address model overfitting by reducing model complexity (i.e., the number of parameters used). For this analysis, the latter was selected since Lasso regression models are able to incorporate both parameter regularization and residual sum of squares into the loss function. The cost function for

Lasso regression $\hat{\beta}_{lasso} = \underset{\theta}{arg\,min}\left(SSE + \alpha \sum_{j=1}^{p}|\beta_j|\right)$ incorporates an L1 regularization term $\alpha \sum_{j=1}^{p}|\beta_j|$, which penalizes the magnitude of the β terms. This penalization tends to shrink coefficients to zero, rendering a more parsimonious model; we use an $\alpha = 0.003$ for defining the impact of the regularization on the regression kernel. Specifically, the penalization acts as a bias, which in turn can reduce overall error due to the bias–variance tradeoff [31].

Standardized variables are passed into Lasso regression to learn a linear model, which relates the input variables to energy. Multiple combinations of input variables were used to train the regression models (more details below). For all models, a randomized (80–20%) split is utilized to partition the preprocessed, standardized data into train and test partitions, respectively.

In addition to individual parameter influences, interactions and temporal factors were incorporated as input features to capture nuances within the datasets. Interaction parameters, which allow the effect of one parameter on the response variable to be weighted by the value of another variable, are introduced by including terms which are the product of two or more predictor variables. For example, Figure 4 shows that the relationship between E and I does vary across C_{DC}. Thus, the inclusion of an I and C_{DC} interaction term may be helpful in predicting the generated energy. The suite of interaction combinations are instantiated using polynomial models up to the third order (i.e., degree $d = 3$). In a model with $d = 2$ and 2 covariates, the initiated regression model would take the following form:

$$y = \beta_0 + \beta_1 x_1^2 + \beta_2 x_2^2 + \beta_3 x_1 x_2 + \beta_4 x_1 + \beta_5 x_2. \tag{3}$$

Notice that a $d = 2$ also includes $d = 1$ parameters (i.e., $\beta_4 x_1$ and $\beta_5 x_2$). This remains true for all values of the polynomial power (e.g., for a model initiated with $d = 3$, terms from $d = 2$ and $d = 1$ are also included). Two interaction polynomial orders are tested: a second-order ($d = 2$) and a third-order ($d = 3$) (Table 2). The particular interaction noted above ($I \times C_{DC}$) is captured in multiple models, including an additive model with a single interaction term (Table 2).

In addition to interactions, temporal factors are used to capture a variable's changing effect on the energy generated over time. For instance, the correlation between I and E changes over the course of the year due to spectral irradiance effects [32,33]. Therefore, allowing the model to capture time-variant nuances may be important for capturing such nonlinearities. Three temporal based conditions were explored: seasonal (four per year), monthly, and hourly. A model with two predictor variables and monthly temporal-based variable conditions would be instantiated as:

$$\begin{aligned} y = a_{jan}1_{t \in jan}x_1 + a_{feb}1_{t \in feb}x_1 + \ldots + a_{dec}1_{t \in dec}x_1 \\ + b_{jan}1_{t \in jan}x_2 + b_{feb}1_{t \in feb}x_2 + \ldots + b_{dec}1_{t \in dec}x_2, \end{aligned} \tag{4}$$

where the a and b parameters are coefficients describing the effect of parameter x_1 and x_2, respectively, when conditioned on a month of the year. For instance, a_{jan} describes the effect of x_1 on the y response variable during the month of January. The indicator function 1_t masks the predictor variable to ensure it is within its timeframe. With the various combinations of interactions and temporal conditions, a total of 13 regression kernels were evaluated (Table 2; see Appendix A for some of the mathematical formulations).

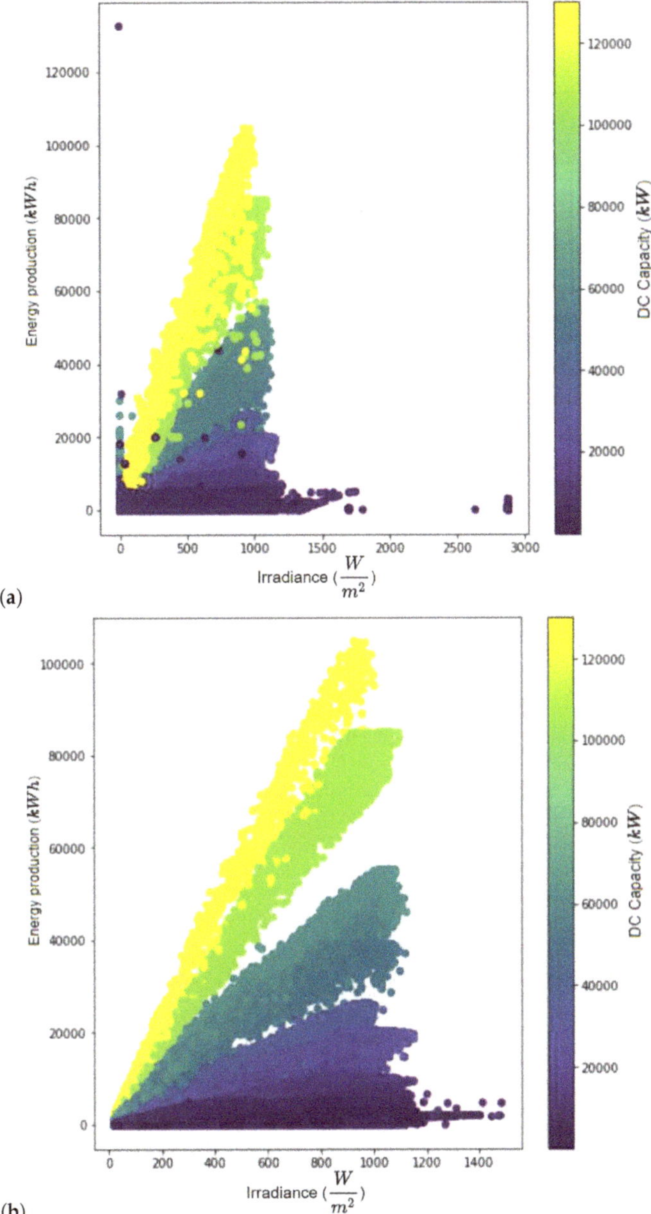

Figure 4. Correlation between energy production and irradiance for raw data (**a**) and preprocessed data (**b**) for different site capacities. Higher correlations in the preprocessed data indicate interaction between DC capacity and irradiance for energy production.

Table 2. Combinations of parameters used to initiate the 13 regression kernels evaluated in this study. The simple additive model attaches a parameter to each predictor variable to evaluate individual effects. * For hourly temporal conditions, only 15 h are used to reflect daytime hours.

Model Category	Time Variables (t)	Interaction Degree (d)	Number Parameters for M Inputs	Number Parameters for 2 Inputs		
Additive			$M + 1$	3		
Additive with interaction		2nd-order (d = 2)	$M + 2$	4		
Additive with Time-weighted	Season (t = 4) Month (t = 12) Hour (t = 15 *)		$tM + 1$	9 25 31		
Polynomial		2nd-order (d = 2) 3rd-order (d = 3)	$\sum_{i=1}^{d} C^R(M, i) + 1$	6 10		
Polynomial with Time-weighted	Season (t = 4) Month (t = 12) Hour (t = 15 *)	2nd-order (d = 2) 3rd-order (d = 3)	$t \sum_{i=1}^{d} C^R(M, i) + 1$	d = 2 t = 4 25 t = 12 73 t = 15 90	d = 3 41 121 151	

2.5. Model Evaluation

Three metrics were used to evaluate the performance of the trained expected energy models: logarithmic root mean squared error (log $RMSE$), coefficient of determination (R^2), and percent error (δ). Both partner-provided expected energy values and those calculated by the leading standardized expected energy model (i.e., IEC 61724) were used as reference values for model evaluations.

The root mean squared error ($RMSE$) is a common goodness-of-fit statistic used for model evaluation. The $RMSE$ is expressed as:

$$RMSE = \sqrt{\frac{1}{n} \sum_{i=1}^{n} (y_i - \hat{y}_i)^2}$$

where y_i and $\hat{y}_i$ are the measured and predicted values of the response variable, and n is the number of samples. $RMSE$ is in the same units as the response variable (i.e., kWh). Lower $RMSE$ values indicate a better, lower predicted error. Because the error can be quite large in magnitude (10^0 to 10^{10}), a logarithmic transform is applied to facilitate evaluations. Because the magnitude of the error is closely connected to a site's capacity, the log $RMSE$ cannot be used to compare model performance between sites unless the sites are similar in size.

The coefficient of determination (R^2), however, can be used to compare model performance across different site sizes. Specifically, R^2 is calculated as:

$$R^2 = 1 - \frac{\sum (y_i - \hat{y}_i)^2}{\sum (y_i - \bar{y})^2}, \tag{5}$$

where $\bar{y}$ is the average of the y values. R^2 denotes the proportion of variability in the response explained by the model with a value of 1, indicating a perfect fit. R^2 was used to compare trained model outputs with partner-generated expected energy values, whose underlying model structures were unknown.

Generally, however, R^2 is not well-suited for comparing models across varying numbers of parameters. Thus, when comparing the 13 trained models to one another, we utilize an adjusted R^2_{adj} metric, which checks whether the added parameters contribute to the explanation of the predictor variable and penalizes models with unnecessary complexity [34]. Low-effect parameters (i.e., $\beta \approx 0$) reduce the model's overall fit score. The adjusted R^2_{adj} is calculated as follows:

$$R^2_{adj} = 1 - \left(\frac{(1 - R^2)(n - 1)}{n - p - 1} \right), \tag{6}$$

where n is the number of samples, and p is the number of predictors.

Finally, δ was used to capture the directionality of error (i.e., overprediction vs. underprediction):

$$\delta = 100 \times \frac{\hat{y} - y}{y}. \tag{7}$$

The log $RMSE$ and R^2 were implemented to evaluate model performance at both site and fleet (i.e., across multiple sites) levels, while δ was only implemented at the fleet level; all metrics were reported on the test dataset. T-tests were used to evaluate significance in performance variations between the trained and reference values.

3. Results and Discussion

Data processing activities generally increased the correlations between the predictor variables (i.e., irradiance and capacity) and the response variable (i.e., energy) (Table A1). The processed data were inputted into a total of 13 trained models—ranging in model complexity (pre-lasso) from 3 parameters for the 'simple additive' model pre-lasso to 151 parameters for the 'third-order-hour' (see Table 3). Generally, the number of parameters were lower for all models post-lasso fit, except for the 'simple additive' and 'additive interaction' models, likely indicating the already sparse construction of these models.

Table 3. This table describes the parameterization and performance of all of the models evaluated in the results of this paper. Wins are summarized by showing the percentage of sites where a given model was the top performer according to the associated goodness-of-fit metric (Adj. R^2 or log $RMSE$); the IEC model was used as the reference value for log $RMSE$ calculations. The third-order interactions model and basic model perform consistently well, a conclusion also found on the heatmaps (Figure 5). Additionally, because lasso regression was leveraged, the models decrease in size after training the model.

Models	Time Variable	Interaction Degree	Number Parameters Pre-Lasso	Number Parameters Post-Lasso	>1000 kW Systems		<1000 kW Systems	
					Wins: Adj. R^2	Wins: logRMSE	Wins: Adj. R^2	Wins: logRMSE
Third-order interactions		3	10	9	47.9	49.0	38.7	38.7
Additive interaction		1	4	4	22.9	17.7	32.3	29.0
Second-order interactions		2	6	5	15.6	10.4	6.5	0.0
Second-order seasonal	season	2	25	18	9.4	15.6	3.2	3.2
Third-order seasonal	season	3	41	27	1.0	4.2	3.2	3.2
Second-order month	month	2	73	43	0.0	0.0	3.2	3.2
Second-order hour	hour	2	90	44	0.0	0.0	0.0	0.0
Third-order month	month	3	121	68	0.0	0.0	0.0	3.2
Third-order hour	hour	3	151	81	1.0	1.0	0.0	6.5
IEC					1.0	1.0	12.9	12.9
hour	hour	1	31	26	0.0	0.0	0.0	0.0
month	month	1	25	25	0.0	0.0	0.0	0.0
seasonal	season	1	9	9	1.0	1.0	0.0	0.0
simple additive		1	3	3	0.0	0.0	0.0	0.0

Initially, the various models were trained using data across all system sizes. However, this approach demonstrated systemic underperformance for low-capacity systems (<1000 kW DC capacity). Specifically, the best trained models (i.e., 'third-order interactions' and 'additive interaction') outperformed the IEC model in terms of log $RMSE$ when tested on every single system above 1300 kW DC capacity; however, 12 of 34 systems below a 1300 kW DC capacity underperformed relative to the IEC model. This result likely reflects the varying relationships between the site DC capacity and the energy generated; systems of higher capacity tend to receive a higher maximum energy generated per DC capacity (Figure A1). To better deal with this varying linearity, two separate sets of models were trained: one for models under 1000 kW DC capacity and another over.

Across both high-capacity and low-capacity systems, models with the $I \times C_{DC}$ interaction term perform better than those without the interaction term (i.e., 'hour', 'month', 'seasonal', and 'simple additive' models). For example, two of the top-performing models (across both high-capacity and low-capacity systems) are the 'simple additive' and the 'third-order interactions' models, both of which contain this interaction term (Table 3 and Figure 5). The 'additive interaction' trained (AIT) model has four parameters:

$$\hat{e} = \begin{cases} 0.07 + 0.69i + 0.65c + 0.42ic & C_{DC} < 1000 \text{ kW} \\ -0.06 + 0.29i + 0.76c + 0.40ic & C_{DC} \geq 1000 \text{ kW} \end{cases} \tag{8}$$

where i and c define the standardized irradiance and capacity variables, respectively, as defined in Table 1. The 'third-order interactions' model, on the other hand, contains these four terms as well as higher-order interactions (e.g., irradiance^2 × capacity, capacity^3). Although the variables within both of these two models are similar to the IEC standard, the inclusion of the interaction term, which highlights that the linear relationship between I and E is moderated by C_{DC} (Figure 4), likely explains the superior performance of these models relative to that standard. The heatmaps of the log $RMSE$ values highlight the evaluation metric's dependence on site capacity (Figure 5a,c), while the adjusted R^2 heatmaps show consistent performance across site capacity (Figure 5b,d). The vertical concentration of dark bars likely reflects data quality issues not addressed by data preprocessing steps. A comparison of the associated partner generated expected energy estimates to those predicted from models also demonstrates that the AIT-derived estimates have lower average percent errors than the other models and the IEC (Table A2). Further evaluation of 2 years of records at a single site demonstrates that the AIT-derived estimates have a lower standard deviation and do not overestimate as much as the IEC-derived estimates (Figure A3). Given its parsimonious nature, the AIT model is subjected to further evaluation for both high- and low-capacity systems.

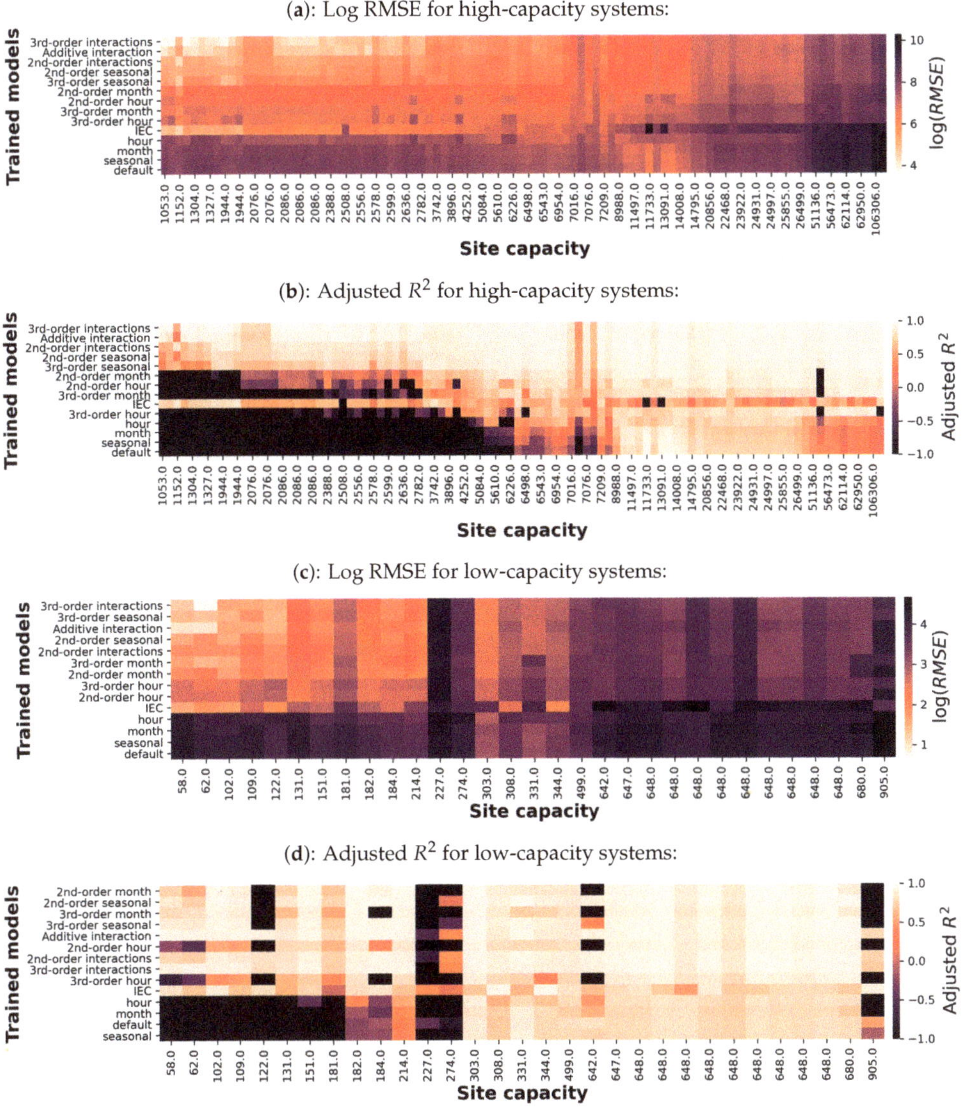

Figure 5. High-capacity (**a**,**b**) and low-capacity (**c**,**d**) site-level model evaluations with test data. Models with lighter colors (i.e., low values for log *RMSE* and values closer to 1 for adjusted R^2) indicate better performance.

3.1. High-Capacity Systems

For high-capacity systems, a significant (almost uniform) difference is found in both the log *RMSE* and R^2 values between the AIT and the IEC reference models (Figure 6a,b). Across the sites, the AIT model improves the goodness of fit by 0.42 in R^2 (IEC: 0.501; AIT: 0.93) and 1.16 in log *RMSE* (IEC: 6.99; AIT: 5.83). Generally, there are very few systems for which the IEC model performs better than the trained models (Table 3). The percent error (δ), on average, of the AIT model (3.65) is significantly lower than the IEC model (20.86) for high-capacity systems. An evaluation of percent error shows that the AIT generally performs well (i.e., $\delta \approx 0$ and thinner standard deviation bars) for most irradiance levels, except at the two extremes (i.e., <200 and $>1100 \frac{W}{m^2}$) (Figure 7).

The difference in performance relative to the IEC model is especially pronounced for larger system sizes (Figure 6a,b). Although the AIT model appears to show a small improvement over the partner-provided values (Figure 6a,b), a t-test concluded that the distributions of both the log $RMSE$ (p-value: 0.78) and R^2 (p-value: 0.48) are not significantly different.

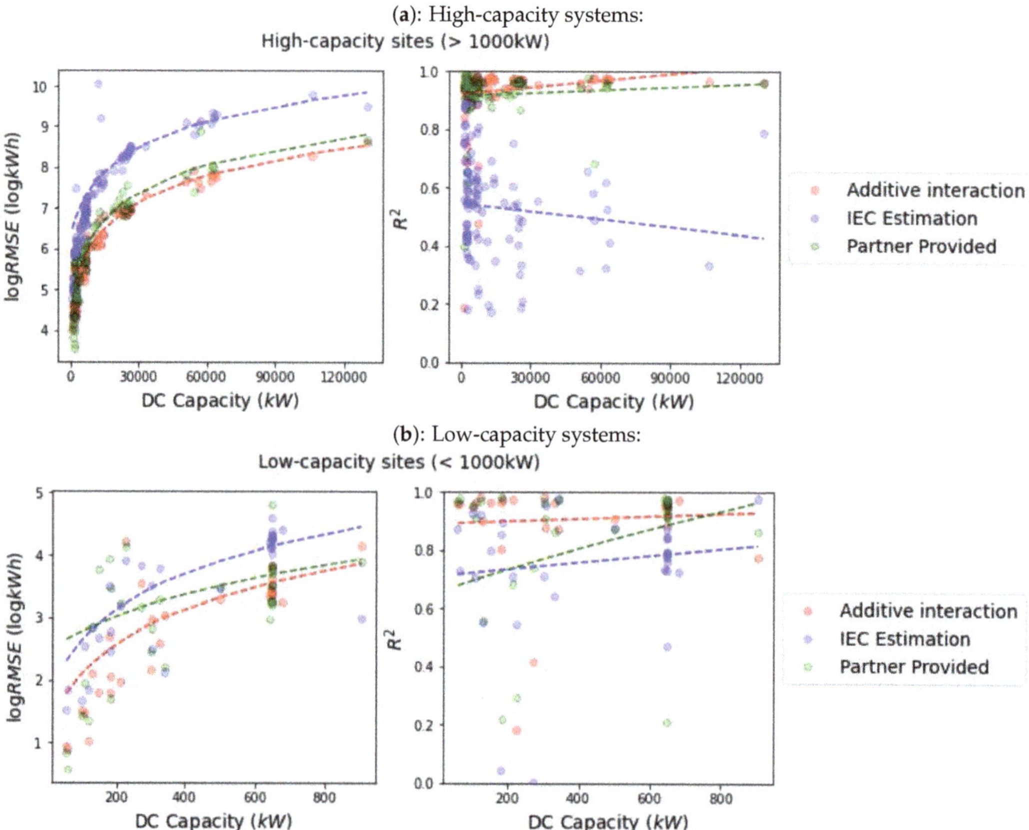

Figure 6. Model evaluation using log $RMSE$ and R^2 metrics for high-capacity systems (**a**) and low-capacity systems (**b**). Data points reflect site-level summaries of associated test data while dotted lines reflect best line fits to support visual pattern identification. The R^2 metric was used for this analysis (vs. adjusted R^2) since partner-provided model architectures are unknown. The 'additive interaction' regression model (in red) is comparable to the partner-provided proprietary values (green) and consistently performs better than the IEC standard (blue), especially at higher capacity values.

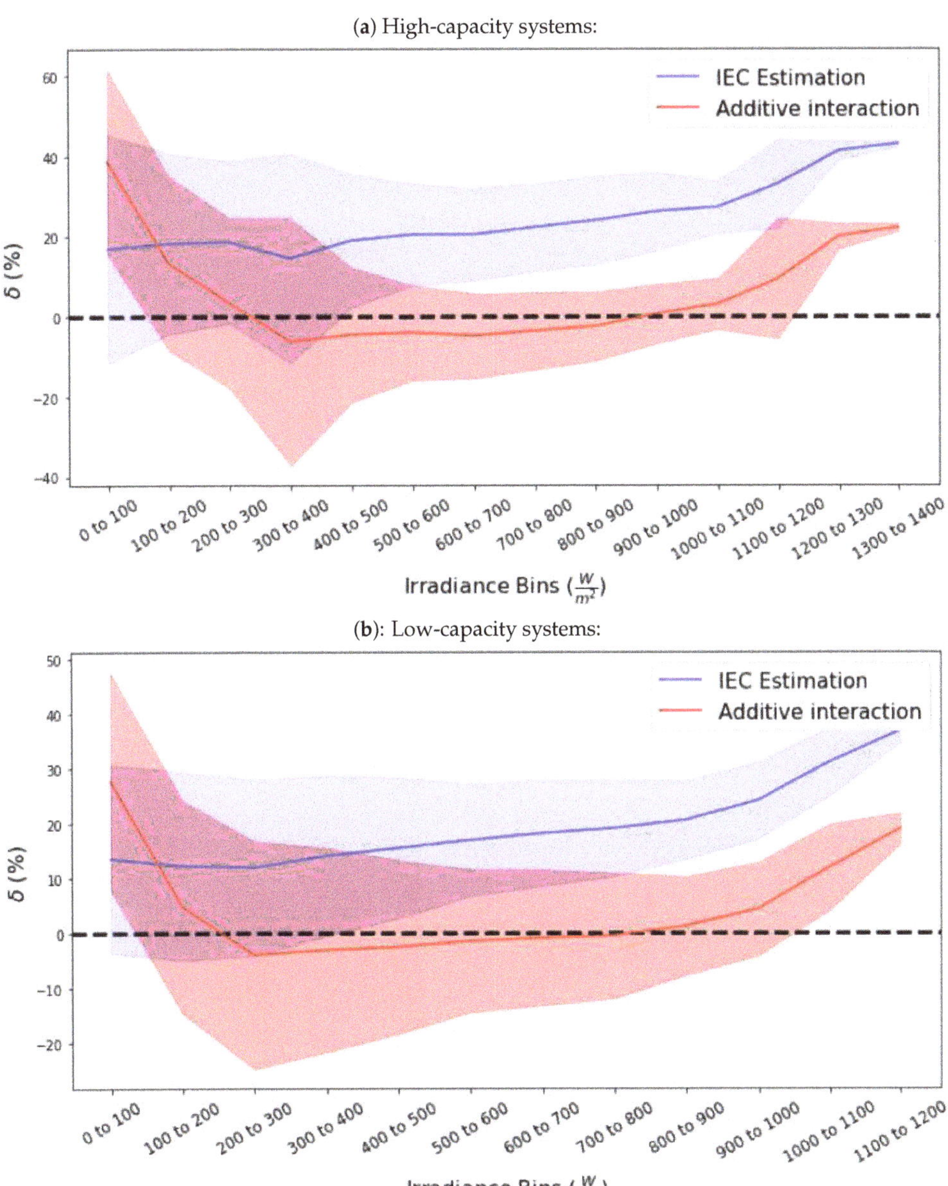

Figure 7. Percent error as a function of irradiance shows that the 'additive interaction' model outperforms the IEC standard across both high-capacity (**a**) and low-capacity (**b**) systems. Lines indicate mean values, while shaded region captures one standard deviation. The 'additive interaction' model performs best ($\delta \approx 0$) at 500–1100 $\frac{W}{m^2}$ and at 200–1000 $\frac{W}{m^2}$ for high-capacity and low-capacity systems, respectively.

3.2. Low-Capacity Systems

The model performance of the AIT for low-capacity systems was generally comparable to that of high-capacity systems, although the improvements were not as high. Across all low-capacity sites, the AIT model's goodness of fit improved by 0.165 in R^2 (IEC: 0.74; AIT:

0.90) and 0.61 in log $RMSE$ (IEC: 3.35; AIT: 2.75) (Figure 6a,b). Out of the 31 low-capacity sites (comprising 50K hours), the IEC-based model outperformed the trained models in 4 systems (Table 3). In some of the low-capacity systems, the measured energy is much higher than expected (Figure 5). The tendency of the IEC model to overpredict likely describes why this model performs better for some of the lower-capacity systems.

The δ, on average, is 4.42 and 15.37 for the AIT model and IEC model, respectively. Similar to the high-capacity system, the percent error values are greater at the extremes (Figure 7b). However, the standard error is generally higher in the low-capacity systems, as evidenced by wider standard deviation bars across the irradiance levels (Figure 7b).

3.3. Limitations and Future Work

The methodological approach of this analysis was strongly guided by available data. However, future work could extend these methods to consider: (1) energy generation at finer resolutions, (2) additional co-variates, and (3) alternate model formulations. For example, scaling could be used to consider alternate frequencies (beyond the hourly intervals considered in this study) post-evaluation. The methods used in this analysis explicitly omitted variables not included in current standard models (e.g., [15]). However, future assessments could more explicitly incorporate co-variates such as temperature, wind speed, and even age of the site. The latter would especially enable active consideration of degradation influence, which can influence long-term energy generation of PV sites [35]. Additional co-variates (such as type of inverters and modules) could also be included in subsequent iterations to capture more subtle impacts associated with differing site designs. Finally, future work could consider alternate model formulations (e.g., splines) to improve model continuity and better capture energy generation for smaller system sizes.

4. Conclusions

This work demonstrates the opportunities for leveraging data-driven, machine learning methods to generate more robust expected energy models. Generally, when compared to partner-provided values, the trained regression models outperform the IEC standard, especially in high-capacity systems. Detailed evaluation of the parsimonious AIT or 'additive interaction' model, in particular, demonstrated significant potential for use as a standardized, fleet-level expected energy model. The specific code used to train the regression models as well as the AIT model have been integrated with pvOps, an open source Python package which supports the evaluation of field data by PV researchers and operators; pvOps can be accessed at https://github.com/sandialabs/pvOps, accessed on 22 December 2021. Although this work presents findings specific to PV systems, the general methodologies can be applied to any domain that uses expected energy models to support site planning and O&M activities. Ongoing evaluations and improvements of these standardized expected energy models will continue to increase the accuracy and precision of site-level PV performance evaluations, which is critical to supporting reliability and economic assessments of PV operations and maintenance.

Supplementary Materials: The following supporting information can be downloaded at https://www.mdpi.com/article/10.3390/app12041872/s1. The supplementary material also includes a subsection with mathematical models for top-performing trained models.

Author Contributions: Conceptualization, T.G.; methodology, M.W.H. and T.G.; software, M.W.H.; validation, M.W.H. and T.G.; data curation, M.W.H.; writing—original draft preparation, M.W.H.; writing—review and editing, T.G.; visualization, M.W.H.; supervision, T.G.; project administration, T.G.; funding acquisition, T.G. All authors have read and agreed to the published version of the manuscript.

Funding: This research was funded by U.S. Department of Energy Solar Energy Technologies Office (Award No. 34172).

Institutional Review Board Statement: Not applicable.

Informed Consent Statement: Not applicable.

Data Availability Statement: The raw data was procured under non-disclosure agreements and thus, cannot be shared. However, an anonymized version of the filtered dataset used in this study analysis can be found within the article and Supplementary Information.

Acknowledgments: The authors would like to thank our industry partners for sharing data in support of this analysis as well as Sam Gilletly for their assistance with reviewing an earlier version of this manuscript. Sandia National Laboratories is a multimission laboratory managed and operated by National Technology and Engineering Solutions of Sandia LLC, a wholly owned subsidiary of Honeywell International Inc. for the U.S. Department of Energy's National Nuclear Security Administration under contract DE-NA0003525. The views expressed in the article do not necessarily represent the views of the U.S. Department of Energy or the United States Government.

Conflicts of Interest: The authors declare no conflict of interest.

Abbreviations

The following abbreviations are used in this manuscript:

E	Energy (kWh)
$\hat{E}$	Expected energy (kWh)
I	irradiance ($\frac{W}{m^2}$)
C_{DC}	DC capacity (kW)
T_{amb}	ambient temperature (°C)
T_{mod}	module temperature (°C)
e	standardized E (see Table 1)
i	standardized I (see Table 1)
c	standardized C (see Table 1)
$RMSE$	Root mean squared error
n	number of samples
p	number of predictors

Appendix A

Appendix A.1. Tables & Figures

Table A1. Correlations with energy generation and regression model parameters pre- and post- data processing. Correlations with energy production are generally comparable with irradiance across raw and filtered data, while correlations for site capacity are significantly higher for the filtered data than the raw data.

	Data Subsets		
Parameters	**Raw**	**Post-Data Quality Filters**	**Post-System Anomaly Filters**
Irradiance (I)	0.46	0.41	0.40
Capacity (C_{DC})	0.55	0.85	0.89

Table A2. Summary of average percent errors for each of the trained models, using the partner-generated values as the reference value.

	Average Percent Error	
	>1000 kW Systems	**≤1000 kW Systems**
Third-order interactions	0.03	0.23
Second-order interactions	0.03	0.21
Third-order seasonal	0.05	0.20
Second-order seasonal	0.04	0.16

Table A2. *Cont.*

	Average Percent Error	
	>1000 kW Systems	≤1000 kW Systems
Third-order month	0.12	0.27
Second-order month	0.08	0.27
Third-order hour	0.13	0.15
Second-order hour	0.10	0.05
Additive interaction	0.03	0.00
simple additive	0.36	1.44
IEC	0.39	1.75
hour	0.28	1.01
month	0.32	1.30
seasonal	0.32	1.55

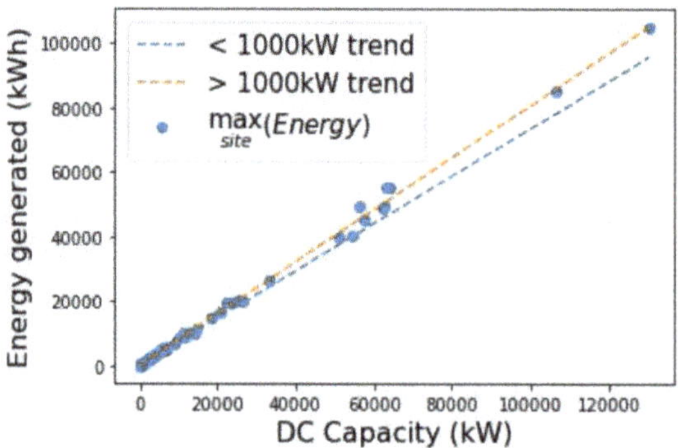

Figure A1. Relationship between DC capacity (kW) and hourly generated energy (kWh). The blue dots show a site's DC capacity versus its maximum recorded energy generated in a single hour. Although the trends are largely linear, the slopes differ for sites smaller than 1000 kW (blue dashed lines) and sites larger than 1000 kW (orange dashed lines). Since slight deviations in slope can render large prediction error, we train two separate model based on site size.

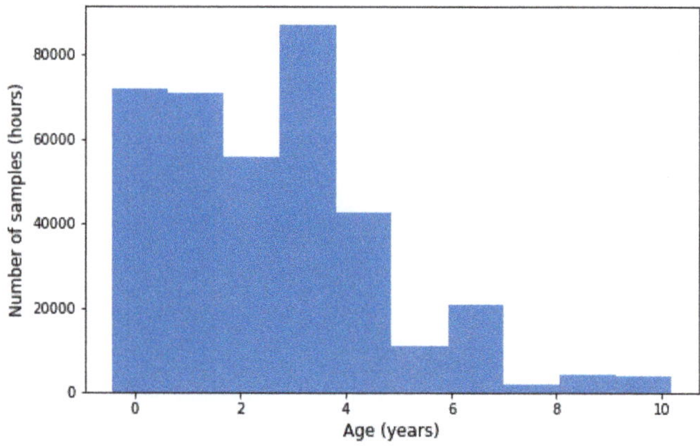

Figure A2. Age of sites within the final dataset. A majority are less than five years of age.

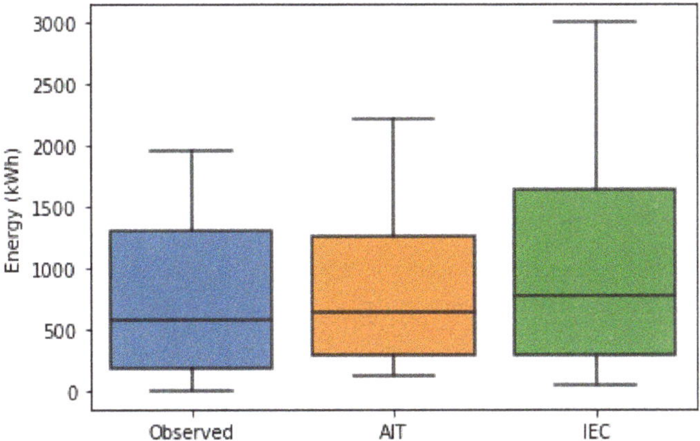

Figure A3. A 2-year comparison of observed energy with expected energy estimates from the trained additive interaction trained (AIT) model and the IEC standard at a single site. The IEC-derived estimates overestimate much more so than AIT-derived estimates. Slightly higher values of AIT-derived estimates relative to observed values likely indicate local failures at the site.

Appendix A.2. Top-Performing Trained Models

Mathematical equations associated with the trained regression models. In general, these model formulations contain more parameters and do not perform as well as the additive interaction model. High-capacity refers to systems with greater than or equal to 1000 kW in C_{DC} while low-capacity refers to systems smaller than 1000 kW in C_{DC}.

High-capacity second-order seasonal:

$$
\begin{aligned}
e = {} & 0.309 i I_{\text{winter}} + 0.292 i I_{\text{spring}} + 0.281 i I_{\text{summer}} + 0.287 i I_{\text{fall}} \\
& + 0.762 c I_{\text{winter}} + 0.734 c I_{\text{spring}} + 0.724 c I_{\text{summer}} + 0.730 c I_{\text{fall}} \\
& - 0.003 i^2 I_{\text{spring}} - 0.009 i^2 I_{\text{summer}} - 0.007 i^2 I_{\text{fall}} + 0.430 i c I_{\text{winter}} \\
& + 0.397 i c I_{\text{spring}} + 0.380 i c I_{\text{summer}} + 0.384 i c I_{\text{fall}} + 0.008 c^2 I_{\text{winter}} \\
& + 0.008 c^2 I_{\text{spring}} - 0.003 c^2 I_{\text{fall}} - 0.054
\end{aligned}
$$

High-capacity third-order interactions:

$$
e = 0.293 i + 0.769 c - 0.019 i^2 + 0.394 i c + 0.021 i^2 c + 0.004 i c^2 + 0.001 c^3 - 0.039
$$

High-capacity third-order seasonal:

$$
\begin{aligned}
e = {} & 0.265 i I_{\text{winter}} + 0.270 i I_{\text{spring}} + 0.252 i I_{\text{summer}} + 0.243 i I_{\text{fall}} \\
& + 0.749 c I_{\text{winter}} + 0.740 c I_{\text{spring}} + 0.722 c I_{\text{summer}} + 0.730 c I_{\text{fall}} \\
& - 0.006 i^2 I_{\text{summer}} - 0.388 i c I_{\text{winter}} + 0.378 i c I_{\text{spring}} + 0.365 i c I_{\text{summer}} \\
& + 0.346 i c I_{\text{fall}} + 0.019 c^2 I_{\text{winter}} + 0.005 c^2 I_{\text{spring}} - 0.002 c^2 I_{\text{fall}} \\
& + 0.016 i^3 I_{\text{winter}} + 0.008 i^3 I_{\text{spring}} + 0.014 i^3 I_{\text{summer}} + 0.017 i^3 I_{\text{fall}} \\
& - 0.013 i c^2 I_{\text{winter}} + 0.008 i c^2 I_{\text{spring}} + 0.007 i c^2 I_{\text{summer}} + 0.019 i c^2 I_{\text{fall}} \\
& - 0.002 c^3 I_{\text{winter}} - 0.054
\end{aligned}
$$

High-capacity second-order interactions:

$$e = 0.295i + 0.745c - 0.018i^2 + 0.404ic + 0.004c^2 - 0.042$$

Low-capacity second-order seasonal:

$$\begin{aligned}
e = {} & 0.711iI_{\text{winter}} + 0.686iI_{\text{spring}} + 0.680iI\text{summer} + 0.714iI_{\text{fall}} \\
& + 0.617cI_{\text{winter}} + 0.576cI_{\text{spring}} + 0.605cI\text{summer} + 0.636cI_{\text{fall}} \\
& - 0.046i^2 I_{\text{winter}} - 0.028i^2 I_{\text{spring}} - 0.046i^2 I\text{summer} - 0.038i^2 I_{\text{fall}} \\
& + 0.401icI_{\text{winter}} + 0.381icI_{\text{spring}} + 0.382icI\text{summer} + 0.427icI_{\text{fall}} \\
& + 0.001c^2 I_{\text{winter}} - 0.006c^2 I\text{summer} - 0.104
\end{aligned}$$

Low-capacity third-order interactions:

$$e = 0.746i + 0.642c - 0.043i^2 + 0.426ic - 0.019i^3 - 0.034i^2 c - 0.017c^3 + 0.115$$

Low-capacity third-order seasonal:

$$\begin{aligned}
e = {} & 0.757iI_{\text{winter}} + 0.684iI_{\text{spring}} + 0.667iI\text{summer} + 0.712iI_{\text{fall}} \\
& + 0.577cI_{\text{winter}} + 0.486cI_{\text{spring}} + 0.567cI\text{summer} + 0.619cI_{\text{fall}} \\
& - 0.023i^2 I_{\text{winter}} - 0.027i^2 I_{\text{spring}} - 0.050i^2 I\text{summer} - 0.033i^2 I_{\text{fall}} \\
& + 0.413icI_{\text{winter}} + 0.382icI_{\text{spring}} + 0.381icI\text{summer} + 0.428icI_{\text{fall}} \\
& - 0.028i^3 I_{\text{winter}} - 0.007i^3 I\text{summer} - 0.022i^2 cI_{\text{winter}} - 0.004i^2 cI_{\text{fall}} \\
& - 0.047c^3 I_{\text{winter}} + 0.068c^3 I_{\text{spring}} + 0.027c^3 I\text{summer} + 0.016c^3 I_{\text{fall}} + 0.099
\end{aligned}$$

Low-capacity second-order interactions:

$$e = 0.719i + 0.630c - 0.063i^2 + 0.411ic - 0.124$$

References

1. Golnas, A. PV system reliability: An operator's perspective. In Proceedings of the 2012 IEEE 38th Photovoltaic Specialists Conference (PVSC) PART 2, Austin, TX, USA, 3–8 June 2012; pp. 1–6.
2. Feng, M.; Bashir, N.; Shenoy, P.; Irwin, D.; Kosanovic, D. SunDown: Model-driven Per-Panel Solar Anomaly Detection for Residential Arrays. In Proceedings of the 3rd ACM SIGCAS Conference on Computing and Sustainable Societies, Guayaquil, Ecuador, 15–17 June 2020; pp. 291–295.
3. Dyamond, W.; Rix, A. Detecting Anomalous Events for a Grid Connected PV Power Plant Using Sensor Data. In Proceedings of the 2019 Southern African Universities Power Engineering Conference/Robotics and Mechatronics/Pattern Recognition Association of South Africa (SAUPEC/RobMech/PRASA), Bloemfontein, South Africa, 29–31 January 2019; pp. 287–292.
4. Harrou, F.; Sun, Y.; Taghezouit, B.; Saidi, A.; Hamlati, M.E. Reliable fault detection and diagnosis of photovoltaic systems based on statistical monitoring approaches. *Renew. Energy* **2018**, *116*, 22–37. [CrossRef]
5. Platon, R.; Martel, J.; Woodruff, N.; Chau, T.Y. Online fault detection in PV systems. *IEEE Trans. Sustain. Energy* **2015**, *6*, 1200–1207. [CrossRef]
6. Blair, N.; Dobos, A.P.; Freeman, J.; Neises, T.; Wagner, M.; Ferguson, T.; Gilman, P.; Janzou, S. *System Advisor Model, Sam 2014.1. 14: General Description*; Technical Report; National Renewable Energy Lab.: Golden, CO, USA, 2014.
7. Deline, C.; Anderson, K.; Jordan, D.; Walker, A.; Desai, J.; Perry, K.; Muller, M.; Marion, B.; White, R. *PV Fleet Performance Data Initiative: Performance Index-Based Analysis*; Technical Report; National Renewable Energy Lab.: Golden, CO, USA, 2021.
8. Gilletly, S.; Jackson, N.; Staid, A. Quantifying Wildfire-Induced Impacts to Photovoltaic Energy Production in the western United States. In Proceedings of the 2021 IEEE 48th Photovoltaic Specialists Conference (PVSC), Online, 20–25 June 2021.
9. Klise, G.T.; Stein, J.S. *Models Used to Assess the Performance of Photovoltaic Systems*; Sandia National Laboratories: Albuquerque, NM, USA, 2009.
10. De Soto, W.; Klein, S.A.; Beckman, W.A. Improvement and validation of a model for photovoltaic array performance. *Sol. Energy* **2006**, *80*, 78–88. [CrossRef]

11. Arab, A.H.; Taghezouit, B.; Abdeladim, K.; Semaoui, S.; Razagui, A.; Gherbi, A.; Boulahchiche, S.; Mahammed, I.H. Maximum power output performance modeling of solar photovoltaic modules. *Energy Rep.* **2020**, *6*, 680–686. [CrossRef]

12. King, D.L.; Kratochvil, J.A.; Boyson, W.E. *Photovoltaic Array Performance Model*; Sandia National Laboratories: Albuquerque, NM, USA, 2004.

13. Boyson, W.E.; Galbraith, G.M.; King, D.L.; Gonzalez, S. *Performance Model for Grid-Connected Photovoltaic Inverters*; Technical Report; Sandia National Laboratories: Albuquerque, NM, USA, 2007.

14. Dobos, A.P. *PVWatts Version 5 Manual*; Technical Report; National Renewable Energy Lab.: Golden, CO, USA, 2014.

15. *ISO/IEC TR 29110-1:2016; Photovoltaic System Performance—Part 1: Monitoring*. International Electrotechnical Commission: Geneva, CH, USA, 2017.

16. Myers, D. *Evaluation of the Performance of the Pvusa Rating Methodology Applied to Dual Junction pv Technology: Preprint (Revised)*; Technical Report; National Renewable Energy Lab.: Golden, CO, USA, 2009.

17. DNV GL. *Narrowing the Performance Gap: Reconciling Predicted and Actual Energy Production*; Technical Report; DNV GL: Oakland, CA, USA, 2019.

18. Hanawalt, S. *The Challenge of Perfect Operating Data*; Technical Report; Power Factors: San Francisco, CA, USA, 2020.

19. Livera, A.; Theristis, M.; Makrides, G.; Sutterlueti, J.; Ransome, S.; Georghiou, G.E. Performance Analysis of Mechanistic and Machine Learning models for Photovoltaic energy yield prediction. In Proceedings of the 36th European Photovoltaic Solar Energy Conference and Exhibition, Marseille, France, 9–13 September 2019; pp. 9–13.

20. Ascencio-Vásquez, J.; Bevc, J.; Reba, K.; Brecl, K.; Jankovec, M.; Topič, M. Advanced PV Performance Modelling Based on Different Levels of Irradiance Data Accuracy. *Energies* **2020**, *13*, 2166. [CrossRef]

21. Wang, S.; Zhang, Y.; Zhang, C.; Yang, M. Improved artificial neural network method for predicting photovoltaic output performance. *Glob. Energy Interconnect.* **2020**, *3*, 553–561. [CrossRef]

22. Bertsimas, D.; Delarue, A.; Jaillet, P.; Martin, S. The price of interpretability. *arXiv* **2019**, arXiv:1907.03419.

23. Klise, G.T. *PV Reliability Operations Maintenance (PVROM) Database: Data Collection & Analysis Insights*; Technical Report; Sandia National Lab.: Albuquerque, NM, USA, 2015.

24. Klise, K.A.; Stein, J.S.; Cunningham, J. Application of IEC 61724 Standards to Analyze PV System Performance in Different Climates. In Proceedings of the 2017 IEEE 44th Photovoltaic Specialist Conference (PVSC), Washington, DC, USA, 25–30 June 2017; pp. 3161–3166.

25. Klise, K.A. *Performance Monitoring Using Pecos*; Technical Report; Sandia National Lab.: Albuquerque, NM, USA, 2016.

26. *IEC TS 61724-3:2016; Photovoltaic System Performance—Part 3: Energy Evaluation Method*. International Electrotechnical Commission: Geneva, CH, USA, 2016.

27. Vining, W.; Hansen, C. PVAnalytics. 2021. Available online: https://github.com/pvlib/pvanalytics (accessed on 22 December 2021).

28. Malik, S.; Dassler, D.; Fröbel, J.; Schneider, J.; Ebert, M. Outdoor data evaluation of half-/full-cell modules with regard to measurement uncertainties and the application of statistical methods. In Proceedings of the 29th European Photovoltaic Solar Energy Conference and Exhibition, Amsterdam, The Netherlands, 22–26 September 2014; pp. 3269–3273.

29. Theil, H. A rank-invariant method of linear and polynomial regression analysis. *Indag. Math.* **1950**, *12*, 173.

30. Tibshirani, R. Regression shrinkage and selection via the lasso: A retrospective. *J. R. Stat. Soc. Ser. B* **2011**, *73*, 273–282. [CrossRef]

31. Geman, S.; Bienenstock, E.; Doursat, R. Neural networks and the bias/variance dilemma. *Neural Comput.* **1992**, *4*, 1–58. [CrossRef]

32. Dirnberger, D.; Blackburn, G.; Müller, B.; Reise, C. On the impact of solar spectral irradiance on the yield of different PV technologies. *Sol. Energy Mater. Sol. Cells* **2015**, *132*, 431–442. [CrossRef]

33. Eke, R.; Betts, T.R. Spectral irradiance effects on the outdoor performance of photovoltaic modules. *Renew. Sustain. Energy Rev.* **2017**, *69*, 429–434. [CrossRef]

34. Arias, A.L. R-Squared, Adjusted R-Squared, the F Test, and Multicollinearity. In *Understanding Regression Analysis*; Chapman and Hall/CRC: Boca Raton, FL, USA, 2020; pp. 185–200.

35. Jordan, D.C.; Deline, C.; Kurtz, S.R.; Kimball, G.M.; Anderson, M. Robust PV degradation methodology and application. *IEEE J. Photovoltaics* **2017**, *8*, 525–531. [CrossRef]

Article

SCADA-Compatible and Scaleable Visualization Tool for Corrosion Monitoring of Offshore Wind Turbine Structures

Joachim Verhelst [1,*], Inge Coudron [1,2] and Agusmian Partogi Ompusunggu [1]

1   Flanders Make, Corelab Decisions, Oude Diestersebaan 133, 3920 Lommel, Belgium;
    inge.coudron@flandersmake.be (I.C.); agusmian.ompusunggu@gmail.com (A.P.O.)
2   Faculty of Applied Engineering, University of Antwerp, Groenenborgerlaan 171, 2020 Antwerpen, Belgium
*   Correspondence: joachim.verhelst@flandersmake.be

Featured Application: Structural corrosion monitoring for offshore wind turbines.

Abstract: The exploitation of offshore windfarms (WFs) goes hand in hand with large capital expenditures (CAPEX) and operational expenditures (OPEX), as these mechanical installations operate continuously for multiple decades in harsh, saline conditions. OPEX can account for up to 30% of the levelised cost of energy (LCoE) for a deployed offshore wind farm. To maintain the cost-competitiveness of deployed offshore WFs versus other renewable energy sources, their LCoE has to be kept in check, both by minimising the OPEX and optimising the offshore wind energy production. As corrosion, in particular uniform corrosion, is a major cause of failure of offshore wind turbine structures, there is an urgent need for corrosion management systems for deployed offshore wind turbine structures (WTs). Despite the fact that initial corrosion protection solutions are already integrated on some critical structural components such as WT towers, WT transition pieces or WT sub-structure (fixed or floating platforms), these components can still be harshly damaged by the corrosive environmental offshore conditions. The traditional preventive maintenance strategy, in which regular manual inspections by experts are necessary, is widely implemented nowadays in wind farm applications. Unfortunately, for such challenging operating environments, regular human inspections have a significant cost, which eventually increase the OPEX. To minimise the OPEX, remote corrosion monitoring solutions combined with supporting software (SW) tools are thus necessary. This paper focuses on the development of a software (SW) tool for the visualisation of corrosion measurement data. To this end, criteria for efficient structural corrosion analysis were identified, namely a scaleable, SCADA-compatible, secure, web accessible tool that can visualise 3D relationships. In order to be effective, the SW tool requires a tight integration with decision support tools. This paper provides three insights: Firstly, through a literature study and non-exhaustive market study, it is shown that a combined visualisation and decision SW tool is currently non-existing in the market. This gap motivates a need for the development of a custom SW tool. Secondly, the capabilities of the developed custom software tool, consisting of a backend layer and visualisation browser designed for this task are demonstrated and discussed in this paper. This indicates that a SCADA-compatible visualisation software tool is possible, and can be a major stepping stone towards a semi-automated decision support toolchain for offshore wind turbine corrosion monitoring.

Keywords: SCADA; visualisation; software; wind-turbine; windfarm; cross-platform; HMI; GUI; corrosion; monitoring

Citation: Verhelst, J.; Coudron, I.; Ompusunggu, A.P. SCADA-Compatible and Scaleable Visualization Tool for Corrosion Monitoring of Offshore Wind Turbine Structures. *Appl. Sci.* **2022**, *12*, 1762. https://doi.org/10.3390/app12031762

Academic Editors: Luis Hernández-Callejo, Maria del Carmen Alonso García and Sara Gallardo Saavedra

Received: 1 December 2021
Accepted: 29 January 2022
Published: 8 February 2022

Publisher's Note: MDPI stays neutral with regard to jurisdictional claims in published maps and institutional affiliations.

1. Introduction

The renewable energy markets in the world have reached a mature state and continue growing at a significant rate [1,2]. Among others, wind-sourced electricity production has the potential to deliver a significant contribution to the electric energy portfolio of the world within the next decades. This attracts many new investors and augments market interactions, thereby increasing the importance of optimisation of operational costs and efficiency gains [1].

Europe is a leading player in this domain, notably for offshore wind turbine manufacturing, as (in 2015) European manufacturers produced around 41 to 50% of annually installed wind power installations worldwide. While this market share had increased up to 61% in 2017 [3], in the last five years, both China and USA have made an unprecedented comeback, partially driven by expiring government subsidies. In 2020 [4], the Chinese wind market dominated the global wind market, totalling 97 GW that year, with a share of 40% of the newly installed production capacity worldwide, both onshore and offshore.

As the world's landmass is rather limited and already under tremendous pressure, offshore installations are becoming ever more popular. The mean cost per produced energy unit for offshore exploitation is rapidly decreasing, thereby closing the gap compared to conventional energy generation methods and onshore wind farms; therefore, many large investors in the domain of windfarm operations move towards larger wind turbines and offshore windfarms. Some sources predict that the global offshore wind farm (WF) market size will continue to grow at a rate of 12% per year [5].

These off-shore investments currently result in a typical levelised cost of (produced) electricity (LCoE) in the range between USD 70 to USD 210 per MWh [2], with a mean around USD 90 per MWh, which is currently still higher than that of most onshore wind turbines estimates (USD 47–60 per MWh$_{produced}$) [6]. These costs are mainly composed of three items: capital expenditures (CAPEX) for power production, conversion and transmission installations; operational expenses (OPEX) and the weighted average cost of capital (WACCs) for financing related costs, such as the cost of depth and risk premiums.

This market growth and shift of deployment area does have consequences: Weather conditions and corrosion effects are more severe in offshore and saline environments, making structural stability more critical, while decreasing maintenance time windows. This leads to increased costs for operation and maintenance (O&M) compared to onshore installations, signalling an economic opportunity for (data-driven or model based) optimisation of maintenance scheduling. Due to the unpredictable nature of corrosion, both for installations that are approaching the end of design lifetime and more recent deployments (which may be engineered with lower tolerances), regular inspection and preventive maintenance is crucial to monitor, track and predict the individual status of structural health of individual wind turbines. Moreover, near the end of the planned turbine lifetime, decommissioning should be scheduled and performed.

The scheduling of these maintenance tasks is a complex planning problem [7], wherein the two key bottlenecks of failure mode analysis are: the availability of suitable measurement data and expert insights to interpret the data correctly. In order to bridge this gap, further research is warranted.

The contributions of this paper are threefold:

- The shifting needs for the wind farm sector with regard to structural corrosion monitoring are identified and listed;
- A market analysis is performed and presented for existing, commercially available software solutions that combine these needs;
- A solution for the identified knowledge domain gap is presented, in the form of an open-source platform based software tool that can combine available SCADA data and web-based datastreams.

In the following sections, the state-of-the-art of corrosion monitoring is presented (Section 2), together with a market analysis (Section 3.1) and a presentation of our custom software (SW) tool (Section 3.3.2) developed to tackle these specific needs.

2. State-of-the-Art

In order to analyse the needs and benchmark available software solutions for structural stability evaluation of wind turbines, objective criteria are needed. Several existing wind farm operation market studies were taken as a guideline for this comparison [5,8,9], together with input from EU-project partners involved in the WATEREYE project [10].

In the case of offshore wind turbines, a large variety of support-structure types are used. The most prevalent design is a monopile structure mounted on the seabed floor, yet other structures such as a tripod structures, jacket type or gravity-based structures are used as well [11], while even floating turbines are in the research phase for the deployment of offshore wind turbines in deep waters. A few examples are given in Figure 1.

(a)   (b)   (c)

Figure 1. Some commercial offshore wind turbine structures: (**a**) monopile wind turbine; (**b**) floating wind turbine; (**c**) lattice type wind turbine. Images are courtesy of [12].

Therefore, especially for new, non-standard installations such as floating wind turbines, expert supervision for structural analysis will likely be required. Visualisation and decision support tools can assist them in this analysis. A breakdown of the needs for these roles, lead to the following standard practices and derived criteria for structural stability monitoring.

2.1. Maturity of the Wind Turbine Market

The collective world experience with regard to offshore wind turbine structural corrosion is rather limited: Most of the experience regarding corrosion of offshore wind turbine structures is condensed in Europe, especially in Denmark, the Netherlands, Germany and the UK, as 90% of the globally installed offshore installations (by capacity) were produced and installed by European companies [13].

The first offshore windfarm was constructed in 1991 (Ørsted), yet these demonstrators contained small wind turbines (0.5 to 2 MW each) installed in shallow water (3–5 m), whereas in the last decade, the trend is to go towards larger sizes and capacities per wind turbine (3 to 12 MW each), and with anchoring on ever deeper seabeds [13]. In the US, the first commercial offshore wind farm was only constructed in 2016, near Rhode Island. In China, the first commercial offshore windpark was deployed in 2010 [14].

2.2. Standard Practices for Structural Stability Monitoring

Most wind turbine manufacturers offer tools for operational management of their wind turbines; however, the focus of these tools is mostly on the rotating components (gearbox, generator, etc.), whereas the evaluation of structural stability of the wind turbine pylons are rarely addressed in these tools [15].

The default inspection mode for structural stability of offshore wind turbines, are visual (RGB-camera-based) inspections and flooded member detection. For submerged parts, this is traditionally performed by divers or remotely operated vehicles, with a frequency between one and five years. These types of inspections and full scale post construction impact evaluation campaigns [16]; however, can be costly in terms of vessel time, and involve hazards to personnel [17].

Therefore, in more recent constructions, manufacturers or operators often embed sensors to detect and quantify structural loading, flooding or corrosion effects, for example accelerometers, saline senors, displacement measurements or acoustic emission systems.

In this data-rich environment, it is not the data capture that is the main difficulty. Rather, the robustification of the measurements, the detection of all failure modes and the subsequent evaluation and analysis of the captured data is what leads to understanding and effective decision making.

2.3. Criteria for Structural Analysis of WT

From existing literature reviews and interviews, including [17–20], multiple features for structural analysis of wind turbines are extracted and subsequently condensed in criteria. A holistic SW tool for structural health monitoring of offshore WF should be:

- Graphical, as structural integrity is related to spatial distribution and location of occurrence.
- Scaleable and modular. WFs are modular and scaleable by nature, consisting of one, up to several hundreds, of WTs. Modular—to include other, non-corrosion wind-turbine failure modes (blade monitoring, gearbox monitoring, inverter monitoring, . . .).
- SCADA compatible, in order to leverage the vast amount of data already captured at the wind turbine (WT) and WF level.
- Web-based and secure, in order to allow data-analysis by experts, independent of the WF location and shielded from external tampering or unauthorised access.
- Maintenance planning-inclusive, as the data-based insights can trigger condition-based maintenance maintenance (CBM) or predictive maintenance (PdM) scheduling decisions, actively reducing OPEX.

These identified criteria are subsequently used to evaluate existing SW solutions in the market.

3. Software Solutions

3.1. Existing Windfarm Visualisation Tools

Both for wind-farm development [21] and for structural monitoring and maintenance planning, visualisation tools are indispensable. Dozens of software packages exist in the market for these applications. The scope of this review is not to be exhaustive, nor to provide a judgement or ranking. Rather, the aim is to classify a subset of existing software tools that are publicly advertised, according to the criteria presented in Section 2.

This exercise results in the non-exhaustive overview shown in Table 1. This table shows that there are multiple software providers; each offering unique, functional pieces of software; however, to the author's knowledge, none of these suppliers currently offer a holistic and tailored solution specifically for the structural monitoring of the windfarm market, including 3D corrosion visualisation and maintenance task planning.

3.2. Existing Visualisation Tools in Other Industries (Oil and Gas)

Beyond the domain of WF, it makes sense to extend the search into an other domain where metal structures are deployed in offshore, saline environments, where corrosion monitoring is important, namely, the offshore oil and gas industry.

Since the economic loss in these industries due to corrosion can be extremely high, monitoring and managing corrosion in the oil and gas industry is paramount importance; therefore, this industry has deployed and is inspecting and maintaining ten-thousands of offshore drilling platforms, undersea pipelines and supporting pylons.

Offshore oil drilling platforms have a long history: the first deep-sea oil-rig was erected in 1947 in the gulf of Mexico, whereas the first platform might have been built around 1890 (Grand Lake, St. Maris) [22].

A similar classification exercise was undertaken for this industry, using the same criteria as stipulated in Section 2.3, in order to carry over suitable approaches and frameworks (where available) to tackle monitoring of renewable energy projects.

Interested readers are directed to existing literature reviews in this field, including (among others) following book [23] and paper [24]. These authors describe common practices for design, maintenance and monitoring of offshore installations.

Table 1. Non-exhaustive list and classification of commercial wind turbine corrosion visualisation and maintenance planning software tools.

Name	Application	Graphical	Modular *	SCADA	Web-Based	Planning	Reference
WindFarmer: Analyst	Design	2Dmap	V (import)	X	X	X	dnv.com accessed on 30 July 2021
Surfacepro 3D	Corrosion	3D	V	V	X	X	eddyfi.com accessed on 28 July 2021
Yokagawa	EMS (Energy Management System)	2D	V	V	V	V	yokagawa.com accessed on 8 October 2021
BraveDigital	Bespoke (EWT)	2D	V	V (OPC)	V	X	bravedigital.com accessed on 28 July 2021
PingMonitor	Blade monitoring	X	V	V	V	V (forecast)	pingmonitor.co accessed on 28 July 2021
Effector Octavis	EMS	X	V	V	V	V (diagnosis)	ifm.com accessed on 14 september 2021
Multilevel Wind SCADA Center	EMS	V	V	V	V	X	siemens.com accessed on 28 July 2021
nCode	Fatigue analysis	V	X	V	X	X	hbm.com accessed on 16 August 2021
General Electric	Digital twin	V	V	V	X	V	ge.com accessed on 12 September 2021

* 1 Scaleability and Modularity.

Table 2 summarises our (non-exhaustive) market analysis of structural corrosion analysis tools used in the oil and gas industry. There are uncountable sources and software packages published online, that list multiple (hundreds) of software packages in this industry [25], that may be applied (with minimal modifications) to WF applications.

Table 2. Non-exhaustive list and classification of commercial oil and gas industry corrosion visualisation and maintenance planning software tools.

Name	Application	Graphical	Modular *	SCADA	Web-Based	Planning	Reference
SmartCET	Design	x (reporting)	V	V	V	X (alarms)	honeywell.com accessed on 14 September 2021
rysco-corrosion	Sensing	x	V	V	X	X	ryscocorrosion.com accessed on 14 September 2021
Beasy	Corrosion	3D	V	V	X	X	nace.org, [15] accessed on 8 October 2021
SGS Corrosion monitoring	Monitoring	x	X	X	X	V (alarms)	www.sgsgroup.com. cn accessed on 14 July 2021
Corrosion Clinic	Manual analysis	V	X	X	X	X	[25]

* 1 Scaleability and Modularity.

From our investigation, most software packages offer one or more of the required functionalities of the listed criteria, yet to our knowledge, no single application offers a holistic solution fitting the needs of the WT structural corrosion monitoring. Furthermore, there is a need for adapting these solutions to the needs of the offshore windfarm industry. Even if one would fit all criteria, it is not certain that it is applicable in plug-and-play fashion to monitor the structural health of wind farms.

3.3. Custom SW Tool

3.3.1. Custom Architecture

In offshore wind turbines, failures of structural components, caused by corrosion at tower including the "tower-platform" junction and the entire splash-zone can cause significant and critical down-times and subsequent loss of electricity production. As no holistic commercially available tool was identified, to the best of the authors' knowledge, in the framework of the ongoing WATEREYE research project [10], a custom SW demonstrator tool was developed. The initial concept and architecture of the custom SW tool was developed in our previous work [26]. In this paper, the concept and architecture are further refined and finally implemented into a demonstrable and deployable SW tool.

In the WATEREYE project [10], corrosion monitoring and O&M planning tools for offshore wind farms and turbine structures are developed and integrated in (a subsequent version of) the custom SW-tool. Further, in the scope of the WATEREYE project, technologies for monitoring, data analytics, modelling, and diagnosis and for wind turbine (WT) and wind farm (WF) O&M advanced control strategies have been investigated and implemented.

The analysis listed in Table 1, resulted in the design and development of an SW-tool that attempts to cover all criteria identified in Section 2.3. For the framework of this SW tool, Qt [27] was chosen. Because of its highly scaleable characteristics, it allows developers to build components that can run on embedded, desktop and mobile computers, and it supports graphical development (inclusion via widgets). Qt supports all of today's user interface paradigms, controls and behaviors, making it easy to design really attractive HMIs that users intuitively understand. Furthermore, Qt uses the latest SSL and TLS implementations to safely encrypt data communications for cyber-secure remote-access.

In order to leverage both web-based and SCADA-capabilities, pvbrowser [28] was chosen, which is an open source C++ GUI framework, built on top of Qt. This application framework provides a specialised browser for the client computer and an integrated development environment for creating servers that interact with data acquisition programs (daemons) for many SCADA protocols [28].

The stock architecture of pvbrowser is shown in Figure 2a. PvBrowser is available under the GPL v2 (program) and GNU LGPLv3 (library) licence frameworks.

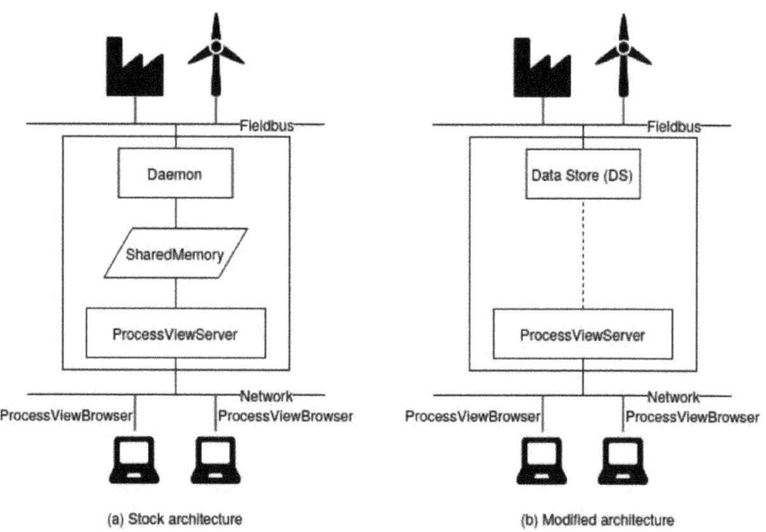

Figure 2. (**a**) Original pvbrowser architecture [28] and (**b**) modified architecture and data flow, to support the WATEREYE project [10] needs.

For 3D-visualisation purposes, the visualisation toolkit (VTK) [29] was used. This C++ based open source toolbox is capable of filtering, processing and visualising large datasets, with numerous user and SCADA interaction capabilities. Pvbrowser provides some support for VTK, but the default browser does not support it out of the box. To setup the VTK-based Qt-widget and to streamline the interaction between the user and the widget, TCL-TK scripts (Graphical toolkit for TCL) were used, available under a BSD-type licence [30].

3.3.2. Custom Visualisation SW Tool

In the WATEREYE project [10], measurements containing corrosion status information are generated using mobile and fixed sensors. These data are stored in a secure database, accessible through a Representational State Transfer application Programming interface (REST-API) . A visualisation tool is needed to support O&M experts to understand the status of an offshore wind turbine, based on gathered measurement data. To this end, a visualisation SW tool was developed according to the criteria developed defined in Section 2.

This framework was used to develop a custom web interface, which takes (maintenance staff) user input, queries a REST-datbase to fetch the measurement and prediction data from a windfarm database and subsequently presents the data on a custom pv-dashboard. The user can interact with these data, either visually (in 3D or 2D), or by downloading the underlaying data and processing it offline.

This combination results in the following architecture, depicted in Figure 3. It was based on pvbrowser, but with modified architecture, as explained in Figure 2b. This SW tool is capable of:

- Querying measurement and geometric data from the database using a REST-API, based on user inputs in the GUI.
- Pre-processing these data, merging the inputs from different sources into key performance indicators (KPI's) such as relative corrosion rate (mm/y).
- Exporting these data toward data-files and image.
- Visualising these data in interactive 3D and 2D visualisations.

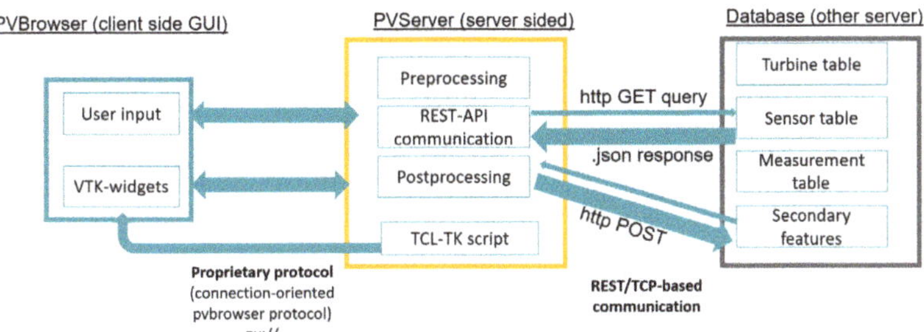

Figure 3. Architecture of custom SW tool, with main components: QT-based (customised) pvserver (orange box) and VTK-enabled pvbrowser (blue box), fetching data from a database (grey box).

Through the browser client, the user can access and interact with the data, using a browser-like application. The user can use the software, agnostic of the architecture or changes at server side.

The server architecture allows us to centrally manage and adapt both the data and visualisation output from different windfarms and users.

Based on the chosen parameters, the server component makes a query to the data-server, parses the data and presents them to the user in predefined widgets in the browser side (see Figure 4). There, the user can interact with the data (zoom, query, rotate) to increase his/her understanding of spatial and time relationships, or adapt the query for additional data.

The core functionalities of this SW tool are:

- Initialisation, with connection to the database, and auto-populating of the dropdown menus for all wind turbines and sensor data available in the database.
- Selection of a wind turbine to present in 2D and 3D views.
- Selection of parameters, such as attributes and sensors to visualise and time period of interest.
- Modularity and expandability, for future functionality (as described in following section).

Any change in the parameters above, triggers the visualisation of:

- The 3D visualisation of the sensor values at each sensor position on the wind turbine, as shown in Figure 5a,b.
- The 2D or 3D visualisation of derived the wall thickness loss/defect features or relative thickness.
- The 2D visualisation of a time series of a selected attribute, at a selected position, over the selected time period, shown in Figure 5c.

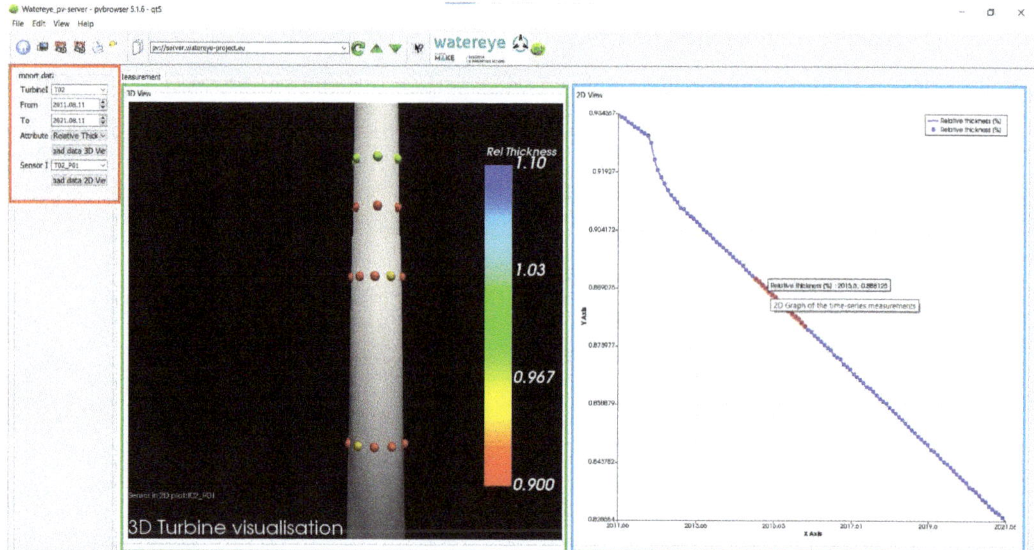

Figure 4. Example of the browser window of our SW-tool. It consists of three areas: user input (red box), a 3D visualisation area (green box) and a 2D time series visualisation (blue box). All widgets are interactive and responsive, as explained in Section 3.3.2.

The user-friendliness and responsiveness of the tool is further enhanced by enabling 3D-panning, rotating and zooming, sensor highlighting, 2D-point picking and informational tooltips, as shown in Figure 5b,c.

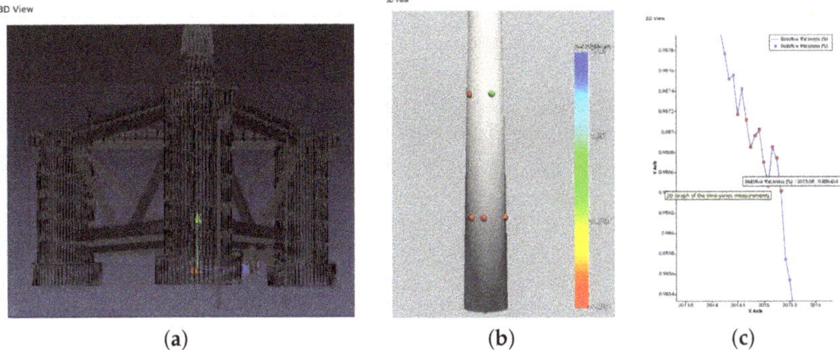

(a)   (b)   (c)

Figure 5. (**a**) Example of the browser 3D window, with a large 3D wind turbine model. (**b**) Example of the browser 3D window, with 3D model, measurement location and status visualisation capability. (**c**) Example of the browser 2D window, with interaction and picking capability.

3.4. Discussion

The presented market analysis shows the potential and opportunities present in this growing market of structural monitoring of offshore wind turbine parks, both for applied research and further commercialisation of dedicated software solutions.

The technological readiness level (TRL) of the SW-tool presented here is relatively low (not production ready), as it is not deployed in the field yet, nor tested by a large userbase; however, it does form a profound starting point for further developments.

Paths toward industrial uptake could focus on security and robustness, as well as on an improved user-friendliness and modularity. Further research includes the extension of this SW-tool to stochastic estimates of corrosion processes and inclusion of risk-assessment and economic impact of different degradation scenarios. Moreover, with minimal effort it is possible to transition the SW tool to other fields where 3D visualisation of corrosion measurement data is relevant.

4. Conclusions and Further Research

Based on the insights presented in Section 2 and the existing tools identified in Section 3.3.2, it is clear that there currently is a lack of commercial tools that consider and address all the needs for corrosion monitoring of offshore wind turbine structures.

Furthermore, as corrosion is a slow, stochastic process influenced by multiple disturbances, it is very hard to quantify theoretically; therefore, data-based and risk aware maintenance planning is crucial to ensure the structural integrity of offshore wind assets throughout their intended service life. This improves the risk–return balance of CAPEX and OPEX, thereby further reducing the weighted average cost of capital (WACC).

The SCADA-compatible software tool presented here, can help wind turbine/wind farm operators to translate corrosion-related information to insights on the wind turbine structural degradation. Furthermore, it can assist the operators with the economic scheduling of preventive maintenance interventions.

Author Contributions: J.V.: Software, Methodology, Writing—original draft, Writing—review & editing; I.C.: Conceptualisation, Software, Methodology, Validation; A.P.O.: Conceptualisation, Supervision, Project administration, Funding acquisition, Writing—review & editing. All authors have read and agreed to the published version of the manuscript.

Funding: This research was carried within the WATEREYE project that has received funding from the European Union's Horizon 2020 research and innovation programme under grant agreement No. 851207.

Institutional Review Board Statement: This study did not involve any testing on animals.

Informed Consent Statement: This study did not involve any testing on humans.

Data Availability Statement: The software described in this paper, developed in the framework of this H2020 WATEREYE project, is defined as a proprietary deliverable. Therefore, the source code of this software cannot be made open source or shared through open access.

Acknowledgments: The authors are grateful for the technical support of Stijn Helsen and Jia Wan during the development of the SW tool.

Conflicts of Interest: The authors declare no conflict of interest. The funding agencies had no role in the design of the study; in the collection, analyses, or interpretation of data; in the writing of the manuscript, or in the decision to publish the results.

References

1. Alliance, C.G. Creating a Road to Market for Renewables. 2021. Available online: https://cleangridalliance.org/about/history (accessed on 1 August 2021).
2. Department of Energy, National Renewable Energy Laboratory Transparent Cost Database. 2021. Available online: https://openei.org/wiki/Transparent_Cost_Database (accessed on 11 August 2021).
3. Lacal-Arantegui, R. Globalization in the wind energy industry: Contribution and economic impact of European companies. *Renew. Energy* **2019**, *134*, 612–628. [CrossRef]
4. BloombergNEF Global Wind Industry Had a Record, Near 100GW, Year. Available online: https://about.bnef.com/blog/global-wind-industry-had-a-record-near-100gw-year-as-ge-goldwind-took-lead-from-vestas/ (accessed on 3 August 2021).
5. Research, C.M. 7th Edition Global Wind Farm Operation Market Report. 2021. Available online: https://www.cognitivemarketresearch.com/service--software/wind-farm-operation-market-report (accessed on 2 August 2021).
6. Lee A. Offshore Wind Power Price Plunges by a Third in a Year: BNEF. Available online: https://www.rechargenews.com/transition/offshore-wind-power-price-plunges-by-a-third-in-a-year-bnef/2-1-692944 (accessed on 3 August 2021).
7. Kovács, A.; Erdős, G.; Viharos, Z.J.; Monostori, L. A system for the detailed scheduling of wind farm maintenance. *CIRP Ann.* **2011**, *60*, 497–501. [CrossRef]
8. U.S. Energy Information Administration. Levelized Costs of New Generation Resources in the Annual Energy Outlook. 2021. Available online: https://www.eia.gov/outlooks/aeo/pdf/electricity_generation.pdf (accessed on 2 August 2021).
9. Ziegler, L.; Gonzalez, E.; Rubert, T.; Smolka, U.; Melero, J.J. Lifetime extension of onshore wind turbines: A review covering Germany, Spain, Denmark, and the UK. *Renew. Sustain. Energy Rev.* **2018**, *82*, 1261–1271. [CrossRef]
10. WATEREYE_H2020. O&M Tools Integrating Accurate Structural Health in Offshore energy. Available online: http://www.watereye-project.eu/ (accessed on 15 November 2021).
11. Fraunhofer Institute for Wind Energy Systems Condition Monitoring of Wind Turbines: State of the Art, User Experience and Recommendations. Available online: https://www.vgb.org/vgbmultimedia/383_Final+report-p-9786.pdf (accessed on 16 August 2021).
12. Shutterstock. 3D Models. Available online: https://www.turbosquid.com/Search/Index.cfm?keyword=offshore+wind+turbine (accessed on 5 October 2021).
13. IEA. Offshore Wind Outlook. 2019. Available online: https://iea.blob.core.windows.net/assets/495ab264-4ddf-4b68-b9c0-51429 5ff40a7/Offshore_Wind_Outlook_2019.pdf (accessed on 11 August 2021).
14. Zhang, D.; Zhang, X.; He, J.; Chai, Q. Offshore wind energy development in China: Current status and future perspective. *Renew. Sustain. Energy Rev.* **2011**, *9*, 4673–4684.
15. Adey, R.; Peratta, C.; Baynham, J. *Corrosion Data Management Using 3D Visualisation and a Digital Twin*; Whitepaper NACE International; Perdido Beach Resort and Wharf Event Center: Orange Beach, AL, USA, 2020; C2020-14535.
16. Dong energy, C. Barrow Offshore Wind Farm (Denmark), Post Construction Monitoring Report. Available online: https://tethys.pnnl.gov/sites/default/files/publications/Barrow_Offshore_Wind_Monitoring_Report.pdf (accessed on 2 August 2021).
17. Giuliani, B. Structural Integrity of Offshore Wind Turbines. In Proceedings of the Earth and Space 2010: Engineering, Science, Construction and Operation in Challenging Environments, Honolulu, HI, USA, 14–17 March 2010. [CrossRef]
18. Wilkinson, M.; Darnell, B.; van Delft, T.; Harman, K. Comparison of methods for wind turbine condition monitoring with SCADA data. *IET Renew. Power Gener.* **2014**, *8*, 390–397.. [CrossRef]
19. Zhu, C.; Li, Y. Reliability Analysis of Wind Turbines. In *Stability Control and Reliable Performance of Wind Turbines*; Okedu, K.E., Ed.; IntechOpen: London, UK, 2018. [CrossRef]
20. Tautz-Weiniter, W. Using SCADA data for wind turbine condition monitoring—A review. In *Special Issue: Wind Turbine Condition Monitoring, Diagnosis and Prognosis*; Institution of Engineering and Technology (IET): London, UK, 2016. [CrossRef]
21. Avolio, F. Visualisation tools for wind farms development. In *Wind Energy and Landscape*; CRC Press: London, UK, 1998; Chapter 6, ISBN 9781003078159.
22. Oil, A.; Society, G.H. Overview of Offshore Oil & Gas History. Available online: https://aoghs.org/offshore-oil-history/ (accessed on 14 September 2021).
23. Wang, G.; Serratella, C.K.S. Current practices in condition assessment of aged ships and floating offshore structures. In *Condition Assessment of Aged Structures*; Elsevier Ltd.: Amsterdam, The Netherlands, 2008; pp. 3–35.

24. Sadeghi, K. An Overview of Design, Analysis, Construction and Installation of Offshore Petroleum Platforms Suitable for Cyprus Oil/Gas Fields. *J. Soc. Appl. Sci.* **2011**, *2*, 1–16.
25. Clinic, C. List of Corrosion Prediction Software. Available online: https://www.corrosionclinic.com/corrosion%20prediction%20and%20corrosion%20modeling.htm#list_of_corrosion_prediction_software (accessed on 29 September 2021).
26. Ompusunggu, A.P.; Coudron, I.; Wan, J. Towards a digital twin for corrosion monitoring and health estimation of offshore wind turbine structures. In Proceedings of the WESC2021 Conference Book, Theme 07: Reliability, Monitoring and Sensing Technology, Hannover, Germany, 25–28 May 2021.
27. Sterz. Modernizing SCADA HMIs. *Industrial Applications, iPhone Expectations, KDAP Whitepaper.* 2019. Available online: https://www.kdab.com/making-industrial-applications-match-iphone-expectations/ (accessed on 11 August 2021).
28. Lehrig, S. Pvbrowser Manual; Lehrig Software Engineering. 2016. Available online: https://github.com/pvbrowser/pvb/commits/master (accessed on 11 August 2021).
29. Kitware Inc. *The VTK User's Guide*; Kitware Inc.: New York, NY, USA, 2010; ISBN 978-1-930934-23-8.
30. Sun Microsystems, University of Carolina, Scriptics Corporation TCL-TK Licence. Available online: https://tcl.tk/software/tcltk/license.html (accessed on 11 August 2021).

applied sciences

Article

Ultrasound-Based Smart Corrosion Monitoring System for Offshore Wind Turbines

Upeksha Chathurani Thibbotuwa [1,*], Ainhoa Cortés [1,2,*] and Andoni Irizar [1,2]

1   CEIT-Basque Research and Technology Alliance (BRTA), Manuel Lardizabal 15, 20018 San Sebastian, Spain; airizar@ceit.es

2   Department of Electronics & Communications, Universidad de Navarra, Tecnun, Manuel Lardizabal 13, 20018 San Sebastian, Spain

*   Correspondence: uthibbotuwa@ceit.es (U.C.T.); acortes@ceit.es (A.C.); Tel.: +34-943212800

Abstract: The ultrasound technique is a well-known non-destructive and efficient testing method for on-line corrosion monitoring. Wall thickness loss rate is the major parameter that defines the corrosion process in this approach. This paper presents a smart corrosion monitoring system for offshore wind turbines based on the ultrasound pulse-echo technique. The solution is first developed as an ultrasound testbed with the aim of upgrading it into a low-cost and low-power miniaturized system to be deployed inside offshore wind turbines. This paper discusses different important stages of the presented monitoring system as design methodology, the precision of the measurements, and system performance verification. The obtained results during the testing of a variety of samples show meaningful information about the thickness loss due to corrosion. Furthermore, the developed system allows us to measure the Time-of-Flight (ToF) with high precision on steel samples of different thicknesses and on coated steel samples using the offshore standard coating NORSOK 7A.

Keywords: corrosion monitoring; FPGA; offshore wind turbines; ultrasound; thickness loss

Citation: Thibbotuwa, U.C.; Cortés, A.; Irizar, A. Ultrasound-Based Smart Corrosion Monitoring System for Offshore Wind Turbines. *Appl. Sci.* **2022**, *12*, 808. https://doi.org/10.3390/app12020808

Academic Editors: Luis Hernández-Callejo, Maria del Carmen Alonso García and Sara Gallardo Saavedra

Received: 10 December 2021
Accepted: 11 January 2022
Published: 13 January 2022

Publisher's Note: MDPI stays neutral with regard to jurisdictional claims in published maps and institutional affiliations.

1. Introduction

Wind energy plays a significant role as a renewable energy source, urging the need for cost-effective renewable energy sources. Compared to onshore wind energy, offshore wind turbines benefit from two main advantages: higher mean wind speeds (modest increases in wind speed can result in doubling the generated power) and steadier wind supply, which makes power generation more reliable. These two factors combined have a dramatic effect on the Return on Investment (RoI) of the farm [1].

In the last decade, we have witnessed a steady increase in the total installed capacity of offshore wind farms. By the end of 2019, 6.1 GW capacity has been newly installed. The total capacity installed by 2019, was 29.1 GW, which is a 10% increase with respect to 2018 [2]. Europe (UK, Germany, Denmark, Belgium) is leading by contributing 75% of total global offshore wind installation, as of the end of 2019, and Asian offshore wind energy production keeps growing significantly, lead by China and followed by Taiwan, Vietnam, Japan, and South Korea. Due to COVID-19, the overall expected wind capacity in 2020 was not met because of the challenges in project construction, execution activities, and investments. However, the Global Wind Energy Council (GWEC) has predicted that offshore wind energy will have a 20% of contribution to the global wind installation by the year 2025.

Offshore structures are installed in relatively deep water and exposed to the harsh marine environment. This leads to an increase of the cost of construction and deployment compared to onshore structures [3] and higher operation and Maintenance (O&M) costs of installations compared to land or coastal-based structures due to the limitations in accessibility, manpower, special equipment requirements, etc. [4]. The added costs in the construction and deployment of offshore turbines were mitigated at the beginning as the turbines were installed mainly in coastal sites with shallow waters. In recent years, there has been a constant increase in the water depth and distance to shore for the offshore wind farms [5], as farms are located further away from shore (60 km) and as depth water reaches 200 m [6]. This leads to higher Capital Expenditure (CAPEX) and O&M costs that reduce the benefits of the farm [7]. Several foundations for wind turbines have been developed trying to reduce the construction and installation costs. Another key to the profitability of the wind farm is to improve the useful life of the wind turbines by maintaining the operational conditions of the towers to minimize the breakdowns or downtime of the turbines. Therefore, improving O&M models and tools and, at the same time, reducing costs have become key to maintain the profitability of offshore wind farms. Numerous research works providing different approaches that support directly or indirectly to improve O&M process of offshore wind turbines/farms have been discussed in Reference [8–12].

1.1. Structural Health Monitoring of Wind Turbines: Corrosion Monitoring

Structural Health Monitoring Systems (SHMSs) for Offshore Wind Turbines (OWTs) are used to detect damage in blades [13,14], tower, and support structure. Since the tower and foundation are two key elements of OWT that support the structural integrity, they must operate as much time as possible. Once the turbine is installed, it should sustain with associated loads, and their partial failure would carry catastrophic consequences. Not much progress has been made in developing robust applications regarding SHMs for operating WTs [15].

Corrosion is one of the main root causes that can produce offshore structural failure. Corrosion monitoring is the process of obtaining very frequent corrosion measures to evaluate the progress of corrosion within a specific environment. Corrosion inspections are done to evaluate the material condition at any given time based on a pre-scheduled routine or the available risks. Usually, inspections are performed much less frequently than corrosion monitoring. Frequent corrosion measures are always advantageous of early detecting the risks, repairing conditions, and, consequently, reducing the operational and maintenance cost associated with corrosion. The rust formation due to corrosion depends on the environment and the type of material, as well. Carbon steels corrode mostly by general corrosion (uniform corrosion), but localized types of corrosion can also take place, e.g., pitting, which usually are not considered in conventional corrosion analysis [16]. Different reasons can be identified behind the growth of corrosion process in an offshore structure, such as temperature, salinity, pH, and coating damage due to tidal fluctuations, dissolved oxygen, variable cyclic load due to wave and wind impact, etc. [17].

Even for a well-designed structure, with a long time of exposure to the harsh marine environment, corrosion causes the degradation of the metal leading to create corrosion fatigue cracks and, consequently, structural failure. This scenario increases the O&M cost because of the possibility of more frequent maintenance, repair activities, and any replacement if needed [18]. Therefore, a successful corrosion monitoring approach can potentially decrease the ongoing O&M cost of wind turbines directed at reducing economic losses and environmental damages. However, corrosion monitoring is one of the major challenges that an SHM system developed for an OWT faces, mainly due to the accessibility difficulties and the area the corrosion monitoring system must cover, which is very large even when one considers that some zones are more affected than others, e.g., the splash zone. Thinking about an automated solution, ideally, the corrosion sensor should work unattended 24/7 for several months or even years. This imposes very hard specifications that the monitoring system must meet. Therefore, prior to the selection of an appropriate corrosion monitoring technique and sensors, the zone of interest (splash zone, atmospheric zone, submerged zone) of the wind tower and the respective environmental conditions must be considered separately [19].

1.2. Corrosion Inspection and Monitoring Techniques

There are different types of corrosion inspection and monitoring techniques to find out the corrosion condition of metals, as is discussed and reviewed in Reference [20,21]. Figure 1 shows a detailed classification of these techniques based on the corrosion detection method and sensing parameters.

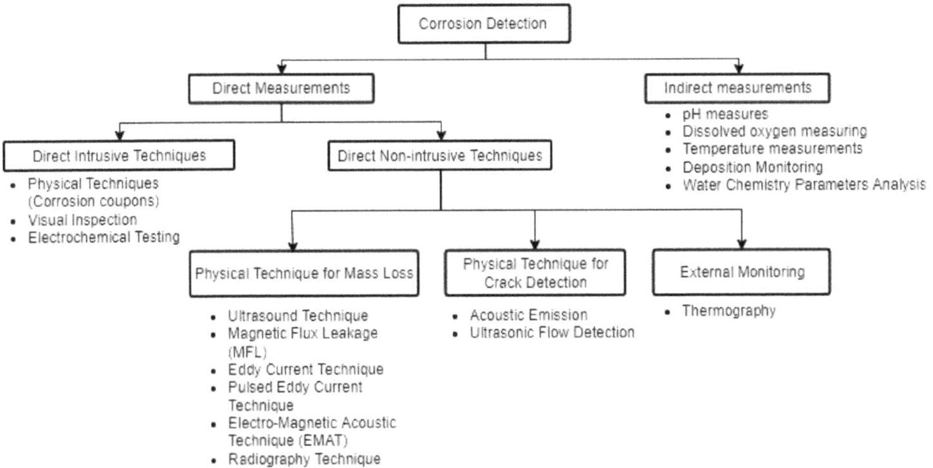

Figure 1. Corrosion detection methods.

In general, corrosion detection measures can be direct measuring metal loss due to corrosion or corrosion rate or indirect measuring any parameter that may be a cause or consequence of the metal loss or corrosion. Some of these techniques need direct contact/exposure to the same corrosive fluids or to access the internal environment where the corrosion process takes place. They are known as intrusive techniques and can alter the corrosion process and create disturbances to the operation during installation and re-installation of measuring probes or inspection processes. Therefore, it is beneficial if corrosion testing can be performed from the outside of the test object. This is possible with non-intrusive techniques through detecting physical and geometry changes due to corrosion (mass loss, crack, or surface discontinuities), and there is no need for any direct contact to the same corrosive fluids or access to the internal environment. On the other hand, it is very important to perform corrosion detection activities in the field applications without destroying the engineering properties of the material and without affecting its long-term future performance during the inspection process. This kind of evaluation can be achieved by using Non-Destructive Testing (NDT) methods for material testing [22]. Among the techniques listed in Figure 1, if this condition is satisfied, they are known as an NDT method. Therefore, we can propose that the most appropriate corrosion detection technique for offshore wind turbines shall be both non-intrusive and non-destructive.

The main objective of this research work is to develop a smart ultrasound solution for corrosion monitoring in offshore wind turbines. Due to the harsh and distant environment they have been installed in, it is a challenge to find the most suitable corrosion detection technique. Based on the expected solution, we have identified the following design specifications to be achieved and feasible with the selected corrosion detection method.

- SHM requires a reliable and permanent acquisition of monitoring data with minimum interruptions.
- Corrosion detection technique shall be non-intrusive and non-destructive.
- The monitoring system is able to detect uniform corrosion.

- The solution is able to detect corrosion in both splash zone and atmospheric zone of the wind turbine tower.
- Data acquisition must run for several years in order to obtain meaningful results about the tower's health.
- Robust wireless communications are needed to rely on the sensor data sent to the tower's central computer in a difficult environment for electromagnetic signals.
- Corrosion detection components must have the capability of integrating with Unmanned Aerial Vehicles (UAVs) as they offer significant potential benefits to the monitoring of large-scale facilities, e.g., atmospheric zone.
- Tower locations which are not easily reached by a mobile platform must be covered by battery powered sensor nodes fixely installed in those areas, e.g., splash zone.

The paper is structured as follows. In Section 2, a review of different ultrasound solutions for corrosion monitoring is introduced. Section 3 describes the concept of the presented approach for corrosion monitoring in offshore wind turbines. Section 4 provides the description of the ultrasound testbed and the design methodology followed to miniaturize the solution. In Section 5, the samples preparation and the experimental setup are presented. Section 6 shows the most relevant results we have obtained through our experiments. Finally, the conclusions are summarized in Section 7.

2. Related Work

In ultrasound techniques, a high-frequency ultrasound wave will travel along with a material that shows elastic properties. The frequency range of ultrasonic sound waves used for NDT of materials is generally 1 MHz to 15 MHz [23]. The velocity of the ultrasound wave propagation through a material is a function of the elastic modulus and density of the material [22,24]. Importantly, ultrasound waves are distinctly reflected at the boundaries when the acoustic impedance change due to a different material or medium properties. Based on that, the thickness loss due to corrosion will be given as a function of propagating velocity, frequency, and energy components [25]. The setup can be designed based on two common techniques [23,26]:

1. Pulse echo method: A short period of ultrasonic energy is introduced into the test piece with regular pause, and the reflected part of the ultrasonic wave is utilized for the analysis. As the sender and receiver probes use the same transducers, either one transducer or two transducers placed on the same side can be used.
2. Through transmission method: The transmitted part of the ultrasound wave is used for the analysis. Separate transducers placed on opposite sides are used as the sender and receiver probes.

It can be said that the best technique for corrosion monitoring would be the pulse-echo method as it gives the depth/location of the defect in comparison to the through transmission method.

Thus, current sensing methods using the ultrasound technique give valuable information about various forms of degradation that occur in all sorts of materials (cracks, deformations, thinning, corrosion, etc.), but, in general, they are classified as inspection devices, rather than monitoring devices. The inspection devices are usually "once-off" measurements that cover a given period of time (according to maintenance schedules), while the monitoring devices involve very frequent measurements that must detect small fluctuations in the advancement of corrosion in order to be used as input for a general assessment of corrosion [27]. Several previous works have been reported to assess corrosion in pipelines using ultrasounds [28].

Although not focused on corrosion detection, Ref. [13] proposes a novel walking robot-based system for NDT in wind turbines. Moreover, commercial systems based on ultrasounds exist in the market to monitor large structures. For example, Eddyfi Technologies has developed the Scorpion2 product [29] for ultrasonic tank shell inspections. To the best of our knowledge, it monitors wall thickness during the inspection using a dry-coupled, remote-access ultrasonic crawler that brings major efficiency and data improvements to tank shell inspections. It scans the tank wall to get an image of the wall status with the aim of estimating the wall thickness. Among other NDT techniques, the walking robot offers ultrasonic testing, but it is aimed at identifying cracks, delamination, etc., in composites (e.g., in blades). Furthermore, this walking robot has already been used in onshore wind turbines.

Ultrasound is a technique with a lot of potentials to deploy a Wireless Sensor Network (WSN). Nodes can be housed in small devices with communication capabilities, drawing very little power from a battery and offering very good corrosion detection capabilities. However, this is not enough when referred to large structures (e.g., offshore WT) where corrosion may develop randomly at a given point. The continuous technological improvements of all kinds of unmanned vehicles (terrestrial, maritime, and aerial) during the last decade make this technology very attractive for many industries that require frequent and costly maintenance operations in areas where access is difficult. Ref. [30] proposes an autonomous ultrasonic inspection using UAVs. This research is very relevant since it demonstrates the influence of the accuracy of the measurements positioning and of the electrical noise produced by control electronics and the propeller motors into the thickness estimation. These uncertainties cause ultrasonic signal coupling issues. One of the main advantages of permanently installed ultrasound sensors is that they achieve a better repeatability since the coupling and probe-positioning errors are removed. Hence, in the case of mobile solutions, there are challenges to solve regarding probe alignment, more robust control strategies, enhanced positioning the accuracy, and novel ultrasonic angular coupling capabilities. The paper is focused on the improvement of the UAV design, and they do not show the accuracy obtained in the thickness estimation. Additionally, in this paper, they claim to detect the loss of thickness over non-coated aluminium samples (11 mm of thickness). It is well known that coating reduces accuracy of wall thickness estimations [31], and wind turbine walls are usually protected with standard coatings of different thicknesses and formulations. Therefore, coating affects the ultrasound signal response, and it is another key challenge to be solved, not only for ultrasound mobile solutions but also for fixed solutions. Another problem is that ultrasound monitoring employs liquid couplants, which makes it necessary to integrate a couplant dispenser into the UAV and a cleaning phase after the ultrasound measurement has been taken.

Corrosion is an extremely slow process, and corrosion rates between 0.1–0.2 mm/year are typical rates. Therefore, with a resolution of 1 μm, we will have to wait, on average, more than 4 days between consecutive measures to get statistically meaningful results. The work presented by References [32,33] proposes the use of curve fitting algorithms of a Gaussian echo model to estimate the time of arrival of ultrasound echoes. In order to improve the precision of the ToF measurements, it is necessary to compensate for temperature variations that affect both the thickness of the material and the speed of sound, digitally processing several back-wall echoes or using adaptive filtering techniques [34]. Other proposals rely on improving the Signal-to-Noise Ratio (SNR) and time resolution of the signals by employing the Wiener filter and spectral extrapolation [35]. An interesting idea to increase the SNR of the ultrasound system is to use coded excitation as in [36], which introduces a code gain in the signal's path. Finally, Ref. [37] proposes a non-standard approach based on cross-correlation, but, instead of performing the correlation with the transmitted signal, they correlate a back-wall echo with a reference signal that has been modified using an iterative algorithm that also involves cross-correlations. Thus, this approach achieves wall-thickness estimations (below 1 μm) on ultrasonic signals. However, this is at the expense of increased data processing complexity that will affect the power consumption and the cost of the solution.

To the best of our knowledge, none of the proposals in the literature develop a low-cost and low-power corrosion monitoring system based on ultrasound able to measure the thickness loss precisely showing that the solution can work for different thicknesses and mainly, for coated samples using the offshore standard coating NORSOK 7A. On top of that, the proposed approach aims to operate unattended inside offshore metallic towers.

3. Corrosion Monitoring System Concept

Our corrosion monitoring system design concept is focused on a novel and smart low-power and low-cost corrosion monitoring system for offshore wind turbines. The zones of interest are the splash zone and the atmospheric zone. The splash zone is the area that is intermittently exposed to seawater due to the action of tide or waves or both. This intermittence produces the highest level of corrosion. The atmospheric zone is the zone that is permanently exposed to marine air conditions. Therefore, the level of corrosion is also very high. Our approach includes two types of deployment, namely (i) a fixed and (ii) a mobile solution. In the case of the splash zone, the fixed sensors are attached to the structure due to the difficulties to access by a mobile platform. However, the atmospheric zone is covered by the mobile solution based on a drone flying inside the tower. Therefore, the main requirements to be achieved by our corrosion monitoring solution can be highlighted as follows.

- The monitoring solution shall be operational under the harsh conditions of offshore platforms.
- The corrosion sensor of the mobile and fixed solutions shall perform one measurement every month. However, this rate must be parameterizable with the aim of being increased when necessary.
- The corrosion sensor must be able to measure the ToF related to the thickness loss of the selected steel with a thickness between 40 mm and 70 mm but also with a thickness of 5 mm or 6 mm, which shall be the thickness of the produced samples to characterize the ultrasound solution.
- Be able to operate unattended for long periods of time.
- Both fixed and mobile solutions shall be able to perform corrosion measurements in the field and send those measurements wirelessly.
- The size of the overall final monitoring solution shall be up to 6×11 cm with a weight of up to 100 g. The sensor head shall be connected to the processing platform through wires.

4. Smart Ultrasound Sensor Design

Being aware of developing a reliable corrosion monitoring solution to decrease the cost of the maintenance and the downtimes during operation in the offshore wind sector, this paper presents a smart and robust corrosion monitoring system based on ultrasound technology and ToF technique that is capable of operating unattended inside an offshore metallic tower. With that aim, a prototype of the ultrasound test device has been designed. This is the Ultrasound Testbed (UT) with the capability of upgrading into a miniaturized solution that can be placed and operate inside of the wind turbine. UT is developed with the aim of performing frequent ultrasound measurements and analyzing the ultrasound responses. More details about sensor selection, excitation signals, and design methodology of the UT system are discussed in the following subsections.

4.1. Probe Selection

The selection of ultrasound probe or transducer needs to be done by considering the sensor characteristics, such as sensor dimensions, weight, dead zone duration, waveform duration, and peak frequency. The transducer operating frequency spectrum and the waveform duration are often provided by the manufacturer. The peak frequency of the transducer controls the penetrating power that affects the possible inspection thickness of test material, e.g., high-frequency ultrasound has low penetration power and might almost attenuate within the material. As at a fixed velocity in a perfectly elastic material the wavelength and frequency are inversely proportional, a change in the probe frequency will affect the sensitivity and the resolution to detect flaws in the material. Due to the diffraction of ultrasonic, the sensitivity is about half of the corresponding wavelength [38]. Thus, sensitivity and resolution are increased when probe frequency increases. The dead zone duration is the interval following the initial pulse where the transducer ringing prevents detection or interpretation of reflected echoes. This is a constraint with thinner size samples as it is better that the time interval between back-to-back echoes has to be out of the dead zone.

As we are going to measure 5 mm thick samples, it is important to check the dead-zone of the selected probe. A 5 mm thick steel sample will introduce a delay between back-to-back echoes of $2 \times 5/5.9 \times 10^6 = 1.69\,\mu s$ (the speed of sound in steel is approximately 5.9×10^6 mm/s). This is the expected value of the ToF before corrosion takes place. This value is important when considering the dead zone of an ultrasonic probe. Dead zone value is highly dependent on the test conditions and instrumentation used. It is not normally given in the technical documentation of the probe and needs to be determined experimentally. The values shown on Table 1 correspond to the time when the ultrasound signal has almost completely settled (see Figure 2). However, it must be noted that the dead zone does not entirely invalidate the ultrasound data received on that interval. By proper filtering, and with the appropriate receiver's gain, it would be possible to extract echoes within the dead zone of the piezo.

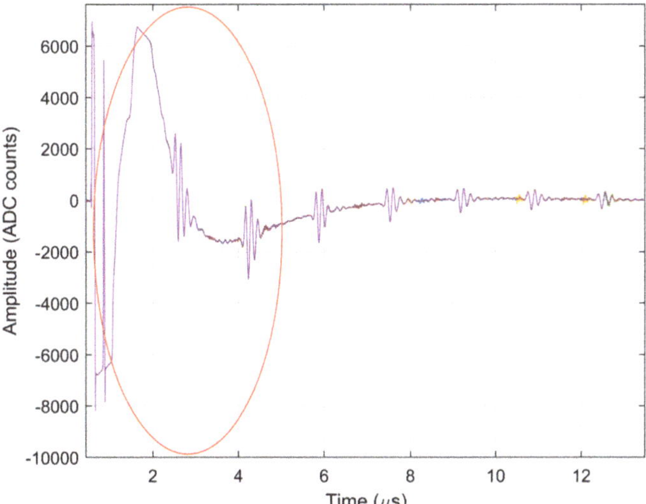

Figure 2. Dead Zone of an ultrasound probe.

Table 1. Performance and features of ultrasound probes.

Part Name	YS15-10	TSB10-06	V111
Manufacturer	YUSHI	OKOndt	Olympus
Diameter (mm)	10	6	13
Peak Frequency (MHz)	7.86	10.2	8.44
-6 dB Bandwidth (MHz)	3	3	6.93
Waveform Duration -20 dB Level (µs)	0.7	0.840	0.272
Dead Zone (µs)	2.6	2.4	3.5
Connector/Position	BNC/Top	Microdot/Side	Microdot/Side
Weight (g)	50	7	20
Price (€)	~100	~100	~600

A comparative analysis of the performance specifications of three sensors has been done to select the most suitable one for the UT system as given in Table 2. As a result, V111 from Olympus [39] has been selected as the most suitable ultrasound probe for our system, specifically prioritizing its lower waveform duration (higher bandwidth) and sound pressure power. Having a higher dead zone compared to the other two sensors will not be a constraint in the real scenario as the system is designed for testing thickness around 40 mm.

Table 2. Applicability to the US system.

Part Name	YS15-10	TSB10-06	V111
Manufacturer	YUSHI	OKOndt	Olympus
Dead Zone	✓✓	✓✓	✓
Waveform Duration	✓	✓	✓✓✓
Bandwidth	✓	✓	✓✓✓
Sound Pressure Power	✓✓	✓	✓✓✓
Testing 5 mm Samples	✓	✓	✓✓✓
Testing 40 mm Samples	✓✓	✓	✓✓✓
Weight and Size	✓	✓✓✓	✓✓
Price	✓✓✓	✓✓✓	✓

4.2. Excitation Signal

The conventional method is using a square wave as the excitation signal. The frequency response of the V111 peaks at 8.44 MHz; therefore, the frequency of the pulse must be similar in order to resonate and optimize the received signal's strength. Adding more pulses increases the signal's strength but at the expense of a larger pulse duration that deteriorates the time resolution of the ToF estimation. In our system, the period and the rise and fall times of the square wave can be controlled using a 125 MHz clock. The square wave pulse is bipolar ($\pm$15 V), and the generated frequency is obtained by dividing the clock frequency by an even integer $125/16 = 7.8$ MHz to guarantee a 50% duty cycle for positive and negative pulses.

4.3. Signal Processing and Time-of-Flight Estimation

Based on ultrasound waves, corrosion is evaluated by estimating the wall thickness loss that happens due to corrosion. The proposed approach is based on the Time of Flight (ToF) technique, which estimates the thickness of the test object by measuring the elapsed time between two consecutive echoes of the ultrasound response. This is based on the reflection of the ultrasound waves in different acoustic impedance conditions. This way, the effect of the couplant is greatly reduced because no changes in the couplant are expected in the time period of a ToF (in our case, less than 5 µs). The main drawback is the additional attenuation of the second echo that reduces the SNR of the signal and, therefore, the quality of the estimation. The determination of the time is done using the cross-correlation operation.

A block diagram of the signal processing operations performed to complete the ToF estimation in our approach is given in Figure 3. To have an accurate time base, it is important that both the ultrasound signal generator (US Generator) and the signal acquisition circuit (ADC Read) work with the same clock input. The filter removes unwanted frequencies components outside the bandwidth of the probe. The Echo Windowing block determines the location and size of the echo signals that will be cross-correlated. Some signal quality measurements are performed on the echo signals (signal level, signal width, echo amplitude and decay rate, separation between echoes, etc.). These measurements will help to detect low-quality signals and discard those ToF measurements in advance. Finally, cross-correlation between two consecutive echoes, as well as peak detection and analysis, produce the ToF result.

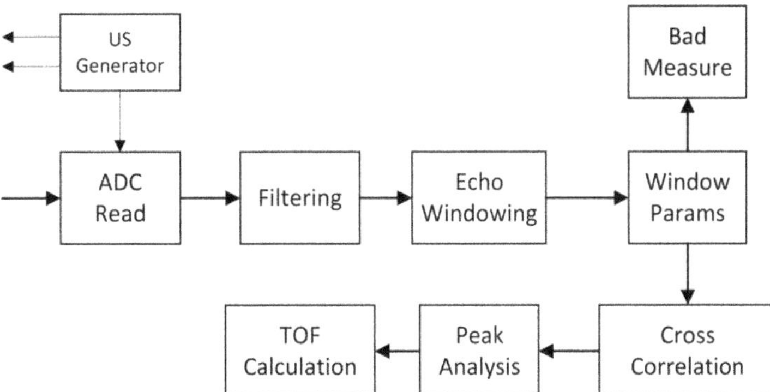

Figure 3. Block diagram of the signal processing chain.

4.4. Design Methodology

The design of a highly integrated monitoring system involves the collaboration between several design teams: analog circuits, digital signal processing, FPGA and microcontroller development, PCB prototyping and design, validation, etc. To coordinate the efforts from different teams, it is very important to have a system design methodology that takes you from the conceptual design to the final device to be deployed.

We have used a prototype-based design methodology in which the system is first designed from large building blocks that allow us to define the system using high-level abstraction languages, such as MATLAB or C/C++. As can be seen in Figure 4, the main blocks of the system prototype include a PC/Laptop and an Ultrasound Testbed. The latter is composed of a high-performance embedded processor that is able to generate and acquire ultrasound signals. The embedded processor runs a Linux system with the Debian distribution and includes an SD card for mass storage, Ethernet and Wi-Fi connections, and access to an internal FPGA that serves the purpose of buffering the ultrasound data from a high-speed 14-bit ADC. The Ultrasound Analog Front End (AFE) is composed of a pulser generator, adaptation circuitry, and a variable gain amplifier.

The system prototype permits us first to set up an experiment and gather raw ultrasound data for a variety of steel samples with different coatings, to test different piezo sensors, and change the experiment conditions, such as temperature, signal frequency, number of pulses, etc. The raw data is stored as a JSON file in a data repository that can be accessed from MATLAB to design and test signal processing and detection algorithms with the aim of estimating the time-of-flight from the ultrasound response.

Secondly, thanks to the embedded system, it is possible to run a C/C++ implementation of the algorithm in the Red Pitaya microprocessor and test it for quantization effects. In this case, the embedded system is responsible for estimating the time-of-flight, and the data obtained can also be uploaded to the data repository as a JSON file.

The hardware architecture of the final implementation is designed by analyzing the computing requirements of the algorithms designed in the prototype phase. We have chosen an architecture that includes a low-power Cortex-M4-based microcontroller (μC), an FPGA that interfaces with the μC using an SPI link and the AFE devices. The μC is in charge of supervising the system, i.e., powering on and off the different devices (FPGA, AFE, memories), launching measurements and gathering the results and raw data from the FPGA using the SPI interface and providing communication interfaces with the external world (UART and wireless communications). The high-speed sampling of the ultrasound signals has called for the use of an FPGA. Ultrasound signal generation is also handled by the FPGA to produce synchronized signals for the pulser. The processing of ultrasound data requires intensive computation that is more efficiently performed in the FPGA.

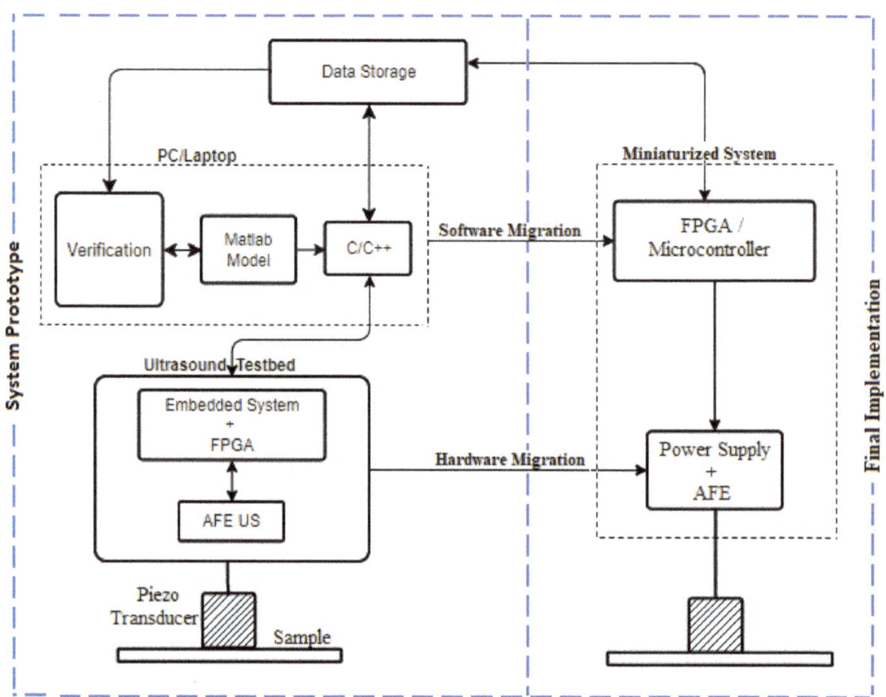

Figure 4. System design methodology.

By having models of the processing of ultrasound signals at different abstraction levels (MATLAB, C/C++, and FPGA), it is possible to perform validations at different stages of the design process as it advances towards the final implementation using the raw data stored in the data repository (see Figure 4). In this way, we can feed the MATLAB model with the real data from the miniaturized system and compare the ToF results obtained for verification.

Finally, after validating the ultrasound test method and algorithm implemented in UT, it has been migrated to a miniaturized system, as shown in Figure 5 (right), being able to place it inside a wind turbine to conduct ultrasound tests in the real scenario. The performance specifications of the miniaturized solution are given in Table 3.

Table 3. Miniaturized ultrasound system performance.

Specification	Value
Dimensions	$(110 \times 60 \times 15)$ mm
Power supply	Battery 3.6 V, 5800 mAh
Average power consumption (Sleep mode)	5.4 µW
Average power consumption (per measurement event)	850 mW
Wake Up Time before US measurement	1 s
Measurement Time	200 µs

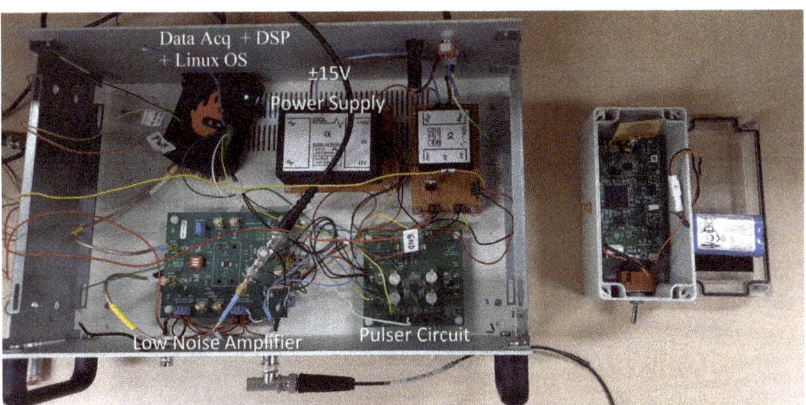

Figure 5. Pictures of the UT (**left**) and miniaturized solution (**right**).

5. Sample Preparation and Experimental Setup

5.1. Sample Preparation

Two types of S355 structural steel samples have been used for the system performance validation as reference samples and test samples. Reference samples have been used mainly to verify the test method and evaluate the precision of the measurements during the algorithm development. On the other hand, test samples have been used to test the long-term system performance. The dimensions of the reference and test samples are shown in Table 4.

Table 4. Reference and test sample dimensions.

Sample Size	(75 × 150 × 5) mm	(75 × 150 × 40) mm
Reference Samples	✓	✓
Test Samples	✓	–

The real thickness of a wind turbine tower is around 40 mm. However, the test samples have been used only with 5 mm thickness due to the practical difficulties in transportation and handling of the samples during the frequent experiments.

Coating layers are applied to protect the substrate from reacting with its environment so that it will not engage in an electrochemical process and corrode. Coatings provide protection against moisture, dissolved gases, acids, and other reactants in the environment. The test coated samples with applied standard coating NORSOK 7A (see Figure 6) were tested by our ultrasound system to observe the changes in the characteristics of the ultrasound signal and to analyze the performance of the developed signal processing algorithm.

Corrosion is a very slow process, and it takes a long time to induce corrosion on coated samples. Therefore, in order to observe a measurable thickness loss with time, the corrosion process has been accelerated by producing a 2 mm-wide and 50 mm-long scribe, cut completely through the coating, which is made in the coated samples, as can be seen in Figure 6. This scribe emulates in some way a coating defect or failure. The prepared coated samples were exposed and conditioned based on cyclic aging resistance processes (cyclic processes of exposure to UV, neutral salt spray, and low-temperature exposure) in agreement with EN ISO 12944-9. The test method consists of 25 cycles of 4200 h of exposure: over three days, the samples are exposed to UV/condensation according to EN ISO 16474-3; during the following three days, the samples are exposed to neutral salt spray according to EN ISO 9227; during the next day, the samples are exposed to low temperature at $(-20 \pm 2\,°C)$.

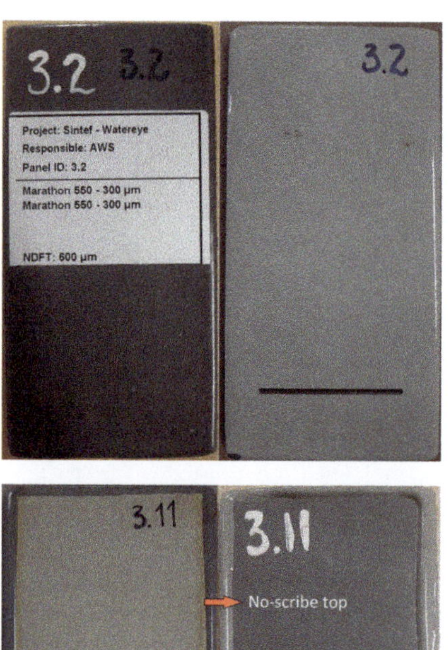

Figure 6. Picture of coated test samples: non-exposed coated test sample (**top**) and three months exposed sample (**bottom**).

In order to observe and measure the corrosion process, the coated samples have been exposed to cyclic aging resistance process for different time durations as 1 month, 3 months (see Figure 6 (bottom)), and 6 months. Testings of coated samples have been performed in three different layers as: on scribe, no-scribe bottom layer, and no-scribe top layer, to get an indication of thickness loss due to corrosion under cyclic aging process.

5.2. Experimental Setup

The developed Ultrasound Testbed (UT) prototype is shown in Figure 5 (left). It consists of a low-noise amplifier, pulser, piezoelectric transducer, and FPGA controller with high-frequency ADC and DAC conversion. The UT system is capable of performing frequent ultrasound testings. In the purpose of performing tests and analyzing results, the UT is connected to a PC/Laptop with MATLAB via Wi-Fi or Ethernet, and the experiments are launched and controlled via PC/Laptop. The main advantages of the UT system are that it allows the user to perform frequent tests, to visualize and compare the results easily, and to develop, analyze, and update the signal processing algorithms for better ToF estimations.

The ultrasound test is performed by placing the sensor on the sample using a couplant applied as shown in Figure 7. Olympus D12-gel, that is recommended for rough surfaces, has been used as the couplant to perform the ultrasound tests. Each test is performed by launching 25 excitation signals in a row and collecting the corresponding response from the test sample. Therefore, the ToF value calculated for a particular location is an average of 25 measures, and this helps to minimize the uncertainty of the results. In addition to that, the precision is calculated as the standard deviation of those 25 measurements.

Figure 7. Non-exposed coated sample: on scribe measurements.

Ultrasonic couplant is used to make good contact between transducer and test piece, facilitating quality sound energy transmission. In NDT, the couplant is used because they do not transmit effectively through air. Even so, the applied couplant layer still can affect to ultrasonic velocity and attenuation of the back wall echoes from the test piece. This depends on the acoustic impedance of the couplant being used and how good the contact is made between the test surface and the transducer. On the other hand, if the amount of couplant is excessive, a couplant layer can act as a wedge and alter the direction of the sound wave affecting the final result.

The measurement of ToF in ultrasound signals propagating through steel is directly related to the variation of the speed of sound with temperature and, to a lesser extent, to thermal expansion. Therefore, we need to measure the temperature of the steel to compensate for the temperature effect in the measurement of ToF. However, all the measures performed using UT were made at room temperature. Therefore, no temperature compensation is needed at this point.

The value of Time of Flight depends on the speed of the sound through the target material. The speed of sound in steel is approximately 5900 m/s [40]. To increase the accuracy of the measurements, the speed of sound has been calculated for the selected steel material (S355J2G3) using the 5 mm reference samples. In the process of calculating the speed of sound, the reference bare steel sample surface was divided into 8 zones, and the average thickness of each zone was measured using a digital micrometer. Considering it as the thickness reference, the average speed of sound for each zone was calculated using Equation (1), based on 5 ToF measures performed by using the UT on each zone. Accordingly, the final calculated average speed of sound at room temperature for bare steel reference samples is 5950 m/s. This value has been set as the speed of sound in steel for the rest of the test samples measured by the ultrasound solution.

$$\text{Speed of sound [m/s]} = \frac{2 \times 10^3 \times \text{Reference thickness of sample [mm]}}{\text{Time of Flight (ToF) [μs]}}. \tag{1}$$

The UT test approach has been validated by comparing the results from UT and micrometer measures from reference samples of both 5 mm and 40 mm thickness. The results showed a good agreement between the thickness estimations done by the UT and by conventional methods, such as using a caliper or a micrometer.

6. Results and Discussion

6.1. Test Samples Measurements

Though the test method of UT system has been validated using reference samples, the long-term system performance has been analyzed by measuring the test samples, as explained in Section 5.1. The following subsections discuss the results obtained by the UT system for non-exposed and exposed coated test samples.

6.1.1. Measurements Precision of Non-Exposed Test Samples

The time response of an ultrasound sensor collected from a test object shows the characteristic behavior of echoes bouncing back and forth the wall's boundaries.

As discussed in Section 4.3, after ADC reading follows a filtering process that includes a bandpass filter that is used to remove the unwanted frequency components from the signal. Figure 8 presents the bandpass filtering results obtained for non-exposed bare steel and coated samples. Figure 8 (top) shows the ultrasound response from the bandpass filter for a bare steel sample that has a well separated and small echo duration, which facilitates the separation of the consecutive echoes for the ToF calculation. In addition, compared with Figure 8 (bottom), it can be seen that the echo duration is longer for coated samples compared to the bare steel samples. This fact makes the echo separation more difficult in the case of coated samples. The estimated echo durations for bare steel and coated samples are around 0.7–0.9 µs and 1.3–1.5 µs, respectively.

The precision of the ultrasound measures gives important information about the system's performance. Having a good precision allows us to take more frequent corrosion measurements, which can potentially be used to estimate the corrosion rate at any instant by using a derivative FIR filter. The latency of the corrosion rate measurements using this method will be in the range of a few hours which, in practical terms, can be considered a real-time signal for an O&M operator. Figure 9 shows the data plotted between the obtained ToF precision (σ_{ToF}) and the signal level (SL) for both non-exposed bare steel and coated samples. Here, the SL refers to the Root Mean Square (RMS) value of the first detected echo measured after the amplifier. This value is calculated digitally after the bandpass filter mentioned above. Because of this, SL is an indication of the SNR of the digitized signal after bandpass filtering.

The precision of ToF values we are plotting here is the standard deviation of 25 measures launched at one location, as discussed in Section 5.2. The data points of the graph for non-exposed coated samples in Figure 9 are related to both on scribe and no-scribe areas. It has been clearly observed, in the Figure 9, that the SL of ultrasound measures has considerable variations, depending on the position of the piezo, and, when the signal level decreases, the precision obtained for those respective measures gets worse.

The data points of signal level versus ToF precision were best fitted with an exponential decay function ($\sigma_{ToF} = A \cdot e^{-SL/B}$). Each fitted curve is represented over the signal level range of 35 mV to 250 mV. Considering the decay constant coefficient associated with each fitted curve (parameter B in mV indicates the increase in SL necessary for a 63% reduction in σ_{ToF}), the ToF precision versus SL behavior is almost the same for the bare steel and coated samples. Both bare and coated samples have good ToF precision results, but the obtained SLs are slightly better in coated samples. The reason behind this behavior may be that the quality of contact made between the transducer and the coated samples was better than in the case of the bare steel samples. Furthermore, the results do not show any significant reduction of the signal level due to the coating layer.

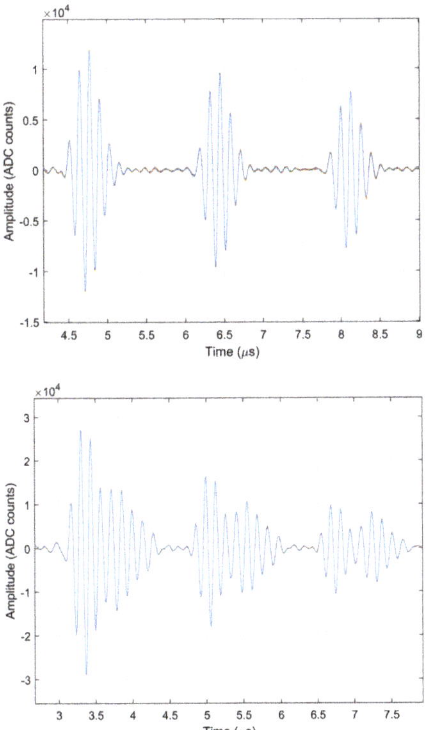

Figure 8. Bandpass filter results for bare steel sample (**top**) and coated sample (**bottom**).

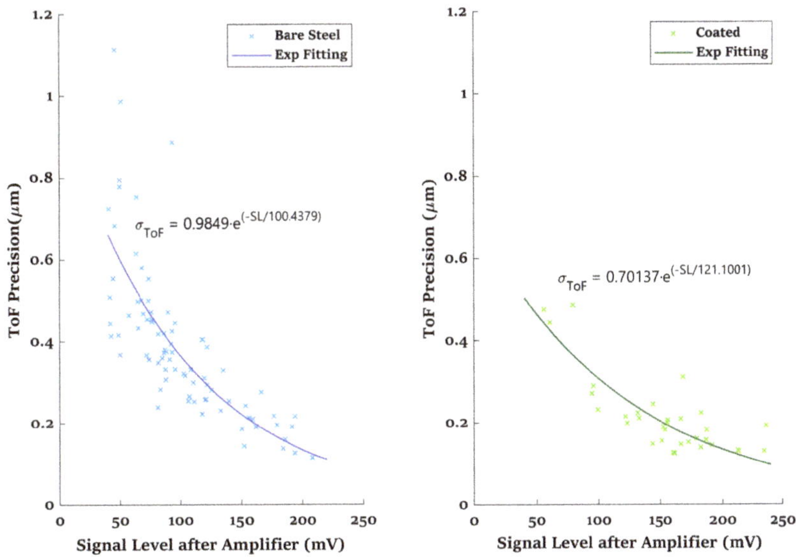

Figure 9. Non-exposed samples: signal level versus precision.

6.1.2. Measurements Precision of Exposed Test Coated Samples

Different sets of numbered coated samples have been exposed to a controlled corrosive environment under the cyclic aging test after 1 month, 3 months (see Figure 6 (bottom)), and 6 months exposure. After the completion of the exposure time, each sample has been tested using the UT. The ultrasound thickness measures have been performed in both on scribe and no-scribe locations of the coated samples, as discussed in Section 5.1.

The precision of measures obtained for the test coated samples exposed under cyclic aging test for different time durations with respect to the signal level has been plotted in Figure 10. Here, also, the SL refers to the Root Mean Square (RMS) value of the first detected echo measured after the amplifier, and precision refers to the standard deviation of 25 measures performed in the same location (σ_{ToF}). Comparing the precision of ToF measures between on scribe and no-scribe areas, it can be seen that, in order to obtain similar precisions in Scribe and Non-Scribe areas, we must increase the SL by $\sim$75 mV (see the B parameter of fitting curve) when measuring Scribe areas.

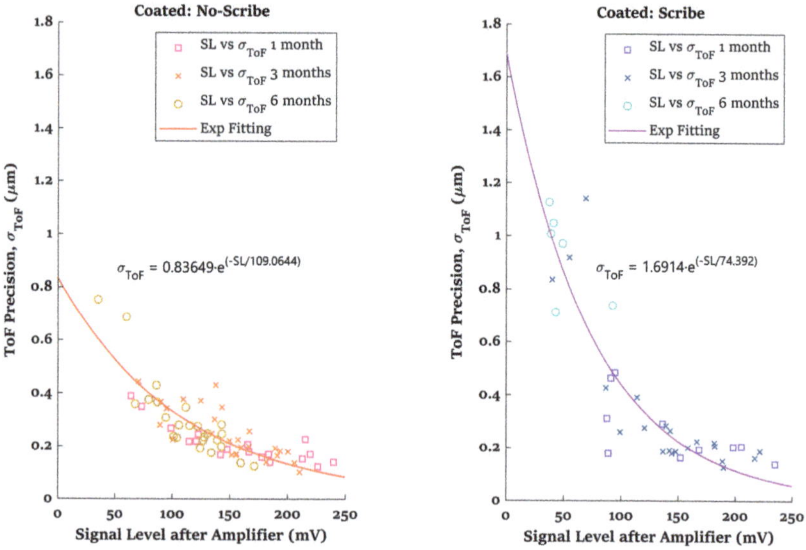

Figure 10. Exposed coated samples: signal level versus precision.

Furthermore, in the case of the on scribe area, it can be observed that the most lower signal level data points belong to the samples with higher exposition time. This may be caused due to the non-uniformity of the corrosion layer developed on the scribe area. In general, we are obtaining a good precision for both bare and coated samples. The ToF precision computed here is the standard deviation of ToF values of 25 measures at the same sensor position (as discussed in Section 5.2). If we place the sensor on different locations of the same sample and calculate the standard deviation of ToF measures, the obtained precision is lower than the given values here. This is because the thickness at different locations of the same sample varies, even for the non-exposed bare steel samples. These small thickness variations (at μm level) are due to the common existing imperfections during the sample production. Therefore, in order to have a good precision of ToF measures, the positioning of the sensor on the same location is very important. As discussed in Section 3, from the two modes of operation (fixed and mobile), the positioning accuracy can be a constraint for the mobile solution, which we have to address in future work.

6.1.3. Thickness Loss Measurements on Exposed Test Coated Samples

Each exposed coated sample has been measured using UT, as mentioned in Section 5.1. Figure 11 shows the thickness measures on scribe and no-scribe locations of coated samples for different exposed time durations. Considering the fact that no significant difference has been observed between the no-scribe bottom and no-scribe top layer thickness measures, only an average of no-scribe top and bottom layer measures has been used as the no-scribe measures.

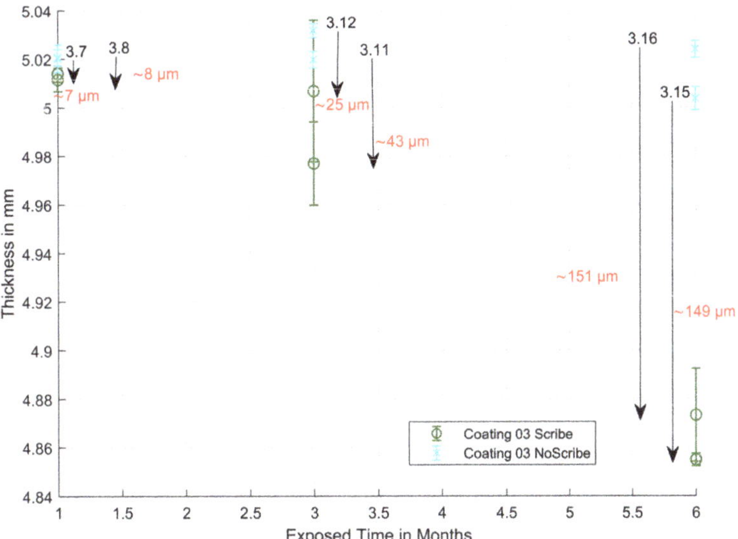

Figure 11. Coated samples: thickness versus exposed time in months.

The detected thickness differences between the scribe and no-scribe locations of the same sample are indicated by the arrows. With the scribe made in the coated samples, the corrosion process has been accelerated and initiated along with the scribe. Apparently, the results show that the thickness losses are higher on the scribe area than on the no-scribe area when the exposure time is sufficient to produce a significant thickness loss due to corrosion, as is the case of 6 months exposure.

7. Conclusions

In this paper, we present a corrosion monitoring system based on ultrasound technology. The size, weight, and other performance parameters, such as precision in thickness loss measurements, power consumption, and wireless connectivity, make the proposed solution very suitable for unattended deployment in offshore wind turbines or installed in mobile platforms, such as drones operating inside the tower, to cover easily large structures. This mobile solution would be able to detect earlier critical failures due to corrosion and, consequently, plan better maintenance actions.

The system design was developed using an Ultrasound Testbed (UT) that allows us to assess the performance of the detection and measuring algorithms in MATLAB and C/C++. The UT can launch ultrasound experiments with different parameters (frequency, waveform duration, gain, etc.) in order to find the optimal values. As a result, we created a large database of ultrasound raw signals that were later used to validate the algorithms implemented in the final miniaturized system. After validation, the UT system has been successfully migrated into a miniaturized system with low power, low cost, and size of $110 \times 60 \times 15$ mm.

All the measurements presented in the paper were done using the UT and analyzed under a MATLAB development environment for test samples of $75 \times 150 \times 5$ mm. However, the system was validated for reference samples of $75 \times 150 \times 40$ mm, as well, demonstrating that the solution is flexible enough to work with different thicknesses closer to a realistic scenario. Further large-scale validation of the proposed solution mainly for sensor placement and alignment in the case of the mobile solution will be done in the near future. Related to that, and taking into account the harsh conditions of these offshore platforms, we pay special attention to the batteries and sensor probe with the aim of reducing the number of replacements and to facilitate those replacements when needed.

Considering the results obtained in this research work, the signal level (SL) and ToF precision measurements show a similar exponential decaying relation for both bare steel and coated samples. In general, we could observe that the signal level is slightly higher for coated samples than for bare samples and that precisions below 1 µm can readily be obtained for SL > 50 mV. The thickness loss could be estimated in the coated (NORSOK 7A system) samples as the difference in the sample thickness between the no-scribe and on scribe areas after running the cycling aging test for several months. The results show, as expected, a growing thickness loss as exposition time increases. The values of the thickness loss reported is an average of the difference between the thickness along the on scribe area and the thickness in many no-scribe areas of the same sample. Other observations are that SL in on scribe areas is normally worse than in no-scribe areas, SL deteriorates as exposition time increases, and ToF precisions are again below 1 µm for SL > 50 mV.

Therefore, the thickness estimation of coated samples exposed under the cyclic aging test shows meaningful information about thickness loss due to corrosion. The variable tolerance of the wall thickness at different locations of the same sample very well obliges sensor positioning for the mobile platform. However, this is not an issue for the fixed solution.

To sum up, it can be said that the main objectives related to achieve a miniaturized system, to get a precision of 1 µm, and to estimate the thickness loss with high precision have been accomplished, taking into account bare and coated samples. The obtained precision allows us to take several thickness measurements per day to estimate the value of the corrosion rate in real-time for practical purposes.

Author Contributions: Conceptualization, A.C. and A.I.; Formal analysis, A.I. and U.C.T.; Funding acquisition, A.C. and A.I.; Investigation, A.C., A.I. and U.C.T.; Methodology, A.I. and U.C.T.; Project administration, A.C.; Software, A.I. and U.C.T.; Supervision, A.C. and A.I.; Validation, A.I. and U.C.T.; Writing—Original draft, U.C.T.; Writing—Review & editing, A.C. and A.I. All authors have read and agreed to the published version of the manuscript.

Funding: This work was supported by the WATEREYE project, which has received funding from the European Union's Horizon 2020 research and innovation program under grant agreement No. 851207.

Institutional Review Board Statement: Not applicable.

Informed Consent Statement: Not applicable.

Data Availability Statement: Not applicable.

Acknowledgments: This work has been possible thanks to the cooperation of CEIT with all the WATEREYE partners, especially in this case with SINTEF Industry, who is responsible for producing the samples, measuring the samples through conventional methods, and corroding the samples in lab.

Conflicts of Interest: The authors declare no conflict of interest.

Abbreviations

The following abbreviations are used in this manuscript:

AFE	Analog Front End
CAPEX	Capital Expenditure
GWEC	Global Wind Energy Council
NDT	Non-Destructive Testing
O&M	Operation and Maintenance
OWT	Offshore Wind Turbines
PCB	Printed Circuit Board
ROI	Return on Investment
SHMS	Structural Health Monitoring Systems
SNR	Signal-to-Noise Ratio
WSN	Wireless Sensor Network
SL	Signal Level
ToF	Time-of-Flight
UAV	Unmanned Aerial Vehicle
UT	Ultrasound Testbed
WT	Wind Turbine

References

1. What Would the Return on Investment Be from a Wind Turbine? 2019. Available online: https://www.renewablesfirst.co.uk/windpower/windpower-learning-centre/what-would-the-return-on-investment-be-from-a-wind-turbine/ (accessed on 12 March 2021).
2. GWEC Global Wind Report 2021. Available online: https://gwec.net/wp-content/uploads/2021/03/GWEC-Global-Wind-Report-2021.pdf (accessed on 12 March 2021).
3. Hevia-Koch, P.; Jacobsen, H.K. Comparing offshore and onshore wind development considering acceptance costs. *Energy Policy* **2019**, *125*, 9–19. [CrossRef]
4. Soh, J.Y.; Lee, M.W.; Kim, S.K.; Kim, D.H. Corrosion Monitoring for Offshore Wind Farm's Substructures by using Electrochemical Noise Sensors. *KEPCO J. Electr. Power Energy* **2016**, *2*, 615–618. [CrossRef]
5. Walsh, C. (Ed.) *Offshore Wind in Europe*; WindEurope. 2019. Available online: https://windeurope.org/wp-content/uploads/files/about-wind/statistics/WindEurope-Annual-Offshore-Statistics-2019.pdf (accessed on 12 March 2021).
6. Keene, M. Comparing Offshore Wind Turbine Foundations. 2021. Available online: https://www.windpowerengineering.com/comparing-offshore-wind-turbine-foundations/ (accessed on 12 March 2021).
7. Castellà, X.T. Operations and Maintenance Costs for Offshore Wind Farm. 2020. Available online: https://upcommons.upc.edu/bitstream/handle/2117/329731/master-thesis-xavier-turc-castell-.pdf (accessed on 12 March 2021).
8. Silva, J.G.; Doekemeijer, B.; Ferrari, R.; van Wingerden, J.W. Active Power Control of Waked Wind Farms: Compensation of Turbine Saturation and Thrust Force Balance. *arXiv* **2021**, arXiv:2104.03894.
9. Hofmann, M.; Sperstad, I.B. NOWIcob—A tool for reducing the maintenance costs of offshore wind farms. *Energy Procedia* **2013**, *35*, 177–186. [CrossRef]
10. Hofmann, M. A review of decision support models for offshore wind farms with an emphasis on operation and maintenance strategies. *Wind Eng.* **2011**, *35*, 1–15. [CrossRef]
11. Brennan, F. Risk based maintenance for offshore wind structures. *Procedia CIRP* **2013**, *11*, 296–300. [CrossRef]
12. Sørensen, J.D. Framework for risk-based planning of operation and maintenance for offshore wind turbines. *Wind Energy* **2009**, *12*, 493–506. [CrossRef]
13. Herraiz, A.H.; Marugán, A.P.; Ramirez, I.S.; Papaelias, M.; Márquez, F.P.G. A novel walking robot based system for non-destructive testing in wind turbines. In Proceedings of the 1st International Conference on Welding & NDT, Athens, Greece, 22–23 October 2018.
14. Pedro García, F.; Arcos Jimenez, A.; Quiterio Gómez Muñoz, C. Non-Destructive testing of wind turbines using ultrasonic waves. In *Non-Destructive Testing and Condition Monitoring Techniques for Renewable Energy Industrial Assets*; Butterworth-Heinemann: Oxford, UK, 2020; pp. 91–101. [CrossRef]
15. Martinez-Luengo, M.; Kolios, A.; Wang, L. Structural health monitoring of offshore wind turbines: A review through the Statistical Pattern Recognition Paradigm. *Renew. Sustain. Energy Rev.* **2016**, *64*, 91–105. [CrossRef]
16. Nawfi, N.M.; Sarusan, N.; Piyathilake, S.; Sivahar, V.; Munasinghe, R. Remote Estimation of Degree of Corrosion Using Ultrasonic Pulse Echo Methods. In Proceedings of the 2018 Moratuwa Engineering Research Conference (MERCon), Moratuwa, Sri Lanka, 30 May–1 June 2018.
17. Khodabux, W.; Causon, P.; Brennan, F. Profiling corrosion rates for offshore wind turbines with depth in the North Sea. *Energies* **2020**, *13*, 2518. [CrossRef]
18. Thomas, D.J. A Life at Sea and the Corrosion Fatigue Lives of Offshore Structures *J. Fail. Anal. Prev.* **2021**, *21*, 707–710. [CrossRef]
19. Ahuir-Torres, J.I.; Bausch, N.; Farrar, A.; Webb, S.; Simandjuntak, S.; Nash, A.; Thomas, B.; Muna, J.; Jonsson, C.; Mathew, D. Benchmarking parameters for remote electrochemical corrosion detection and monitoring of offshore wind turbine structures. *Wind Energy* **2019**, *22*, 857–876. [CrossRef]

20. National Association of Corrosion Engineers. *Techniques for Monitoring Corrosion and Related Parameters in Field Applications*; Technical Report 24203; NACE International: Houston, TX, USA, 1999.
21. Bardal, E.; Drugli, J. Corrosion detection and diagnosis. *Mater. Sci. Eng.* **2004**, *3*, 1–8.
22. Forsyth, D.S. Non-destructive testing for corrosion. In *Corrosion Fatigue and Environmentally Assisted Cracking in Aging Military Vehicles (RTO-AG-AVT-140)*; NATO, Distribution of Unclassified RTO Puplications: Neuilly-Sur-Seine, France, 2011.
23. Berger, H.; Brackett, R.; Mittleman, J. Underwater Inspection Of Naval Structures. In Proceedings of the Proceedings OCEANS'83, San Francisco, CA, USA, 29 August–1 September 1983; pp. 555–559.
24. Stegemann, D.; Raj, B.; Bhaduri, A. NDT for Analysis of Microstructures and Mechanical Properties of Metallic Materials. In *Encyclopedia of Materials: Science and Technology*; Elsevier: Amsterdam, The Netherlands, 2016. [CrossRef]
25. Du, C.; Owusu Twumasi, J.; Tang, Q.; Guo, X.; Zhou, J.; Yu, T.; Wang, X. All-optical photoacoustic sensors for steel rebar corrosion monitoring. *Sensors* **2018**, *18*, 1353. [CrossRef] [PubMed]
26. Manjula, K.; Vijayarekh, K.; Venkatrama, B. Weld flaw detection using various ultrasonic techniques: a review. *J. Appl. Sci.* **2014**, *14*, 1529–1535. [CrossRef]
27. Roberge, P.R., Corrosion Monitoring or Corrosion Inspection? In *Corrosion Inspection and Monitoring*; Wiley Series in Corrosion; Wiley Online Library: Hoboken, NJ, USA, 2007; Chapter 3.6.3.
28. Canales, R.V.; Furukawa, C.M. Signal processing for corrosion assessment in pipelines with ultrasound PIG using matched filter. In Proceedings of the 2010 9th IEEE/IAS International Conference on Industry Applications-INDUSCON, Sao Paulo, Brazil, 8–10 November 2010.
29. Scorpion2 Ultrasonic Tank Inspection. 2021. Available online: https://eddyfi.com/en/product/scorpion-2 (accessed on 25 March 2021).
30. Zhang, D.; Watson, R.; Dobie, G.; MacLeod, C.; Pierce, G. Autonomous Ultrasonic Inspection Using Unmanned Aerial Vehicle. In Proceedings of the 2018 IEEE International Ultrasonics Symposium (IUS), Kobe, Japan, 22–25 October 2018.
31. Mihaljevic, M.; Cajner, H.; Markucic, D.; Kozic, K. Coating Influence on Ultrasonic Thickness Measurement Result. In *Proceedings of the 29th DAAAM International Symposium*; Katalinic, E.B., Ed.; DAAAM International: Vienna, Austria, 2018; pp. 341–346. [CrossRef]
32. Ramazan Demirli, J.S. Model-Based Estimation of Ultrasonic Echoes Part I: Analysis and Algorithms. *IEEE Trans. Ultrason. Ferroelectr. Freq. Control* **2001**, *48*, 787–802. [CrossRef] [PubMed]
33. Ramazan Demirli, J.S. Model-Based Estimation of Ultrasonic Echoes Part II: Nondestructive Evaluation Applications. *IEEE Trans. Ultrason. Ferroelectr. Freq. Control* **2001**, *48*, 803–811. [CrossRef] [PubMed]
34. Rommetveit, T.; Johansen, T.F.; Johnsen, R. A combined approach for high-resolution corrosion monitoring and temperature compensation using ultrasound. *IEEE Trans. Instrum. Meas.* **2010**, *59*, 2843–2853. [CrossRef]
35. Honarvar, F.; Sheikhzadeh, H.; Moles, M.; Sinclair, A.N. Improving the time-resolution and signal-to-noise ratio of ultrasonic NDE signals. *Ultrasonics* **2004**, *41*, 755–763. [CrossRef]
36. Zhang, D.; Watson, R.; Cao, J.; Zhao, T.; Dobie, G.; MacLeod, C.; Pierce, G. Dry-Coupled Airborne Ultrasonic Inspection Using Codded Excitation. In Proceedings of the 2020 IEEE International Ultrasonics Symposium (IUS), Las Vegas, NV, USA, 7–11 September 2020. [CrossRef]
37. Gajdacsi, A.; Cegla, F. The effect of corrosion induced surface morphology changes on ultrasonically monitored corrosion rates. *Smart Mater. Struct.* **2016**, *25*, 115010. [CrossRef]
38. Centre for Nondestructive Evaluation ISU. Wavelength and Defect Detection. Available online: https://www.nde-ed.org/Physics/Waves/defectdetect.xhtml (accessed on 25 November 2021).
39. Olympus. Contact Transducers-Olympus-IMS.com. Available online: https://www.olympus-ims.com/en/ultrasonic-transducers/contact-transducers/#!cms[focus]=cmsContent10861 (accessed on 25 November 2021).
40. Magda, P.; Stepinski, T. Corrosion assessment using ultrasound. *Diagnostyka* **2015**, *16*, 15–17.

 *applied sciences*

Article

Diagnosis of Broken Bars in Wind Turbine Squirrel Cage Induction Generator: Approach Based on Current Signal and Generative Adversarial Networks

Yuri Merizalde Zamora [1], Luis Hernández-Callejo [2,*], Oscar Duque-Pérez [3] and Víctor Alonso-Gómez [4]

1   PhD School of University of Valladolid (UVA), Faculty of Chemical Engineering, University of Guayaquil, Clemente Ballén 2709 and Ismael Perez Pazmiño, Guayaquil 593, Ecuador; yuri.merizaldez@ug.edu.ec
2   ADIRE-ITAP-Department of Agricultural Engineering and Forestry, University of Valladolid (UVA), Campus Universitario Duques de Soria, 42004 Soria, Spain
3   ADIRE-ITAP-Department of Electrical Engineering, University of Valladolid (UVA), Escuela de Ingenierías Industriales, Paseo del Cauce 59, 47011 Valladolid, Spain; oscar.duque@eii.uva.es
4   Department of Phisical, University of Valladolid (UVA), Campus Universitario Duques de Soria, 42004 Soria, Spain; victor.alonso.gomez@uva.es
*   Correspondence: luis.hernandez.callejo@uva.es; Tel.: +34-975-129-213

Citation: Zamora, Y.M.; Hernández-Callejo, L.; Duque-Pérez, O.; Alonso-Gómez, V. Diagnosis of Broken Bars in Wind Turbine Squirrel Cage Induction Generator: Approach Based on Current Signal and Generative Adversarial Networks. *Appl. Sci.* **2021**, *11*, 6942. https://doi.org/10.3390/app11156942

Academic Editor: Daniel Villanueva Torres

Received: 28 June 2021
Accepted: 26 July 2021
Published: 28 July 2021

Publisher's Note: MDPI stays neutral with regard to jurisdictional claims in published maps and institutional affiliations.

Abstract: To ensure the profitability of the wind industry, one of the most important objectives is to minimize maintenance costs. For this reason, the components of wind turbines are continuously monitored to detect any type of failure by analyzing the signals measured by the sensors included in the condition monitoring system. Most of the proposals for the detection and diagnosis of faults based on signal processing and artificial intelligence models use a fault-free signal and a signal acquired on a system in which a fault has been provoked; however, when the failures are incipient, the frequency components associated with the failures are very close to the fundamental component and there are incomplete data, the detection and diagnosis of failures is difficult. Therefore, the purpose of this research is to detect and diagnose failures of the electric generator of wind turbines in operation, using the current signal and applying generative adversarial networks to obtain synthetic data that allow for counteracting the problem of an unbalanced dataset. The proposal is useful for the detection of broken bars in squirrel cage induction generators, which, according to the control system, were in a healthy state.

Keywords: wind turbine; faults diagnostic; artificial intelligence; unbalanced datasets; synthetic data

1. Introduction. Fault Diagnosis in Wind Turbines by Means of the Current Signal

Due to the remote locations where wind farms are installed and the considerable height of the wind turbines (WTs), condition-based maintenance predominates in the wind industry [1–3]. The detection and diagnosis of failures of WT components is usually performed using signal processing techniques applied to the vibration signal [4]. However, according to [5], the vibration signal has some disadvantages that can be overcome using the current signal, since with this signal it is possible to detect not only electrical faults, but also mechanical faults.

When a current flows through the stator of the induction machine, a flux is created in the air gap that depends on the design parameters of the motor. This flux induces currents in the rotor bars, which will create their own field. According to [6], when the rotor is in good condition, there is only one field that rotates in the same direction as the rotor, at the slip frequency. However, when there is an asymmetry in the rotor, a current and a field appear in the opposite direction of the normal current. The electromotive force due to the fault current induces a current in the stator, whose frequency is given by $f_{bb} = (1 - 2ks)f$ Hz [7].

The broken rotor bar causes a torque pulsation at twice the slip frequency (*2sf*), in addition to a speed oscillation that is also a function of the inertia of the rotor. The speed oscillation can reduce the magnitude of the sideband $(1 - 2s)f$, but instead an upper sideband current component is induced at $(1 + 2s)f$ in the stator winding, which is reinforced by the magnetic core nonlinearity, that is, as a summary it can be said that the failure due to broken bars or other oscillations induces additional frequencies in the stator current given by Equation (1).

$$f_{bb} = [(1 \pm 2ks)]f \tag{1}$$

Equation (1) is what is known as a double slip frequency sideband due to broken rotor bars, which modulates the amplitude and phase of the line current. According to [8], as the oscillations of the load torque are also manifested in the spectrum around the fundamental frequency, the components of Equation (1) would not be useful to diagnose broken bars. For this, it is necessary to use the higher order components given by Equation (2).

$$f_{bb} = \left[\left(\frac{k}{p} \right) (1 - s) \pm s \right] f \tag{2}$$

Having analyzed the way in which the components associated with failures are produced, such as broken bars, now the problem is to find the appropriate method for their detection and diagnosis. For the detection and diagnosis of faults in rotating electrical machines, the prevalent approach is based on working in the frequency domain, identifying the frequency components of failure in the vibration spectrum. Although many proposals in this regard can be found in the specialized literature, there are still challenges to overcome, especially in the detection of incipient faults and in operation at low load. Besides, if the slip varies due to change in speed, load, or grid instability, there is no constant spectrum, and signal processing techniques must be applied, such as: time-frequency transforms, filtering techniques to eliminate the components that are not of interest, and algorithms to obtain the spectrum corresponding to a specific speed [9,10].

Thus, in relation specifically to the use of the current signal for the fault detection of the electric generator of WTs and prime mover coupled to its shaft, some examples can be mentioned. In [11], the faulty gears of the gearbox are detected using power spectrum density (PSD), [12] proposes a model that allows for detecting the broken teeth of gears applying fast Fourier transform (FFT), [13] also detects the broken teeth of gears, using FFT prior to the application of SVM, [14] uses PSD to extract the characteristics of the signal that, after passing through a model called particle filtering, feeds an adaptive neuro Fuzzy inference System (ANFIS) capable of detecting broken teeth in gears, [15] proposes the use of a multiphase imbalance separation technique (EMIST) together with FFT to detect faults in the inner and outer race bearings of a gearbox, [16] proposes the detection of roughness in bearings, decomposing the current signal by means of wavelets. In [17], faults of the stator and rotor windings are detected by analyzing the spectrum obtained with the FFT. Through a comparative study, [18] determined that, to detect bearing failures, broken bars, and eccentricity, the wavelet-based methodology is superior to the Welch periodagram, PSD and Short Time Fourier Transform (STFT), while according to [19], Hilbert transform (HT) is superior to Park transform and Teager energy operator when it comes to detecting generator bearing failures. As a summary, once the signal processing techniques have been applied, the identification of the components associated with the failures is done according to the equations, such as those included in Table 1.

Table 1. Frequency components associated with faults [5].

Fault	Mathematical Model
Inter-turns short circuit	$f_{st} = \left[\frac{n}{p}(1 - s) \pm k \right] f$
Eccentricity	$f_{ecc} = \left[1 \pm k \frac{(1-s)}{p} \right] f$

From a historical point of view, methodologies based on signal processing techniques were the first to be used regarding the detection and diagnosis of faults in rotating electrical machines. After the appearance of the AI models, it would not take long before they were applied to the diagnosis and detection of faults in rotating induction machines. At present, signal processing techniques constitute a previous step for the application of AI models.

AI has been the subject of a huge number of publications and research papers, which have allowed for the construction of sophisticated equipment to control, monitor and detect the presence of faults, especially using ANN, Fuzzy logic and Neuro-Fuzzy systems [20] and [21]. As an example of the basic methodology used, we can mention the publication made by [22], approximately twenty-two years ago. According to this proposal, the frequency components that are relevant for the diagnosis depend on the slip or speed, so it is necessary to make two measurements, a sampling at high frequencies to determine the slip by means of the harmonic of the first slot and another sampling of greater duration at low frequencies to find the components associated with the faults. The inference machine of an expert system compares the spectrum obtained with a database containing the components associated with different types of faults and filters out the part of the spectrum that is not of interest. A simple way to carry out the diagnosis is by comparing the values obtained against a previously established threshold. The other alternative is through ANFIS with a Sugeno-type first-order inference system. The set of faults obtained are the input of the membership functions that form the adaptive nodes of the first layer of a multilayer neural network fed forward. The first layer is associated with the membership functions of linguistic variables (small, medium, and large), while the last layer provides the diagnosis expressed as: no failure, incipient failure, one broken bar, and two broken bars. To diagnose broken bars, the input variables are the components associated with these faults, whilst the negative and positive sequence components are used to detect short circuits between turns.

According to [21], as the ANFIS model only has one output, it can detect only one fault. For this reason, [21] proposes the co-active ANFIS (CANFIS) model capable of detecting several faults at the same time. The previously filtered and demodulated signal is decomposed by wavelet transform to obtain the coefficients that are affected by the faults. These coefficients are the input of the CANFIS model, which by having multiple outputs, can detect several faults at the same time. For [23,24], when fault diagnosis is done applying wavelet transform, it is not necessary to use slip. However, according to [25], as wavelets have the drawback of energy loss and edge distortion, it is preferable to use empirical mode decomposition (EMD) to obtain intrinsic mode functions (IMF). Each IMF represents a frequency band in which a type of failure can be found using Support Vector Machines (SVM), but after the optimization of the parameters by means of a genetic immune algorithm.

Although most of the proposals are based on models that have been previously trained with data containing the faults to be detected, according to the proposal of [26], this requirement would not be necessary. According to [27,28], the diagnosis can be made based only on data from systems such as Supervisory Control and Data Acquisition (SCADA). In [29], the signal is filtered to obtain the root mean square (RMS), kurtosis, skewness, standard deviation and crest factor. These parameters are the input data of a neural network trained with the Bayesian Regularization algorithm; the ANN has a hidden layer with 46 neurons and for the output layer it uses the tan-sigmoid transfer function.

According to [30], once the rotor current signal of a doubly fed induction generator (DFIG) has been sampled, the instantaneous rotation frequency of the shaft is obtained and the signal is demodulated using HT to obtain its envelope, but as the speed and shaft rotation frequency are not constant, an angular resampling algorithm is applied to obtain a constant envelope. Once this constant envelope is obtained, its PSD can be obtained to detect faults. In each phase, the amplitudes of the rotation frequencies corresponding to the input shaft, pinion and output shaft are obtained, plus RMS, kurtosis, peak value, and signal-to-noise ratio (SNR). These variables become the input of a deep learning model called a stacked autoencoder. Finally, the characteristics obtained from the learning of the ANN feed an SVM algorithm that classifies the failures. In [13], the frequency components that correspond to the faults are derived as a function of the interaction between the rotor and stator currents. When the gearbox suffers a failure, the vibration produces a torque on the shaft, Equation (3), altering its speed and frequency, which will be reflected in the current spectrum. Thus, when the FFT is applied to rotor and stator signals, characteristic frequencies of gearbox failures can be detected. In total, 16 variables of the rotor and 9 of the stator are extracted, of which 19 are in the frequency domain and 6 are in the time domain (RMS, kurtosis, and peak value of both signals). These variables feed an SVM whose output offers a diagnosis of the failures expressed in probability form, unlike the traditional SVM model that only classifies the failures.

$$T = T_0 + \sum T_i \cos(2\pi f_i t + \varphi_i) \tag{3}$$

Another way to visualize the state of a component is by means of its remaining useful life (RUL), as applied in [14] to analyze the state of the bearings of a gearbox. According to this study, whilst gears cause torsional vibrations, bearings do so radially. For this reason, in a healthy state the phase current will contain the fundamental component f and sidebands fi caused by the normal vibrations of the gearbox. However, as the faults will be reflected in the amplitude of the frequency components, if there are bearings with localized faults, new components will appear in the current signal, according to Equation (4). Once the signal has been sampled and before obtaining the PSD, the high frequency noise is eliminated by means of a forward-backward filter. The variable used to predict the state of the multiplier and its RUL is the SNR calculated according to Equation (5). The SNR values are the input of a five-layer ANFIS model, with an inference system formed by a set of 16 Fuzzy rules of the IF-THEM form based on a first order Sugeno model and optimizing the parameters using an ANN.

$$\begin{aligned} i_{sa} = i_{sT0} \cos(2\pi f t) + \sum_i \tfrac{1}{2} A_{sT_i} [\cos(2\pi(f - f_i)t - \varphi_{Ti}) + \cos(2\pi(f + f_i)t + \varphi_{Ti})] \\ + \sum_j \tfrac{1}{2} A_{sT_j} [\cos(2\pi(f - f_j)t - \varphi_{Tj}) + \cos(2\pi(f + f_j)t + \varphi_{Tj})] \end{aligned} \tag{4}$$

$$NSR = \frac{P_{noise}}{P_{signal}} = \frac{P_{total} - P_{signal}}{P_{signal}} \tag{5}$$

According to [31], although the rotor imbalance is manifested in the amplitude of the lateral harmonics of the fundamental frequency, the FFT and STFT fail to detect faults when the SNR is low, and the speed is not constant. As a single ANN would not be sufficient to cover the entire range of the electric generator speed, the authors propose dividing the interval in which the speed varies by several ranges and using one ANN for each range. The variables used correspond to wind speed, wind direction, pitch angle, turning speed and power output for one year, all of them obtained from the SCADA. To test the proposal, [31] simulates in Simulink a WT whose current signal is sampled at 5 kHz for 300 s. Applying FFT to 2 s signal segments, the spectrum formed by 250 components that become the input of the ANN is obtained. Having trained the model first with the signal in a healthy state and then with the signal containing the fault, the author states that it is possible to detect the frequency components associated with rotor eccentricity, which according to classical spectral analysis are given by Equation (6).

$$f_{ecc} = [1 \pm (2k - 1/p)]f \tag{6}$$

In [32], through the equations that relate voltage, current, flux, mutual inductance and torque between stator and rotor, a WT with DFIG is simulated in Matlab, but unlike [31], the faults studied are the one-phase fault and the inter-turns short circuit of the stator, while the model used is Fuzzy logic. The current signals from the three phases of the stator feed the Fuzzy system, which interprets them as linguistic variables (zero, small, medium, and large). From the database obtained with the measurements, the membership functions and 14 Fuzzy rules are built. In another simulation proposed in [33], the current signal from a DFIG and ANN are used to monitor islanding events of WTs that are part of a wind farm connected to the grid. Other models used include the detection of broken bars and inter-turns short circuits, using equations from Table 1 [34].

In relation to the publications on the monitoring, detection, and diagnosis of failures in induction motors, using current signal and AI models, there is less research that deals with WTs. Some proposals are shown in Table 2. The components that have received the most attention are: bearings, gearbox, blades and electric generator. It can also be observed that among the models used, the following predominate: SVM, ANN, fuzzy logic and ANFIS. However, at the time of writing this research, only the works of [15,35–37] deal with the use of the current signal based on data from WTs in operation, although none of them apply AI models. By way of summary, it can also be stated that:

- Generally, signal processing and analysis techniques are combined with AI;
- Not just one AI model is used, but a combination of them;
- The component of the WTs that has received the most attention is the gearbox;
- The types of failures most analyzed are broken gears and broken teeth;
- In most cases, the current signal is from a DFIG; and
- Stator, rotor or both currents signal can be used.

In most of the proposals, the AI model used is trained with the signal from the fault-free machine, to later use a sample of the signal that contains some type of failure caused on purpose (so the type of failure is known in advance). In these conditions, diagnosis is relatively easy, but due to the related costs, in the wind industry it would be very difficult to proceed in this way, so only the signal during the WT operation is available. However, if we only have the signal from the electric generator of TWs during its operation, the diagnosis process is complicated, since, in this case, and assuming that there are few faults or the faults are in an incipient state, the data are unbalanced. Since the efficiency of generative adversarial networks (GANs) is so high that it is difficult to distinguish between real and synthetic data, samples obtained by GANs could be used to compensate for unbalanced datasets.

Considering what has been stated previously, the main objectives of this work are to investigate the detection and diagnosis of faults of the SCIG installed in an operating WT (on which there are very few field studies), using electrical current signal and GANs, which is a methodology little explored thus far. The rest of this investigation is organized as follows: in Section Two, a brief conceptual analysis is made on the use of the current signal and GANs for fault detection and diagnosis in WTs. Section Three includes the methods and materials. In Section Four, the results obtained when applying the proposed methodology are shown and discussed. Conclusions and recommendations are included in Section Five.

Table 2. Proposals on fault detection in WTs using current signal and AI models.

Reference	Generator Type	Signal Source	Component	Failure	Models Used
[30]	Doubly fed induction generators	Rotor	Gearbox	Cracks gear Chipped gear one-tooth-missing two-tooth-missing	Hilbert transform Power spectrum density Support vector machine Artificial neural networks Stacked autoencoder
[33]	Doubly fed induction generators	Stator	Stator	Islanding	Artificial neural networks
[31]	Permanent magnet synchronous generator	Rotor	Generator	Rotor imbalance	Artificial neural networks
[32]	Doubly fed induction generators	Stator	Generator	Inter-turns short circuit Open phase	Fuzzy Logic
[36]	Permanent magnet synchronous generator	Stator	Gearbox	Remaining useful life	Particle filtering Adaptive neuro Fuzzy inference System
[13]	Doubly fed induction generators	Rotor Stator	Gearbox	chipped gear cracked gear one-tooth-missing two-tooth-missing	Fast Fourier transform Support vector machine
[38]	Squirrel cage induction generator	Stator	Rotor	Broken bars	Fast Fourier transform Hilbert transform Wavelets transform STFT

2. Fault Diagnosis by Means of GANs

From a computational point of view, there are several methods to improve the efficiency of AI models trained with an unbalanced dataset, [39]. However, according to [40], statistical, regression, clustering and reconstruction models are not efficient when it comes to unbalanced datasets with very few outliers. Non-parametric models require large amounts of data and computational resources, while proximity-based models are affected by the volume and dimensionality of the data. Therefore, to overcome the drawbacks of unbalanced datasets and the lack of information caused by the course of dimensionality, [40] proposes Artificially Generating Potential Outliers (AGPO), whose main idea is to apply generative adversarial networks (GANs) to detect outliers of unbalanced datasets.

Among the most widely used AI models are those called generative modeling, whose main objective is to learn the exact distribution of the data with which they are trained, so that new data similar to the original can be generated, simulated or predicted. Although the main application of these models has been the treatment of images, they have also been used in the field of video games, cinema, graphic design, audio analysis and body language. However, the main difficulty has been finding the function that allows modeling the input data, and for this purpose, these models use the Monte Carlo method based on Markov chains, which is computationally very expensive. To overcome this drawback, several proposals, such as the Variational Autoencoders (VAEs) model have incorporated the use of ANNs capable of obtaining a powerful approximation function, through backpropagation. VAEs obtain the probability function using Bayesian statistics and two generative networks. The first ANN generates a probability function and random values on the studied phenomenon. The second ANN performs the discriminator function and provides a model only for the variables labeled conditional on the observed variables [41].

Until a few years ago, VAEs were among the most powerful and popular models used for the unsupervised autonomous learning of complicated distributions; however, they have the drawback that to determine probability distributions, they use Bayesian networks and Markov chains. However, in 2014, Ian Goodfellow [42] proposed the GANs, which is a model composed of two multilayer perceptron (MLP) ANNs. The first ANN plays the role of generator (NNG), since, after obtaining the distribution function of the dataset, it is capable of generating synthetic data very similar to the originals. The second ANN works as a discriminator (NND), because it determines whether a sample is real or is generated by the NNG. The two ANNs compete with each other until they find the Nash equilibrium for the non-cooperative game between two players trying to minimize their cost function. When the optimal point has been reached, the synthetic samples are so similar to the real ones that the NNG is not able to determine if a sample is real or fake. All this is achieved without using Markov chains and inference systems, and both ANNs are trained simultaneously with the same dataset, unlike models such as VAE, in which the NNG and NND are trained separately, [42].

According to [43], the minimization technique based on the gradient to lower the cost of each player simultaneously, which is used in [42], fails in convergence, so it is preferable to train the generator to match the value that it should have for an intermediate layer of the discriminator. This strategy, called GAN coincident characteristic, is excellent as a semi-supervised learning classifier, since, according to [40], its strength is based on the fact that: *"Instead of directly maximizing the output of the discriminator, the new objective requires the generator to generate data that matches the statistics of the real data, where we use the discriminator only to specify the statistics that we think are worth matching."*

Currently, one of the most recent alternatives to solve the problem of classifying outliers from an unbalanced dataset is to generate synthetic data using semi-supervised and unsupervised models based on GANs [44]. The applications of GANs are no longer limited to image processing, but they are also applied to tabular data. In addition, to improve efficiency, proposals can be found that combine GANs with other models [45], while other investigations even modify the structure of the original GANs proposal. In [40], the Single-Objective Generative Adversarial Active Learning (SO-GAAL) algorithm is proposed to detect outliers from an unbalanced tabular dataset. The SO-GAAL model, which is based on the Generative Adversarial Active Learning (GAAL) proposed by [46], is basically a GAN that performs the classification when the NND separates the real data from those potential synthetic outliers generated by NNG. However, as the training progresses and the min-max game reaches Nash equilibrium, the information about the potential outliers is too close to the real data and the NND fails to distinguish between real data and outliers, causing the accuracy of SO -GAAL to drop dramatically.

The SO-GAAL model is unable to obtain a distribution function that represents the whole dataset and fails to detect the outliers because it does not stop the training when the outliers provide the necessary information, and this is due to fact that the GAN proposed by [42] has no prior information. To correct this problem, [40] proposes to modify the structure of the GANs, adding multiple NNGs with different objectives (MO-GAAL). The real data are divided into subsets of affine samples and each subset will feed an NNG, which will learn to generate outliers similar to the real data. In this way, a set of distributions is obtained that represents the whole dataset, which allows the NND to classify the outliers.

Regarding the detection and diagnosis of faults in rotating electrical machines, based on unbalanced datasets, several approaches are available, such as: [47], which proposes the Adaptive Boosting (AdaBoost) method to detect broken bars, and [48], which proposes a multiclass support vector machine (SVM) based on the one-vs-one strategy to detect broken bars in the case of speeds close to synchronism. In general, it can be said that, when it comes to fault detection and diagnosis, proposals based on an unbalanced dataset combine various statistical and AI models, but there is no standardized methodology. The application of GANs is still limited and there are not many references: [49] uses the current signal and synthetic data generated by GANs to correct the overtraining of a deep neural network used to detect the faults of an induction motor, while, to detect incipient faults in a gearbox, [50] obtains a synthetic dataset through GANs that are added to the original dataset to properly train a Stacked Denoising Autoencoder (SDAE) using the vibration signal. In general, it can be said that no references have been found on the use of the current signal together with GANs for fault detection in WTs, and especially in squirrel cage induction generators (SCIG) of WTs. Furthermore, the mentioned proposals for induction motors have been demonstrated on test benches and using small power electric motors. The GANs could also be applied to other types of signals, in such a way that the results can be compared with other proposals, such as [51].

3. Materials and Methods

For this study, measurements were made in a wind farm located in the region of Castilla y León, Spain. The wind farm has 33 WTs (NEG Micon brand) that use two winding SCIG, one of 750 kW that operates when the wind speed exceeds 6.5 m/s and another 200 kW winding for wind speeds between 3 and 6.5 m/s (see Table 3). As it is necessary to turn off the WTs to install the measurement equipment, permission was only obtained to perform the measurements at four WTs (WT-3, WT-4, WT-16 and WT-25). It must also be noted that the tests were done on the highest power winding, since the wind speed was relatively high at the time the signal was sampled.

Table 3. Technical characteristics of the SCIG.

Manufacturer	ABB
Type	2 speeds
Cooling	by water
Rated power	750/200 kW
Rated current	697/204 A
Poles	4/6
Synchronism speed	1500/1000 rpm
Speed at rated power	1510/1007 rpm
Slip at rated power	0.67/0.7%
Voltage	3×690 V
Frequency	50 Hz
Winding connection	Δ

For measurements, three Fluke i3000s FLEX-36 current clamps (one for each phase) were connected to the main panel of the WT. The other end of the current clamp was connected to a PicoScope® 4424, which must necessarily be connected to a computer where the software has previously been installed to configure the acquisition of the signal (see Figure 1). Although generally to obtain a good resolution in the frequency domain the sampling rate used fluctuates between 2 and 5 kHz, in this work a 10 kHz sampling rate was applied, since the frequencies that could be found were unknown. The total measurement time in each WT was approximately 8.5 min and to reduce the effects of spectrum variation, the sampled signal is recorded every two seconds. The software to sample the signal simultaneously records a file in *mat* format for each phase. Under these conditions and considering the four WTs, a total of 1006 signal files are obtained for each phase and 3018 files if the three phases are considered.

The signal records are processed in Matlab to obtain the power spectral density. The difference in magnitude between the fundamental frequency and its sidebands is then calculated, and depending on the magnitude of this difference, the data are labeled as 0 (healthy) or 1 (broken bars) [52]. Through Equation (2), the lateral components of the fifth, seventh, eleventh and thirteenth harmonics are also obtained.

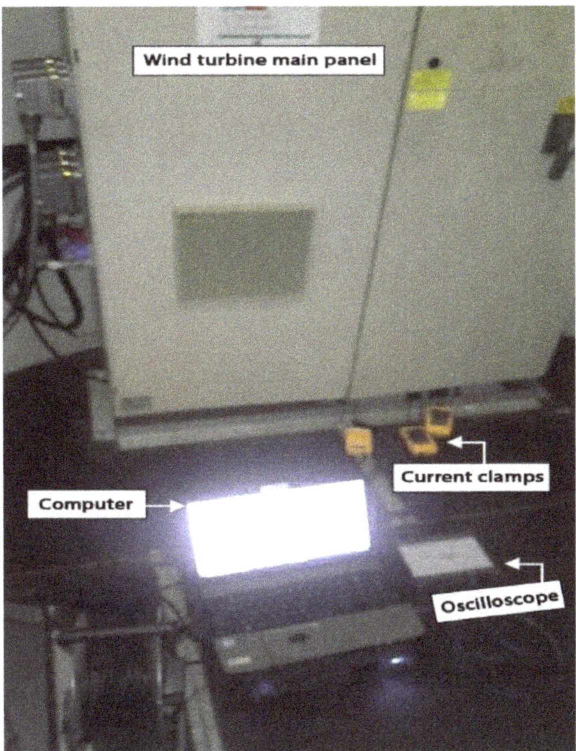

Figure 1. Generator electrical signal sampling.

According to what was seen in Section Two, the identification of the frequency components associated with the faults must be done at a fixed speed and slip. However, as the wind speed has a stochastic behavior, the spectrum of the generator will vary, and it is necessary to apply a method that allows the analysis in the frequency range at which the faults occur [53]. To solve the inconvenience described, one of the most accepted alternatives is the wavelet transform, since it allows for analyzing the signal in both the time and frequency domain [23]. Using the wavelet transform, a signal can be represented as a sum of small waves or wavelets throughout the time domain, which is known as a continuous wavelet transform (CWT). Each wavelet is a wavelet function that represents the original signal but scaled and shifted. However, as CWT involves too many calculations, another alternative is to apply the discrete wavelet transform (DWT), which can be seen as a downsampling process to decompose a signal into two sequences called cA_1 (approximation coefficients) and cD_1 (detail coefficients). cA_1 corresponds to the lower frequency range, while cD_1 consists of the high-frequency noise of the original signal. If we decompose cA_1, a second level of decomposition formed by cA_2 and cD_2 will be obtained. Decomposing cA_2, we will obtain cA_3 with cD_3, and so on until the frequency level we are trying to analyze is reached [54].

Applying Equation (1) for broken bars, a frequency component of 49.33 Hz is obtained. Then, following the methodology proposed in [55], the signal is decomposed using discrete wavelet transform (DWT) in 8 levels (see Figure 2). Since the 49.33 Hz frequency is contained within the frequency range of level $d8$ (38-78.16 Hz), signal power is obtained from level d8. In addition, the maximum and minimum values of the signal power are obtained, as well as the median, mean, mode, standard deviation, and variance. With all these data obtained in Matlab, a file type *csv* is created, which becomes the dataframe to work in Python and tensorflow. The data are scaled to the range 0 to 1 for the best behavior of the neural networks.

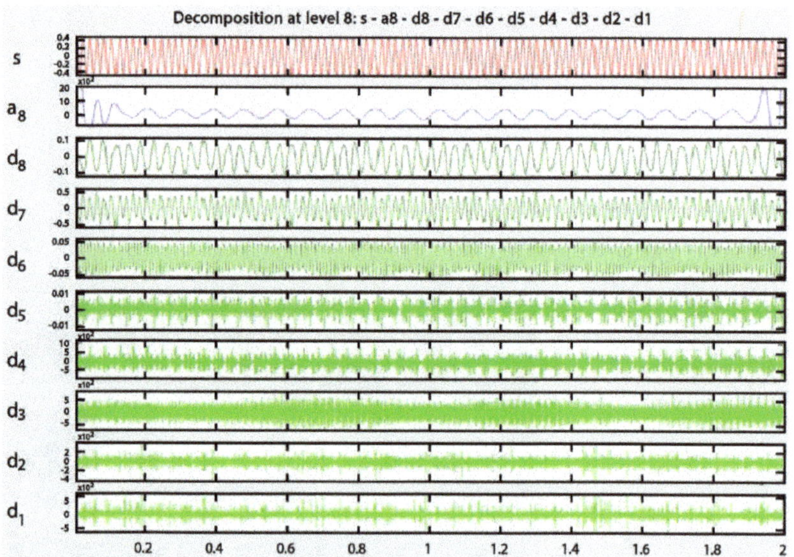

Figure 2. Signal decomposition using DWT, [55,56].

In Python, Kmeans is first applied to obtain three clusters, in such a way that the dataset is separated into healthy and faulty samples. Once the failure samples have been identified, and because they are relatively few, then the GANs are used to generate synthetic samples with faults which can compensate the unbalanced dataset. The synthetic data obtained in this way are put together with the original samples to retrain an ANN (see Figure 3).

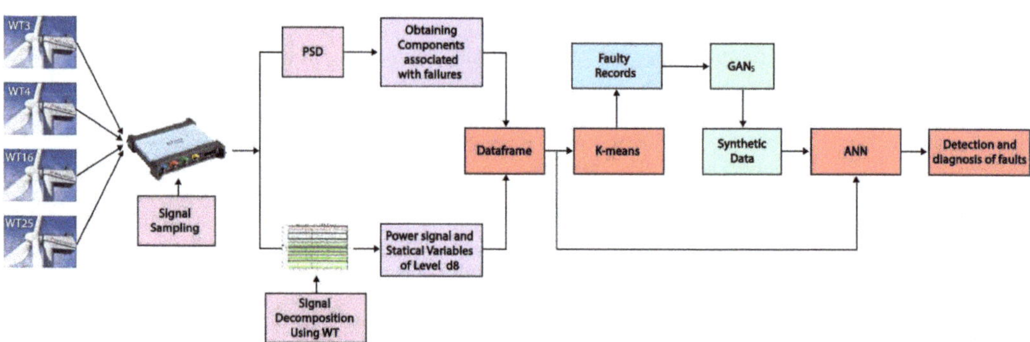

Figure 3. General scheme of the proposed model.

The GANs are MLP with backward propagation and are fed with the same variables used to apply Kmeans but are from the samples with failures. The structure of the ANNs is:

- 1 input layer with 17 neurons, with ReLU activation function;
- 2 hidden layers of 15 neurons each; and
- 1 output layer.

The hyper-parameters of the ANNs are: 100 training cycles (epochs) and the samples presented to the network each time are 10 (batch size). Accuracy is defined as the initial results metric. To guarantee the independence of the training and test data, ANNs are trained using 5-fold CrossValidation. The implementation of the ANN models was done with tensorflow and Keras. The computer used was the same one that was used to sample the signal, that is, a Toshiba laptop with an Intel Core i3-3120M processor, 2.50 GHz, 8 GB of RAM and an Intel HD Graphics 4000 graphics card.

To compare the results obtained with the proposed procedure, the proposal of [40] is applied, in which, in addition to generating synthetic data using GANs to compensate unbalanced tabular data, it is also proposed to use several generating neural networks (GNN), instead of a single GNN that is used in the original proposal of GANs. The research code from [40] is available in the GitHub repository, but since it is written in previous versions of Python and Tensorflow, it is necessary to create another virtual environment in Anaconda Powershell.

4. Results and Discussion

Applying some basic AI models, it can be observed that for all the models, the convergence is excellent, the RMSE is very small, and the accuracy value is very high (see Table 4 and Figure 4). The same does not happen with the Presiccion, Recall and F1 metrics, since the uniformity of the data and the existence of very few outliers make it difficult to identify the outliers. Applying kmeans and segmenting the dataset into three clusters (see Figure 5), the uniformity of the data can be appreciated, but also in one of the clusters the dispersion of the outliers can be observed. Outliers reduce the effectiveness of kmeans, as can be seen in the confusion matrix in Figure 6.

Table 4. Root mean square (RMSE) error for AI models trained with real data.

AI Model	RMSE
Linear Regression	0.122
Logistic Regression	0.10
Decision Tree Classification	0.012
SVM-Linear	0.122
Nearest Neighbors	0.122
Neural Network	0.138
AdaBoost Classifier	0.018

Out of 3018 samples, 2953 are free of faults and only 65, approximately 2% of samples, have any indications of a fault. The distribution of the asymmetries found according to each WT is summarized in Table 5, where 78% of the failures indicated correspond to WT-3 and WT-4. The small number of samples which have failures causes the set of data available to train the AI model to be unbalanced, which would also likely cause the metrics of the first Kmeans model to be improved.

The low number of outliers (incomplete dataset) reduces the efficiency of the AI models. So, to improve the quality of the prediction, we apply the strategy of generating synthetic data using GANs. First, we tested with the methodology proposed by [40]. When the SCIG signal is processed using the algorithm proposed to generate synthetic samples using GANs with only one GNN, the precision in the detection of outliers is shown in Figure 7a and the area under the curve (AUC) is 0.52. When the GANs model with multiple GNNs is used, which according to [40] is superior to models such as: kNN, FastABOD, Parzen and k-means, the precision in the detection of outliers improves considerably (see Figure 7b) and the value AUC increases to 0.84.

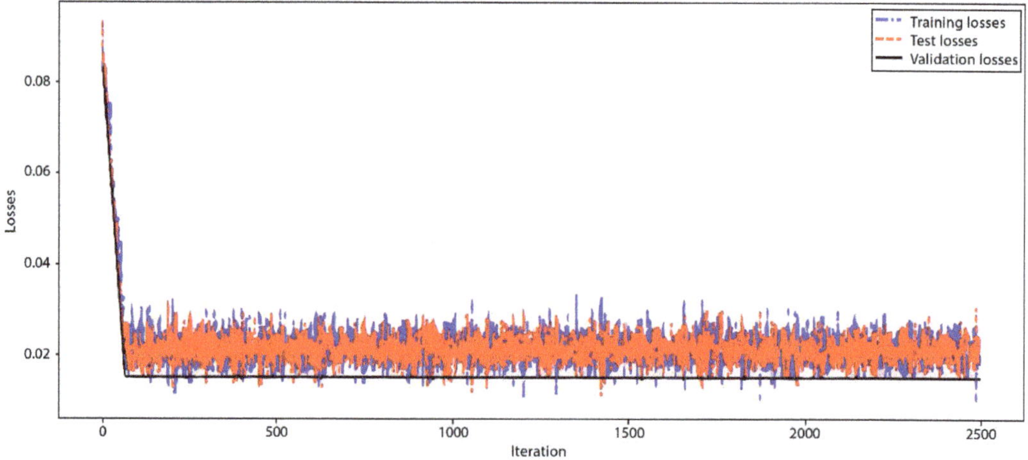

Figure 4. ANN convergence trained with real data.

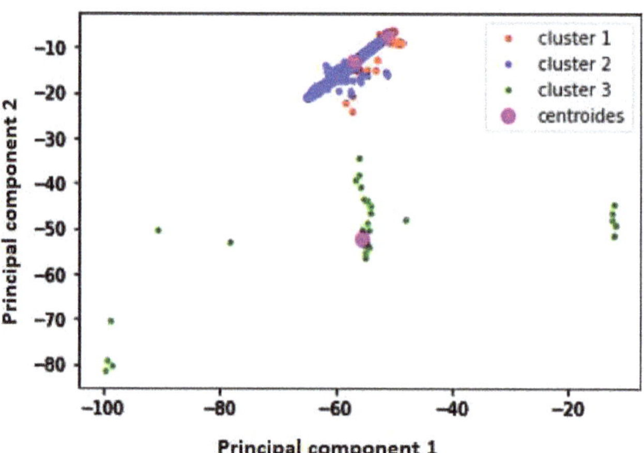

Figure 5. Kmeans (k = 3) model trained with real data.

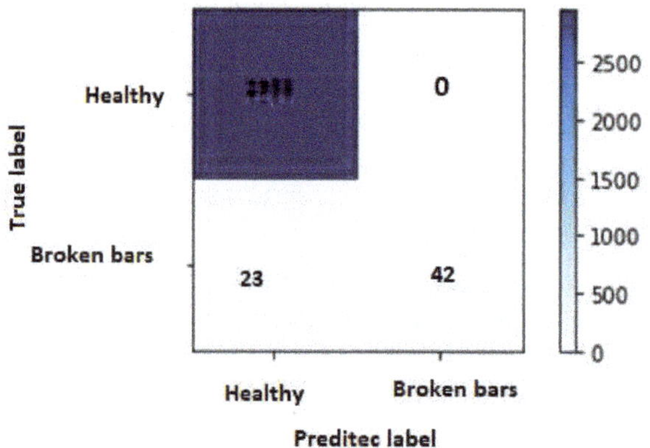

Figure 6. Confusion matrix obtained with original dataset and Kmeans.

Table 5. Number of samples with trace of failure.

Fault	WT-3	WT-4	WT-16	WT-25
Broken bar	30	21	12	2

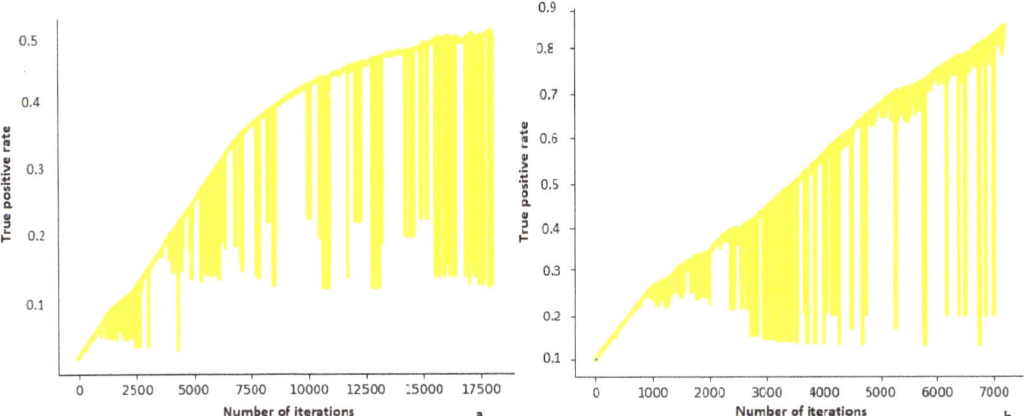

Figure 7. Precision in the detection of broken bars of the SCIG, using the proposal of [40]. (**a**) GANs with one GNN; (**b**) GANs with multiple GNNs.

As can be seen in Figure 7, the ROC curves are very unstable and the predictions fall below the non-discrimination line, which means that the model has difficulties converg and is ineffective at predicting failures. When only one GNN is used, the instability of the model is maintained despite reaching 17,000 iterations (see Figure 7a), while, using several GNNs, the model stabilizes after 7000 iterations (see Figure 7b), and the sensitivity also improves markedly.

Applying the methodology proposed in this research, as described in the fourth section, with the samples showing signs of failure (see Table 5), we train the GANs to generate 100 synthetic data, which is added to the original data, to retrain a neural network. Proceeding in this way, the Receiver Operating Characteristics (ROC) curve (see Figure 8) and the value of 0.95 for the AUC are obtained. Compared with the ROC curve obtained by applying the proposal of [40] (see Figure 7), the ROC curve in Figure 8 is much smoother and indicates a better convergence of the models used in this research. In fact, the AUC is also higher.

Another way to visualize the efficiency of the proposed model is through the confusion matrix (see Figure 9). From t 9 it can be seen that the proposed model is capable of appropriately classifying the synthetic data and in this way improves the accuracy of the prediction.

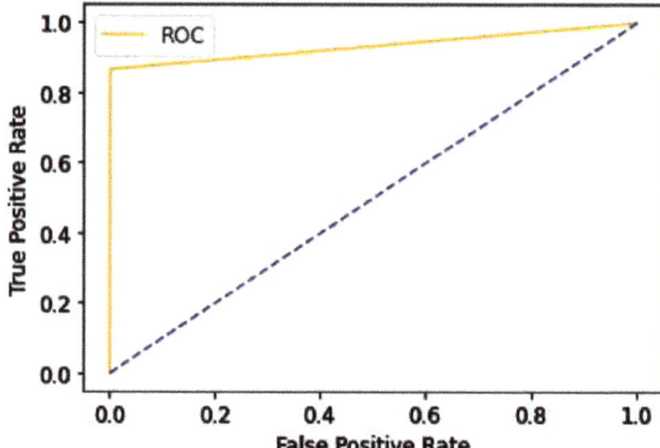

Figure 8. Receiver Operating Characteristic curve of the K-means model trained with data generated by GANs.

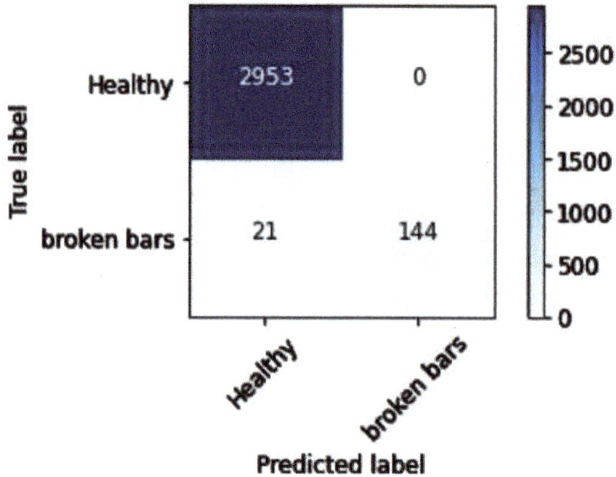

Figure 9. Confusion matrix of the ANN trained with synthetic data.

5. Conclusions

The signal processing techniques have represented an important advance regarding the detection and diagnosis of faults; however, in many cases they are not enough and must be combined with other mathematical models that generally assume some idealizations. Another alternative is artificial intelligence (AI) models, which from a conceptual point of view are characterized by their ability to adapt to uncertainty and to work with incomplete data. These AI models are used individually or together with signal processing techniques; however, when only the current signal of a WT in operation is available, but the signal has not been previously sampled in a healthy state or with some type of fault, the detection and diagnosis is complicated. It must also be noted that when failures are incipient and there are few failure records, the efficiency of AI models will be reduced.

In the usual procedures, the diagnosis algorithms are trained using data from tests with healthy equipment and with equipment in which failures have been caused. However, these situations may not be extrapolated to real situations, in addition to decreasing the performance of the classification algorithms when working with unbalanced datasets.

To improve the efficiency of AI models and wind turbine fault diagnosis procedures in cases of unbalanced data, this research proposes generating synthetic data using GANs. The methodology has shown its effectiveness for the early detection of failures due to broken bars in SCIG, in addition to allowing for improving the metrics of the AI model used.

Although this study has focused on broken bars, the proposed model could be applied to detect other faults. At the output of the proposed model, another model based on ANN or Fuzzy logic could be added to obtain a more precise diagnosis of the failure studied (half section broken bar, one broken bar, two broken bars, many broken bars). It would also be advisable to continue with the research trying to build GANs, not only with MLP, but also with other AI models, and to use GANs not to generate a synthetic dataset, but to carry out the diagnosis exclusively through GANs. In this study, tabular data have been used; however, the proposed methodology could be tested with images. In addition, other types of signals could also be used, such as: vibration, acoustic and thermal.

Author Contributions: Conceptualization, Y.M.Z. and L.H.-C. methodology, Y.M.Z.; validation, Y.M.Z., L.H.-C. and O.D.-P.; formal analysis, Y.M.Z., L.H.-C. and O.D.-P.; resources, Y.M.Z. and L.H.-C.; writing—original draft preparation, Y.M.Z.; writing—review and editing, L.H.-C. and O.D.-P.; visualization, V.A.-G.; supervision, L.H.-C.; project administration, O.D.-P. All authors have read and agreed to the published version of the manuscript.

Funding: This research was funded by University of Guayaquil and CETASA.

Institutional Review Board Statement: Not applicable.

Informed Consent Statement: Not applicable.

Data Availability Statement: Not applicable.

Acknowledgments: The authors would like to thank the University of Valladolid and University of Guayaquil for the assistance in the preparation of this research. We would also like to thank the company CETASA for allowing the acquisition of the signals and providing the necessary equipment. Thanks also to the anonymous reviewer for their assistance in the improvement of the study.

Conflicts of Interest: The authors declare no conflict of interest.

Nomenclature

i_s	stator current
i_r	rotor current
θ_r	angular displacement of the rotor
ω_r	rotor electric field speed
T	total torque
T_0	average or constant torque
T_i	torque component induced by torsional vibration
f_i	characteristic frequency induced by torsional vibration
φ_i	T_i offset
f	fundamental frequency
f_i, f_j	characteristic frequencies of gears and bearings respectively
f_{ecc}	component due to rotor eccentricity
f_{bb}	component due to broken bars
f_{st}	component due to inter-turn short circuit
s	slip
p	polos pairs
k	$1, 2, 3, \ldots \ldots$
n	$1, 2, 3, \ldots \ldots$

References

1. Wang, K.; Sharma, V.; Zhang, Z. SCADA data based condition monitoring of wind turbines. *Adv. Manuf.* **2014**, *2*, 61–69. [CrossRef]
2. Coronado, D.; Fischer, K. *Condition Monitoring of Wind Turbines: State of the Art, User Experience and Recommendations*; Project number: VGVB-Nr.383; Fraunhofer Institute for Wind Energy and Energy System Technology: Kassel, Germany, 2015. Available online: https://www.vgb.org/vgbmultimedia/383_Final+report-p-9786.pdf (accessed on 29 May 2020).
3. Tchakoua, P.; Wamkeue, R.; Ouhrouche, M.; Slaoui-Hasnaoui, F.; Tameghe, T.A.; Ekemb, G. Wind Turbine Condition Monitoring: State-of-the-Art Review, New Trends, and Future Challenges. *Energies* **2014**, *7*, 2595–2630. [CrossRef]
4. Qiao, W.; Lu, D. A Survey on Wind Turbine Condition Monitoring and Fault Diagnosis—Part II: Signals and Signal Processing Methods. *IEEE Trans. Ind. Electron.* **2015**, *62*, 6546–6557. [CrossRef]
5. Merizalde, Y.; Hernández-Callejo, L.; Duque-Pérez, O.; Alonso-Gómez, V. Diagnosis of wind turbine faults using generator current signature analysis: A review. *J. Qual. Maint. Eng.* **2019**, *26*, 431–458. [CrossRef]
6. Kliman, G.; Koegl, R.; Stein, J.; Endicott, R.; Madden, M. Noninvasive detection of broken rotor bars in operating induction motors. *IEEE Trans. Energy Convers.* **1988**, *3*, 873–879. [CrossRef]
7. Kral, C.; Kapeller, H.; Gragger, J.V.; Haumer, A.; Kubicek, B. Phenomenon Rotor Fault-Multiple Electrical Rotor Asymmetries in Induction Machines. *IEEE Trans. Power Electron.* **2010**, *25*, 1124–1134. [CrossRef]
8. Didier, G.; Ternisien, E.; Razik, H. Detection of incipient rotor cage fault and mechanical abnormalities in induction motor using global modulation index on the line current spectrum. In Proceedings of the International Symposium on Diagnostics for Electric Machines, Power Electronics and Drives, Vienna, Austria, 7–9 September 2005. [CrossRef]
9. Henao, H.; Capolino, G.-A.; Fernandez-Cabanas, M.; Filippetti, F.; Bruzzese, C.; Strangas, E.; Pusca, R.; Estima, J.; Riera-Guasp, M.; Kia, S.H. Trends in Fault Diagnosis for Electrical Machines: A Review of Diagnostic Techniques. *IEEE Ind. Electron. Mag.* **2014**, *8*, 31–42. [CrossRef]
10. Hameed, Z.; Hong, Y.; Cho, Y.; Ahn, S.; Song, C. Condition monitoring and fault detection of wind turbines and related algorithms: A review. *Renew. Sustain. Energy Rev.* **2007**, *13*, 1–39. [CrossRef]
11. Sreenilayam, R.; Azariana, M.; Kimb, N.; Pechta, M. Effect of Multiple Faults and Fault Severity on Gearbox Fault Detection in a Wind Turbine using Electrical Current Signals. *Chem. Eng. Trans.* **2013**, *33*, 79–84. [CrossRef]
12. Lu, D.; Gong, X.; Qiao, W. Current-based diagnosis for gear tooth breaks in wind turbine gearboxes. In Proceedings of the IEEE Energy Conversion Congress and Exposition, Raleigh, NC, USA, 15–20 September 2012. [CrossRef]
13. Cheng, F.; Peng, Y.; Qu, L.; Qiao, W. Current-Based Fault Detection and Identification for Wind Turbine Drivetrain Gearboxes. *IEEE Trans. Ind. Appl.* **2017**, *53*, 878–887. [CrossRef]
14. Cheng, F.; Qu, L.; Qiao, W. Fault Prognosis and Remaining Useful Life Prediction of Wind Turbine Gearboxes Using Current Signal Analysis. *IEEE Trans. Sustain. Energy* **2018**, *9*, 157–167. [CrossRef]
15. Zhang, P.; Neti, P. Detection of Gearbox Bearing Defects Using Electrical Signature Analysis for Doubly Fed Wind Generators. *IEEE Trans. Ind. Appl.* **2014**, *51*, 2195–2200. [CrossRef]
16. Gong, X.; Qiao, W.; Zhou, W. Incipient bearing fault detection via wind generator stator current and wavelet filter. In Proceedings of the 36th Annual Conference of the IEEE Industrial Electronics Society, Glendale, CA, USA, 7–10 November 2010; pp. 2615–2620.
17. Stefani, A.; Yazidi, A.; Rossi, C.; Filippetti, F.; Casadei, D.; Capolino, G.-A. Doubly Fed Induction Machines Diagnosis Based on Signature Analysis of Rotor Modulating Signals. *IEEE Trans. Ind. Appl.* **2008**, *44*, 1711–1721. [CrossRef]
18. Al Ahmar, E.; Choqueuse, V.; Benbouzid, M.; Amirat, Y.; El Assad, J.; Karam, R.; Farah, S. Advanced signal processing techniques for fault detection and diagnosis of a wind turbine induction generator drive train: A comparative study. In Proceedings of the IEEE Energy Conversion Congress and Exposition, Atlanta, GA, USA, 12–16 September 2010. [CrossRef]
19. Amirat, Y.; Choqueuse, V.; Benbouzid, M.E.H. Wind Turbine Condition monitoring and fault Diagnosis Using Generator current amplitude demodulation. In Proceedings of the IEEE International Energy Conference, Manama, Bahrain, 18–22 December 2010. [CrossRef]
20. Helmy, K.; Prabhakar, M. *Complex System Maintenance Handbook*; Springer: London, UK, 2008.
21. Yang, H.; Mathew, J.; Ma, L. Intelligent diagnosis of rotating machinery faults—A review. In Proceedings of the 3rd Asia-Pacific Conference on Systems Integrity and Maintenance, Cairns, Australia, 25–27 September 2002. Available online: https://eprints.qut.edu.au/17942/1/17942.pdf (accessed on 6 July 2020).
22. Filippetti, F.; Franceschini, G.; Tassoni, C.; Vas, P. Recent developments of induction motor drives fault diagnosis using AI techniques. *IEEE Trans. Ind. Electron.* **2000**, *47*, 994–1004. [CrossRef]
23. Kia, S.H.; Henao, H.; Capolino, G.-A. Diagnosis of Broken-Bar Fault in Induction Machines Using Discrete Wavelet Transform Without Slip Estimation. *IEEE Trans. Ind. Appl.* **2009**, *45*, 1395–1404. [CrossRef]
24. Kim, K.; Parlos, A. Induction motor fault diagnosis based on neuropredictors and wavelet signal processing. *IEEE/ASME Trans. Mechatron.* **2002**, *7*, 201–219. [CrossRef]
25. Chen, F.; Tang, B.; Chen, R. A novel fault diagnosis model for gearbox based on wavelet support vector machine with immune genetic algorithm. *Measurement* **2013**, *46*, 220–232. [CrossRef]
26. Schoen, R.; Lin, B.; Habetler, T.; Schlag, J.; Farag, S. An unsupervised, online system for induction motor fault detection using stator current monitoring. *IEEE Trans. Ind. Appl.* **1995**, *31*, 1280–1286. [CrossRef]

27. Cambell, P.; Adamson, K. Identification of blade vibration causes in wind turbine generators. In Proceedings of the 4th International Confernce on Data Mining Including Building Applications for CRM & Competitive Intelligence, Rio de Janeiro, Brazil, 1–3 December 2003. Available online: https://www.witpress.com/Secure/elibrary/papers/DATA03/DATA03015FU.pdf (accessed on 22 October 2018). [CrossRef]

28. Kusiak, A.; Verma, A. A Data-Driven Approach for Monitoring Blade Pitch Faults in Wind Turbines. *IEEE Trans. Sustain. Energy* **2011**, *2*, 87–96. [CrossRef]

29. Dhomad, T.A.; Jaber, A. Bearing Fault Diagnosis Using Motor Current Signature Analysis and the Artificial Neural Network. *Int. J. Adv. Sci. Eng. Inf. Technol.* **2020**, *70*. [CrossRef]

30. Cheng, F.; Wang, J.; Qu, L.; Qiao, W. Rotor current-based fault diagnosis for DFIG wind turbine drivetrain gearboxes using frequency analysis and a deep classifier. *IEEE Trans. Ind. Appl.* **2017**, *54*, 1062–1071. [CrossRef]

31. Ibrahim, R.; Tautz-Weinert, J.; Watson, S. Neural networks for wind turbine fault detection via current signature analysis. In Proceedings of the WindEurope Summit 2016, Hamburg, Germany, 27–29 September 2016. Available online: https://www.researchgate.net/publication/308991040_Neural_Networks_for_Wind_Turbine_Fault_Detection_via_Current_Signature_Analysis (accessed on 18 June 2020).

32. Merabet, H.; Bahi, T.; Halem, N. Condition Monitoring and Fault Detection in Wind Turbine Based on DFIG by the Fuzzy Logic. *Energy Proc.* **2015**, *74*, 518–528. [CrossRef]

33. Abd-Elkader, A.G.; Allam, D.F.; Tageldin, E. Islanding detection method for DFIG wind turbines using artificial neural networks. *Int. J. Electr. Power Energy Syst.* **2014**, *62*, 335–343. [CrossRef]

34. Thomson, W.; Fenger, M. Current signature analysis to detect induction motor faults. *IEEE Ind. Appl. Mag.* **2001**, *7*, 26–34. [CrossRef]

35. Artigao, E.; Honrubia-Escribano, A.; Gomez-Lazaro, E. Current signature analysis to monitor DFIG wind turbine generators: A case study. *Renew. Energy* **2018**, *116*, 5–14. [CrossRef]

36. Cheng, F.; Qu, L.; Qiao, W.; Wei, C.; Hao, L. Fault Diagnosis of Wind Turbine Gearboxes Based on DFIG Stator Current Envelope Analysis. *IEEE Trans. Sustain. Energy* **2018**, *10*, 1044–1053. [CrossRef]

37. Artigao, E.; Koukoura, S.; Honrubia-Escribano, A.; Carroll, J.; McDonald, A.; Gómez-Lázaro, E. Current Signature and Vibration Analyses to Diagnose an In-Service Wind Turbine Drive Train. *Energies* **2018**, *11*, 960. [CrossRef]

38. Merizalde, Y.; Hernández-Callejo, L.; Duque-Perez, O.; López-Meraz, R. Fault Detection of Wind Turbine Induction Generators through Current Signals and Various Signal Processing Techniques. *Appl. Sci.* **2020**, *10*, 21. [CrossRef]

39. Chawla, N.V.; Bowyer, K.W.; Hall, L.O.; Kegelmeyer, W.P. SMOTE: Synthetic Minority Over-sampling Technique. *J. Artif. Intell. Res.* **2002**, *16*, 321–357. [CrossRef]

40. Liu, Y.; Li, Z.; Zhou, C.; Jiang, Y.; Sun, J.; Wang, M.; He, X. Generative Adversarial Active Learning for Unsupervised Outlier Detection. *IEEE Trans. Knowl. Data Eng.* **2020**, *32*, 1517–1528. [CrossRef]

41. Doersch, C. Tutorial on Variational Autoencoders. *arXiv* **2016**, arXiv:1606.05908v2. Available online: https://arxiv.org/abs/1606.05908 (accessed on 26 July 2020).

42. Goodfellow, I.; Pouget-Abadie, J.; Mirza, M.; Xu, B.; Warde-Farley, D.; Ozair, S.; Courville, A.; Bengio, Y. Generative Adversarial Nets. *arXiv* **2016**, arXiv:1406.2661. Available online: https://arxiv.org/abs/1406.2661v1 (accessed on 26 July 2020).

43. Salimans, T.; Goodfellow, I.; Zaremba, Z.; Cheung, V.; Radford, A.; Chen, X. Improved Techniques for Training GANs. *arXiv* **2016**, arXiv:1606.03498. Available online: https://arxiv.org/abs/1606.03498v1 (accessed on 26 July 2020).

44. Xu, L.; Skoularidou, M.; Cuesta-Infante, A.; Veeramachaneni, K. Modeling Tabular data using Conditional GAN. *arXiv* **2019**, arXiv:1907.00503. Available online: http://arxiv.org/abs/1907.00503 (accessed on 10 December 2020).

45. Bian, J.; Hui, X.; Sun, S.; Zhao, X.; Tan, M. A Novel and Efficient CVAE-GAN-Based Approach with Informative Manifold for Semi-Supervised Anomaly Detection. *IEEE Access* **2019**, *32*, 88903–88916. [CrossRef]

46. Zhu, J.; Bento, J. Generative Adversarial Active Learning. *arXiv* **2017**, arXiv:1702.07956. Available online: http://arxiv.org/abs/1702.07956 (accessed on 5 April 2021).

47. Díaz, I.M.; Morinigo-Sotelo, D.; Duque-Perez, O.; Romero-Troncoso, R. Early Fault Detection in Induction Motors Using AdaBoost With Imbalanced Small Data and Optimized Sampling. *IEEE Trans. Ind. Appl.* **2017**, *53*, 3066–3075. [CrossRef]

48. Gangsar, P.; Tiwari, R. A support vector machine based fault diagnostics of Induction motors for practical situation of multi-sensor limited data case. *Measurement* **2019**, *135*, 694–711. [CrossRef]

49. Lee, Y.O.; Jo, J.; Hwang, J. Application of deep neural network and generative adversarial network to industrial maintenance: A case study of induction motor fault detection. In Proceedings of the IEEE International Conference on Big Data, Boston, MA, USA, 11–14 December 2017. [CrossRef]

50. Wang, Z.; Wang, J.; Wang, Y. An intelligent diagnosis scheme based on generative adversarial learning deep neural networks and its application to planetary gearbox fault pattern recognition. *Neurocomputing* **2018**, *310*, 213–222. [CrossRef]

51. Glowacz, A. Fault diagnosis of electric impact drills using thermal imaging. *Measurement* **2021**, *171*. [CrossRef]

52. Soto, N.; De La Torre, F. Diagnóstico de problemas de asimetrías rotóricas en un motor de inducción de gran potencia. *Ing. Mecánica* **2007**, *10*, 47–50. [CrossRef]

53. Gong, X.; Qiao, W. Imbalance Fault Detection of Direct-Drive Wind Turbines Using Generator Current Signals. *IEEE Trans. Energy Convers.* **2012**, *27*, 468–476. [CrossRef]

54. Misiti, M.; Misiti, Y.; Oppenheim, G.; Poggi, J. Wavelet ToolboxTM 4 User's Guide Product Enhancement Suggestions Wavelet ToolboxTM User's Guide. 1997. Available online: www.mathworks.com%0Awww.mathworks.com/contact_TS.html (accessed on 13 April 2021).
55. Riera-Guasp, M.; Antonino-Daviu, J.A.; Pineda-Sanchez, M.; Perez-Cruz, J.; Puche-Panadero, R. A General Approach for the Transient Detection of Slip-Dependent Fault Components Based on the Discrete Wavelet Transform. *IEEE Trans. Ind. Electron.* **2008**, *55*, 4167–4180. [CrossRef]
56. Kar, C.; Mohanty, A. Monitoring gear vibrations through motor current signature analysis and wavelet transform. *Mech. Syst. Signal Process.* **2006**, *20*, 158–187. [CrossRef]

Article

A Learning-Based Methodology to Optimally Fit Short-Term Wind-Energy Bands

Claudio Risso [1,*] and **Gustavo Guerberoff** [2]

1   Computer Science Institute, Engineering Faculty, University of the Republic, Montevideo 11200, Uruguay
2   IMERL, Engineering Faculty, University of the Republic, Montevideo 11200, Uruguay; gguerber@fing.edu.uy
*   Correspondence: crisso@fing.edu.uy

Abstract: The increasing rate of penetration of non-conventional renewable energies is affecting the traditional assumption of controllability over energy sources. Power dispatch scheduling methods need to integrate the intrinsic randomness of some new sources, among which, wind energy is particularly difficult to treat. This work aims at the optimal construction of energy bands around wind energy forecasts. Complementarily, a remarkable fact of the proposed technique is that it can be extended to integrate multiple forecasts into a single one, whose band width is narrower at the same level of confidence. The work is based upon a real-world application case, developed for the Uruguayan Electricity Market, a world leader in the penetration of renewable energies.

Keywords: wind power; non-conventional renewable energy; forecasting; energy bands; combinatorial optimization

---

Citation: Risso, C.; Guerberoff, G. A Learning-Based Methodology to Optimally Fit Short-Term Wind-Energy Bands. *Appl. Sci.* **2021**, *11*, 5137. https://doi.org/10.3390/app11115137

Academic Editor: Luis Hernández-Callejo

Received: 21 April 2021
Accepted: 18 May 2021
Published: 31 May 2021

Publisher's Note: MDPI stays neutral with regard to jurisdictional claims in published maps and institutional affiliations.

1. Introduction

Whether due to economic pressure or environmental concerns, the rate of penetration of non-conventional renewable energies has been increasing rapidly over recent years, and it is expected to grow even faster in the years to come. Short-term operation and maintenance of electrical systems relies on optimal power dispatch scheduling methods.

Either renewable or not, conventional energy sources are dispatchable on request, i.e., authorities can control when and how much power will be provided from each source. Conversely, non-conventional renewable energies are not controllable, are intermittent and uncertain, even within a few hours period ahead. The intrinsic stochastic nature of the new energy sources turns out the short-term dispatch of the grid into a much harder challenge, which necessarily must coexist with randomness coming from significant portions of the installed power plant. This work regards with the optimal crafting of wind-energy bands (in a sense precisely defined in Section 3). It is based on an application case of Uruguay, a worldwide leader in the usage of renewable energies. The Uruguayan case was chosen as reference because: (i) Uruguay is the country of these authors, so the case is very proximate to our research group; (ii) the country counts an immense relative penetration of renewable generation of diverse sources; and (iii) there is plenty of open access information available. Although this document is guided by that reference application case, the methodology is general and its results extendable, so it can be ported to another system or country.

1.1. Literature Overview

In the context of forecast of daily scenarios there exist a vast literature and plenty of methods proposed to accurately predict a likely behavior for the stochastic process involved. Several of them use purely statistics techniques—parametric, semi-parametric and nonparametric—to infer forecasts. Particularly, there are standard and well-known techniques coming from the analysis of time series for the daily forecasting electricity prices and electricity demand (see [1]) that can be adapted to wind power inference. Statistical methods like those, perform well to forecast time series without very strong fluctuations, like: electricity prices, electrical demand time series, hydraulic contributions to water reservoirs, or even to forecast mid and long-term availability of wind power. When the application regards with short-term wind power generation, the accuracy of such statistical methods degrades notably, even over short periods of time that range from a few hours till two or three days ahead. This is because wind power is much more volatile than electricity prices and electrical demand time series (and the other examples mentioned before).

Complementarily, there are approaches for short-term wind power forecasting based on numerical simulations of atmosphere's wind flows (see [2,3]). For a couple of days ahead period, or even larger time windows, numerical simulations are usually more accurate than purely statistical models. Such models are deterministic, while the underlying physical phenomena is chaotic by nature. So, they perform better than purely statistical methods to follow the process whereabouts at early stages but are far from being trustworthy in what respects to the construction of likely scenarios at larger times. Summarizing: the scheduling of short-term dispatch must coexist with randomness, so, even though wind simulations provide valuable information, they must be enriched in order to account the intrinsic stochasticity of the process.

In the last decade, the concept of prediction interval associated to probabilistic forecasts was introduced (see [4]). The construction is based on a nonparametric approach to estimate—at once, for all instant of time within a selected grid over a forecast horizon- the interdependent quantiles for the (unknown) distribution probability of the wind power process. For each time, the corresponding prediction interval provides an estimate of the expected accuracy of predictions with respect to what the actual value of the wind power will be. Therefore, this technique crafts bands inside which the wind power process is expected to stay with a given probability (the nominal coverage rate [5,6]). Complementing the previous line of work, in [7] other methods that also use deterministic forecasts as input are introduced, and through a subtle analysis that involves historical data to estimate nonparametric forecast error densities, for some relevant times chosen appropriately, the authors have succeeded in generating wind power scenarios with their respective probabilities.

Referred to the construction of confidence bands, several papers have been issued. In other works (see [8]), parametric processes guided by stochastic differential equations (SDEs) are studied. These processes involve a drift term, which acts as a force that tends to attract the trajectories towards the forecast (which is known as an input), and also involve a diffusive term modulated by a Wiener process factor as usual. Different parametric models are studied by considering specific forms for the drift and diffusion terms. The basic idea of the mentioned work consists in approximate these parametric non-Gaussian stochastic processes by Gaussian processes with the same mean, variance and covariance structure. Such approximations allow the estimation of the parameters through maximum likelihood techniques. Following the ideas of these authors and using the same data set for wind power in Uruguay as in the present work, the article [9] synthesizes multiple realizations of the calibrated process in order to build confidence bands. As a subsequent and tightly coupled step towards tackling the crafting of stochastic optimal short-term dispatches schedules of the grid, the previous research team uses those SDEs as supply for a Continuous-Time Stochastic Optimal Control (CTSOC) [10], with very promising results. However, scalable numerical techniques to solve optimal control problems derive from dynamical programming [11], what limits the number of state variables to integrate to the problem. The reference [10] is a pretty good example of what can be done when the number of states is manageable, but many times, the intrinsic structure of the problem makes it untreatable through such approaches. Some practical applications inevitably require crafting scenarios to use other optimization techniques, where the availability of wind-energy bands is essential.

Energy bands crafted in this article were used as a supply for an optimal short-term dispatch problem for the reference application case, which combines generation units with complex commitments, temporal dependencies among them, and other intrinsic characteristics that makes the problem too hard to be tackled when approached with dynamical programming or related techniques. We suggest [12] as an illustrative reference for practical applications of this work.

In recent years a considerable effort was put into controlling strong variability of weather conditions through the incorporation of what is called in the literature Ensemble Weather Forecasting (see for instance [13], focused on wind-power predictions in Japan); a strategy significantly different from that present in Section 4 of this paper. Methodologies aside, objectives are in fact quite similar: to control rare events, in that case through numerical studies over an area-averaged, added to a rigorous probabilistic analysis. Of course, the particularities of the Uruguayan case are notably different to the Japanese case due to the frequent existence in the eastern Pacific region of strong climatic anomalies as the passing of extratropical cyclones, that the authors called wind ramp events. Their probabilistic wind-power prediction achieved a good statistical reliably through confidence interval for the wind-power variability. Numerical Weather Predictions and Ensemble Weather Forecasting are also used in [14].

Finally, researchers from South Korea introduce an interesting ensemble [15] through different machine learning techniques, combining multilayer perceptron (MLP), support vector regression (SVR), and CatBoost to improve power forecasting of renewable sources. As we see later on, the article here presented focuses upon wind energy rather than power forecasts, and in fact, uses power forecasts as a supply. Furthermore, instead of using existing learning techniques as in [15], this work introduces a novel one, conceptually simpler, and yet promising according on its results.

1.2. Particulars of the Reference Application Case

As we previously mentioned, the methodology elaborated in this work is general and it can be applied to other cases. However, practical relevance of these results relies upon the magnitude and diversity of renewable energy sources in the particular application. Uruguay is one the countries of the world with highest penetration of renewable energies. Nowadays, over 98% of the annual energy consumed by the country or exported to neighbors (i.e., Argentina and Brazil) comes from renewable sources. Table 1 presents the installed power plant and annual generation by type of energy source (Information regarding capacity and energy is available at: https://portal.ute.com.uy/composicion-energetica-y-potencias while geographic distribution is in: https://www.ute.com.uy/institucional/infraestructura/fuentes-de-generacion Other historical or real-time data is available at: https://www.adme.com.uy and https://adme.com.uy/mmee/infanual.php. A few years ago, when data of this work was acquired, that fraction was 96%, slightly lower but still remarkable high.

Table 1. Details of the installed power plant and energy by type of source [ADME: 2019, UTE: 2019].

Energy by Type of Source	Installed Power Plant (MW)	Relative Subtotal	Produced Energy Total 2019 (GWh)	Relative Subtotal
Hydroelectric	1534	31.5%	6134	55.6%
Wind-power	1506	30.9%	3690	33.5%
Solar-Photovoltaic	254	5.2%	314	2.8%
Biomass (thermal)	413	8.5%	660	6.0%
Fossile (thermal)	1170	24.0%	225	2.0%

Figure 1 on the other hand shows the geographical distribution of renewable units.

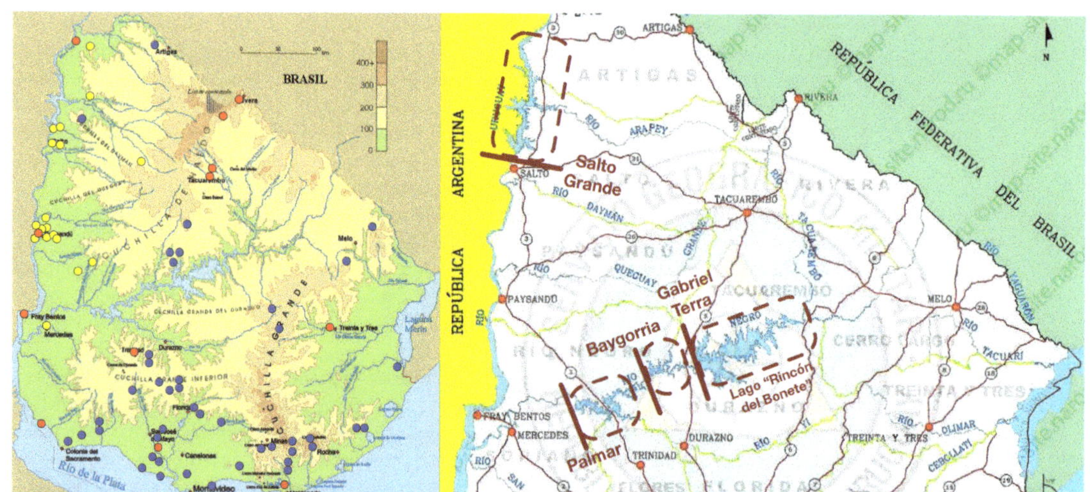

Figure 1. Geographical distribution of renewable sources. Leftmost: wind-power (blue), solar-photovoltaic (yellow) and biomass thermal units (red). Rightmost: hydroelectric dams. [source UTE: 2019 and [12]].

The information previously presented is public and accessible through the provided URLs. Detailed historic information about actual wind-power and related forecasts was also public by the time this work was realized, but unfortunately is no longer so.

1.3. Main Goals of This Work

Regardless of the technique used to narrow uncertainty with bands, those works mentioned in Section 1.1 share a common characteristic: the electric-power is the magnitude to be captured. For some applications and/or contexts, the *energy* coming from wind sources—which is a derivative of the wind's power anyway—is an important magnitude itself, and since it is the outcome of integrating the previous, results less noisy and easy to capture. This work aims on crafting bands representative of the energy to be produced along some periods, rather than focusing on accurate measures of the instant power. It is worth mentioning that since both magnitudes are strictly dependent, a fitted power-process cannot be outside energy-bands for too long. Thus, though this kind of bands could resemble the other in their shape, strictly speaking, they are different.

Regarding the process to craft such wind-energy bands, the idea goes by designing a learning model that is fed from: wind-power forecasts as a mainstream of what to expect for the days to come, and actual wind-power data to incorporate information related to the historical deviations of the process. The area inside the band is a measure of the quality of such calibration. The smaller the area, the better the quality of the fitting. So, crafting energy-bands of minimal area is our objective. Assuming persistent behavior, an optimal fitting over a training set is expected to replicate reasonably well over other independent instance, so a historical calibration could be used to estimate energy-bands in the near future.

Summarizing, our interest is not focused on particular power trajectories and their probabilities but aims on crafting optimal area energy-bands around wind-power forecasts, in such a way that the energy outside those bands be bounded. At this respect, our article presents a novel approach. The method is purely nonparametric, since it makes no assumptions on the physical phenomena, nor any hypothesis about the involved random processes (conditions on homogeneity in time, seasonal behavior of the temporal series or any kind of markovian hypotheses, are not necessary for this work). Our proposal is based upon a mixed integer optimization problem, which aims on getting the narrowest average band around a set of forecasts, or a combination of forecasts, that keeps the off-band aggregated energy below a given threshold. As an innovation, to allow the optimization to go as farther as possible, the model enables to discard up to a given percentage of training samples that are treated as *atypical profiles*.

At first glance this last feature might look risky, in the sense that, in advance, one cannot tell whether the day to come will match or not an atypical profile. However, as we see later, when the method is trained with more than one forecast, whenever they are conditionally independent or weakly dependent, a combination of those forecasts and their bands allows to regain the lost confidence; the result is a more accurate band than those of constructions computed by separate. This is another remarkable point of the present work, since it helps to improve band's quality by taking the best of more than one forecasts provider.

1.4. Structure of This Document

This work uses data coming from two independent wind-power forecast providers for the Uruguayan Electricity Market: Garrad Hassan and Meteológica, which was available during the period. Complementarily, a third and purely probabilistic forecast was constructed up from historical wind-power realizations, by closely following other documented ideas (see [16]). Regarding actual wind-process realizations, we also have used power records measured over the Uruguayan grid. Therefore, the experimental evaluation here presented for training and test sets is based upon real-world data.

The structure of this article matches the stages of the novel technique. In Section 2 we describe the main characteristics of wind-power in Uruguay, together with the forecasts used as supplies for computations and analysis. Section 3 presents the optimization model to create likely bands of minimum width as well as results from experimental evaluation; while Section 4 shows how after filtering atypical days, a combination of bands calibrated up from independent forecasts performs better than any of them by separate. Finally, Section 5 summarizes the main results and possible applications.

2. Wind-Power Uncertainty and the Use of Forecasts

This section shows how variable wind-power is -when described as a stochastic process—and it presents the forecasts that are used to anticipate power realizations. The historical of wind-power data in Uruguay has a few years and along this period the installed power plant has been firmly growing. So, instead of expressing power in term of MW, we use the Plant Load Factor (PLF), which corresponds to the actual power generated at each time divided by the sum of the installed power capacity of each wind turbine in the system at that moment (i.e., the wind-power plant). Therefore, PLF is a dimensionless quantity that takes values between 0 and 1. Over an hour time-slot basis we consider the average PLF, i.e., the average power along each hour divided by the power plant; this is the main variable we use along this work. In this way the information is normalized, and we can disregard of changes in the installed capacity during the period of analysis. In what follows we present a summary of the behavior of wind-power in Uruguay. The data sample used involves around 730 days of the years 2014 to 2016. We decided to use this period as reference because of the homogeneity of forecast providers and the availability of open data sources.

2.1. Seasonal Regularity

Figure 2 shows the average energy and relative deviations of the daily PLF with respect to the historical mean. Daily PLF is the cumulated value of hourly PLFs over a day, so it ranges from 0 to 24. The temporal horizon varies from 1 day to 730 days ahead.

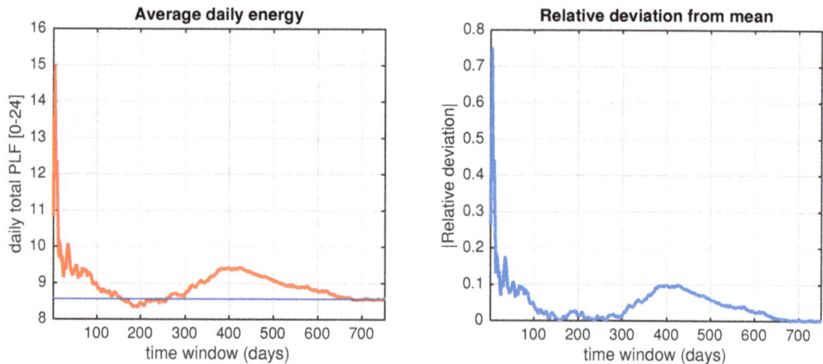

Figure 2. Total and relative deviations of actual PLF with respect to the historical average (leftmost horizontal blue line) as a function of the horizon (730 days).

The intermediate rebound in the deviation is due to a seasonal phenomenon; this effect decreases considerably if the records are limited to a single season. For instance, Figure 3 shows the equivalent plot when only summer days of the same period are considered.

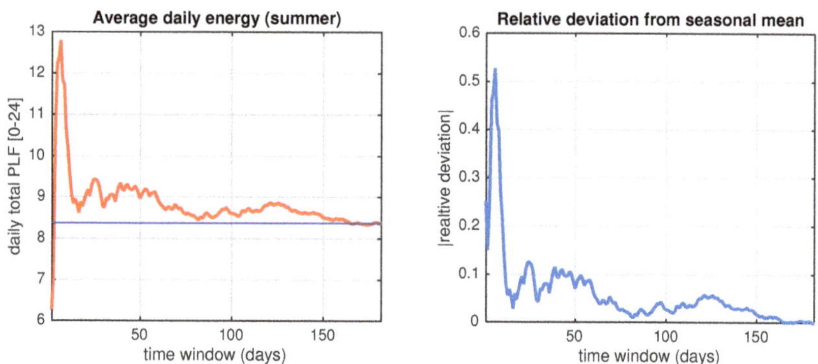

Figure 3. Total and relative deviations of actual PLF with respect to the historical average for summer season (leftmost horizontal blue line).

The previous figure shows how after a week or two, the process goes inside the 10% error band, respect to the average value for that season. We must conclude that wind-power is fairly regular when used in medium-term planning. For shorter periods of time, the situation is quite the opposite.

2.2. Changing Daily Behavior

Managing the electric grid of a country is a challenging task that must be carried out carefully and optimally. To accomplish that, multiple problems are to be solved, spanning different scales of time and components. Medium-term planning usually refers to the valuation of intangible resources, such as the height of the lake in an electric dam accounted as an economic asset. Seasonal regularity as that observed in Section 2.1 is an advisable characteristic to develop mid-term optimization planning models. Short-term planning consists in crafting optimal dispatch schedules for some days ahead, and its aim is upon efficiently coordinate the usage of available resources. The object of this work is on suppling energy-bands for the short-term power dispatch of the Uruguayan grid, whose outcome sets the prices of energy in the electricity market. Due to its short scale of time (a few days ahead) and time-step (of an hour each) short-term planning requires accurate PLF estimations hour-by-hour over some days to come.

A histogram of daily cumulated PLFs along the available two years of samples is shown over the leftmost of Figure 4. Observe that many days within the period have daily cumulated PLFs below 5 (which is approximately 20% of the power plant). The number of samples whose cumulated PLFs are above 19 (80% of the power plant) are lesser, and yet there are days where the average wind-power was pretty close to the power plant. To get an insight about hourly behavior, the rightmost of Figure 4 shows the hour-by-hour mean marked with asterisks. The mean PLF is around 20% higher in the night hours compared with the hours of sun. Complementarily, the rightmost image plots actual wind-power samples, concretely those 30% with higher distance ($||.||_2$) respect to the average over the first year. Observe how divergent are these samples when compared with the mean trajectory. We are not going further in the direction of standard statistical descriptive, since it exposes the predictions of wind-power to important errors and thus is seldom used.

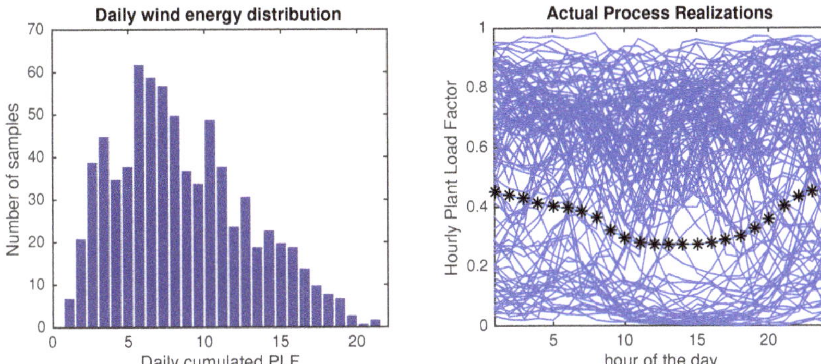

Figure 4. Histogram of daily wind energy samples [daily PLF upon the leftmost] and 30% most atypical realizations for Uruguayan wind-power over a year [rightmost].

2.3. Additional Accuracy Coming from Numerical Forecasts

This section experimentally analyzes the benefits of using short-term wind-power forecastings based on numerical simulations of atmosphere's wind flows. As a measure of the error incurred we use the $||.||_1$ distance between forecasts and actual realizations. Hence, the total energy error for the period starting on day d (denoted err^d) is computed as: $err^d = \sum_{t=1}^{T} |w_t^d - p_t^d|$, being w_t^d and p_t^d, respectively, the actual power (PLF) for the day d as seen t hours ahead and its corresponding forecasted value. On the other hand, T is the time horizon of the forecasts: 72 h in our examples. In Section 3 we explain the convenience of using such an error measure.

This work used information for the Uruguayan grid, which was of public domain by the time computations were realized. Two independent forecast providers are considered: Garrad Hassan and Meteológica. Their common samples span around 300 days, starting in early 2016. A third forecast -about which we elaborate later on—that uses a purely statistics analysis (following PSF ideas [16]) is built to benchmark statistical and numerical forecast. In spite of its lower performance as an isolated technique, we see in Section 4 that this final forecast (referred to as PSF44) increases the overall quality of a convex combinations of filtered forecasts.

Regarding statistical moments of the series, the mean value of the PLF error samples is 6.80 for Garrad Hassan and 5.99 for Meteológica, with respective variances of values 2.48 and 2.21. On the other hand, those figures for the PSF44 are: 13.99 and 4.52; notoriously worse than numerical forecasts. Complementarily, histograms in Figure 5 represent error distributions for each case, reinforcing the idea that Meteológica's forecast slightly outperforms Garrad Hassan's, while both are much better than PSF44.

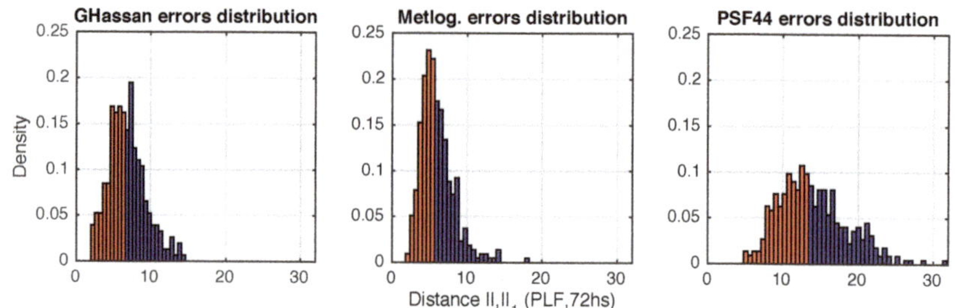

Figure 5. Histograms for total deviations of forecasts within a horizon of 72 h ahead. Red and blue areas concentrate 50% of probability in all cases [leftmost: Garrad Hassan, center: Meteológica, rightmost: PSF44].

2.4. PSF-Like Forecast

Pattern Sequence-based Forecasting algorithm (PSF) is a novel nonparametric approach to infer forecasts. This method has provided promising results when applied to an assortment of time series forecasting in several international markets, at a horizon of one or a few days ahead. The main idea of the PSF algorithm -and more recent variants- involves three parts working sequentially:

1. The historical of time series is clustered into groups that have similar temporal daily profiles;
2. The time series of daily profiles is converted into a discrete sequence of labels for the cluster each day belongs to;
3. Given the label corresponding to the current day, a window of days of fixed length is used to seek along the whole past sequence of labels for those that match the window ending at present time. The profile for the day to be predicted is constructed averaging profiles that follow to each one of those identical windows in the past.

A remarkable advantage of the PSF method is its reduced number of parameters. There are only two main parameters to adjust: the number of clusters K and the historical time window W.

Figure 6 shows the centroids computed over the actual wind-power data-set when using different numbers of clusters. For example purposes, assume $K = 3$ and $W = 4$; so, a sequence of D days translates into a sequence $\{s_d\}$ ($d \in D$) of digits, with $s_d \in \{1, 2, 3\}$. Here, s_d is the index of the closest centroid to the realization of the day d.

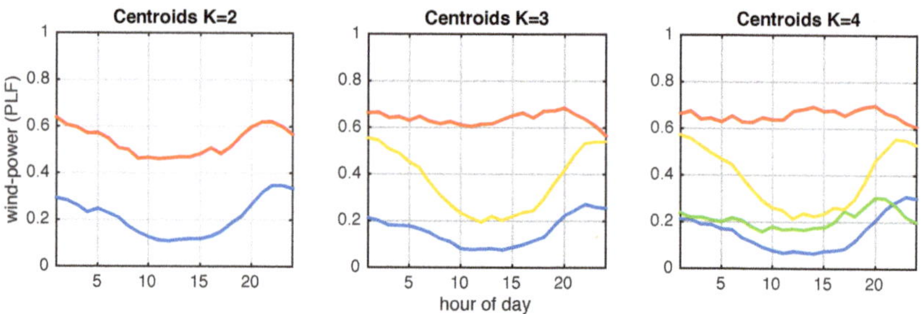

Figure 6. Centroids of actual wind-power samples for different number of clusters.

Within such sequence, we aim now on finding subsequences with $W = 4$ symbols, for instance, the sequence 1231 (we are assuming that current day belongs to cluster 1, given by the last symbol in this sequence). Figure 7 shows examples where that subsequence could be found. The outcome of this predictor (one day ahead) is the average of the actual PLF realizations, among those samples immediately following the subsequences registered. To extend the construction to a two-days ahead forecast, one could repeat the process seeking for the subsequence composed of the previous $W - 1$ days plus the new forecasted one, and so on.

$$\cdots \quad 1 \quad \boxed{1 \quad 2 \quad 3 \quad 1} \quad 3 \quad \cdots \quad 1 \quad \boxed{1 \quad 2 \quad 3 \quad 1 \quad 2 \quad 3 \quad 1} \quad \cdots$$

Figure 7. Example of matching subsequences within a historical of symbols.

The purpose behind the development of this forecast is not devaluating statistical methods. On the contrary, this work shows an example of how such a simple method, may contribute to the overall quality when combined with forecasts coming from complementary techniques.

3. Optimization of Wind-Energy Bands

Optimization of wind-energy bands is in the core of this framework. We provide an expression to compute a band around any forecast, and, for that concrete formula, we seek for the narrowest band that satisfies a set of constraints, which imposes limits to the actual process in its deviations in accordance with the historical behavior. A traditional approach would go the way of setting constraints to keep the power deviation under certain boundaries. Conversely, this work aims on minimizing the expected off-band energy.

The previous is explained by the particulars of the Uruguayan electricity installed plant, but it is also justified by trends of new technologies. Around 98% of the electricity annually consumed in this country comes from renewable sources (see [17]). In average, 50% is from hydroelectric sources, while 35% is from wind-power. All of the hydroelectric dams in Uruguayan have water reservoirs; two of them (Bonete and Salto Grande, see Figure 1) are particularly huge. Almost 40% of the hydroelectric capacity is located after the greater lake (Bonete's), which would take 5 months to empty at full-power. Therefore, in fact, the hydroelectric plant also constitutes an accumulator, i.e., a kind of battery that can plenty compensate short-term fluctuations of the power coming from wind turbines. Hence, regarding Uruguayan short-term planning concerns, an accurate prediction of energy boundaries is more convenient than a power forecast of limited punctual quality. In a complementary manner, smart-grids capabilities are rapidly advancing towards active applications, capable of dynamically adjusting portions of the demand to adapt them to fit system needs (see [18]), while electricity storage units based on batteries are just around the corner (read [19] and also see the "Neoen & Tesla Motors" project in Australia). Therefore, in the near future, this work could be a useful experience for other countries.

The information required to determine an instance of our optimization problem comprises the following data sets. At first place, we need a historical of wind-power forecasts. We consider a collection $\mathcal{P}$ of deterministc registers that involves short-term point forecasts over a horizon of a few days ahead; i.e., a family of vectors $p^d \in [0, 1]^T$, with fixed T, which is set by the number of samples along the time horizon. Here, d is the index for each day on which the construction of a band begins; $d \in D$, being D the set of indices for days with historical observations. Wind-power forecasts usually span from one up to three days, i.e., from 24 to 72 h, and time is discretized at a rate of one sample per hour. Let $T - 1$ be the limit of hours ahead available for each forecast. We assume that all forecasts share the same time horizon, and that in $t = 0$ the current power is the only data known for sure. As we mentioned earlier, for simplicity the wind-power is expressed as the PLF, which corresponds to the actual power generated divided by the sum of the installed power capacity of wind turbines in the system at each moment. Thus $p_t^d \in [0, 1]$ is the normalized point forecast of the wind-power t hours ahead, within the vector associated to the forecast issued on the day d.

The second part of the input data set comprises the actual historical wind-power time series samples, grouped into a collection: $\mathcal{W}$, whose elements $w^d \in [0, 1]^T$ are also assumed normalized. Hence, $w_t^d \in [0, 1]$ is the actual PLF measured t hours after the beginning of the day d. For consistency, since the current state can be measured rather than forecasted, $p_0^d = w_0^d$ for each day d. Observe that the set $\mathcal{W}$ usually has duplicated records, for instance:

$w_{24}^d = w_0^{d+1}$. Despite that, we have chosen this format to simplify those expressions that link with forecast information. Regarding forecasts, however, the previous equality doesn't hold. In fact, p_{24}^d (a sample, forecasted 24 h ahead) is different from p_0^{d+1} (the actual value measured a day later).

It is clear that, given any two bands containing the real process inside of them at the same instants, the narrower band is of better quality. Wind-power generation is a process hard to anticipate, and violations to computed bands is a fact we must coexist with. However, not every violation has the same severity in terms of its impact to the power grid. In the context of the short-term energy dispatch, how much cumulated energy falls down outside the band is a convenient metric to assess the confidence of the pair: forecast plus computed band. In this work, we define the following expression as a metric for the reliability (The expression on the right hand side corresponds precisely to the anti-reliability, which of course is the complement of the reliability; hence the notation for the left hand side) $\mathcal{R}^d$, of a band around a given forecast p^d:

$$1 - \mathcal{R}^d(w, lb, ub) = \frac{1}{T} \sum_{t=0}^{T-1} \max\left[w(t) - ub\left(p^d, t\right), 0\right] + \frac{1}{T} \sum_{t=0}^{T-1} \max\left[lb\left(p^d, t\right) - w(t), 0\right] \quad (1)$$

where ub and lb respectively are the functions that determine upper and lower limits for the bands along the forecasted period, and w is the actual generation, unknown until the near future where reality is revealed. Functions lb and ub take a forecast (p^d) and an instant (t) as their inputs, while their outputs are the respective bounds to expect.

As mentioned, the feasible region of the optimization model imposes limits to the severity of violations to the band. Besides, in order to improve the quality as further as possible, the model allows to discard up to a limit of elements in the training set, which are *atypical*, specially bad forecasts that whether included would either: deteriorate the accuracy of the result, or force us to use too broad bands. So, to complete an instance we must set values to those quantities. The parameter $\theta \in [0, 1]$ limits the amount of energy allowed to fall down outside the band along the optimization horizon. The parameter $\lambda \in [0, 1]$ sets a minimum fraction of *regular* (i.e., not atypical) forecasts to be used in the effective training set or, in other terms, $(1 - \lambda)$ is the maximum fraction of atypical days allowed to be discarded. It is worth mentioning that the limit for off-band energy only accounts over regular forecasts.

3.1. Minimal Relative Width of Bands

This work considers those bands defined by relative deviations with respect to forecasted values, which are simple to calculate and optimize, and yet lead to accurate results. Let $\{x_t \geq 0\}$ be a set of coefficients associated to the time series analyzed, which delimits the width of the band. That is, for any instant t within the time horizon of the forecast issued on day d, we take p_t^d and compute the lower and upper limits of the band using the expressions $lb_t^d = \max\left[0, (1 - x_t)p_t^d\right]$ and $ub_t^d = \min\left[1, (1 + x_t)p_t^d\right]$ respectively. Hence, $\{x_t : 0 \leq t \leq T - 1\}$ comprises the first set of control variables that modulates the relative width of the band for a given forecast $p^d \in \mathcal{P}$. Figure 8 sketches about how these variables and derivatives are related, through a hypothetical forecast (centroid of the band, highlighted in blue), its correspondent energy-band (shaded in grey), and the actual power-process registered afterwards (red curve).

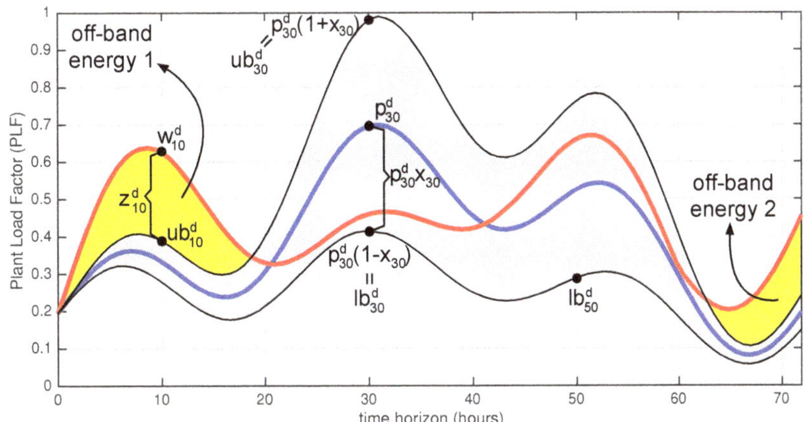

Figure 8. A wind-power band (grey) crafted after a forecast (blue), for some day d, and the actual process (red).

The objective function of this optimization is $\sum_{t=0}^{T-1} \hat{w}_t x_t$, where $\hat{w}_t = \left(\sum_{d \in \overline{D}} w_t^d \right) / |\overline{D}|$ is the average PLF at time t over a historical record of observations $\overline{D}$, eventually different from that of the training set. In other words, $\hat{w}_t$ corresponds to the sequence of asterisks in the rightmost of Figure 4, while $\sum_{t=0}^{T-1} \hat{w}_t x_t$ matches the average grey area in Figure 8. Whenever forecasts are statistically reliable, the objective function corresponds with the expected absolute PLF area of the band along the period T.

Defined so, the optimization is not instant-to-instant greedy, in the sense that it could deteriorate the performance at some points in order to surpass the overall performance by gaining more in others. That differentiates this work from related ones (like [7]), whose intention is to track power rather than energy. In fact, this model doesn't need conventional hypotheses about stochastic processes, such as homogeneity or markovianity.

The second group of control variables is composed by those who determine which are the regular forecasts. The variable $y_d \in \{0,1\}$ indicates whether the forecast issued on day d should be considered regular ($y_d = 1$), or atypical ($y_d = 0$). Unlike the $\{x_t\}$ variables, these new ones are boolean. We denote D to the set of days for which the optimization problem is implemented (i.e., the training-set). The complete combinatorial optimization model is that in (2).

$$
\begin{cases}
\min \sum_{t=0}^{T-1} \hat{w}_t x_t & \\[2mm]
p_t^d x_t - y_d + z_t^d \geq |w_t^d - p_t^d| - 1, 0 \leq t \leq T-1, d \in D, & (i) \\[2mm]
\sum_{t=0}^{T-1} z_t^d \leq T(\theta + 1 - y_d), \ d \in D, & (ii) \\[2mm]
\sum_{d \in D} y_d \geq \lambda D, & (iii) \\[2mm]
y_d \in \{0,1\}, 0 \leq x_t, 0 \leq z_t^d \leq 1. &
\end{cases}
\qquad (2)
$$

The auxiliary variables (z_t^d) account by how much power the process (w_t^d) violates the band around the forecast (p_t^d), either at the top or the bottom, for those days classified as regular (i.e., when $y_d = 1$). For instance, if $y_d = 1$ and $w_t^d \geq p_t^d(1 + x_t)$, then it must be held $z_t^d \geq w_t^d - p_t^d(1 + x_t) \geq 0$ to satisfy equation (i) in (2) for that day d at time t. When $y_d = 1$ and $w_t^d \leq p_t^d(1 - x_t)$, then z_t^d should verify $z_t^d \geq p_t^d(1 - x_t) - w_t^d \geq 0$ to satisfy equation (i). For a graphical reference about both situations, please see Figure 8. The optimization process pushes down the z_t^d values, which ultimately are to be set to

$\max\left[0, w_t^d - p_t^d(1 + x_t), p_t^d(1 - x_t) - w_t^d\right]$, the anti-reliability of (1). That equation is always satisfied when $y_d = 0$ simply by choosing $z_t^d = 0$ for every t; therefore, atypical days are disregarded for violations.

Given any day d, when $y_d = 1$ (an effective day of the training set), the second equation guarantees that the time-normalized cumulated off-band energy along the forecasted period T is below θ. That is, in terms of the reliability: $1 - \mathcal{R}^d = \left(\sum_{t=0}^{T-1} z_t^d\right)/T \leq \theta$; so θ bounds the energy that lies outside the band to a fraction of the installed power plant. As it happens with (i), equation (ii) is automatically satisfied when $y_d = 0$. Coming back to Figure 8 as a reference instance, by combining equations (i) and (ii) inside an optimization process, we are forcing the total off-band energy (the result of adding up both yellow areas) to be under a desired threshold. Finally, equation (iii) forces the problem to select at least λD days to be regular, which combined with the persistence hypothesis conveys likelihood to the result.

3.2. Experimental Evaluation

The experimental evaluation of this work is based upon a later open data from the Uruguayan Electricity Market. From that past repository, we chose two independent forecast sources: Garrad Hassan and Meteológica. The data were pre-processed using a power assimilation methodology, which fits forecasts along the first 6 h in order to match the starting state (w_0^d). The exact process is described in paper [3]. The used forecasts from Garrad Hassan were those issued at 1AM between 5 April 2016, and 10 March 2017. Within this period there are 302 days where both, forecast and actual data, are complete. Regarding the other provider (Meteológica), the number of complete records is 394, with dates of issue ranging from 1 January 2016, to 10 March 2017. Regarding our own forecast (PSF44), synthesized up from a series of actual power registers, we used the same 730 days of between years 2014 to 2016 that were used upon the first part of Section 2.1. Best performance was found by using $K = 4$ and $W = 4$ (acronym PSF44 refers to those parameters).

Throughout this work, we relied upon `IBM(R) ILOG(R) CPLEX(R) Interactive Optimizer12.6.3` as the optimization solver. The server was an `HP ProLiant DL385 G7`, with `24 AMD Opteron(tm) Processor 6172` with 64 GB of RAM. After running model (2) over a training set comprising around 30% of Meteológica's days, we find bands like those sketched in Figure 9.

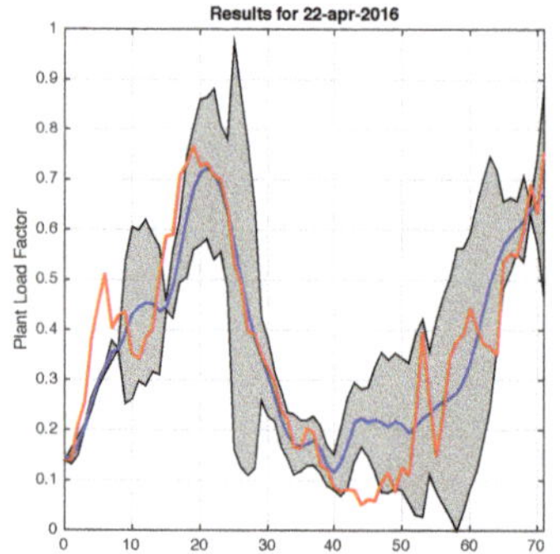
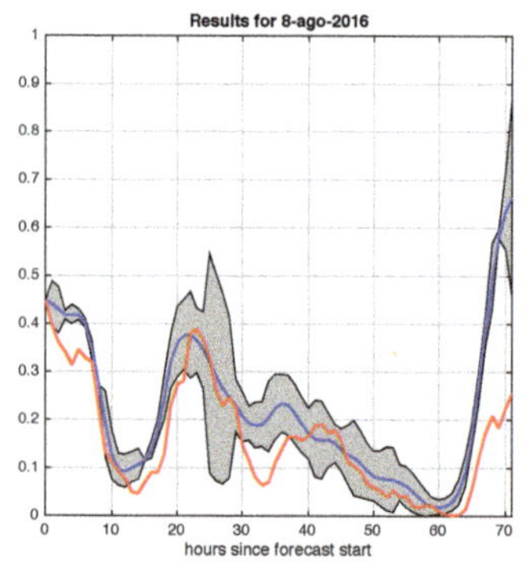
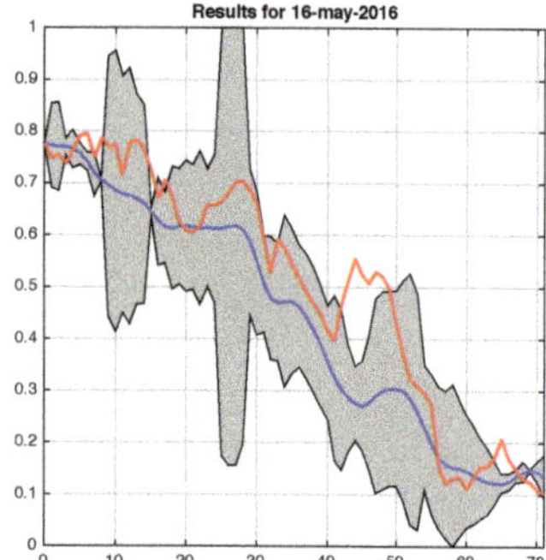

Figure 9. Wind-energy bands for three random days within Meteológica's training set [$\lambda = 1$, $\theta = 0.05$].

The x-axis represents the number of hours ahead for each forecast, while the y-axis corresponds to the PLF. Blue curves are associated with power forecasts while red ones are the actual values. Finally, the grey area represents the wind-power band for $\theta = 0.05$ and $\lambda = 1$. Since λ equals 1, every day within the training set must be effectively included; that is, $y_d = 1$ for each $d \in D$, so all days are treated as regular. Furthermore, when $\lambda = 1$ then (2) turns out to be a pure linear programing problem, and running times are within the second. Fixing λ to 1, it is of interest to explore how θ modifies the bands. Figure 10 shows the result over the same training set when $\theta = 0.01$.

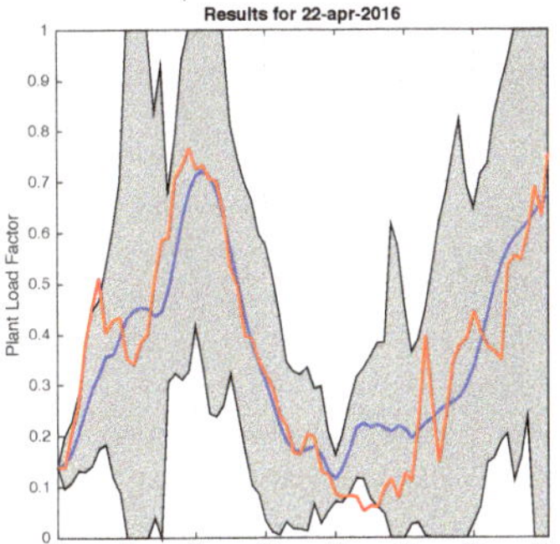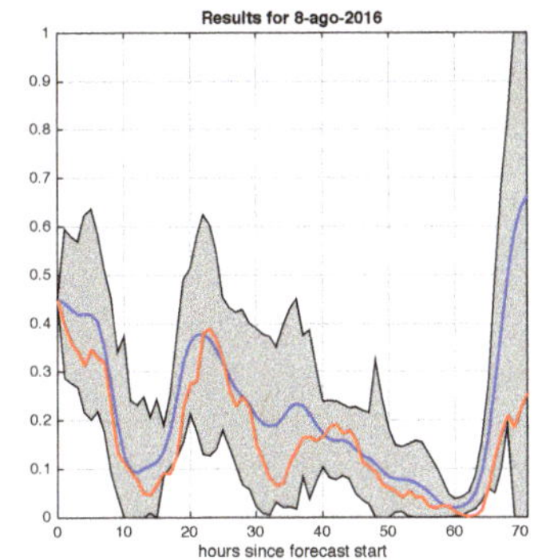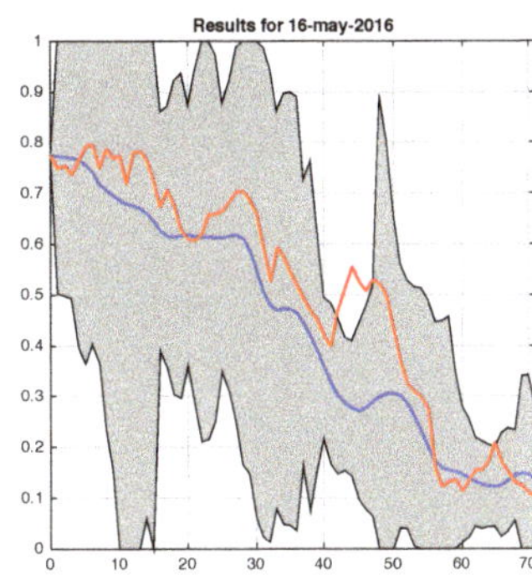

Figure 10. Wind-energy bands for the same days when $\theta = 0.01$ instead of $\theta = 0.05$ [$\lambda = 1$].

Observe that bands in Figure 10 are wider than in Figure 9, which is expected since we are less tolerant respect to how much energy lies outside those bands. In order to balance reliability and thickness, it is of interest to compute how much area do bands cover as we change θ while keeping $\lambda = 1$.

The training in all of the previous cases was performed over a set D of 120 randomly selected days out of a set of 300 days in common for all providers. The common complement, i.e., the set of (180) days shared by these three forecasts and not being in the *training-set*, is used as the *test-set*. The calibration of PSF44 was crafted using the set of 430 contiguous days previous to those of training and test sets. Experimental evaluation (see leftmost of Figure 11) verifies that the average width of the bands, when trained over the entire training set of forecasts ($\lambda = 1$), falls down rapidly to 0, which is reached upon both companies when θ is close to 0.2. Although similar, Meteológica's bands (blue) are always better than Garrad Hassan's (red). PSF44 (green) requires much wider bands to achieve the same grades of reliability. The middle plot shows the relative difference between widths of original bands (those of the training set), and widths computed over the test-set using the corresponding x vector found for each θ. It is worth mentioning that widths are always similar (divergence is low), so the objetive function in (2) when evaluated over the training set is representative of what happens outside it. This is sustained even for relatively higher θ values right below 0.2, where widths tend to zero and the relative deviation makes no sense to be accounted.

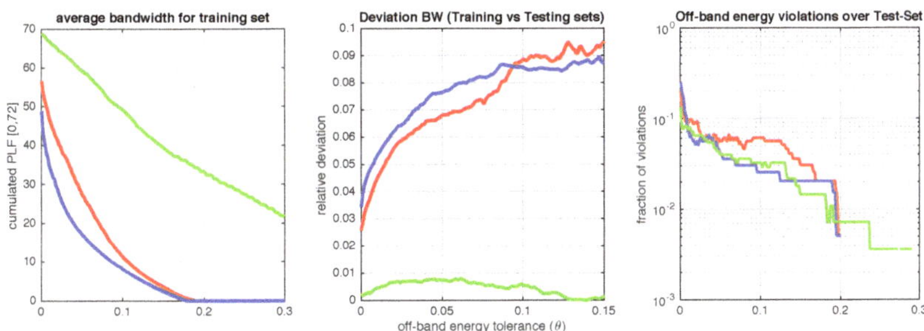

Figure 11. Average width of bands found for the training set as θ changes while λ is fixed in 1 [leftmost], relative deviation between average widths register for training and testing sets [middle], and violations of off-band energy limits over testing set. [red samples correspond to Garrad Hassan, blue ones to Meteológica and green to PSF44].

Regarding off-band energy violations to the limit θ when computed over the test-set (they do not happen in the training-set because of (2)(*ii*)), the rightmost of Figure 11 shows the fraction of those violations, i.e., the fraction of samples where the off-band energy surpasses $T\theta$. They are also low for all forecasts and are particularly lower as values of θ get apart from zero. The previous exercise experimentally justifies the persistence hypothesis this technique is based on.

The goal of this work is providing stochastic short-term optimal power dispatch schedulers, with accurate wind-energy bands, in the context of the Uruguayan Electricity Market. In particular, our interest is keeping off-band energy below 10% of the average PLF, which is around 0.35; so we consider $\theta = 0.035$. In Uruguay, 35% of electricity comes from wind-power sources, thus that value of θ corresponds to 1.23% of the average energy consumed, what is ambitious. That value is used as reference during this work.

The other parameter to consider is λ which attends to the fact that, whatever accurate a family of forecasts may be, there will always be samples that degrade the overall quality of the whole. Table 2 shows how some attributes of the bands change as λ decreases from 1 to 0.6, while θ remains fixed in 0.035 (our target off-band violation).

Table 2. Experimentally estimated attributes for confidence $\theta = 0.035$ bands as λ decreases.

λ	% Anomalous	$\overline{BW}$	%$\overline{BW}$	t (s)	% ANOMALOUS	$\overline{BW}$	%$\overline{BW}$	t (s)	% Anomalous	$\overline{BW}$	%$\overline{BW}$	t (s)
1.00	6.6%	21.3	29.6%	<1	5.6%	34.3	47.6%	<1	5.4%	62.3	86.5%	<1
0.95	14.8%	16.0	22.2%	117	12.2%	26.8	37.2%	129	8.0%	56.6	78.6%	77
0.90	20.4%	13.8	19.2%	138	16.3%	24.8	34.4%	428	11.2%	51.6	71.7%	66
0.85	27.0%	12.5	17.4%	265	22.5%	22.1	30.7%	272	13.4%	48.4	67.2%	292
0.80	32.1%	11.7	16.3%	664	25.5%	19.7	27.4%	1518	16.3%	45.6	63.3%	421
0.75	36.7%	10.8	15.0%	1296	37.2%	17.4	24.2%	1097	20.7%	41.8	58.1%	657
0.70	45.4%	9.9	13.7%	2020	45.4%	15.6	21.7%	1950	28.3%	38.5	53.5%	842
0.65	49.0%	8.9	12.3%	1225	49.0%	13.1	18.2%	1480	28.6%	35.9	49.9%	1468
0.60	54.6%	8.3	11.5%	1390	52.6%	11.9	16.5%	1560	37.0%	34.2	47.5%	2425
	Meteológica				**Garrad Hassan**				**PSF44**			

The first three four columns correspond to Meteológica forecasts, the second part does to Garrad Hassan's and the last one to PSF44 forecasts. These metrics were computed over the test-set by using optimal x coefficients for each λ over the samples in the training-set. Columns labeled as *%anomalous* indicate the percentage of the samples, in each case, whose off-band energies surpasses the 0.035 of the total plant factor ($T\theta$). We decided to use different adjetives to distinguish between *atypical days*: samples intentionally excluded from the training-set, and *anomalous days*: samples in the testing-set that by chance surpass the off-band energy limit. The columns $\overline{BW}$ and %$\overline{BW}$ respectively show the average absolute and relative areas of the wind-power bands over the test set, using 72 as the full plant factor for the time horizon. Finally, the number of seconds spent by the solver to find the optimal solution for each case appears in the column labeled as *t(s)*. Observe that as λ decreases so it does the expected width for energy bands, because the solver is allowed to select down to λD days during the training, and the optimization ends up by crafting bands for the best subset with a λ fraction of the original number of days. Computation times ascend, because (2) becomes combinatorial for $\lambda < 1$. Conversely, the percentage of *anomalous days* (i.e., those whose off-band energy falls outside the limit) increases, since the calibration performed over a partial/elite training-set is no longer representative over the complement (i.e., the test-set). A second goal of this work is keeping the percentage of *anomalous days* below 10%, which translates into attaining the target θ at least 90% of the times. The final goal is over the allowed variance for wind-power. Until now, we have focused upon energy rather than power. Keeping the process within narrower bands is equivalent to expect lower power variations. According to official sources, the total electricity produced by Uruguay during 2017 (to meet internal demand plus energy exports to Argentina and Brazil) was of 12,600 GWh. The equivalent hourly average power is 1438 MW. The total wind-power plant by late 2017 was of 1437 MW (the fact these final figures match is just a coincidence). Hence, aiming on having energy bands whose relative width is below 20% is equivalent to expect average power fluctuations (either upwards or downwards the centroid) below 10% of the installed wind-power plant, which matches the average power consumption. In summary, our targets are: $\theta \leq 0.035$, $\%\overline{BW} \leq 20\%$ and $\%anomalous \leq 10\%$. Observe that no record in Table 2 fulfills all these goals simultaneously. Along Section 4 we see how to deal with that issue.

4. Combining Forecasts

At first sight, we might think that a convex combination of forecasts and their bands would inherit the width of each one, and that we cannot improve bands quality by means of combining them. The only mechanism we have seen that can get narrower bands goes by reducing λ. As a drawback, this also increases the percentage of anomalous days. However, we might regain confidence if anomalous days -for the different forecasts- were somehow independent, since a combination of anomalous situations in all bands would be rarer than in any of them by separate. That's the idea behind this section. To check the consistency of this idea we analyze how independent anomalous days are, by its correlation matrices. Table 3 recapitulates figures of anomalous days for different values of λ with $\theta = 0.035$.

Table 3. Correlation matrices for anomalous days for different λ's and $\theta = 0.035$.

	MT	GH	PS		MT	GH	PS		MT	GH	PS
MT	1.0000	0.2621	−0.0178	**MT**	1.0000	0.2854	0.0147	**MT**	1.0000	0.2797	−0.0093
GH	0.2621	1.0000	−0.0648	**GH**	0.2854	1.0000	−0.0720	**GH**	0.2797	1.0000	−0.0308
PS	−0.0178	−0.0648	1.0000	**PS**	0.0147	−0.0720	1.0000	**PS**	−0.0093	−0.0308	1.0000
	$\lambda = 0.60$				$\lambda = 0.65$				$\lambda = 0.70$		
	MT	GH	PS		MT	GH	PS		MT	GH	PS
MT	1.0000	0.2450	0.1127	**MT**	1.0000	0.2488	0.0581	**MT**	1.0000	0.2507	0.0109
GH	0.2450	1.0000	0.0817	**GH**	0.2488	1.0000	0.0810	**GH**	0.2507	1.0000	0.0929
PS	0.1127	0.0817	1.0000	**PS**	0.0581	0.0810	1.0000	**PS**	0.0109	0.0929	1.0000
	$\lambda = 0.75$				$\lambda = 0.80$				$\lambda = 0.85$		
	MT	GH	PS		MT	GH	PS		MT	GH	PS
MT	1.0000	0.3246	0.0262	**MT**	1.0000	0.2830	−0.0926	**MT**	1.0000	0.2921	−0.0707
GH	0.3246	1.0000	0.1119	**GH**	0.2830	1.0000	0.0281	**GH**	0.2921	1.0000	0.0243
PS	0.0262	0.1119	1.0000	**PS**	−0.0926	0.0281	1.0000	**PS**	−0.0707	0.0243	1.0000
	$\lambda = 0.9$				$\lambda = 0.95$				$\lambda = 1.00$		

These numbers were computed over the test-set for Bernoulli's random variables, $Mt(d)$, $Gt(d)$ and $Pt(d)$, indicator of the event of finding an anomalous day: they evaluate to 1 (respectively 0) if and only if the forecast for day d of the respective corresponding -Metológica, Garrad Hassan and PSF44- classifies as anomalous (resp. regular). From these correlation values, we infer that anomalous days of Metológica and Garrad Hassan are positively but weakly correlated. By running simple simulations with two sets of dependent Bernoulli's random variables with the same expected value, we observe that the correlations values of the table appeared when 1 out of between 3 to 4 samples of one set copy the value of the other. PSF44 is basically independent of the others providers, and in fact can either be positively or negatively correlated with them.

Given the three sets of forecasts: $\mathcal{P}_{mt}$, $\mathcal{P}_{gh}$ and $\mathcal{P}_{ps}$, and their corresponding functions to compute bands (lower and upper bounds): $bdMT_\lambda(p) \rightarrow [0,1]^{T \times 2}$, $bdGH_\lambda(p) \rightarrow [0,1]^{T \times 2}$ and $bdPS_\lambda(p) \rightarrow [0,1]^{T \times 2}$, we explore convex combinations of them: $bdMX = \alpha \cdot bdMT_{\lambda 1} + \beta \cdot bdGH_{\lambda 2} + (1 - \alpha - \beta) \cdot bdPS_{\lambda 3}$, with $0 \leq \alpha, \beta \leq \alpha + \beta \leq 1$, for different combinations of λ's. The goal is on finding the combination that is closest to satisfy the targets: $\theta \leq 0.035$, $\%\overline{BW} \leq 20\%$ and $\%anomalous \leq 10\%$. This second stage of training was performed over the half of the test-set (90 days). The other half remains as the definite test-set to check results. The most convenient combination over the new training-set was found for values: $\alpha = 0.66$, $\beta = 0.25$, $\gamma = 0.09$, $\lambda_1 = 0.85$, $\lambda_2 = 0.70$ and $\lambda_3 = 0.65$. After checking over the now reduced test-set we verify that for $\theta = 0.035$ as limit for off-band energy, 8.9% of the days fall into the anomalous condition, while the value for the average bandwidth is $\%\overline{BW} = 21.7\%$. This final figure does not attain our original goal (i.e., 20%), but it is pretty close to it.

Performance of Optimally Combined Bands

The lecture of the previos figures indicates that the most performant family of forecasts (Metológica) contributes with 66% of the weight when is calibrated using 85% of its better forecast samples. Despite having similar performance (recall Figures 5 and 11), Garrad Hassan's forecasts only contributes with 25% of the weight, and that is after filtering 30% of its samples. Unexpectedly, being the worst by far, PSF44 contributes with almost 10% to the final result, although after purging 35% of its samples. Probably, the higher weight of PSF44 comes from its almost independence (small correlation) with respect to the other forecasts, rather than its quality.

To analyze the performance of the combined band we present qualitative and quantitive evidence. Figure 12 sketches random bands, its centroid and the corresponding actual power over six days within the test-set. The last two figures (middle and rightmost plots in the bottom row) correspond to two of the eight anomalous days found. Although the off-band energy surpasses the $T\theta$ limit, overall, the performance of those bands doesn't look that bad either.

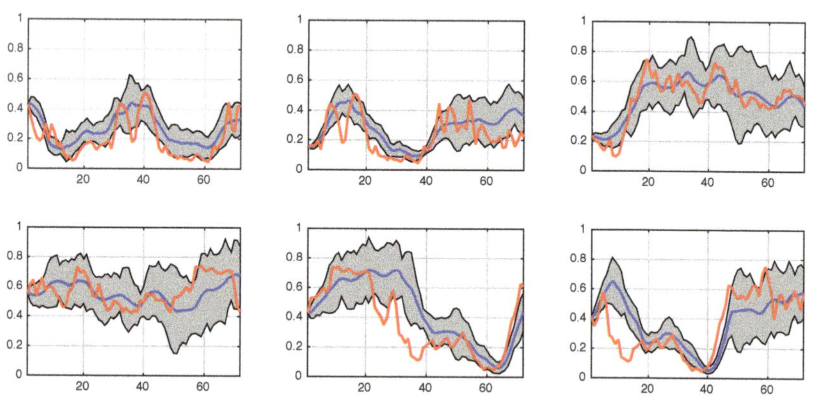

Figure 12. Hybrid bands for six random days in the test-set [blue is the centroid, the red one is the actual power].

Figure 13 shows other group of six random bands. This case does not include anomalous samples, but there are a couple of samples where the area of the band is above the target value. The most notorious case is that on the leftmost of the bottom row.

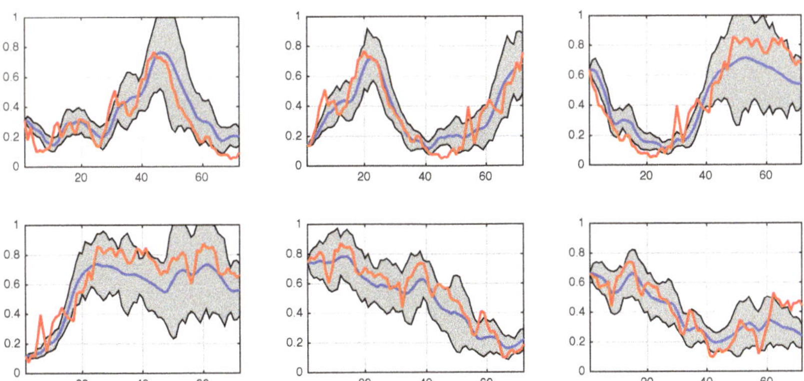

Figure 13. Hybrid bands for six random days in the test-set [blue is the centroid, the red one is the actual power].

It is worth wandering how much energy lies outside the band when violations happen, and how narrow confidence bands are. The residual test-set is so small (90 samples) that, although biased, we decided to use the old one to craft histograms. Figure 14 shows histograms computed up from the original (180 samples) test-set. The leftmost corresponds to the distribution of the off-band energy normalized by the total PLF along the period (72). It is observed that no sample disagrees in more than 7.3%, while in 50% of the samples (those colored with red) that percentage is lower than 1.6%. The rightmost represents the distribution of normalized widths ($\%\overline{BW}$). As in the previous case, samples colored in red add up to 50% and all of them are lower than 14.6%.

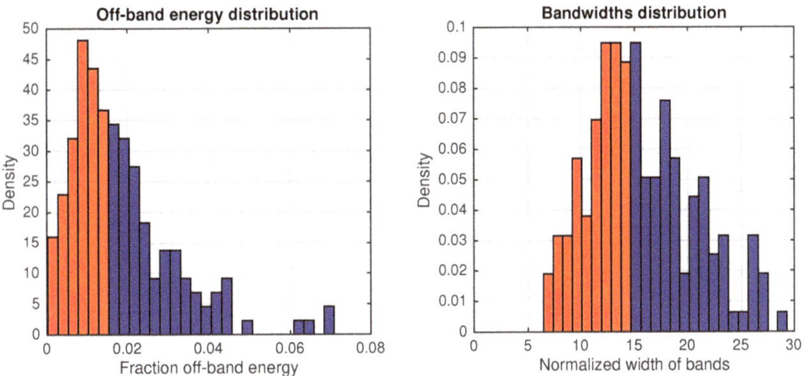

Figure 14. Histograms for relative off-band energy and widths of the bands.

Complementing the previous figure, Figure 15 marks with green the quantiles where the values of either: off-band energy [leftmost] or normalized widths [rightmost] satisfy original targets. The cumulated probability of samples in the first totalizes 90.17%, while those over the rightmost add up to 79.8%. These results reflect the quality of the forecasts and bands computed by this method. We conclude then, that the result is not only satisfactory regarding our initial average performance goals ($\theta \leq 0.035$, { but it is pretty good in terms of the overall quality of the bands and specially in terms of the energy confidence of them.

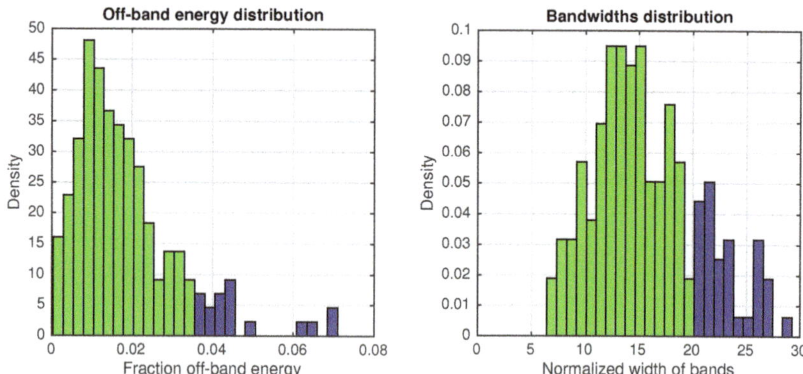

Figure 15. Histograms for relative off-band energy and widths of the bands.

5. Conclusions and Future Work

In this work we introduce a novel learning technique for crafting wind energy bands around forecasts of wind-power generation over a horizon of 72 h ahead. The analysis is based on a historical data set provided by the Uruguayan Electricity Market. The technique allows to discard a portion of atypical days in the training-set, while controls the average cumulated energy that lies outside bands. With an appropriate choice of the parameters involved in the analysis, the model has succeeded in providing bands satisfying natural requirements on confidence and width.

A remarkable conclusion of this work is that the use of an optimal convex combination of conditionally independent (or weakly dependent) forecasts and its corresponding bands improves significantly the performance of the model. For instance, the experimental evaluation of Section 3 suggests that Meteológica forecasts performs, in average, better than Garrad Hassan's and PSF44. However, an appropriate convex combination of all of them (even when the performance of PSF44 is rather bad) provides better results. While most of the weight of the combination goes to Meteológica, the inclusion of Garrad Hassan and PSF44 forecasts conveys stability to the result, compensating the fact that some anomalous days for one forecast are regular according to the others. This idea could be extended of course including more forecasts providers.

A drawback of the analysis that we mention here is that the available data set at the moment this work was developed was not too large (around 300 days). Regarding the quality of bands, we expect the performance of the technique will work even better with a more extensive data-set, perhaps spanning a few years. Nevertheless, this size introduces a challenge: increasing the training-set significantly increases computation times. Notice that after adding up computation times reported in Table 2, the total time is above 6 h, which is pretty good for the purposes of these experiments. However, those times are expected to be much higher as the training data-set increases in size, so, in such situations is necessary the introduction of specific algorithms to solve the optimization problem in (2), i.e., not to rely upon standard solvers. A line of future work precisely goes the way of experimenting with other exact methods or derivatives thereof and the exploration of Metaheuristics, in order to find more efficient algorithms to tackle the problem.

Complementarily, the current model uses a single set of x's variables to delimit bands around forecasts, which results in symmetric widths either upwards or downwards. It is worth testing this hypothesis by including two sets of x's, on per each direction, and letting the optimization to find solutions over a larger search space. A previous clusterization of forecasts might also improve the performance. Since the training-set indistinguishably comprises both: samples for windy and not-windy days, the relative deviation at a time t necessary to reposition the process within a band shall be greater for forecasts of low prospected energy than the necessary for high energy ones. The previous over-penalizes widths of bands in forecasts with higher expected energy. Training different bands for different seasons might also improve the quality. Most of these ideas however, require historical data sets much larger than the one currently available.

Regarding the application of bands as those developed in this work. They may be particularly important to craft scenarios in stochastic optimization problems where the complexity of state variables does not allow using other techniques, such as dynamical programming. Examples of such situations arise from a combination of: generation units with complex commitments (limit to minimum power, a slow starting/stopping process, a minimum uptime operation once started); temporal correlation between generation units (e.g., dam water reservoirs where water influxes come from another hydroelectric dam); control deferrable consumption (e.g., electrical residential water heating that must be accounted within certain time windows); large scale energy storage to be later returned to the grid; among others. That results in a wide spectrum of potential application cases.

Author Contributions: Both authors have contribute equally. Both authors have read and agreed to the published version of the manuscript.

Funding: This research received no external funding.

Institutional Review Board Statement: None.

Informed Consent Statement: None.

Data Availability Statement: None.

Acknowledgments: This work was partially supported by PEDECIBA-Informática and PEDECIBA-Matemática (Uruguay), by the STIC-AMSUD project 15STIC-07 DAT (joint project Chile-France-Uruguay), and by ANII (Agencia Nacional de Investigación e Innovación, Uruguay)-Fondo Sectorial de Energía 2015, ANII-FSE_110454.

References

1. Weron, R.; Misiorek, A. Forecasting spot electricity prices: A comparison of parametric and semiparametric time series models. *Int. J. Forecast.* **2008**, *24*, 744–763. [CrossRef]
2. Cazes, G.; Ortelli, S. Minimum-Cost Numerical Prediction System for Wind Power in Uruguay, with an Assessment of the Diurnal and Seasonal Cycles of its Quality. *Ciência Nat.* **2018**, *40*, 205–210.
3. De Mello, S.; Cazes, G.; Gutiérrez, A. Operational wind energy forecast with power assimilation. In Proceedings of the 14th International Conference on Wind Engineering, Porto Alegre, Brazil, 21–26 June 2015.
4. Pinson, P. Estimation of the Uncertainty in Wind Power Forecasting. Ph.D. Thesis, École des Mines de Paris, Paris, France, 2006.
5. Pinson, P.; Nielsen, H.A.; Møller, J.K.; Madsen, H.; Kariniotakis, G.N. Non-parametric probabilistic forecasts of wind power: Required properties and evaluation. *Wind Energy* **2007**, *10*, 497–516. [CrossRef]
6. Pinson, P.; McSharry, P.; Madsen, H. Reliability diagrams for non-parametric density forecasts of continuous variables: Accounting for serial correlation. *Q. J. R. Meteorol. Soc.* **2010**, *136*, 77–90. [CrossRef]
7. Staid, A.; Watson, J.P.; Wets, R.J.B.; Woodruff, D.L. Generating short-term probabilistic wind power scenarios via nonparametric forecast error density estimators. *Wind Energy* **2017**, *20*, 1911–1925. [CrossRef]
8. Møller, J.K.; Zugno, M.; Madsen, H. Probabilistic Forecasts of Wind Power Generation by Stochastic Differential Equation Models. *J. Forecast.* **2016**, *35*, 189–205. [CrossRef]
9. Elkantassi, S.; Kalligiannaki, E.; Tempone, R. *Inference and Sensitivity in Stochastic Wind Power Forecast Models*; UNCECOMP: Rhodes Island, Greece, 2017.
10. Caballero, R.M. Stochastic Optimal Control of Renewable Energy. Master's Thesis, King Abdullah University of Science and Technology, Tuwa, Saudi Arabia, 2019. [CrossRef]
11. Bertsekas, D.P. *Dynamic Programming and Optimal Control*; Athena Scientific: Belmont, MA, USA, 1995; Volume 1.
12. Risso, C. Benefits of demands control in a smart-grid to compensate the volatility of non-conventional energies. *Rev. Fac. Ingeniería Univ. Antioq.* **2019**, 19–31. [CrossRef]
13. Nohara, D.; Ohba, M.; Watanabe, T.; Kadokura, S. Probabilistic Wind Power Prediction Based on Ensemble Weather Forecasting. *IFAC Pap.* **2020**, *53*, 12151–12156.
14. Foley, A.M.; Leahy, P.G.; Marvuglia, A.; McKeogh, E.J. Current methods and advances in forecasting of wind power generation. *Renew. Energy* **2012**, *37*, 1–8.
15. Khan, P.W.; Byun, Y.C.; Lee, S.J.; Kang, D.H.; Kang, J.Y.; Park, H.S. Machine Learning-Based Approach to Predict Energy Consumption of Renewable and Nonrenewable Power Sources. *Energies* **2020**, *13*, 4870.

16. Álvarez, F.M.; Troncoso, A.; Riquelme, J.C.; Ruiz, J.S.A. Energy Time Series Forecasting Based on Pattern Sequence Similarity. *IEEE Trans. Knowl. Data Eng.* **2011**, *23*, 1230–1243 [CrossRef]
17. REN21 Secretariat. *Renewables 2018 Global Status Report*; REN21 Secretariat: Paris, France, 2018. [CrossRef]
18. Jeon, W.; Lamadrid, A.J.; Mo, J.Y.; Mount, T.D. Using deferrable demand in a smart grid to reduce the cost of electricity for customers. *J. Regul. Econ.* **2015**, *47*, 239–272. [CrossRef]
19. Mo, J.Y.; Jeon, W. How Does Energy Storage Increase the Efficiency of an Electricity Market with Integrated Wind and Solar Power Generation? A Case Study of Korea. *Sustainability* **2017**, *9*, 1797. [CrossRef]

 *applied sciences*

Article

Use of Ecofriendly Glass Powder Concrete in Construction of Wind Farms

Eva M. García del Toro [1,*], Daniel Alcala-Gonzalez [1], María Isabel Más-López [2], Sara García-Salgado [1] and Santiago Pindado [3]

1 Departamento de Ingeniería Civil, Hidráulica y Ordenación del Territorio ETSI Civil, Universidad Politécnica de Madrid Alfonso XII, 3, 28014 Madrid, Spain; d.alcalag@upm.es (D.A.-G.); sara.garcia@upm.es (S.G.-S.)
2 Departamento de Ingeniería Civil, Construcción, Infraestructura y Transporte ETSI Civil, Universidad Politécnica de Madrid Alfonso XII, 3, 28014 Madrid, Spain; mariaisabel.mas@upm.es
3 Instituto Universitario de Microgravedad "Ignacio Da Riva" (IDR/UPM), ETSI Aeronáutica y del Espacio, Universidad Politécnica de Madrid, Pza. del Cardenal Cisneros 3, 28040 Madrid, Spain; santiago.pindado@upm.es
* Correspondence: evamaria.garcia@upm.es

Abstract: Silicon is the main element in the composition of glass and it has been seen that it can be used as a partial replacement for cement in the manufacture of concrete. Different dosages of glass powder and cement were applied to manufacture the concrete mixes. Initially, the characteristics of fresh concrete were studied, such as consistency, air content, apparent density and workability. Secondly, compressive strength tests were performed on the different concrete mixtures produced. The consistency tests allowed us to classify these concretes within the group of fluids. The air content of these concretes increased with the rate of substitution of cement by glass powder, resulting in lighter concretes. Density tests showed that this parameter decreased as the rate of substitution of cement increased. A coefficient k has been calculated for the substitution of glass powder by cement in the binder, using the Bolomey formula. Also, a mathematical model has been proposed to further analyze the experimental data. Major contributions of this work were to study the possible application of this concrete in different dispersions as a surface protection layer against the action of corrosion, in wind turbine foundations as well as the stabilization of the wind farm roads.

Keywords: sustainability; compressive strength; Bolomey formula; sustainable concrete; glass powder

Citation: García del Toro, E.M.; Alcala-Gonzalez, D.; Más-López, M.I.; García-Salgado, S.; Pindado, S. Use of Ecofriendly Glass Powder Concrete in Construction of Wind Farms. *Appl. Sci.* **2021**, *11*, 3050. https://doi.org/10.3390/app11073050

Academic Editors: Luis Hernández-Callejo, Maria del Carmen Alonso García and Sara Gallardo Saavedra

Received: 27 February 2021
Accepted: 24 March 2021
Published: 29 March 2021

1. Introduction

By designing the 2030 Agenda, the United Nations (UN) established 17 primary global objectives, the so-called SDGs (Sustainable Development Goals) [1], mainly focusing on eradicating poverty, protecting the planet by fighting climate change and defending the environment. It is a commitment and a challenge that must be addressed jointly, seriously and responsibly from all areas of society. Since ancient times, civil engineering has promoted the development of society through the construction of different types of infrastructures [2]. However, this development has caused severe environmental damage due to the large amount of natural resources demanded, as well as the pollution produced [3]. Current trends in the field of civil engineering are aimed at adapting to the SDGs by achieving resilient and sustainable infrastructures that contribute in some way to the circular economy, where the value of products and materials is kept as long as possible. Waste and the use of resources are minimized, as these resources are kept within the economy when a product has reached the end of its useful life in order to be repeatedly reused and continue creating value [4], and contributing to achieve innovative products that represent an economic benefit and a higher quality of life for people [5]. Therefore, it is important to carry out an evaluation of the efficiency and sustainability of the works to determine the degree of efficiency of the materials and construction methods [6].

Within the framework of the circular economy, the role of glass is worth highlighting [7]. Glass is a material that is easily recyclable due to its physical-chemical characteristics [7]. All types of glass waste are used in the recycling process, coming from the selective recovery of containers and packaging from the glass and ceramic industry [8].

Although in most industrialized countries the percentage of annually recycled glass is increasing, there is still a high percentage of glass that is disposed in landfill [9], which involves an important problem due to the accumulation of non-degradable waste, especially in highly populated areas [10].

Some of the problems to increase the recycling rate of glass waste come from the combination of different colors of glass waste, as well as difficulties in removing dirt, paper or other contaminants from glass products [11]. These wastes that cannot be recycled will be reused for certain uses.

Currently, there are many studies that show the good properties of glass waste, which cannot be recycled, as substitutes for certain materials in the preparation of mortars and concrete. They are considered indeed one of the most suitable substitutes for sand and cement, due to their physical characteristics and chemical composition [12–14]. This reuse of waste materials becomes a viable strategy to reduce the use of Portland cement and natural aggregates in the preparation of mortars and concretes, reducing environmental and energy impacts. Among these, the reduction of CO_2 emissions is significant [14,15], as well as of areas destined for controlled landfills [13,16]. In this context, the so-called eco-efficient concretes arise, which comply with the characteristics outlined, but, in some cases, they have some worse properties, such as compressive strength or durability, when compared to those made with natural materials [16].

The use of finely grinded glass powder in the manufacture of mortars and concretes has been widely studied, especially the optimum particle size. Most of the studies have focused on assessing how the properties of concrete vary depending on the substitution percentage of cement by glass powder, as well as its particle size, which has been shown to play a vital role in the alkali–silica reaction (ASR) [17–19]. At this regard, and according to Idir et al. [19], with a particle size between 0.9 and 1 mm and a substitution percentage of 20% with glass powder, the classic contractions due to ASR do not occur. Corinaldesi et al. [20] stated that up to a substitution percentage of the aggregates by glass powder of 70% can be reached, provided that a particle size between 36 and 50 μm is used. This showed that by reducing the particle size of glass powder, the pozzolanic properties of the binders manufactured increase. In addition, a greater long-term strength resistance of the pastes manufactured with this type of cement was obtained, due to the higher presence of C-S-H gels [19,20]. Also, these gels have a self-repairing property when they are used in the stabilization of rolling track soils by prolonging their setting over time [21].

Liu [22] reported that self-compacting concrete produced with 10% glass powder to replace cement had good properties when fresh. They indicated that workability decreased as the glass powder content of the concrete increased due to the geometry of the glass dust. In the same line, Parghi et al. [13] indicated that the sharper edge and the more angular shape of the glass powder (GP) particles reduced the fluidity of the cement mortars and concretes.

Nassar and Soroushian [23] reported pozzolanic activity when glass powder with a particle size of 13 μm was used as a fractional replacement for cement in concrete. Schwarz et al. [24] studied the properties of ground glass powder and reported that up to a certain percentage substituting ground glass for cement was a viable solution for fabrication in concrete.

Studies carried out by Sahyan et al. [25] have shown that at constant water-to-binder mass ratio, the addition of 20% of glass powder significantly reduced the chloride ion permeability of concrete, which was confirmed by Schawrz et al. [21]. This property confers protective properties against corrosion to concretes made with glass powder and cement.

Shaoa et al. [23] observed an increase in compressive strength of 120% at the cure age from 3 to 90 days, when concrete was produced with ground glass powder with particle size up to 38 μm.

According to Pengwei et al. [24], in the traditional concrete used in civil construction that is generally subjected to large changes in temperature, its durability is clearly affected, and in extreme cases can leave the concrete out of service. To avoid these consequences in traditional concretes, air-entraining additives are incorporated into the mix. In the case of concrete made with glass powder as a binder, the higher the percentage of substitution of glass powder for cement, the higher the air content, so it is not necessary to include additives.

In 1935, Bolomey gave a formula to predict the compressive strength of cement mortar, which expresses a linear relationship between the water–cement ratio and compressive strength. This expression indicates that compressive strength of cement-based materials is mainly dependent on the water–cement ratio among all the other factors. Therefore, it is seen as a mathematical form of water–cement ratio law. In this regard, based on Fernández Cánovas studies [25], a calculation of theoretical compressive strengths at 28 and 90 days was reported using Bolomey's dosage, introducing a coefficient k that represented the replacement of cement CEM I52.5 R by glass powder.

Considering all the above-mentioned points, our initial research hypothesis was that the glass powder used in this work, with its characteristics and particle size, will allow us to produce an ecofriendly concrete, whose mechanical properties will not be adequate to use it as a structural concrete, but it may be useful as a surface protection layer to avoid or reduce corrosion phenomenon. Therefore, the aim of this work consisted of studying the compressive strength of cements produced with different substitution percentages of a certain glass powder. A mathematical model has been proposed to fit the experimental compressive strength results. Also, Bolomey's formula was applied for simulation of the relationship between the water–cement ratio and 28- and 90-day compressive strength.

In summary, major contributions of this work were to study the possible application of concrete made with cement and glass powder in different dispersions as a surface protection layer against the action of corrosion, in wind turbine foundations as well as the stabilization of the wind farm roads, since it is a sustainable and environmentally friendly material. On the other hand, the mathematical model developed has resulted in an appropriate simulation tool, since errors between real and simulated final stable values of compressive strength were lower than 3.3%. Finally, it has been proved that glass powder exerted an important activity in increasing the long-term compressive strength of concretes, according to results obtained by the application of Bolomey's formula. Also, the use of glass powder as a binder in the concrete would be beneficial from the point of view of the circular economy and environmental footprint because a final waste, which cannot be further recycled and whose destiny would be a landfill, may have another useful application.

2. Materials and Methods

2.1. Materials

CEM I 42.5-R Portland cement (Cementos Portland Valderrivas, Morata de Tajuña, Madrid, Spain) was used. This cement had a density of 3.12 g/cm^3, a specific surface of 4.440 cm^2/g and a green–gray color. Its chemical composition was as follows—CaO (65%), SiO$_2$ (19%), Al$_2$O$_3$ (5.5%), Fe$_2$O$_3$ (2.65%), SO$_3$ (2%), MgO (2%), Na$_2$O (0.15%), K$_2$O (0.7%).

The aggregate in mortar was essentially siliceous and non-reactive. It was sand with a granulometry <4 mm, gravel 4–12 mm and gravel 12–20 mm.

The glass powder used came from the grinding of waste from the ceramic industry, as well as from containers and packaging from the selective rubbish collection, which cannot be reused due to their characteristic (high percentage of paper, cork or plastic attached). They have been ground in a bar mill equipped with 15 bars of 3 different diameters and with different grinding times.

d50 glass powder of 16 μm was used (dimension of sample particles for which 50% of them have a diameter lower than a certain value) [8], which provides interesting mechanical results at a cost energy clearly lower than that necessary to obtain smaller granulometries of glass powder. This makes the use of this material more sustainable in the field of Civil Engineering [26].

Chemical composition of glass powder was given by the manufacturing company. It is composed by 71.00% SiO_2, 11.80% Na_2O, 11.28% CaO, 2.20% Al_2O_3, 1.60% Fe_2O_3, 1.40% MgO, 0.60% K_2O, 0.10% MnO, 0.07% TiO_2 and 0.05% P_2O_5, with 0.90% volatile LF (lost of the Fire).

Physical and mechanical characteristics of the waste used were obtained by means of different analysis techniques—laser granulometry, X-ray Diffraction and Scanning Electron Microscopy. It should be noted that all the aforementioned ground glass powder comes from the same batch of waste. Only their granulometries vary.

2.2. Characterization of Fresh Concrete

Consistency tests were carried out using the UNE-EN 12350-5 standard [27], by means of the settlement test, which is sensitive when the mean settlement is between 10 and 200 mm. The air content of prepared concrete was determined by the UNE-EN 12350-7 standard [28] by pressure methods. The procedure followed for the calculation of density and porosity was based on the corresponding standard [29].

2.3. Sample Preparation

In order to evaluate the different characteristics of concrete, 6 series of test pieces were manufactured in accordance with the UNE-EN 12390-2 standard [30]. The only difference between these specimens was the amount of glass powder used to replace the CEM I 52.5 R cement, as described in Table 1.

Table 1. Summary of experimental conditions for the samples prepared.

Concrete Composition	Sample ID					
	Control	G15	G30	G45	G60	G80
Cement substitution rate for glass powder (%)	0	15	30	45	60	80
Cement CEM I 52,5 R (kg/m^3)	330	280.5	231	181.5	132	66
Glass Powder (kg/m^3)	0	49.5	99	148.5	198	264
Equivalent binder (kg/m^3)			330			
Plasticizer (% binder)			0.35			
Total water (L)			184			
Arid <4 mm			740			
Gravel 4–12 mm			310			
Gravel 12–20 mm			850			

All the mixtures were completely homogenized and poured into 10×30 cm cylindrical molds. They were compacted and after 24 h they were removed from the mold and kept in a humid curing chamber at 20 °C, for 2, 7, 28, 90 and 180 days. After this time, the test tubes were broken in accordance with the instructions of the UNE 83-304-84 standard [31], and the properties of compressive strength of manufactured concretes were determined.

3. Results and Discussion

3.1. Characterization of Glass Powder

Three grinding processes were carried out on the waste glass, each one for a different time, in order to obtain three samples of glass powders with different particle sizes.

Three dimensions have characterized the glass powders—d10, d50 and d90. They represent, respectively, the diameter of the sample particles for which 10%, 50% and 90% of the particles have a diameter smaller than that dimension, as can be seen in Table 2. In this work, the value of d50 was used for the characterization of the different batches of ground glass.

Table 2. Granulometric characteristics of glass powders produced as a function of grinding time.

Glass Powder Used	Griding Duration	d_{10}	d_{50}	d_{90}
T1	2 h 30	2.92 ± 0.01 μm	33 ± 1 μm	110 ± 3 μm
T2	4 h 15	1.96 ± 0.01 μm	16 ± 1 μm	59 ± 2 μm
T3	5 h	1.65 ± 0.01 μm	11 ± 1 μm	43 ± 2 μm

The cumulative granulometric curves of the three samples (Figures 1 and 2) revealed considerable differences in the distribution of particle size.

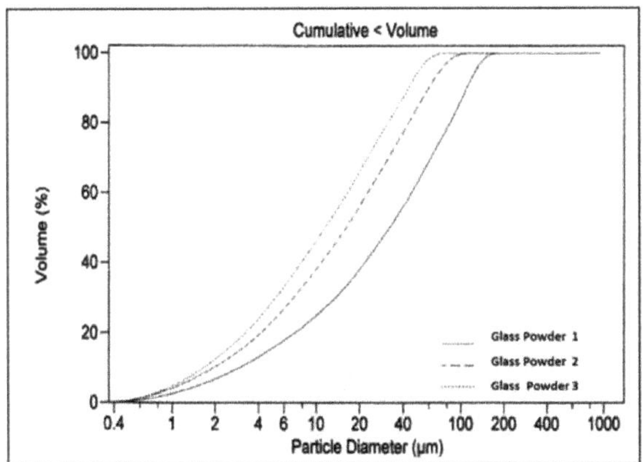

Figure 1. Distribution of the accumulated granulometry of the 3 batches of glass powder [32].

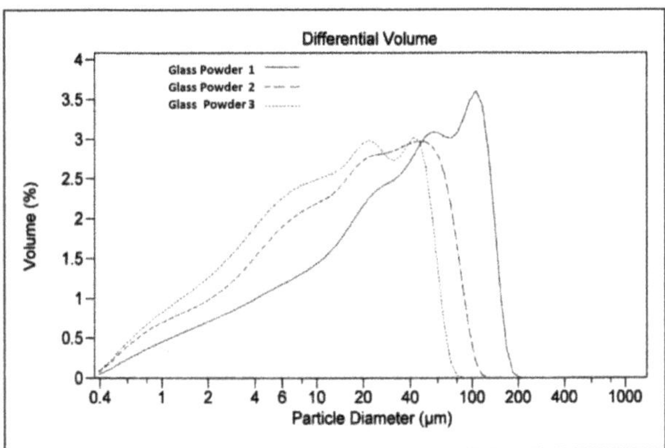

Figure 2. Differential particle size distribution of the 3 batches of glass powder [32].

The glass powder subjected to X-ray diffraction tests resulted in a broad diffraction band between 15 and 45 degrees, which corresponded to its amorphous and disordered structure. The X-ray diffractogram of the glass powder of d50 = 16 μm is shown in Figure 3.

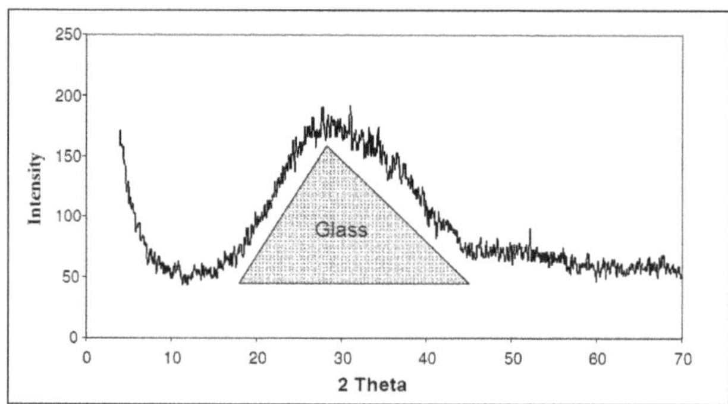

Figure 3. X-ray diffractogram of the glass powder of d50 = 16 μm. The dotted area shows the bulging of the baseline indicating the presence of amorphous phases [32].

By means of the observation through the Scanning Electron Microscope (SEM), glass grains with particle size between 1 and 20 μm can be observed. These grains showed acicular shape and cone-shaped fractures. It should be noted the absence of fine elements attached to these glass particles and their low porosity. A frontal and side views of a glass powder particle, obtained by SEM, are shown in Figure 4.

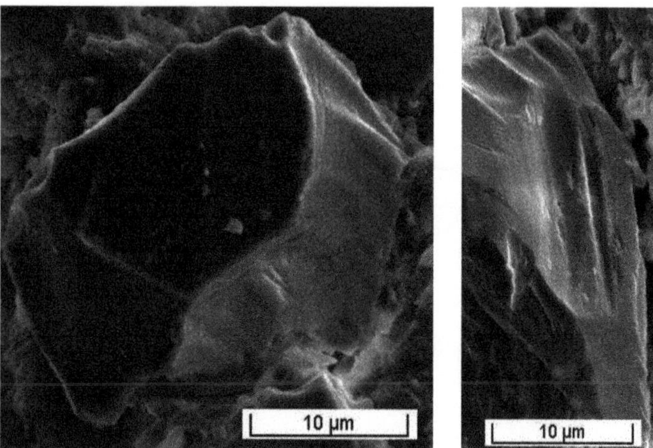

Figure 4. View of a glass powder particle (scanning electron microscopy (SEM) cliche in secondary electrons). Frontal view (**left**) and side view (**right**) [32].

3.2. Results of the Characterization of Concrete

Results obtained for consistency, air content, apparent density and workability tests are shown in Table 3.

Table 3. Results obtained in the characterization of fresh concrete and concrete substituted by glass powder.

	Control	G15	G30	G45	G60	G80
Consistency (mm)	11	16	16	16	14	13
Air content (%)	2.8	3.9	5.0	5.0	6.8	7.5
Apparent density (kg/m^3)	2390.1	2351.0	2338.5	2330.5	2292.0	2282.0
Workability (easy/difficult)	easy	easy	easy	medium	medium	difficult

According to the results obtained for the consistency tests, carried out by means of a settlement test, concretes studied can be classified within the group of fluid concretes, regardless of their dosage. It can be observed that the penetration values decreased as the amount of glass powder added to the mixture increased, which is in agreement with the results obtained by Liu [22]. Regarding the air content, it varied from 3.9% to 7.5%, so air content increased with the substitution rate of the CEM I 52.5 R cement by glass powder. Therefore, it can be said that replacing cement with glass powder led to a higher air content. This phenomenon is caused by the retention of air bubbles by the surface energy of the glass, which is greater the finer the glass grains are. For this reason, the volumetric mass of the different concretes varied, decreasing when the substitution of cement for glass powder increased. In addition to air, a second cause intervened in the decrease in the volumetric mass of the concretes, which was the volumetric mass difference between glass and cement. This is because the substitutions of CEM I 52.5 R cement for glass powder were mass substitutions that were made without the volumetric corrections due to the presence of aggregates.

Workability decreased when the amount of glass powder in the binder increased, also in accordance with the tests carried out by Liu et al. [22].

The densities of the different concretes manufactured for this study were between 2351 and 2282 kg/m^3, so density decreased as the rate of substitution of cement for glass powder increased. This is not in agreement with some authors, for example, Parghi et al. [13], who reported that the density was higher the higher the percentage of glass contained in the mixture was. This may be due to the volumetric mass of the cement used, which was 3.12 g/cm^3, while that of the glass powder was 2.54 g/cm^3.

Regarding workability, the results obtained for the manufactured concretes did not show significant changes. This result is in agreement with those obtained by Pereira-de-Oliveira et al. [33] and Taha et al. [34].

3.3. Mechanical Properties of Concrete

Table 4 presents the compressive strengths of manufactured concretes, with and without cement replacement with glass powder (see Table 1 for dosages), for all test ages (2, 7, 28, 90 and 180 days).

Table 4. Compressive strength of concrete manufactured with cement replaced by glass powder.

t (Days)	Compressive Strength (MPa)					
	Control	G15	G30	G45	G60	G80
2	22.4	17.3	12.4	8.6	2.7	0.5
7	29.9	25.9	20.4	15.9	5.0	1.5
28	34.5	32.0	23.9	21.2	10.0	3.7
90	39.6	32.6	29.8	25.2	14.2	8.6
180	40.4	33.8	30.7	26.9	15.8	10.3

As can be observed, for concrete prepared with cement replaced by glass powder, the compressive strength of the concrete decreased when the quantity of glass powder in the binder increased. This can be attributed to the fact that glass powder has a low pozzolanic activity at the early ages. These results are in agreement with the results obtained by Tan et al. [35] and Mizahosseini et al. [36].

3.3.1. Mathematical Analysis of the Experimental Results

A mathematical model has been proposed to further analyze the experimental data. The results obtained fit the following equation:

$$Cs = Cs_0\left(1 - exp\left(-\left(\frac{t}{\tau}\right)^n\right)\right) \tag{1}$$

close to the one that defines a first order system in classical mechanics (a similar model has been used by Fiol et al. [37] In the above equation, Cs is the compressive strength, Cs_0 is the final stable value of the compressive strength, t stands for the number of days after setting, n is a shape constant, and τ is a characteristic time that represents the time when the compressive strength reaches a 63% of its final value. The results of the fittings, together with the Root Mean Squared Error, RMSE:

$$\text{RMSE} = \sqrt{\frac{1}{N}\sum_{i=1}^{N}\left(Cs|_{measured} - Cs|_{calculated}\right)^2} \tag{2}$$

are presented in Table 5. The results of the modeling have been plotted in Figure 5 with dashed lines.

Table 5. Coefficients of Equation (1), Cs_0, τ, and n, fitted to the measured values of the compressive strength, for each one of the studied cases (see Table 4), together with the Root Mean Squared Error (RMSE) of each fitting.

	Control	G15	G30	G45	G60	G80
Cs_0 (MPa)	41.89	33.237	31.978	26.964	16.618	11.011
τ (d)	1.5979	2.0105	2.4732	3.3509	9.3934	59.627
n	4.1999	3.4191	9.4455	10.810	32.400	62.276
RMSE (MPa)	0.608	0.364	1.025	0.634	0.107	0.226

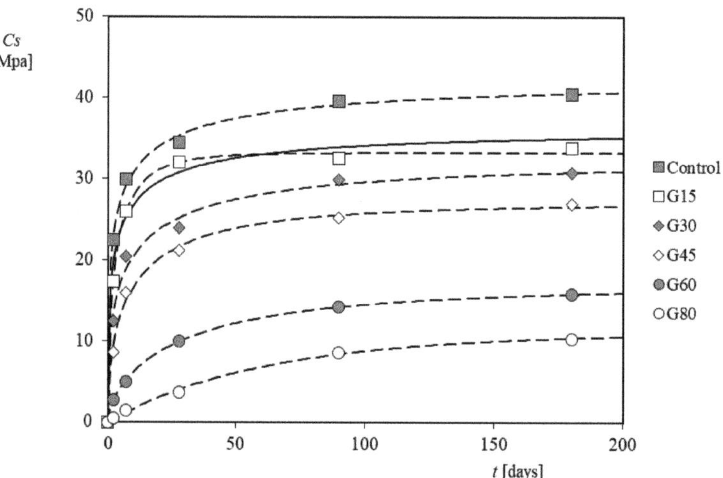

Figure 5. Evolution over time after setting of concrete compression strengths, Cs, for different cases studied (see Tables 1 and 4).

The model has proven to simulate quite well the results, as the errors were between 0.65% and 3.21% in relation to the calculated values of Cs_0. Results of the model fitting to the G_{15} sample were slightly different, as the shape of its curve seemed to detach from the general pattern shown by the curves fitted to the data from the other cases. In Figure 6, the coefficients of Equation (1) extracted from the different samples data are plotted in relation to the amount of glass powder used to replace the cement, Gp. The following equations were fitted to the results (see the dashed lines in graphs from Figure 5):

$$Cs_0 = 41.54 - 38.06C_s \left(R^2 = 0.967 \right) \tag{3}$$

$$\tau = 2.8477 + 46.875C_s^2 + 73.28C_s^4 \left(R^2 = 0.986 \right) \tag{4}$$

$$n = 0.3266 + 1.0277C_s^{2.172} \left(R^2 = 0.857 \right) \tag{5}$$

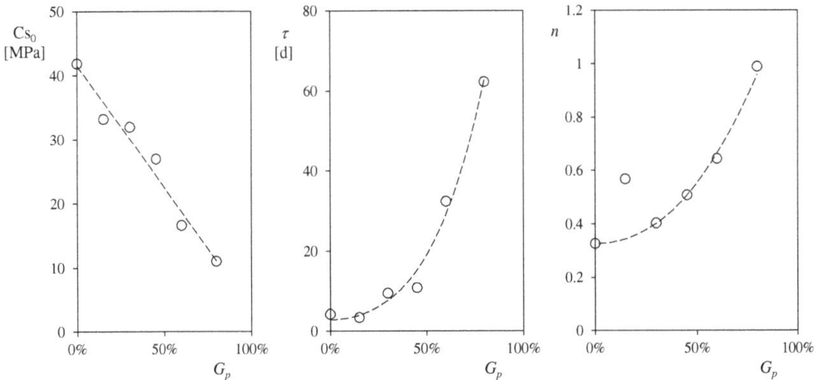

Figure 6. Coefficients of Equation (1), Cs_0, τ, and n, plotted in relation to the amount of glass powder used to replace the cement, Gp. The corresponding lines of tendency (Equations (3)–(5)) have been plotted as dashed lines.

From Figure 5, it can be seen that the coefficient n from the G_{15} sample did not fit Equation (5), making the fitting of Equation (1) to this sample data (see Figure 5) slightly different from the general pattern. This might indicate that the normal errors present in the process of manufacturing the samples has larger effects for lower quantities of glass powder replacing the cement (that is, for lower values of Gp). Finally, with the coefficients extracted from the above equations for $Gp = 15$, the modeling of the sample G15 has been calculated and plotted in Figure 5 with a solid line. The fitting was worse to the experimental data than the one obtained from the direct fitting of Equation (1), with the error being larger (RMSE = 4.32%).

3.3.2. Simulation by Bolomey's Formula

As previously mentioned, Bolomey's formula allows us to predict the compressive strength by means of a linear relationship between the water–cement ratio and compressive strength. According to Bolomey's law, by introducing a coefficient for a certain admixture, the compressive strength values of a concrete in which this admixture has been included as part of the binder can be estimated. From this equation, the effect of such addition on the compressive strengths of concrete can be measured:

$$Cs = Csm \times G \left(\frac{C + k + A}{V} - 0.5 \right) \tag{6}$$

where:

Cs: Compressive strength of concrete (MPa);

Csm: Compressive strength of a mortar of the same age (MPa);

G: Granular coefficient;

C: Amount of cement per m^3 of concrete (kg/m^3);

k: additive coefficient; and

V: volume of water per m^3 of concrete (l/m^3)

Using the above Equation (6), the theoretical values, at 28 and 90 days, of the compressive strengths of concrete in which a part of the CEM I 52.5 R cement has been replaced by an inert additive ($k = 0$), were calculated. Table 6 shows the so-called 'Bolomey compressive strengths-Bolomey Cs' obtained. In this table, Bolomey's 'apparent' compressive strengths and 'apparent' coefficients of the additive correspond to values calculated without correcting for the secondary effect of air entrainment caused by glass dust. Bolomey's 'real' compressive strengths and 'real' coefficients of the additive indicate values calculated after correcting for this effect.

Table 6. Results of Bolomey's 'apparent' and 'real' compressive strengths.

Bolomey Compressive Strengths (MPa)	Control	G15	G30	G45	G60	G80
Cs 28 (Bolomey) apparent ($k = 0$)	34.5	23.7	14.9	8.1	0	/
Cs 28 (Bolomey) real ($k = 0$)	34.5	25.7	18.1	10.6	3.0	/
Coefficient considering the apparent k addition at 28 days	/	1.01	0.57	0.59	0.36	0.35
Coefficient considering the real k addition at 28 days	/	0.67	0.30	0.42	0.20	0.24
Compressive strengths and coefficients considering the addition						
Cs 90 (Bolomey) apparent ($k = 0$)	39.6	27.2	17.1	9.3	0	/
Cs 90 (Bolomey) real ($k = 0$)	39.6	29.5	20.8	12.2	3.4	/
Coefficient considering the apparent k addition at 90 days	/	0.5	0.71	0.63	0.20	0.25
Coefficient considering the real k addition at 90 days	/	0.20	0.42	0.45	0.15	0.34

Figure 7 shows the evolution over time of the compressive strengths of the concrete specimens, at a test age of 28 days, compared to Bolomey's 'apparent' and 'real' compressive strengths. As can be seen, the use of glass powder in the binder increased the compressive strengths of concrete in the long term, increasing their values slightly between 28 and 90 days. It seems that the particle size of glass powder was one important factor responsible for the increased reactivity in the long term [38].

Figure 8 shows the evolution over time of the compressive strengths of the concrete specimens, at a test age of 90 days, compared to Bolomey's 'apparent' and 'real' compressive strengths.

The differences between the 'apparent' and 'real' Bolomey compressive strengths (Figures 7 and 8) were due to the effect of air entrainment caused by glass powder. It is observed that the compressive strengths of concretes containing glass powder were clearly greater than those so-called Bolomey ones, whatever the amount of glass powder that the binder contained. It can be concluded, therefore, that glass dust exerted an important activity in increasing the long-term compressive strength of concretes. This activity can be represented by a coefficient k, at 28 and 90 days, shown in Figures 9 and 10, which present the evolution of the value of 'k apparent' and 'k real' depending on the amount of glass powder contained in the binder, respectively.

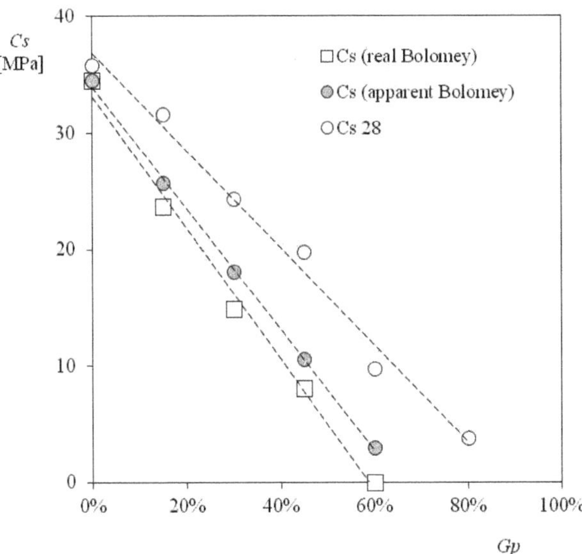

Figure 7. Compressive strength, Cs, 28 days after setting, with regard to the amount of glass powder used to replace the cement (*Gp*). The results are compared to the Bolomey (real and apparent) corresponding ones. The determination coefficients of the linear fittings are $R^2 = 0.988$ (C_S 28); $R^2 = 0.999$ (C_S apparent Bolomey); $R^2 = 0.993$ (C_S real Bolomey).

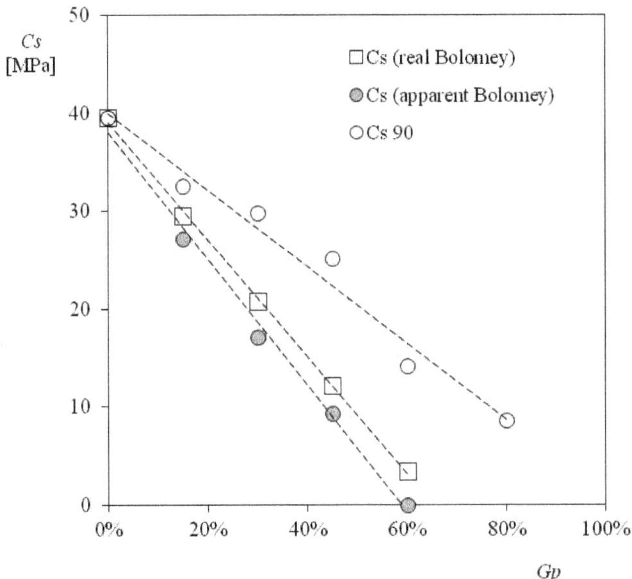

Figure 8. Compressive strength, Cs, 90 days after setting, with regard to the amount of glass powder used to replace the cement (*Gp*). The results are compared to the Bolomey (real and apparent) corresponding ones. The determination coefficients of the linear fittings are $R^2 = 0.973$ (C_S 90); $R^2 = 0.993$ (C_S apparent Bolomey); $R^2 = 0.999$ (C_S real Bolomey).

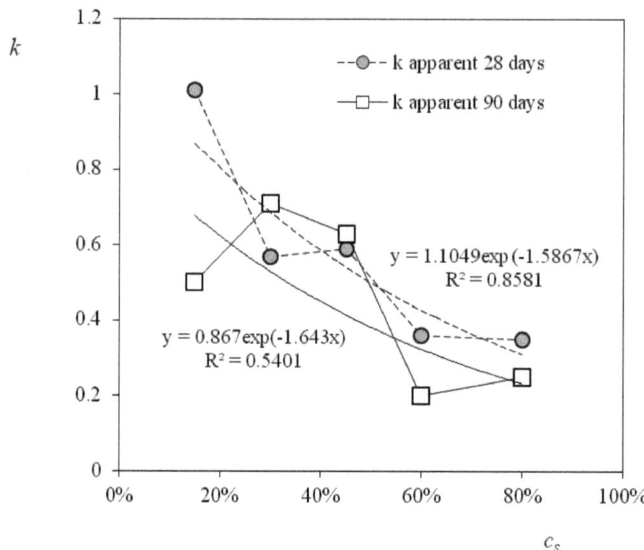

Figure 9. Evolution of the value of '*k* apparent' depending on the amount of glass powder contained in the binder (*Gp*).

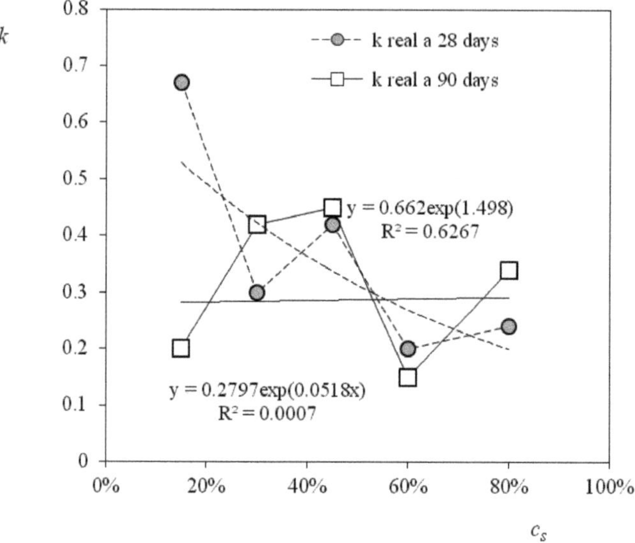

Figure 10. Evolution of the value of '*k* real' depending on the amount of glass powder contained in the binder (*Gp*).

Taking into account the precision of these measurements, two average values of '*k*' can be obtained, one for substitutions lower than 50% of CEM I 52.5 R cement for glass powder, and the other for substitutions higher than 50%.

In this way, it can be considered that '*k* apparent' was equal to 0.6, for replacements of cement for glass powder below 50%, and equal to 0.3 for those higher. Similarly, for '*k* real' a value of 0.4 can be considered for substitutions of cement for glass powder below 50%, and 0.3 for those higher. These results are interpreted so that if *k* was 0.6, for example, to obtain the same compressive strength in a concrete produced with 60 kg of CEM I 52.5 R cement, it would be necessary to use 100 kg of glass powder.

By obtaining the so-called Bolomey coefficients, the percentage of cement that can be replaced for glass powder in concrete to obtain a certain compressive strength can be calculated. According to Figures 9 and 10, for up to 50% replacement of CEM I 52.5 R cement by glass powder, the compressive strength values of concretes are sufficiently important to classify these concretes in the group of building concretes. It was observed that the decrease in compressive strength occurred linearly as a function of the percentage of glass powder in the binder. In this way, from the experimental results, taking into account the percentage of glass powder contained in the binder and the compressive strength obtained in the control specimen, Equation (7) has been deduced. Such an equation makes it possible to predict the compressive strength of the concrete, at a certain setting time, according to the amount of glass powder used to replace the CEM I 52.5 cement:

$$Cs = -31V + Cs_{cem} \tag{7}$$

where:

V is the percentage of glass powder contained in the binder; and
Cs_{cem} is the compressive strength of the concrete whose binder is only CEM I 52.5 R.

Figure 11 shows the variation of compression strengths, for test ages of 2, 7, 28 and 90 days, of concrete manufactured with different replacement percentages of cement by glass powder. At above 50% replacement of CEM I 52.5 R cement by glass powder, the previous law is not applicable. The experimental values of compressive strength are lower than those obtained by this expression. From the values obtained for compressive strength, concretes studied can be classified in the group of concretes for paving roads and highways. Also, a possible use of concrete with a substitution percentage of glass powder for cement lower than 50% may be in the construction of wind farms, as a surface layer in foundations, since it may reduce the effects caused by corrosion because it significantly reduces the chloride ion permeability of concrete due to exposure to meteorological phenomena [25,26].

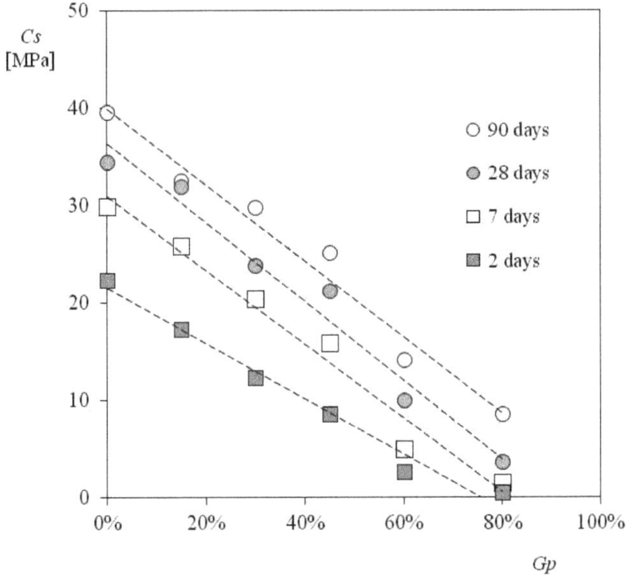

Figure 11. Variation of compression strengths, for test ages of 2, 7, 28 and 90 days, of concrete manufactured with different replacement percentages of cement by glass powder (*Gp*).

On the other hand, concretes manufactured with a replacement percentage higher than 50% of glass powder could be involved in the construction of park roads, especially in those located in areas with a high ecological value. From an environmental point of view, these concretes do not emit pollutant leachate [26] and adopt the color of the aggregate used, thereby minimizing the visual impact caused [32].

4. Conclusions

The results obtained in the tests carried out led us to confirm the initial research hypothesis, and the following conclusions:

On the properties of fresh concrete:

- The addition of glass powder, as a substitute for cement with particles of dimension d50 = 16 µm, to the manufacture of the concrete, whatever its dosage, did not cause any improvement in terms of workability. In fact, the higher the glass powder replacement percentage, the worse the workability of the concrete;
- The air content was higher with higher substitution of cement for glass powder. This fact increased the resistance of the concrete to the atmospheric phenomena of freeze–thaw cycles, without the need to include additives to achieve this effect;
- Replacing cement with glass powder caused a higher air content. For this reason, the density of the different concretes varied, decreasing when the substitution of cement for glass powder increased; and
- The consistency tests by means of a settlement test allowed us to classify these concretes within the group of fluid concretes, regardless of the dosage of glass powder used.

In terms of compressive strength:

- The use of glass in the binder increased the long-term compressive strengths of concrete;
- The long-term compressive strengths of concretes containing glass powder were higher than Bolomey's ones, whatever the amount of glass powder that the binder contained. Therefore, glass powder exerted an important activity in increasing the long-term compressive strength of concretes. At up to 50% replacement of CEM I 52.5 R cement by glass powder, the compressive strength values obtained were important enough to classify these concretes in the group of building concretes. At above 50% substitution of CEM I 52.5 R cement by glass powder, the resistance values obtained allowed these concretes to be classified in the group of concretes for paving roads and highways;
- A mathematical model has been proposed to further analyze the experimental data. The model has proven to simulate quite well the results, as the errors were between 0.65% and 3.21% in relation to the calculated values of the final stable value of the compressive strength; and
- Due to all the characteristic described, this concrete is proposed as ideal as a surface protection layer against the action of corrosion in wind turbine foundations as well as the stabilization of the wind farm roads. The proposed use of this final waster as a binder in ecofriendly concrete has a global impact on circular economy and may reduce the emission of greenhouse gases, because the destiny of this waste would be landfill.

Author Contributions: Conceptualization, E.M.G.d.T. and M.I.M.-L.; methodology, D.A.-G., E.M.G.d.T., M.I.M.-L., S.G.-S. and S.P.; software, M.I.M.-L.; validation, S.G.-S., E.M.G.d.T. and M.I.M.-L.; formal analysis, E.M.G.d.T., M.I.M.-L. and D.A.-G.; investigation, D.A.-G., E.M.G.d.T., M.I.M.-L., S.G.-S. and S.P.; data curation, E.M.G.d.T. and M.I.M.-L.; writing—original draft preparation, D.A.-G., E.M.G.d.T., S.G.-S., M.I.M.-L. and S.P.; writing—review and editing, E.M.G.d.T. and M.I.M.-L.; supervision, S.G.-S., M.I.M.-L. and S.P. All authors have read and agreed to the published version of the manuscript.

Funding: This research received no external funding.

Institutional Review Board Statement: Not applicable.

Informed Consent Statement: Not applicable.

Data Availability Statement: Not applicable.

Conflicts of Interest: The authors declare no conflict of interest.

References

1. ONU Objetivos del Desarrollo Sostenible. Available online: https://www.un.org/sustainabledevelopment/es/ (accessed on 18 October 2020).
2. ASCE. *La Visión para la Ingeniería Civil en 2025*; ASCE: Reston, VA, USA, 2006.
3. Rodríguez, F.; Fernández, G. Ingeniería sostenible: Nuevos objetivos en los proyectos de construcción. *Rev. Ing. Constr.* **2010**, *25*, 147–160. [CrossRef]
4. Fogarassy, C.; Finger, D. Theoretical and practical approaches of circular economy for business models and technological solutions. *Resources* **2020**, *9*, 76. [CrossRef]
5. Fundación Conama; Council Green Building; RCD Asociación. Economía circular en el ámbito de la construccion. In Proceedings of the Congreso Nacional del Medio Ambiente 2018, Madrid, Spain, 26–29 November 2018; pp. 1–63.
6. Švajlenka, J.; Kozlovská, M. Evaluation of the efficiency and sustainability of timber-based construction. *J. Clean. Prod.* **2020**, *259*. [CrossRef]
7. Vidrio España. *Contribución Económica, Ambiental y Social del Sector del Vidrio en España (2014/1016)*; Vidrio España: Madrid, Spain, 2017; p. 44.
8. Marco, L.J.; García, E.; Más, M.I.; Alcaraz, V.; Luizaga, A. Estudio de la resistencia a compresión de morteros fabricados con conglomerante compuesto de polvo de vidrio. *Inf. Constr.* **2012**, *64*, 529–536. [CrossRef]
9. Hogland, W. Remediation of an old landfill site: Soil analysis, leachate quality and gas production. *Environ. Sci. Pollut. Res. Int.* **1994**, *9*, 49–54. [CrossRef] [PubMed]
10. Slomski, V.G.; Lima, I.C.S.; Slomski, V.; Slavov, T. Pathways to urban sustainability: An investigation of the economic potential of untreated household solid waste (HSW) in the city of São Paulo. *Sustainability* **2020**, *12*, 5249. [CrossRef]
11. Rashidian-Dezfouli, H.; Afshinnia, K.; Rangaraju, P.R. Efficiency of Ground Glass Fiber as a cementitious material, in mitigation of alkali-silica reaction of glass aggregates in mortars and concrete. *J. Build. Eng.* **2018**, *15*, 171–180. [CrossRef]
12. Serpa, D.; Santos Silva, A.; De Brito, J.; Pontes, J.; Soares, D. ASR of mortars containing glass. *Constr. Build. Mater.* **2013**, *47*, 489–495. [CrossRef]
13. Aliabdo, A.A.; Abd Elmoaty, A.E.M.; Aboshama, A.Y. Utilization of waste glass powder in the production of cement and concrete. *Constr. Build. Mater.* **2016**, *124*, 866–877. [CrossRef]
14. Thomas, C.; de Brito, J.; Corinaldesi, V. Special issue high-performance eco-efficient concrete. *Appl. Sci.* **2021**, *11*, 1163. [CrossRef]
15. De Castro, S.; De Brito, J. Evaluation of the durability of concrete made with crushed glass aggregates. *J. Clean. Prod.* **2013**, *41*, 7–14. [CrossRef]
16. Idir, R.; Cyr, M.; Tagnit-Hamou, A. Use of fine glass as ASR inhibitor in glass aggregate mortars. *Constr. Build. Mater.* **2010**, *24*, 1309–1312. [CrossRef]
17. García Del Toro, E.M.; Más López, M.I. Study of new formations of C-S-H in manufactured with glass powder as binder mortar. *Ing. Investig.* **2018**, *38*, 24–32. [CrossRef]
18. Liu, M. Incorporating ground glass in self-compacting concrete. *Constr. Build. Mater.* **2011**, *25*, 919–925. [CrossRef]
19. Parghi, A.; Shahria Alam, M. Physical and mechanical properties of cementitious composites containing recycled glass powder (RGP) and styrene butadiene rubber (SBR). *Constr. Build. Mater.* **2016**, *104*, 34–43. [CrossRef]
20. Nassar, R.U.D.; Soroushian, P. Strength and durability of recycled aggregate concrete containing milled glass as partial replacement for cement. *Constr. Build. Mater.* **2012**, *29*, 368–377. [CrossRef]
21. Schwarz, N.; Cam, H.; Neithalath, N. Influence of a fine glass powder on the durability characteristics of concrete and its comparison to fly ash. *Cem. Concr. Compos.* **2008**, *30*, 486–496. [CrossRef]
22. Shayan, A.; Xu, A. Performance of glass powder as a pozzolanic material in concrete: A field trial on concrete slabs. *Cem. Concr. Res.* **2007**, *36*, 457–468. [CrossRef]
23. Shao, Y.; Lefort, T.; Moras, S.; Rodriguez, D. Studies on concrete containing ground waste glass. *Cem. Concr. Res.* **2000**, *30*, 91–100. [CrossRef]
24. Guo, P.; Meng, W.; Nassif, H.; Gou, H.; Bao, Y. New perspectives on recycling waste glass in manufacturing concrete for sustainable civil infrastructure. *Constr. Build. Mater.* **2020**, *257*, 119579. [CrossRef]
25. Fernández Cánovas, M. *Hormigon: Adaptado a la Instruccion para la Recepcion de Cementos y a la Instruccion de Hormigon Estructural Ehe*, 10th ed.; CICCP: Girona, Spain; Garceta: Madrid, Spain, 2013.
26. Mas, M.I.; García, E.M.; Marco, L.J.; De Marco, J. Análisis de la Viabilidad Ambiental de la Utilización de Morteros Fabricados con Polvo de Vidrio en la Estabilización de Suelos. *Inf. Tecnol.* **2016**, *27*, 77–86. [CrossRef]
27. ANEFHOP UNE-EN 12350-5:2020. *Ensayos de Hormigón Fresco. Parte 5: Ensayo de la Mesa de Sacudidas*; UNE: Madrid, Spain, 2020; pp. 1–15.

28. *Comité Técnico CEN/TC 104 Hormigón y Productos Relacionados UNE-EN 12350-7: Ensayos de Hormigón Fresco. Parte 7: Contenido de aire Métodos de Presión*; UNE: Madrid, Spain, 2020.

29. *CTN 83—HORMIGÓN UNE-EN 1015-6:1999: Densidad Aparente del Hormigón Fresco 1999*; UNE: Madrid, Spain, 1999.

30. AENOR UNE-EN 12350-2. *Ensayos de Hormigón Fresco. Parte 2: Ensayo de Asentamiento*; UNE: Madrid, Spain, 2006; pp. 1–9.

31. Hormigón, C. *Técnico 83 UNE 83507 Rotura a Compresión*; UNE: Madrid, Spain, 2004; p. 1.

32. Más-López, M.I.; del Toro, E.M.G.; Patiño, A.L.; García, L.J.M. Eco-friendly pavements manufactured with glass waste: Physical and mechanical characterization and its applicability in soil stabilization. *Materials* **2020**, *13*, 3727. [CrossRef]

33. Pereira-De-Oliveira, L.A.; Castro-Gomes, J.P.; Santos, P.M.S. The potential pozzolanic activity of glass and red-clay ceramic waste as cement mortars components. *Constr. Build. Mater.* **2012**, *31*, 197–203. [CrossRef]

34. Taha, B.; Nounu, G. Properties of concrete contains mixed colour waste recycled glass as sand and cement replacement. *Constr. Build. Mater.* **2008**, *22*, 713–720. [CrossRef]

35. Tan, K.H.; Du, H. Use of waste glass as sand in mortar: Part I—Fresh, mechanical and durability properties. *Cem. Concr. Compos.* **2013**, *35*, 109–117. [CrossRef]

36. Mirzahosseini, M.; Riding, K.A. Influence of different particle sizes on reactivity of finely ground glass as supplementary cementitious material (SCM). *Cem. Concr. Compos.* **2015**, *56*, 95–105. [CrossRef]

37. Fiol, F.; Thomas, C.; Muñoz, C.; Ortega-López, V.; Manso, J.M. The influence of recycled aggregates from precast elements on the mechanical properties of structural self-compacting concrete. *Constr. Build. Mater.* **2018**, *182*, 309–323. [CrossRef]

38. Lu, J.-X.; Duan, Z.-H.; Poon, C.S. Combined use of waste glass powder and cullet in architectural mortar. *Cem. Concr. Compos.* **2017**, *82*, 34–44. [CrossRef]

Article

Fault Detection of Wind Turbine Induction Generators through Current Signals and Various Signal Processing Techniques

Yuri Merizalde [1,*], Luis Hernández-Callejo [2,*], Oscar Duque-Perez [3]
and Raúl Alberto López-Meraz [4]

[1]   PhD School of the University of Valladolid (UVA), Faculty of Chemical Engineering,
      University of Guayaquil, Clemente Ballen and Ismael Perez Pazmiño, Guayaquil 593, Ecuador
[2]   Department of Agricultural Engineering and Forestry, University of Valladolid (UVA),
      Campus Universitario, Duques de Soria, 42004 Soria, Spain
[3]   Department of Electrical Engineering, University of Valladolid (UVA), Escuela de Ingenierías Industriales,
      Paseo del Cauce 59, 47011 Valladolid, Spain; oscar.duque@eii.uva.es
[4]   Unidad de Ingeniería y Ciencias Químicas, Universidad Verzcruzana, Circuito Universitario Gonzalo
      Aguirre Beltrán, Zona Universitaria, Xalapa 91000, Mexico; meraz_raul@hotmail.com
*   Correspondence: yuri.merizaldez@ug.edu.ec (Y.M.); luis.hernandez.callejo@uva.es (L.H.-C.);
     Tel.: +34-975-129-418 (L.H.-C.)

Received: 12 August 2020; Accepted: 19 October 2020; Published: 22 October 2020

Abstract: In the wind industry (WI), a robust and effective maintenance system is essential. To minimize the maintenance cost, a large number of methodologies and mathematical models for predictive maintenance have been developed. Fault detection and diagnosis are carried out by processing and analyzing various types of signals, with the vibration signal predominating. In addition, most of the published proposals for wind turbine (WT) fault detection and diagnosis have used simulations and test benches. Based on previous work, this research report focuses on fault diagnosis, in this case using the electrical signal from an operating WT electric generator and applying various signal analysis and processing techniques to compare the effectiveness of each. The WT used for this research is 20 years old and works with a squirrel-cage induction generator (SCIG) which, according to the wind farm control systems, was fault-free. As a result, it has been possible to verify the feasibility of using the current signal to detect and diagnose faults through spectral analysis (SA) using a fast Fourier transform (FFT), periodogram, spectrogram, and scalogram.

Keywords: wind turbine; electric generator; spectral analysis; fault diagnosis

---

1. Introduction

Regardless of the maintenance strategies and models applied in the wind industry (WI) to detect and diagnose faults, the use of signals, such as vibration, acoustic, temperature, magnetism, and electrical signals, is an indispensable requirement. Each of these types of signal have their advantages and disadvantages. However, because all moving equipment produces some type of vibration, in the WI, the use of vibration signals predominates [1–3]. Even though the current signal does not use intrusive methods, the equipment used is inexpensive, easy to install, and also, according to reference [4,5], both the vibration and current signal can be used to detect failures of the electric generator and loads coupled to its axis. However, according to published reports, in the WI, current signals have rarely been used, and the existing research is based predominantly on laboratory studies.

The processing of the signals used for the detection and diagnosis of faults is carried out using a variety of models in the time, frequency, and time–frequency domains. All signal processing techniques have their own advantages and disadvantages, and the same applies to the domain in which the analysis is carried out [1–4]. Furthermore, the lack of ideal conditions to apply a specific technique directly prompts us to develop mathematical models that allow the detection and diagnosis of a specific type of fault that occurs in a particular component and in certain specific conditions of operation [6–8]. The combination of all these factors has given rise to a huge number of proposed methods, some of which are analyzed in more detail in the following sections.

Based on the foregoing, the purpose of this research is the detection and diagnosis of electrical generator faults by means of the current signal of real wind turbines (WTs) in operation and the application of various techniques for processing and analyzing existing signals to:

- Analyze the models used to detect the frequency components associated with faults.
- Obtain the spectrum of the current signal of an operating turbine.
- Study the effectiveness of signal processing techniques in detecting WT failures.
- Check the effectiveness of the WT control system to determine the status of the generator.
- Compare the results obtained with those of previously published studies.

Due to the variety of stresses to which the rotary induction machine is subjected, there are a variety of failures that can occur in the stator, rotor, and bearings, as described in reference [9]. The objective of this research is not to focus on a specific fault, but rather, applying the different signal processing techniques, try to detect and diagnose the faults that will be described in sections two and three. As one of the main objectives of this research is to use data from WTs in operation, and the only wind farm (WF) available to make the measurements was integrated by WTs that use SCIG, then the study will focus on this type of electric generator. The remainder of this original research is organized as follows. Section 2 analyzes the mathematical models used to determine the frequency components associated with faults in the SCIG using current signal analysis. In Section 3, the fundamentals and application of various signal analysis techniques are discussed, emphasizing the published techniques for fault detection in WTs using the SCIG current signal. Section 4 details the methodology and materials used for the experimental part of this research. Section 5 includes the results obtained by applying the techniques described in Section 3. Finally, the conclusions and recommendations are included in Section 6.

2. Modeling Electrical Generator Faults Using the Current Signal

Due to its design, durability, and low cost, the use of the squirrel-cage induction machine predominates at the industrial, commercial, and domestic levels [10,11]. Although there are several types of generators, according to reference [12], in the WI, the doubly fed induction generator (DFIG), and the squirrel-cage induction generator (SCIG) predominate.

The voltage signals, current, magnetic field, magnetomotive force (MMF), torque, and power of an induction machine are characterized by its sinusoidal behavior [13]. Since the speed of the rotor depends on the coefficients of the associated differential equations, which vary with time, the behavior of the materials used in the construction of the motor is not constant over time but depends on the position of the rotor. Under these conditions, it is hard to analyze signals in a spatial system, which is why in-plane analysis is preferred. For this purpose, using the Clarke and Park transforms, a change of variables is made. With the first transformation, we go from a 3D system (abc) to a 2D plane (alpha-beta) that varies with the stator, while with the second one, we obtain a plane dq0 equivalent to a 2D plane that rotates at the same rotor speed but is offset by an angle θ. Since the three-phase induction machine generally does not use a neutral line, the main current does not have a homopolar component, and the three phases can be represented in the dq plane. In this plane, the stator remains fixed (direct axis d) in relation to a rotor plane (quadrature axis q) that rotates at speed ω_x [13–15].

The transformation between the abc space system and the dq0 plane, when the latter is oriented at an angle θ with reference to the axis that remains fixed, can be performed directly using Equations (1) and (2). When θ is zero, these Equations become (3) and (4), respectively [15–18].

$$\begin{bmatrix} i_{qs} \\ i_{ds} \\ i_{0s} \end{bmatrix} = \frac{2}{3} \begin{bmatrix} \cos\theta & \cos(\theta - \frac{2\pi}{3}) & \cos(\theta + \frac{2\pi}{3}) \\ \sin\theta & \sin(\theta - \frac{2\pi}{3}) & \sin(\theta + \frac{2\pi}{3}) \\ 0.5 & 0.5 & 0.5 \end{bmatrix} \begin{bmatrix} i_a \\ i_b \\ i_c \end{bmatrix} \tag{1}$$

$$\begin{bmatrix} i_a \\ i_b \\ i_c \end{bmatrix} = \begin{bmatrix} \cos\theta & \sin\theta & 1 \\ \cos(\theta - \frac{2\pi}{3}) & \sin(\theta - \frac{2\pi}{3}) & 1 \\ \cos(\theta + \frac{2\pi}{3}) & \sin(\theta + \frac{2\pi}{3}) & 1 \end{bmatrix} \begin{bmatrix} i_{qs} \\ i_{ds} \\ i_{0s} \end{bmatrix} \tag{2}$$

$$\begin{bmatrix} i_{qs} \\ i_{ds} \end{bmatrix} = \frac{2}{3} \begin{bmatrix} 1 & -\frac{1}{2} & -\frac{1}{2} \\ 0 & \frac{\sqrt{3}}{2} & -\frac{\sqrt{3}}{2} \end{bmatrix} \begin{bmatrix} i_a \\ i_b \\ i_c \end{bmatrix} \tag{3}$$

$$\begin{bmatrix} i_a \\ i_b \\ i_c \end{bmatrix} = \begin{bmatrix} 1 & 0 \\ -\frac{1}{2} & -\frac{\sqrt{3}}{2} \\ -\frac{1}{2} & \frac{\sqrt{3}}{2} \end{bmatrix} \begin{bmatrix} i_{qs} \\ i_{ds} \end{bmatrix} \tag{4}$$

According to reference [16], the phase current of the DFIG can be expressed as a function of the flow and torque vectors (the torque angle is 90° over the flow), Equations (5) and (6), respectively, which can be represented in the dq plane, according to Equation (2). Since the variables of Equations (5) and (6) rotate at speed 2πfs, they cannot be measured directly, so it is necessary to apply the inverse Park transform to obtain the phase currents according to Equations (7) to (9). As described previously [19], the different stresses that cause a torque on the rotor include coupled loads; unbalanced dynamic forces; torsional vibration; transient torques; magnetic forces caused by leakage flux over the slots, making them vibrate at twice the frequency of the rotor; air gap eccentricity; centrifugal forces; thermal stresses caused by heat in the short-circuit ring and heat in the bars during starting (skin effect); residual forces due to casting; machining and welding. Under normal operating conditions, the spectrum of the signal has defined components. However, the asymmetries of the generator and the loads coupled to it (gearbox, blades) transmit torsional vibrations that act on the rotor, causing variations in the speed, torque, air gap magnetic flux and current bars. In this way, both mechanical and electrical faults manifest as lateral components of the fundamental wave. The number of harmonics and their amplitude depend on the magnitude of the fault [20,21].

$$i_{sM} = i_{sM_0} + \sum A_{sM_i} \sin(2\pi f_v t + \varphi_{M_i}) \tag{5}$$

$$i_{sT} = i_{sT_0} + \sum A_{sT_i} \cos(2\pi f_v t + \varphi_{T_i}) \tag{6}$$

$$\begin{aligned} i_a(t) = {} & i_0 \sin(2\pi f_s t + \varphi_0) \\ & + \tfrac{1}{2}\{A_{sMi} \cos[2\pi(f_s - f_v)t - \varphi_M] \\ & + A_{sT_i} \cos[2\pi(f_s - f_v)t - \varphi_T]\} \\ & - \tfrac{1}{2}\{A_{sM_i} \cos[2\pi(f_s + f_v)t + \varphi_M] \\ & - A_{sT_i} \cos[2\pi(f_s + f_v)t + \varphi_T]\} \end{aligned} \tag{7}$$

$$i_0 = \sqrt{i_{sM0}^2 + i_{sT0}^2} \tag{8}$$

$$\varphi_0 = tg^{-1} \frac{i_{sT0}}{i_{sM0}} \tag{9}$$

As described in reference [22], in a fault-free machine, the rotor and stator currents should be balanced. However, due to small differences in the winding geometry and the nonlinearity of the materials, an asymmetry arises that causes axial flow dispersion. Under these conditions, the distribution of the harmonics in the air gap undergo alterations that can be easily detected, so that we can detect broken rotor bars, one-phase failure, dynamic eccentricity, a negative sequence phase, and short circuits in the rotor and stator windings. According to the same author, in a three-phase machine, with a full pole-pass and fed by a balanced frequency ω_s, the spatial distribution of the harmonics of the MMF about the stator and as a function of the air gap flux is given by Equation (10). To apply these Equations to the rotor, θ is given by (11) or (12). Substituting these Equations in the general term of (10) and expanding it to obtain the first terms, we obtain Equation (13), which provides the components of the frequency spectrum of the current induced in the rotor by the harmonics of the air gap. That is, the stator current spectrum (CS) includes the components of the supply current and those of the rotor. The presence of short circuits between turns produces an MMF with its own frequency spectrum that is superimposed on the main one to give rise to a new spectrum that is expressed by (14) and whose main term is (15).

$$\Phi_s = \Phi_1 \cos(\omega t - p\theta_s) + \Phi_5 \cos(\omega t + 5p\theta_s)$$
$$-\Phi_7 \cos(\omega t - 7p\theta_s) + \Phi_{11} \cos(\omega t + 11p\theta_s) \ldots\ldots \Phi_n \cos(\omega t + np\theta_s) \tag{10}$$

$$\theta = \theta_r + \theta_{sr} = \theta_r + \omega_r t \tag{11}$$

$$\omega_r = \omega(1-s)/p \tag{12}$$

$$\Phi_s = \Phi_1 \cos(s\omega t - p\theta_r) + \Phi_5 \cos((6-5s)\omega t + 5p\theta_r)$$
$$-\Phi_7 \cos((7s-6)\omega t - 7p\theta_s) + \Phi_{11} \cos((12-11s)\omega t + 11p\theta_s) \ldots\ldots \tag{13}$$

$$\Phi_s = 0.5 \sum \sum \Phi_n \cos\left[(k_1 \pm k_2(\frac{(1-s)}{p})) \pm k_2\theta_r\right] \tag{14}$$

$$f = \left[(k_1 \pm k_2(\frac{(1-s)}{p})) \pm k_2\theta_r\right] \tag{15}$$

As described in reference [14], the faulty and healthy squirrel-cage induction motor current is given by (16) and (17), respectively. According to reference [23], when there is a short circuit or static eccentricity in the stator, a negative sequence component appears, the amplitude of which depends on the percentage of shorted turns. As described in reference [24,25], the components due to stator failures (f_{sf}) are given by Equation (18), while according to reference [26], in the case of a healthy motor, the main components are the first and fifth harmonics. In the case of an unbalanced voltage, regardless of slip, this fault shows itself mainly in the first and third harmonics. According to reference [27], short circuits cause the components given by Equation (19). As described in reference [28], in a symmetrical stator, the CS contains the harmonics given by Equations (20) and (21), for which the harmonics determined by Equations (22) through (24) should be added, in case of asymmetry. As described in reference [29], another simple alternative for the early detection of stator faults depends on the magnitude of the negative sequence of the current, which allows us to obtain the negative impedance to be compared with the average winding impedance.

$$i_a(t) = i_A(t) = i_a(t)\left[1 + k_m \cos(\omega_f t)\right] \tag{16}$$

$$i_A(t) = I \cos\left(\omega_s t - \varphi - \frac{\pi}{6}\right) + \frac{k_m I_L}{\sqrt{2}} \{\cos\left[(\omega_s + \omega_f)t - \varphi - \frac{\pi}{6}\right] + \cos\left[(\omega_s - \omega_f)t - \varphi - \frac{\pi}{6}\right]\} \tag{17}$$

$$f_{sf} = \left[2k_0(\frac{1-s}{p}) \pm k_1\right]f_s \tag{18}$$

$$f = f_s\left[\frac{2k}{p}(1-s) \pm k_1\right] \tag{19}$$

$$f_{s1} = f_s \left| (1 - k_3) - \frac{2k_4 Q_r}{p}(1 - s) \right| \tag{20}$$

$$f_{s2} = f_s \left| (1 - k_3) + (-\frac{2k_4 Q_r}{p} + 2 + 6k_0)(1 - s) \right| \tag{21}$$

$$f_{s3} = f_{s1} - (n_1 + 1)sf_s \big|_{n1=1} = f_{s1} - 2sf_s \tag{22}$$

$$f_{s4} = f_{s2} - (n_1 + 1)sf_s \big|_{n1=1} = f_{s2} - 2sf_s \tag{23}$$

$$f_{s5} = f_s \tag{24}$$

As described in reference [30], rotor faults generate components below the supply frequency in the stator spectrum, according to Equation (25), where the first term does not contribute to increasing the supply current because it induces an MMF of zero sequence. However, the second term induces a set of three-phase currents at the supply frequency and contains a component displaced by twice the slip frequency, $2sf_s$. The fault causes a $2sp\omega_{mr}$ variation in rotor speed, causing a displacement of the lower component $(1 - 2s)f_s$ and the appearance of an upper component at $(1 + 2s)f_s$ modulated by the third harmonic of the stator flux. Other components that may appear are given by Equation (26). Because of the static, dynamic, or mixed eccentricity, the air gap is not uniform, and the forces applied to the shaft become unbalanced, causing friction between the stator and rotor. According to references [26,31,32], the eccentricity causes the appearance of harmonics whose sequence is given by (27). If the eccentricity is static, n_d is zero, while if it is dynamic, it is $1, 2, 3, \ldots$ However, according to reference [33], a difference between static and dynamic eccentricity does not always exist for all motor configurations, in addition to the fact that there are components that are not easily detectable.

$$f_s = \frac{N_r I_2}{2} \{\cos[(3 - 2s)\omega_s t - 3p\theta_1] - \cos[(1 - 2s)\omega_s t - p\theta_s]\} \tag{25}$$

$$f_s = (1 \pm 2k_0 s)\omega_s \tag{26}$$

$$f_{ecc} = \left[2(k_0 Q_r + n_d)\left(\frac{1-s}{p}\right) \pm v \right] f_s \tag{27}$$

As described in reference [34], mechanical faults can be classified as those that cause air gap eccentricity, load torque oscillations, or a combination of both. The first effect is due to unbalanced loads, shaft misalignment, and gearbox and bearing failures. The second effect is due to wear or failure of the bearings and to rotor imbalance caused by poor assembly, for example. The load torque oscillation component, Equation (28), affects the rotor position and stator current. The length of the air gap affects the permeance, flux density of the air gap, and MMF, which ultimately modulates the stator current signal. When there is air gap eccentricity, that failure can vary with time and the angle of the circumference θ, so, in the case of dynamic eccentricity, the length of the air gap is a function of θ and t, according to Equation (29). The use of $\omega_r t$ in the last equation provides an expression for the static eccentricity. From this last expression, it is deduced that the dynamic eccentricity produces components given by (30) in the signal spectrum. In addition, the modulation of the phase and the amplitude occur at the same rotor frequency.

$$T_T(t) = T_o + T_{osc} \cos(\omega t) \tag{28}$$

$$g_{de}(\theta, t) \approx g_0(1 - \delta_d \cos(\theta - \omega_r t)) \tag{29}$$

$$f_s \pm f_r \tag{30}$$

3. Signal Processing Techniques Applied to Wind Turbine Failure Detection

Initially, the study of the signals was carried out in the time-amplitude plane and was based on the variation of the waveform in addition to the calculation of parameters such as the average value, peak value, interval between peaks, standard deviation, crest factor, root mean square value, kurtosis, and skewness [35,36]. According to reference [37], synchronizing the sampling of the vibration signal with the rotation of a gear and evaluating the average of several revolutions provide a signal called the synchronized time average, which is expressed by Equation (31). This equation makes it possible to accurately approximate a periodic signal and obtain the vibration pattern (including any modulation effects) of the gear teeth of a gearbox. However, according to the author, this method requires repeating the analysis for each gear. As described in reference [38], the most advanced proposed methods for the analysis in the time domain apply time series models to the signal, among which are those of auto regression (AR) and the autoregressive moving average (ARMA), which are applied in references [39–41]. Currently, the parameters used in the analysis of the current signal in the time domain can be used for the detection and diagnosis of faults using artificial intelligence models [42]. Thus, in reference [35], the eight parameters of the current signal in the time domain are the input variables of a three-layer artificial neural network (ANN) used to predict the remaining useful life (RUL) of the bearings of the gearbox of a WT.

$$g(t) = \sum_{k_0=0}^{k_0} A_{k_0}(1 + a_{k_0}(t)) \cos(2\pi k_0 f_t e t + \varnothing_{k_0} + b_{m_0}(t)) \tag{31}$$

As described in reference [43], the complex sinusoidal and cosine signals in the time domain can be better analyzed using their frequency components obtained with the Fourier transform (FT). According to references [38,44], the analysis in the frequency domain allows us to obtain information that is not available in the time domain, for example, knowing the origin of the signal, the phase modulation, and the moment at which the components arise. When the signals are stationary, their frequency spectrum is constant over time, so the analysis can be performed using the FT. For example, reference [45] determines the frequencies associated with the DFIG faults of WTs by applying the fast Fourier transform (FFT) to the current signal of the electric generator, but during periods of steady-state, that is, when the speed is constant.

When the signals are transient and not periodic, such as during startup or load variation or under wind speeds with stochastic behavior, the spectrum is oscillatory. Under these conditions, the FFT and analysis in the frequency domain are not sufficient, so it is necessary to resort to other techniques that allow analysis in the time-frequency domain, such as the short-time Fourier transform (STFT), wavelet transform, Wigner–Ville distribution (WVD), and Hilbert transform (HT) [42,46,47].

One of the first alternatives used to overcome the disadvantages of the FT was the windowing technique proposed in 1946 by Dennis Gabor, which consisted of applying the FT to only a small section of the signal at a time. This methodology is the origin of the STFT, which allows the signals to be represented as a function of time and frequency. For this, the total signal time is divided into shorter time intervals [48,49]. The signal is multiplied by a window function, Equation (32), and for each resulting interval, the discrete time FT (DTFT) is given by (33) and (34). For a fixed time (n) of analysis, the DTFT is known as the STFT, where (32), which is a sequence of DTFTs, is a periodic function of frequency ω and period 2π. There is a different spectrum in each window, and by analyzing all the intervals as a whole, the variation in frequency over time can be observed. When, instead of studying the spectrum at a certain time, one wishes to study a specific frequency, then the windowing process is carried out in the frequency domain. The quality of the results depends on selecting the window that minimizes losses due to spectral leakage and reduces the amplitude of the main and secondary lobes [50].

$$x_{\bar{t}}(m) = x(m)h(\bar{t} - m) \tag{32}$$

$$X_{\bar{t}}(e^{j\bar{\omega}}) = \frac{1}{\sqrt{2\pi}} \int e^{-j\bar{\omega}m} x_m(m)h(\bar{t} - m)dm \tag{33}$$

$$X_{\bar{t}}(e^{\overline{j\omega}}) = \sum_{m=-\infty}^{\infty} x(m)h(\bar{t}-m)e^{-\overline{j\omega}m} \tag{34}$$

$$X_{\bar{t}}(e^{\overline{j\omega}}) = (x[t]e^{-j\omega t}) * \omega[t]\big|_{t=\bar{t}} \tag{35}$$

As described in reference [38], the STFT has resolution problems due to signal segmentation; one of the alternatives to overcome this limitation is the WVD. This distribution is one of the most popular ones since, unlike the STFT, it is not based on signal segmentation, providing better resolution in both the time domain and the frequency domain. According to references [47,51], given a signal *s(t)* in the time domain, the WVD is defined by (36), while starting from the frequency spectrum of *s(t)*, it is given by (37). It can be assumed that using the WVD means dividing the signal into two equal parts in relation to a time *t*, with the right part overlaid over the left part, which means that when the signal is null before or after *t*, then the signal is zero in *t*. The correct average is obtained only when the signal can be separated into a component that is a function of time only and another function of frequency. A signal that is not zero at time zero or a signal with frequencies without the existence of a spectrum indicates the presence of interference or cross terms. As described in reference [34], a sinusoidally modulated amplitude current signal has the same components as in (30), which means that when $f = fr$, in the case of faults due to torque oscillations and eccentricity, the use of classic spectral analysis (SA) does not allow us to distinguish between amplitude and phase modulation since the modulation indices are small. However, the use of the WVD does allow us to distinguish these faults using Equation (38).

$$W(t,\omega) = \frac{1}{2\pi} \int_{-\infty}^{\infty} s^*(t - \frac{1}{2}\tau)s(t + \frac{1}{2}\tau)e^{-j\tau\omega}d\tau \tag{36}$$

$$W(t,\omega) = \frac{1}{2\pi} \int_{-\infty}^{\infty} S^*(\omega + \frac{1}{2}\theta)S(\omega - \frac{1}{2}\theta)e^{-jt\theta}d\theta \tag{37}$$

$$f_s \pm \frac{f_r}{2} \tag{38}$$

The small magnitude of the components associated with the faults makes their extraction difficult. To overcome this drawback, one of the most suggested techniques is to demodulate the signal amplitude. For this, according to reference [52], the best option is to use HT due to its strength to handle noise. According to that report, when there are no faults, the amplitude of the envelope is constant over time, and its variance is zero. Otherwise, if the variance is greater than a pre-established threshold level, some type of asymmetry exists.

According to reference [53], the HT of a signal is the relationship between the real and imaginary parts of the FT of said signal. Similarly, the function of time obtained by the Fourier inverse is a complex function called the analytical signal, the imaginary part of which is the HT. The analytical signal can be represented as a phasor whose amplitude and rotation speed vary over time, according to Equation (39), implying that, given a function in the time domain, in addition to the amplitude, one can also obtain the components that modulate the frequency or phase, $\widetilde{a}(t)$. In the HT, the amplitude function is the envelope of both the real and the imaginary parts and represents the modulated signal plus *dc* compensation. In the case of oscillating functions, the direct analysis in the frequency domain of the periodic variations over time of the components introduced by some type of anomaly does not provide enough information. However, if a bandpass filter is used in the region containing the components that modulate the CS and its envelope is obtained, the frequencies associated with the faults can be easily identified. Additionally, since the magnitude of the envelope can be plotted on a logarithmic scale, exponential decays can be converted into straight lines to detect low-level peaks.

This is the reason why the demodulation of the amplitude of the current signal is one of the most used techniques for the detection and diagnosis of faults in rotating electrical machines using SA.

$$A(t)e^{j\omega(t)} = a(t) + j\widetilde{a}(t) = a(t) + \frac{1}{\pi}\int_{-\infty}^{\infty} a(\tau)\frac{1}{t-\tau}d\tau \tag{39}$$

According to reference [54], mathematically, the HT along with its FFT are given by (40). The composition of the analytical signal and its amplitude, phase of the envelope, and instantaneous frequency can be obtained from Equation (39). Both positive and negative components have a 90° offset, and Equation (41), according to the Park transform, takes the same form after applying HT as Equation (17). From this, it follows that the characteristic frequencies are f_m and $2f_m$, there being a dc component in f_m, $2f_m$, $2(f_m + f_1)$, etc. According to reference [55], when some type of failure occurs in the multipliers, a new impulse appears in the original spectrum or phase spectrum. For this reason, prior to using the FFT to obtain the spectrum, demodulation is applied (using the HT) to the current signal, demonstrating the effectiveness of this technique in the detection and diagnosis of pinions and broken teeth in the gearbox of a WT.

$$\overline{x}(t) = \frac{1}{\pi}\int_{-\infty}^{\infty}\frac{x(\tau)}{t-\tau}d\tau = x(t) * h(t) = x(t) * \frac{1}{\pi t} \tag{40}$$

$$i_{sq} = i_{sq0} + i_{sqv}\sin\left(\omega_m t + \varphi_{sqv}\right) \tag{41}$$

According to reference [56], due to the operating characteristics of the WTs, the resolution of the STFT, in both the time and the frequency domains, is limited. The wavelet transform has the capacity to analyze variations in the signal in the coupled time-frequency domain. However, this process depends strongly on the chosen function and requires prior knowledge of the signal used, while empirical mode decomposition (EMD) lacks a theoretical foundation and requires extreme interpolation. In this context, the author proposes a new method to detect the failures of the gearbox of the WTs, which is based on first demodulating the stator current signal of a DFIG using the HT and then applying a signal resampling algorithm based on the generator rotation frequency, such that the resampled envelope has a constant phase angle range.

While the FT decomposes a signal into a set of waves of different frequencies, the wavelet transform transforms a signal contained in space to a time-scale region using an infinite set of functions called wavelets and defined according to Equation (42), called the wavelet mother. Similar to the FT, the continuous wavelet transform (CWT) represents the sum of the products of the signal multiplied by each of the wavelets, as shown in Equation (43). The characteristics and properties vary according to the type of mother wavelets or wavelet families, among which we mention the Haar wavelet, Daubechies wavelets, symlets, coiflets, biorthogonal wavelets, reverse biorthogonal wavelets, Meyer wavelets, discrete approximations of Meyer wavelets, Gaussian wavelets, Mexican hat wavelets, Morlet wavelets, complex Gaussian wavelets, Shannon wavelets, frequency B-spline wavelets, and complex Morlet wavelets. Unlike the STFT, wavelet transformation allows the window measurement to be varied in such a way that a wide window can be used when information about low frequencies is required and narrow windows when it is necessary to analyze high frequencies since the latter are detected better in the time domain, while low frequencies are more accurately analyzed in the frequency domain. This property means that wavelets can be used in the SA at different frequencies and resolutions of both stationary and transient signals [48].

$$\Psi_{a,b}(t) = \frac{1}{\sqrt{a}}\Psi\left(\frac{t-b}{a}\right) \tag{42}$$

$$C(a,b) = \int_{-\infty}^{\infty} f(t)\Psi(a,b)dt \tag{43}$$

As described in reference [57], techniques such as the CWT and discrete FT have disadvantages when the analysis is carried out with small loads and close to the synchronism speed. According to this author, a more effective method for detecting faults and analyzing the evolution of their severity is to use a Kalman filter, which is computationally more efficient. Since the CWT requires considerable computational effort and generates too much data, an alternative is to filter the signal iteratively in such a way that each frequency band obtained is again decomposed and so on until we obtain several high-resolution frequency components, on which we perform the analysis. In other words, the alternative is to discretize the parameters of both scale and time, leading to the discrete wavelet transform (DWT), also called multiresolution analysis [58]. According to reference [59], calculating the coefficients throughout the scale increases the calculation time, while according to reference [60], it can be shown that the CWT is not useful for detecting faults or torque variations; therefore, the use of the DWT is preferable. According to that report, the signal must first be decomposed by the CWT with a Daubechies 8 (Db8) mother function, and then the FFT is used to analyze the spectrum components.

In reference [61], the components of the fundamental frequency and harmonics due to eccentricity, slots, and other unknown causes, including environmental noise, determine the CS. However, these components are not those related to the generalized bearing roughness; therefore, to detect this type of failure, that work proposes to eliminate the mentioned components by filtering the generator stator signal from a WT using the DWT based on the coiflet function. In addition, the signal is also broken down by wavelets into several segments until the components associated with bearing failures are obtained. As described in reference [62], in regard to detecting broken bars, the main problem with steady-state analysis is that the frequency separation depends on inertia, which means that for small loads, the separation decreases to a point where the frequencies associated with the broken bars cannot be distinguished, so those authors propose processing the signal using wavelets. The current signal of an induction motor is also broken down by wavelets for the detection of broken bars under different load conditions. According to the authors, the high-order Daubechies family behaves as an ideal filter and partially avoids overlap between frequency bands. In reference [63], the current signal of an induction motor is also segmented by wavelets to detect broken bars under different load conditions. According to the authors, the high-order Daubechies family behaves as an ideal filter and partially avoids overlapping between frequency bands. According to reference [16], the use of wavelets makes it unnecessary to know the slip, and in reference [20], a diagnosis of broken bars is proposed based on the current signal and the transformation of wavelets, without using the slip.

As described in reference [64], it is possible to identify the incipient presence of broken bars by applying the DWT with the Daubechies-44 family. Furthermore, since the transient state of the induction motor can offer very useful information for the detection of electromechanical faults, such as the dynamic eccentricity, the signal is sampled during startup. The detection of failures of loads coupled to the induction machine using the MCSA has also been extensively studied, such as in reference [65], where this methodology is used to detect gearbox failures caused by broken gears or broken teeth. The use of GCSA for this purpose in WTs has not received the same attention, especially in regard to studies based on real data.

In the time-frequency analysis, both variables are dependent, and according to the Heisenberg uncertainty principle, it is not possible to know exact values but only intervals [66]. The autocorrelation and power spectrum function does not provide all the necessary information, such as phase coupling or bicoherence, whereas the STFT has the drawback of temporal resolution. For this reason, in the case of non-Gaussian and nonlinear signals and signals whose spectrum is made up of a large number of frequencies, SA must be implemented using high-resolution or higher-order-spectrum (HOS) techniques [67]. Among these techniques is an approach using the bispectrum, which, being a complex number, consists of magnitude and phase. The bispectrum can be used to analyze the relationship between the frequencies of two sinusoids and the result obtained due to the modulation between them. For each set of three frequencies, the signal power and phase are calculated; if the phase shift between the two sinusoids tends to zero, then both have the same origin. Otherwise, the phase shift provides an indication of failure [68].

According to reference [69], given a signal $X(K)$, its second-order statistical characterization can be represented by the autocorrelation and power spectrum function. For the same signal, with zero mean, its third-order moment is given by (44). If the signal is not stationary, Equation (44) depends on three parameters (k, τ_1, τ_2), while for stationary signals, the function contains only τ_1 and τ_2. The FT of the second-order momentum is the power spectrum that we have seen previously, while the bispectrum is the double FT of the third-order momentum and is defined by Equation (45). The degree of coupling between frequencies of different phases is measured by the bicoherence index, Equation (46), whose magnitude varies from zero to one. The greater the value, the greater is the coupling [70]. According to reference [26], the bispectrum technique allows us to represent the FFT of both the phase and the amplitude of the signal. Since the magnitude of the dominant component is a function of the level of the fault, when this technique is applied to the current of induction motors, the spectrum of the current signal allows the detection of electrical faults. The theoretical and mathematical foundations of HOSs are addressed in reference [71–75].

$$c_{3,x}(k, \tau_1, \tau_2) = c_{3,x}[x(k)x(k+\tau_1)x(k+\tau_2)] = E[x(k), x(k+\tau_1), x(k+\tau_2)] \tag{44}$$

$$B(f_1, f_2) = E[X(f_1)X(f_2)X \times (f_1 + f_2)] \tag{45}$$

$$bic(f_1, f_2) = \frac{B(f_1, f_2)}{\sqrt{P(f_1)P(f_2)P(f_1 + f_2)}} \tag{46}$$

Other HOS techniques used are frequency estimators based on eigen analysis. This methodology divides the RM autocorrelation matrix into two vector subspaces, one representing the signal and another representing noise, as shown in Equation (47). The order of the matrix and its eigenvalues are given by (48). Among the frequency estimators developed based on this methodology, multiple signal classification (MUSIC) and root MUSIC can be mentioned, which, as addressed in [76], are high-resolution models that allow for the detection of frequencies in signals with low signal-to-noise ratios.

$$d(n) = \sum_{k=1}^{s_i} A_k e^{(j2\pi n f_k + \varnothing_k)} + e(n) \tag{47}$$

$$M = \{\lambda_1 + \sigma^2, \lambda_2 + \sigma^2, \ldots\ldots, \lambda_L + \sigma^2, \sigma^2, \ldots\ldots\sigma^2\} \tag{48}$$

As described in reference [76], although the relevant techniques generally deal with the detection of a fault, in induction motors, it is most likely to find the presence of several faults, and for its detection, a high-resolution model is proposed that combines a bank of infinite impulse responses and MUSIC. According to the authors, this method is capable of detecting broken bars, imbalance, and defects in the outer bearing race. Another approach that MUSIC uses is described in reference [77]. Although signal sampling is generally performed during machine operation and in some approaches during startup, in that work, signal sampling is performed when the machine is disconnected from the network since, according to reference [75], as the terminal voltage is produced by the rotor currents, the presence of broken bars is reflected in the spectrum of the stator voltage. On the other hand, according to reference [78], the disadvantage of MUSIC is that by increasing the correlation matrix to find more frequencies, the required computational effort increases. To overcome this drawback, that work proposes applying an algorithm similar to the zoom-FFT (ZFFT) method that focuses on certain frequencies regardless of the total frequency range while applying zoom-MUSIC (ZMUSIC). According to the authors, very good results are obtained with the proposed method, comparable to those obtained with ZFFT but requiring less sampling time and less memory capacity.

Other models used for HOS fault monitoring and detection include estimation of signal parameter via rotational invariance technique (ESPRIT) and PRONY. ESPRIT belongs to the subspace parametric spectrum estimation methods expressed by Equations (52) and (53) [79], whereas according to reference [80], the PRONY method is used to model the sampled data of a signal using a linear system of complex exponential functions. As described in reference [81], the extensive use of the power converter when the DFIG works below the synchronous speed causes the current to have a high content of interharmonics that can cause resonance, in addition to damage to capacitors, insulation, control elements, and protection. According to the authors, the identification of these harmonics can be performed using the PRONY and ESPRIT methods, although the latter has a lower resolution than the former.

According to reference [82], to overcome spectral leakage, high-resolution analysis should be applied, but since this implies a longer sampling time, the spectrum varies both in frequency and in amplitude, making diagnosis difficult. From this, it can be deduced that there are no stable conditions that are required to apply the FFT and that its use does not guarantee the identification of the frequency components. The DWT allows for better spectral resolution. However, in general, the proposed signal techniques are not efficient at low slips, such as 1%. Based on the foregoing, reference [82] proposed using ESPRIT in combination with an improved Hilbert's modulus method, which was successful in detecting broken bars even with a slip as small as 0.33%, using only the one-phase signal and short sampling time. When the same experiments were performed using MUSIC, satisfactory results were not obtained, demonstrating the superiority of ESPRIT.

Another technique that has been widely used to diagnose faults in electrical machines is Park's vectors. According to Fortescue's theorem, a triphasic system can be represented as the sum of the components as a zero or homopolar, positive sequence and negative sequence, as expressed in Equation (49) [83]. For three-phase induction motors, the three phases can be represented in the 2D dq plane with Equations (3) and (4), known as Park vectors or Concordia patterns. In the absence of faults, the Park vectors have components given by (50) and (51), whose graph is circular and centered on the origin, while when there are faults in the stator and/or rotor, the graph is deformed and takes on an elliptical shape [84–89].

$$
\begin{bmatrix} I_a \\ I_b \\ I_c \end{bmatrix} = \begin{bmatrix} 1 & 1 & 1 \\ 1 & a^2 & a \\ 1 & a & a^2 \end{bmatrix} \begin{bmatrix} I_a^0 \\ I_a^+ \\ I_a^- \end{bmatrix}
\tag{49}
$$

$$
i_d = \frac{\sqrt{6}}{2} I_s \sin \omega t
\tag{50}
$$

$$
i_q = \frac{\sqrt{6}}{2} I_s \sin\left(\omega t - \frac{\pi}{2}\right)
\tag{51}
$$

4. Materials and Methods

This research work has two parts. In the bibliographic part, emphasis has been placed on the presentation of the theoretical foundations, mathematical models, and existing proposals for some of the most used methodologies for the detection and diagnosis of failures in WTs. The second part is a field investigation, with the purpose of verifying the effectiveness of the analyzed models to determine the status of WTs in operation. The mentioned WFs were installed approximately 20 years ago, and the one where the measurements were made consists of 33 WTs of brand NEG Micon. The electric generator used by the WTs is a SCIG with two windings, one of small power (200 kW) for low speed and the other of a higher power (900 kW) for higher wind speeds (see Table 1). As at the time of testing, the wind speed was high, and measurements were made on the highest-power generator (see Figure 1).

Table 1. Technical characteristics of the WT and electric generator.

Brand	**NEG Micon**
Model	NM 52/900
Rotor diameter	52.2 m
blades	3
Power	900 kW
Power control	Stall control
Drive train	Gearbox type: Planetary-Parallel Transmission ratio: 1:67.5 Main bearing: spherical rollers Cooling system: refrigerant, heat exchanger and pump
Electric Generator	Type: SCIG Speeds: 750/500 rpm Poles: 4/6 Power: 900 kW/200 kW Voltage: 690 V/50 Hz Cooling system: water
Coupling to the power grid	Smooth, using thyristors

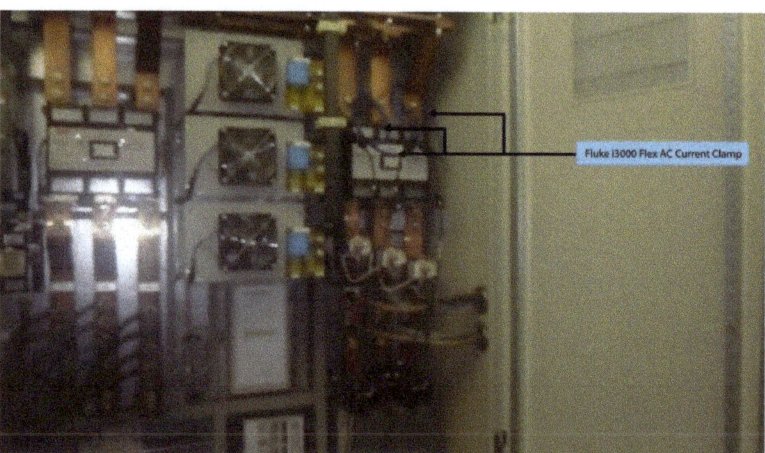

Figure 1. Location of the current clamps on the WT power panel.

For measurements, in addition to the Fluke i3000s flexible clamps, a Pico Technology unit, model PicoScope®4424, was used, which must necessarily be connected to a computer in which software has previously been installed to be able to acquire the signal. Data were acquired by applying a sampling rate of 10 kHz over 2s (representing 20,000 data points per sample). This measurement was made continuously for approximately 10 min. The processing of the data and the application of the various SA techniques were performed in MATLAB r2019b software.

5. Results and Discussion

By applying the FT to the current signal of the generator under study, Figure 2a,b are obtained for one and three phases, respectively. In the graphs, harmonics with frequencies of 40, 42, 46 and 56 Hz can be distinguished, which, according to references [42,76,90], are related to stator failures, broken bars, and phase imbalance. In addition, as described in reference [42], rotor failures are manifested by harmonics of the fundamental frequency (3, 5, 7, etc.), which reinforce the indications of bar failure.

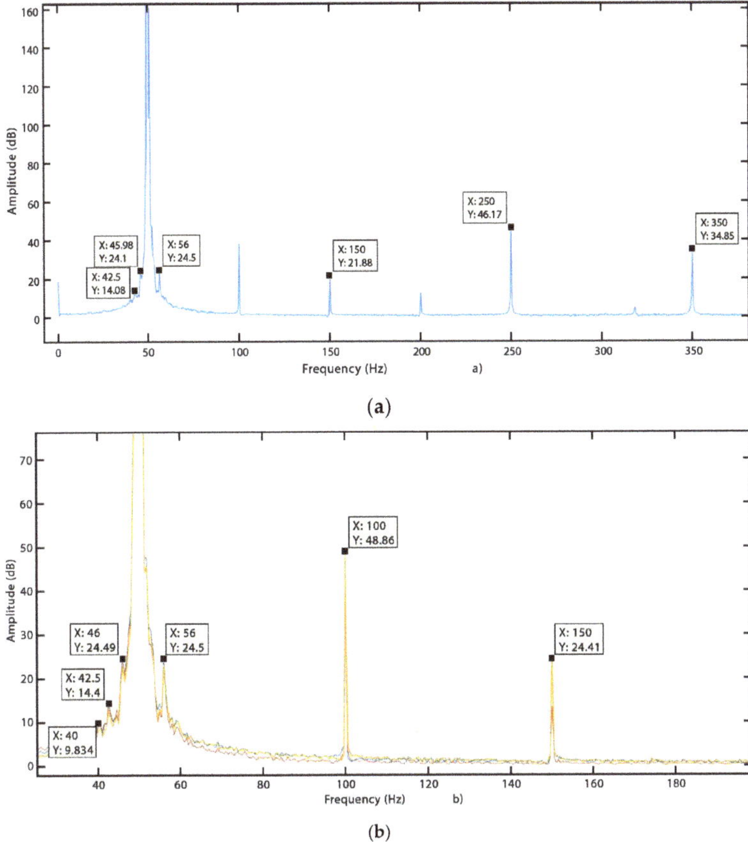

Figure 2. FT of the WT current signal. (**a**) One phase and (**b**) three phases.

The frequency components are very close to the central frequency, and depending on the window used, the width of the lobes can be very large, and the identification of faults by means of the FFT is difficult, making it necessary to resort to other variables and techniques such as the power spectral density (PSD) [91]. In MATLAB, we obtain an improved version of the PSD, which is shown as a Welch periodogram (see Figure 3). In this approach, by definition, MATLAB applies the Hamming window and displays the part of the graph corresponding to the real values. Unlike the case of the classic FT, in the spectrum obtained by the Welch periodogram, a greater number of frequency peaks can be distinguished, such as 30, 46, 54, 63, 124, 165, 261, 451, 534, 575, and 781 Hz. Although the composition of the spectrum is uniform, there are differences in the magnitudes of the components. These differences can also be observed when comparing the three phases of the generator (see Figure 3b).

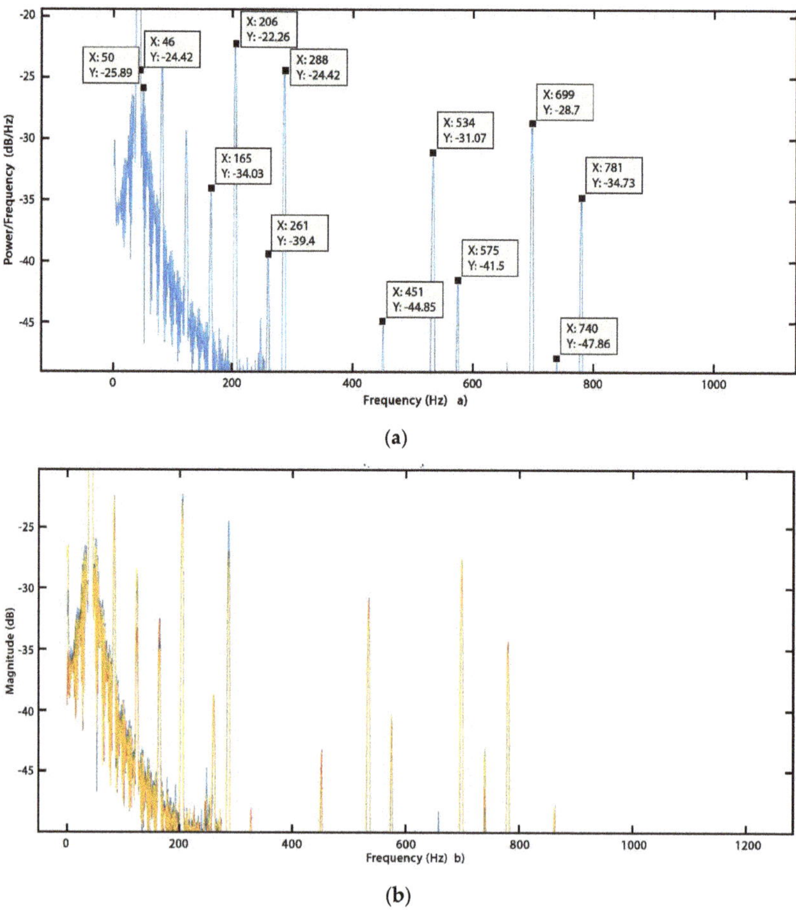

Figure 3. PSD applying Welch's periodogram. (**a**) Phase A and (**b**) three phases.

Another alternative to overcome the drawbacks mentioned so far is the proposed methods to eliminate frequencies that are not of interest so that it is easier to identify the components sought. In this context, one option is the technique known as cepstrum analysis, which calculates the inverse FT of the signal spectrum on a logarithmic scale. Failures alter the rotor speed and magnitude of the components, creating new frequency components that have their own harmonic families. The cepstrum provides an average of each of these families displayed as a single line with their respective harmonics bands. Identifying the frequency of each frequency also allows for the separation of signals that have been combined due to convolution [53,92].

By applying cepstrum analysis and selecting the appropriate scale for the axes, Figure 4 is obtained. Several families of components close to the fundamental wave can be more clearly distinguished than before. For our case, the harmonic families separated by 200 ms that are equivalent to 5 Hz are easily visible. These frequencies are consistent with a fault attributed to broken bars, as was deduced with the previous techniques. Other types of mechanical failures associated with these frequencies are imbalances of the blades and bearings of the generator, which, according to reference [93], are manifested by frequencies of 10 and 5 Hz, respectively. Although there is considerable similarity in the spectrum of the three phases obtained with this technique, differences in the composition of the spectrum and the magnitude of the components can also be highlighted, providing another indication of failure.

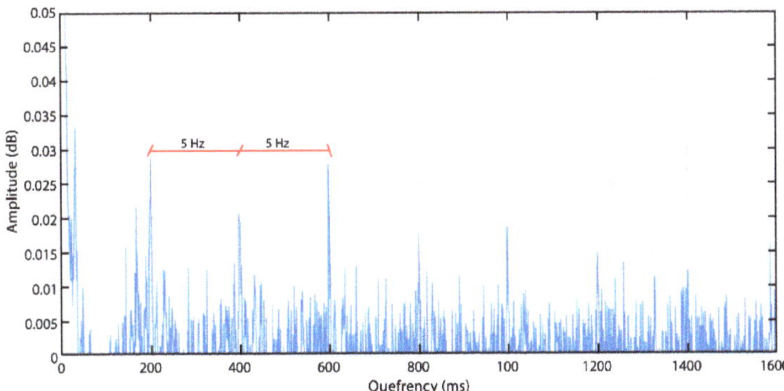

Figure 4. Signal cepstrum of the electric generator.

As we have been able to verify from the techniques used so far, time-domain analysis does not provide the frequency spectrum, while the analysis in the frequency domain does not provide the moment at which the components are produced. By applying the algorithm of reference [94] to calculate the STFT in MATLAB, Figure 5 is obtained. Parts a and b of this figure emphasize how both the fundamental frequency and its components, which remain invariant over time, stand out in terms of their energy (yellow color). Part b reveals harmonics very close to the fundamental (green and orange color), which correspond to the frequencies of 10 and 5 Hz mentioned above.

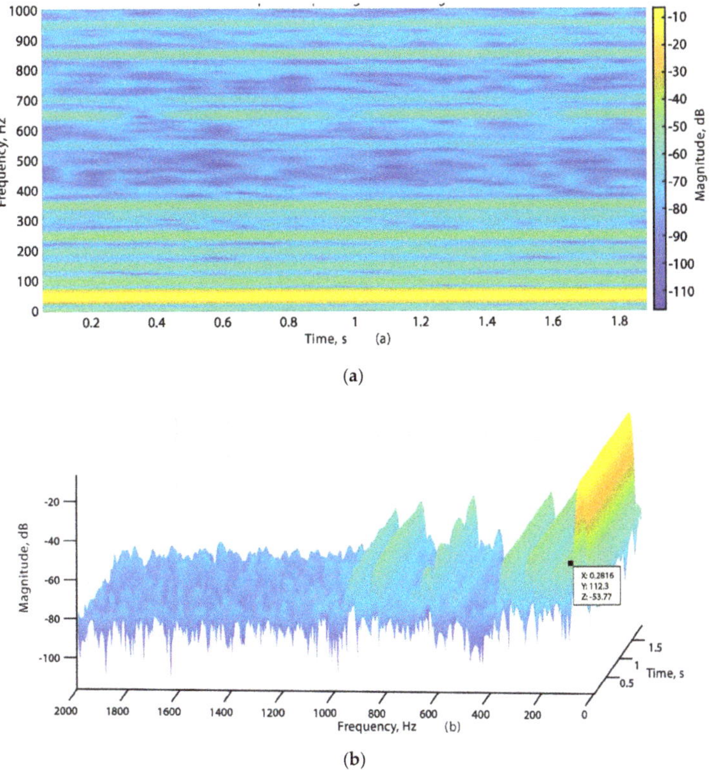

Figure 5. Amplitude spectrogram of the SCIG. (**a**) 2 dimensional STFT, (**b**) 3 dimensional STFT.

From a conceptual point of view, the STFT is one of the most important techniques and serves as the basis for other signal processing techniques. However, its main disadvantage is that it uses the same window width for the entire signal, resulting in the resolution in the time and frequency domain being constant and only one frequency band being known. If the window is wide, a good resolution is obtained in time, but a poor resolution is obtained in the frequency domain, whereas when the window is narrow, the opposite occurs. Therefore, if the frequency components are well separated, a good resolution over time may be preferred, whereas when the components are close together, the frequency resolution is prioritized. The STFT is suitable for the analysis of quasistationary signals (stationary at the window scale), which do not precisely represent the behavior of real signals. Another disadvantage is that there are no orthogonal bases for the STFT, so it is difficult to find a quick and effective algorithm to calculate it [45,65].

According to reference [47], the STFT is positive in all parts and fulfills the positivity requirement, but regardless of the selected window, it does not provide adequate resolution to distinguish the components, nor does it manage to show the instantaneous frequency that can be obtained by the WVD. However, the WVD does not meet the positivity, global average, and finite support requirements. As described in reference [38], one of the main disadvantages of bilinear distributions, such as the WVD, is the interference terms formed by the transformation, which makes interpretation difficult and prevents identifying the true components. To correct this disadvantage, distributions such as the Choi–Williams distribution, pseudo-Wigner–Ville distribution (PWVD), and smooth pseudo-Wigner–Ville distribution (SPWVD) are used [95].

Applying the SPWVD to the WT signal under study, Figure 6 is obtained. The 50 Hz frequency (yellow color) and its harmonics stand out, and—as with the previous techniques—components including 5 Hz can be distinguished along with the fundamental wave. Despite this advantage, in Figures 5 and 6, the disadvantages of the STFT and WVD mentioned by references [38,47] can respectively be seen.

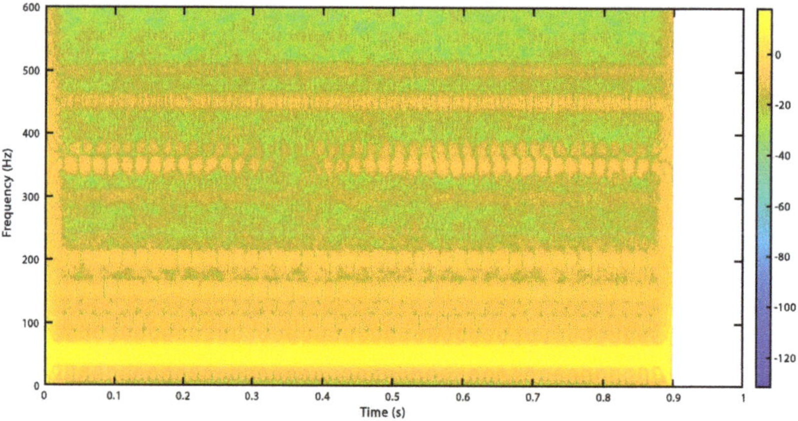

Figure 6. Smoothed PWVD.

According to reference [96], when the carrier signal frequency is on the order of kHz, bearing failures cannot be detected. However, by applying envelope analysis at the lower frequencies, detection is possible. This is the foundation on which techniques such as the shock pulse meter (SPM) and spike energy are based. HT demodulation, either directly to the current signal or to the Park transformation, is used to detect various types of rotor and stator faults, such as broken bars and inter-turn short circuits. In general, signal demodulation is usually the first phase, prior to the application of other mathematical models that are used to improve detection and diagnosis [84,85,97,98]. By applying the HT to our signal, Figure 7a is obtained, and we can distinguish the real and imaginary parts.

Taking this last part and calculating the PSD, the spectrum of Figure 7b is obtained, which is similar to those obtained using the FFT and Welch's periodogram. However, the components associated with broken bars cannot be distinguished, as it was done with the previously applied techniques.

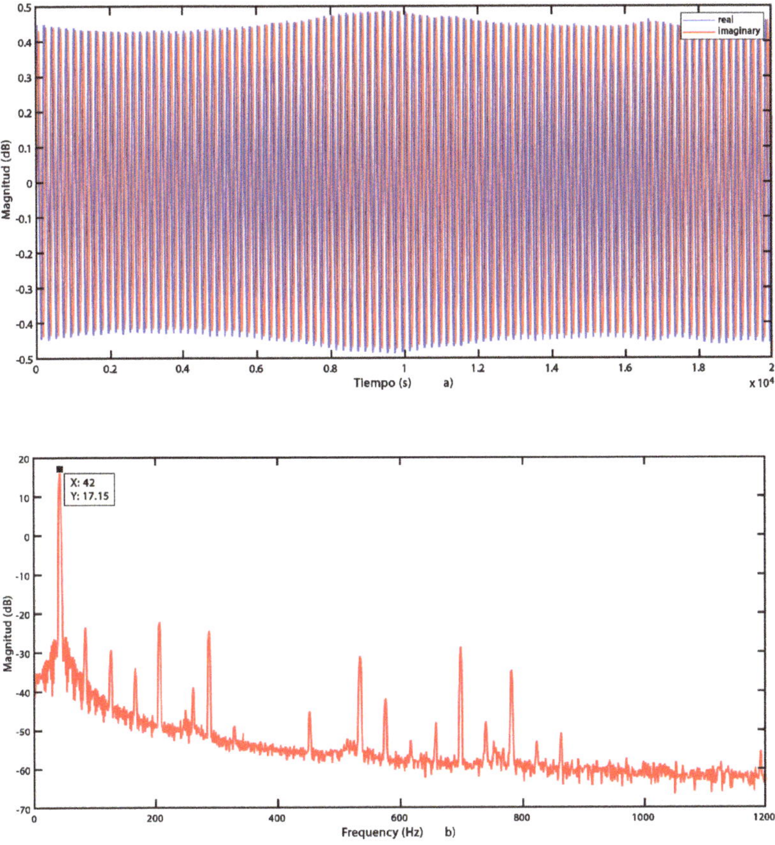

Figure 7. HT for the electrical generator signal. (**a**) Real and imaginary part of the signal, (**b**) HT PSD using the FFT.

According to reference [99], in regard to systems with multiple components, such as WTs, the approach described in reference [54] does not work because the noise processed by the HT generates spurious amplitudes at negative frequencies. To avoid this drawback, according to reference [99], the HT should not be applied directly to the signal but to each of the members of an empirical decomposition of the signal in the IMF, obtained by means of the method called sieving. By applying this methodology to the SCIG signal, Figure 8 is obtained. For our case, the MATLAB algorithm breaks down the signal into seven IMIs (Figure 8a), and proceeding as in reference [55], the HT of the first IMF is obtained (Figure 8b), to which other techniques can be applied, such as the FFT (Figure 8c). In this last figure, the components associated with broken bars are much more evident.

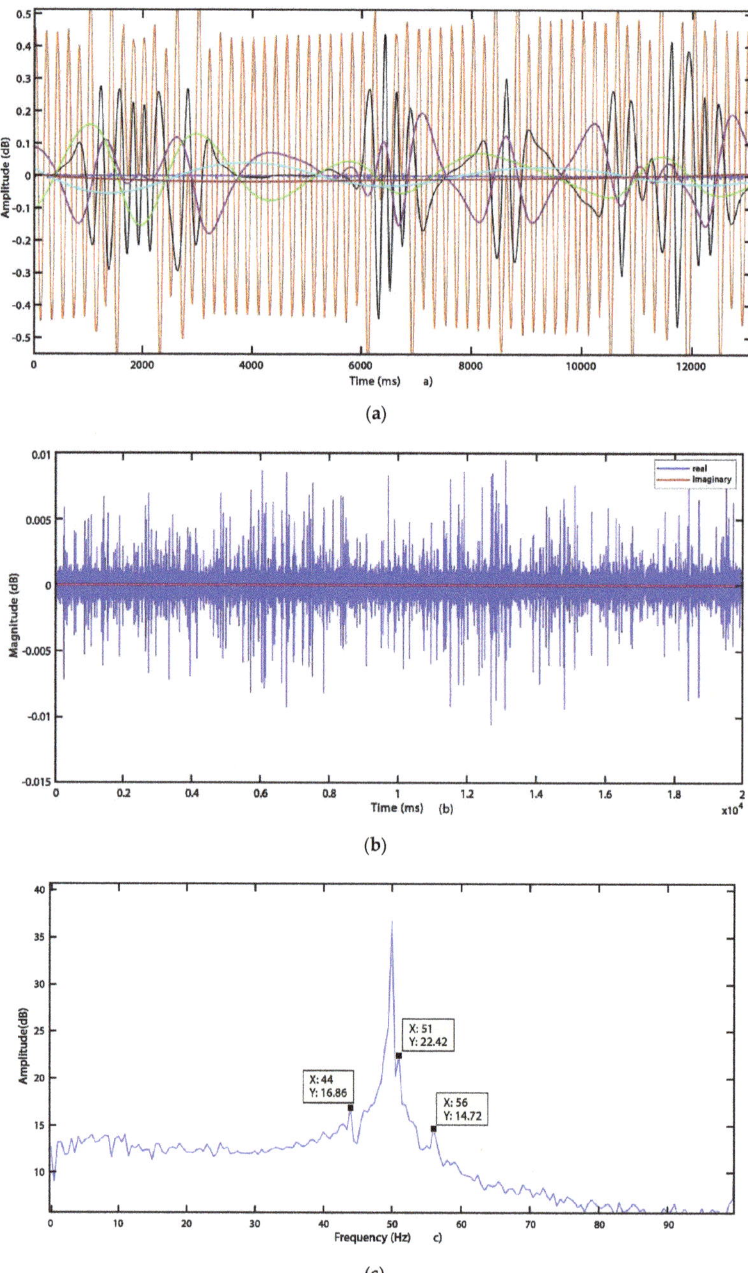

Figure 8. (**a**) Signal IMF, (**b**) HHT for the 1st IMF, (**c**) FFT of the 1st IMFs.

According to reference [42], when the generator is directly coupled to the grid using closed-loop controllers, there is no manual control over the frequency or the terminal voltage. This control system affects the behavior of the generator signal, and for fault diagnosis, it is necessary to use techniques for transient states. For this reason, the author proposes to first filter and decompose the current signal of a SCIG using the DWT and then use the STFT to analyze the evolution over time of the frequency of interest. This allows detecting not only stator and rotor failures but also their location and identification. A similar approach to detect SCIG failures using the current signal is presented in [100]. According to this study, due to its flexibility in the analysis of the evolution of the different frequencies of a signal during transient phenomena, the wavelet transform is the most used signal processing technique for fault diagnosis. However, the author agrees in stating that the DWT cannot analyze the evolution over time of each frequency band in which the signal decomposes, which can be solved by applying the STFT to the obtained frequency bands of interest.

To apply the wavelets to our signal, we proceed in a similar way to reference [60]. First, by means of the DWT and the Daubechies family we decompose the signal into 8 levels (see Figure 9a), in such a way that, at level d7 the frequency range is from 0 to 78 Hz and this is where the frequency components could be found associated with the types of failures mentioned so far. However, observing the d7 level in Figure 9a, it is not enough to make a diagnosis, and in these cases, it is necessary to apply another type of analysis or use other methodologies, as described in reference [101]. Later, CWT with a scale of 1:100, it is applied and whose 2D graph is shown in Figure 9b. Here, in addition to the periodicity of the signal, it can also be seen how as the scale factor increases towards the last low-pass filters, the wavelets compress more and the number of low-frequency components associated with faulty bars or eccentricity becomes more evident. The 3D graph is included in Figure 9c, displaying the periodicity and uniformity or composition of the signal as a function of time. In this last graph, the peaks of the signal occur in the last scales; therefore, by calculating in MATLAB the frequency equivalent of scale 100, the value of 6 Hz is obtained, which is consistent with the results of other signaling techniques.

According to reference [102], one of the disadvantages of classic SA using the FFT is the loss of information when the signal is segmented. This can be compensated by the weighting of the windows. However, this incurs a decrease in spectral resolution. As described in references [45,65], despite the benefits of analysis in the time-frequency domain, since the components associated with the faults may be very close to the fundamental frequency, their identification is complicated, so it is also necessary to determine the frequency at which the analysis should be performed. At high frequencies, a good resolution is obtained in the time domain, while at low frequencies, the resolution is better in the frequency domain.

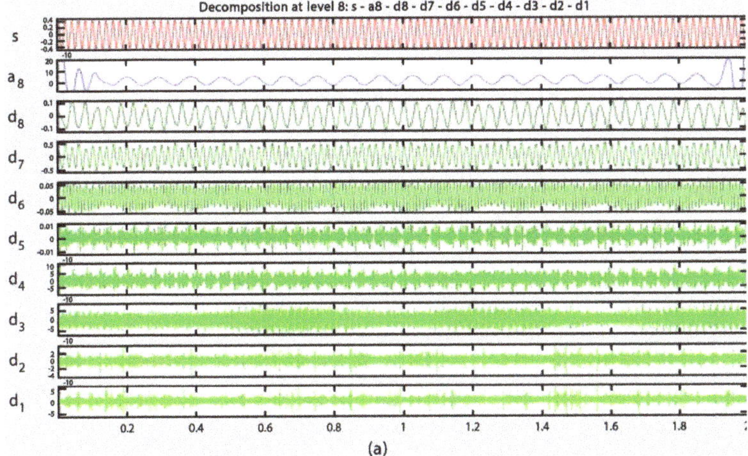

Figure 9. *Cont.*

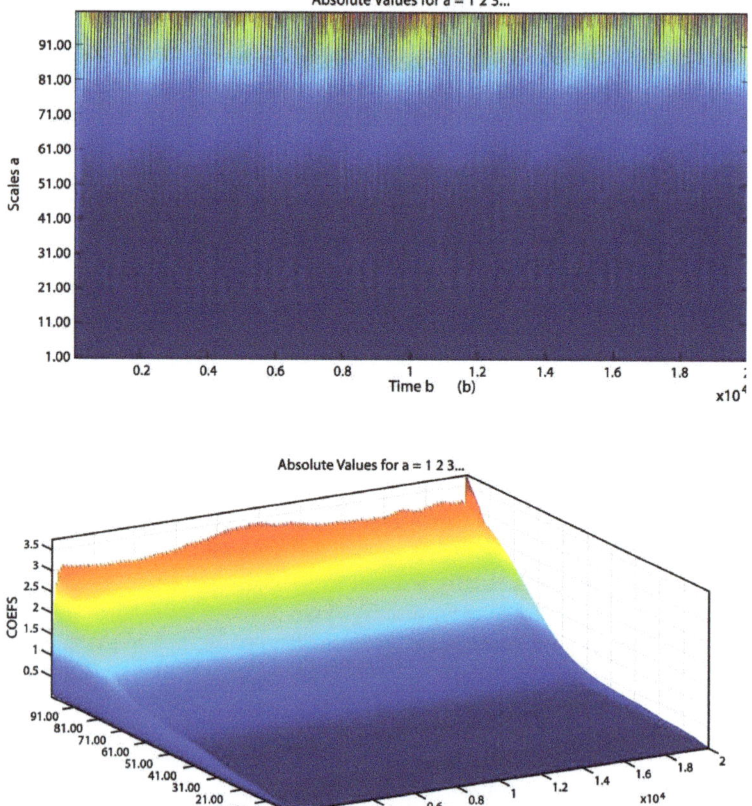

Figure 9. Application of the wavelet transform to the SCIG signal. (**a**) Decomposition of the current signal using the DWT, (**b**) 2-dimensional continuous wavelet transform, (**c**) 3-dimensional continuous wavelet transform.

According to reference [99], the wavelet transform has the disadvantage of overlap between the frequency bands in which the signal has been separated and the need for an optimal selection of the mother wavelet. To overcome these drawbacks, that report proposes to carry out the analysis by applying both the HT and the wavelet transform to the current signal during startup. According to the authors, based on the experimental studies and in contrast to the classic Fourier analysis, the DWT-based approaches are simple and allow clear and reliable patterns to be obtained. The Hilbert–Huang transform (HHT) has the advantage of avoiding dyadic decomposition, allowing greater security in the study of high-frequency components located on the right-hand side, and IMFs allow a more secure theoretical representation of the waveform composed of the left sidebands on the supply frequency, which cannot be achieved with the DWT. Among the disadvantages are the introduction of signal overlap problems, although this effect is negligible during startup. The patterns obtained are not as clear as in other methods and are more difficult to interpret. There is no a priori relationship between IMFs and frequency bands, making it difficult to select the appropriate number of IMFs to consider for the detection of lateral components. As discussed in reference [99], these conclusions have to be verified in field studies so that the results can be generalized for different operating conditions.

Finally, by applying the Park transform to the generator signal, Figure 10 is obtained. According to reference [97], regardless of the slip, the short circuits between turns or broken bars produce an alteration in the envelope repeats cyclically to the same supply frequency, causing the elliptical shape of the Park transform graph. Additionally, the comparison of Figure 10 with the results obtained in reference [103] for the diagnosis of broken bars verifies substantial similarity.

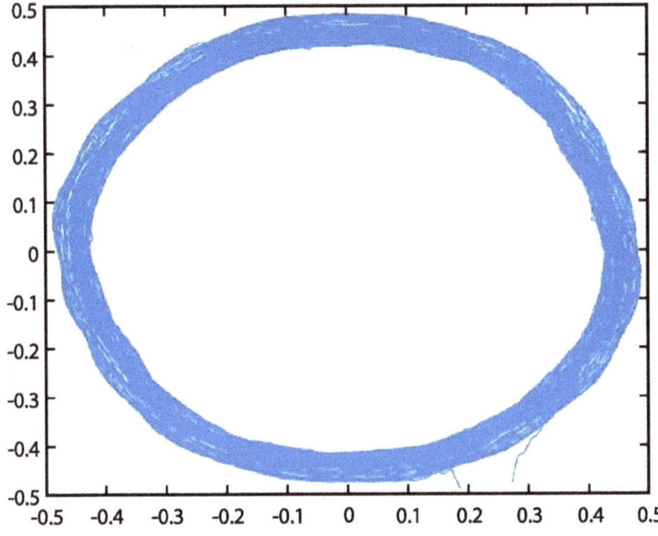

Figure 10. Park vectors of the current signal.

6. Conclusions and Recommendations

Through this research, it has been possible to demonstrate the feasibility of detecting and diagnosing faults in a WT generator using SA of the current signal. According to the models described in the theoretical part of this research, there is at least one indication of failure due to broken bars in the generator under study. The analysis was carried out using various signal processing techniques, obtaining similar results with all techniques. However, the magnitude of the failure has not been included in this investigation. To check the results obtained and to carry out a more in-depth investigation, a periodic sampling could be done to analyze the evolution of the spectrum of the generator current signal.

Although the diagnosis was obtainable with all the signal processing and analysis techniques used, there are some differences. The FT indicates which frequencies exist in the spectrum of a signal, but it does not provide the time at which these frequencies occur, nor does it provide the modulation of the phase. Aliasing and leakage problems can also occur, and in general, this technique is not recommended for transient states. The STFT allows information to be obtained in both the time domain and the frequency domain. However, since it uses a fixed observation window for all frequencies, it cannot adapt to rapid signal changes and cannot eliminate noise. In contrast, the DWT cannot analyze the evolution over time of the frequency composition of each frequency group. The power spectrum does not provide phase information, and the autocorrelation sequence does not provide evidence of nonlinearity. Furthermore, since the power spectrum variance does not tend to zero as the number of samples increases, the second-order periodogram or moment is not a consistent estimator, and it is necessary to resort to third-order estimators such as the bispectrum, MUSIC, and root MUSIC.

Signal processing techniques are a very powerful tool. However, in many cases, especially when conditions are not ideal, the use of these methodologies in isolation is not sufficient, and it is necessary to use other, complementary models to increase the effectiveness of diagnosis. The use of the current signal for the detection and diagnosis of faults in WTs is an area still to be explored, especially through field work. However, based on the few existing references on field studies carried out with WTs in operation, it can be said in general that when the current signal is used, the diagnostic process is based on the models analyzed in Sections 2 and 3 of this investigation.

According to what was seen in the introduction, in this research we have concentrated on the SCIG. However, the most widely used electric generator in the wind industry is the double feed induction generator (DFIG) which has many characteristics in common with the SCIG. However, since the DFIG has a wound rotor that is feed independently, it has an electrical signal from the stator and another signal from the rotor, which could be studied independently or in combination to detect both rotor and stator faults. Another important aspect to consider is that WTs with DFIG use a power converter to control the rotor current, which modifies the spectrum of the signals and increases the difficulty of diagnosis. To limit the research, we have preferred not to delve into the differences that we would have with the DFIG, since it would be preferable to do another specific field study on this type of generator.

Several of the models seen so far require knowledge of the rotor mechanical speed and/or slip and generator design parameters, among other variables, which are generally not available. Additionally, when signal processing and analysis techniques are used, it is necessary to perform the study for each phase and for each record at the same time, so, considering the three phases of each generator and the total WTs of a WF, the work is very complicated and can lead to diagnostic errors. Besides, almost all the reports used as references rely on a signal that includes an explicitly provoked failure, which was not possible to obtain for this investigation. Despite these aspects, a very useful technique at present is to combine signal processing techniques with artificial intelligence models.

Author Contributions: Conceptualization, Y.M. and L.H.-C. methodology, Y.M.; validation, Y.M., L.H.-C., and O.D.-P.; formal analysis, Y.M., L.H.-C., and O.D.-P.; resources, Y.M. and L.H.-C.; writing—original draft preparation, Y.M.; writing—review and editing, L.H.-C. and O.D.-P.; visualization, R.A.L.-M.; supervision, L.H.-C.; project administration, O.D.-P. All authors have read and agreed to the published version of the manuscript.

Funding: This research received no external funding.

Acknowledgments: The authors thank the University of Valladolid and University of Guayaquil for assistance in the preparation of this research. We also thank CETASA for allowing signal acquisition and providing the necessary equipment. Finally, we thank the anonymous reviewer for their assistance in improving this study.

Conflicts of Interest: The authors declare no conflict of interest.

Nomenclature

A_T	torque amplitude
A_{sMi}, A_{sTi}	amplitude of magnetization and torque components
a	$1(120°) = 1e^{j\frac{2\pi}{3}} = -0.5 + j0.866$
a^2	$1(240°) = 1e^{j\frac{4\pi}{3}} = -0.5 - j0.866$
$\vec{B}$	flux density
$e(n)$	sampled noise
$f_r(Hz)$	rotor frequency
f_{mr}	rotor mechanical frequency
f_{te}	gear frequency
f_v	vibration frequency of bearing failure
$g\,(\theta_r, \theta_{sr})$	air gap function (g in the case of a uniform air gap)
g_o	constant air gap length
$g(t)$	mean air gap length as a function of time
I_L	line current
i_0	average or constant component of the current
i_{sM}, i_{sT}	magnetization and torque components of the stator current
i_{sM0}, i_{sT0}	constant value of magnetization and torque components
J	inertia
k_m	failure modulation index
k_{wh}	winding factor for harmonic h

k_0	$0, 1, 2, 3, 4, 5, \ldots$
k_1	$1, 3, 5, 7, 9, \ldots$
k_2	$1, 2, 3, \ldots, (2p-1)$
k_3	$0, \pm2, \pm6, \pm10, \ldots$
N	number of turns per coil
N_r	number of turns of the rotor winding
n_d	0 for static eccentricity, $1, 2, 3, 4, 5, \ldots$ for dynamic eccentricity
MMF	magnetomotive force
P_i	input power
p	pole pairs
p_d	bearing diameter
P_{Fe}	iron losses
Q_r	rotor slots
R_r	rotor resistance
R_s	stator resistance
S	arbitrary contour surface
s_i	number of complex sinusoids
s	slip per unit
T_0	constant torque component
T_{em}	electromechanical torque
T_T	total torque
T_d	damping torque due to failure
T_{osc}	blade torque under normal conditions
θ_s	angular displacement with reference to the stator
θ_r	rotor angular displacement, rotor surface
θ_{sr}	angular displacement between rotor and stator reference position
φ	phase angle, load or power factor
φ_s	phase shift between the stator and rotor MMFs
φ_d	phase angle of the fault
ω	angular velocity of the feed current
ω_{ro}	constant component of the angular speed of the rotor
ω_s	stator field angular velocity
ω_{mr}	rotor mechanical speed
ω_r	rotor magnetic field speed
ω_f	angular velocity of the fault
Λ^2	Laplace operator
ξ	temporal variable
σ^2	variance
δ_d	dynamic eccentricity index

References

1. Qiao, W.; Lu, D. A Survey on Wind Turbine Condition Monitoring and Fault Diagnosis—Part I: Components and Subsystems. *IEEE Trans. Ind. Electron.* **2015**, *62*, 6536–6545. [CrossRef]
2. Qiao, W.; Lu, D. A Survey on Wind Turbine Condition Monitoring and Fault Diagnosis—Part II: Signals and Signal Processing Methods. *IEEE Trans. Ind. Electron.* **2015**, *62*, 6546–6557. [CrossRef]
3. García Márquez, F.P.; Tobias, A.M.; Pinar Pérez, J.M.; Papaelias, M. Condition monitoring of wind turbines: Techniques and methods. *Renew. Energy* **2012**, *46*, 169–178. [CrossRef]
4. Merizalde, Y.; Hernández-Callejo, L.; Duque-Pérez, O.; Alonso-Gómez, V. Diagnosis of wind turbine faults using generator current signature analysis: A review. *J. Qual. Maint. Eng.* **2019**. [CrossRef]
5. Jin, X.; Cheng, F.; Peng, Y.; Qiao, W.; Qu, L. A comparative study on Vibration- and current-based approaches for drivetrain gearbox fault diagnosis. In Proceedings of the IEEE Industry Applications Society Annual Meeting IAS 2016, Portland, OR, USA, 2–6 October 2016; pp. 1–8. [CrossRef]

6. Henao, H.; Capolino, G.A.; Cabanas, M.N.; Filippetti, F.; Bruzzese, C.; Strangas, E.; Pusca, R.; Estima, J.; Riera-Guasp, M.; Hedayati-Kia, S. Trends in Fault Diagnosis for Electrical Machines: A Review of Diagnostic Techniques. *IEEE Ind. Electron.* **2014**, *8*, 31–42. [CrossRef]

7. Pires Leite, G.; Araújo, A.M.; Carvalho, P.A. Prognostic techniques applied to maintenance of wind turbines: A concise and specific review. *Renew. Sustain. Energy Rev.* **2018**, *81*, 1917–1925. [CrossRef]

8. Seera, M.; Peng Lim, C.; Nahavandi, S.; Kiong Loo, C. Condition monitoring of induction motors: A review and an application of an ensemble of hybrid intelligent models. *Expert Syst. Appl.* **2014**, *41*, 4891–4903. [CrossRef]

9. El Bouchikhi, E.H.; Choqueuse, V.; Benbouzid, M. Induction machine faults detection using stator current parametric spectral estimation. *Mech. Syst. Signal Process.* **2015**, *52–53*, 447–464. [CrossRef]

10. AETS Sudamérica. Estudio de Mercado de Motores Eléctricos en Chile. Santiago de Chile. 2010. Available online: http://dataset.cne.cl/Energia_Abierta/Estudios/Minerg/10.Estudio%20Motores%20El%C3%A9ctricos%20en%20Chile_Final%20(1045).pdf (accessed on 8 October 2019).

11. Stroker, J.J. What's the real cost of higher efficiency? *IEEE Ind. Appl. Mag.* **2003**, *9*, 32–37. [CrossRef]

12. Stavrakakis, G.S. Electrical parts of wind turbines. *Compr. Renew. Energy* **2012**, *2*, 269–328. [CrossRef]

13. Chen, S. Induction Machine Broken Rotor Bar Diagnostics Using Prony Analysis. Master's Thesis, University of Adelaide, Adelaide, Australia, 2008.

14. Pyrhönen, J.; Jokinen, T.; Hrabovcová, V. *Design of Rotating Electrical Machines*; John Wiley & Sons: Chichester, UK, 2008.

15. Bose, B.K. *Modern Power Electronics and AC Drives*; Prentice Hall PTR: Upper Saddle River, NJ, USA, 2002.

16. Yacamini, R.; Smith, K.S.; Ran, L. Monitoring Torsional Vibrations of Electro-mechanical Systems Using Stator Currents. *J. Vib. Acoust.* **1998**, *120*, 72–79. [CrossRef]

17. Toliyat, H.A.; Campbell, S.G. *DSP-Based Electromechanical Motion Control*; CRC PRESS: Boca Raton, FL, USA, 2003.

18. Orille, A.L.; Sowilam, G.M.A.; Valencia, J.A. New simulation of symmetrical three phase induction motor under transformations of park. *Comput. Ind. Eng.* **1999**, *37*, 359–362. [CrossRef]

19. Bonnett, A.H. Root cause AC motor failure analysis with a focus on shaft failures. *IEEE Trans. Ind. Appl.* **2000**, *36*, 1435–1448. [CrossRef]

20. Kia, S.; Henao, H.; Capolino, G. Diagnosis of broken-bar fault in induction machines using discrete wavelet transform without slip estimation. *IEEE Trans. Ind. Appl.* **2009**, *45*, 1395–1404. [CrossRef]

21. Mohanty, A.R.; Kar, C. Fault detection in a multistage gearbox by demodulation of motor current waveform. *IEEE Trans. Ind. Electron.* **2006**, *53*, 1285–1297. [CrossRef]

22. Penman, J.; Seddling, H.G.; Fink, W.T. Detection and Location of Interturn Short Circuits in the Stator Windings of Operating Motors. *IEEE Trans. Energy Convers.* **1994**, *9*, 652–658. [CrossRef]

23. Filippetti, F.; Franceschini, G.; Tassoni, C.; Vas, P. Recent developments of induction motor drives fault diagnosis using AI techniques. *IEEE Trans. Ind. Electron.* **2000**, *47*, 994–1004. [CrossRef]

24. Jung, J.H.; Lee, J.J.; Kwon, B.H. Online diagnosis of induction motors using MCSA. *IEEE Trans. Ind. Electron.* **2006**, *53*, 1842–1852. [CrossRef]

25. Joksimovic, G.M.; Penman, J. The detection of inter-turn short circuits in the stator windings of operating motors. *IEEE Trans. Ind. Electron.* **2000**, *47*, 1078–1084. [CrossRef]

26. Benbouzid, M.E.H. A review of induction motors signature analysis as a medium for faults detection. *IEEE Trans. Ind. Electron.* **2000**, *47*. [CrossRef]

27. Thomson, W.T.; Fenger, M. Current signature analysis to detect induction motor faults. *IEEE Ind. Appl. Mag.* **2001**, *7*, 26–34. [CrossRef]

28. Drozdowski, P.; Duda, A. Influence of magnetic saturation effects on the fault detection of induction motors. *Arch. Electr. Eng.* **2014**, *63*, 489–506. [CrossRef]

29. Bellini, A.; Filippetti, F.; Tassoni, C.; Capolino, G. Advances in diagnostic techniques for induction machines. *IEEE Trans. Ind. Electron.* **2008**, *5*, 4109–4126. [CrossRef]

30. Tavner, P.J. Review of condition monitoring of rotating electrical machines. *IET Electr. Power Appl.* **2008**, *2*, 215–247. [CrossRef]

31. Vas, P. *Parameter Estimation, Condition Monitoring, and Diagnosis of Electrical Machines*; Clarendon Press: New York, NY, USA, 1993.

32. Cameron, J.R.; Thomson, W.T.; Dow, A.B. Vibration and current monitoring for detecting airgap eccentricity in large induction motors. *IEE Proc. B Electr. Power Appl.* **1986**, *133*, 155–163. [CrossRef]

33. Nandi, S.; Toliyat, H.A. Detection of rotor slot and other eccentricity related harmonics in a three phase induction motor with different rotor cages. *IEEE Trans. Energy Convers.* **2001**, *16*, 253–260. [CrossRef]
34. Blödt, M.; Regnier, J.; Faucher, J. Distinguishing load torque oscillations and eccentricity faults in induction motors using stator current wigner distributions. *IEEE Trans. Ind. Appl.* **2009**, *45*, 1991–2000. [CrossRef]
35. Teng, W.; Zhang, X.; Liu, Y.; Kusiak, A.; Ma, Z. Prognosis of the remaining useful life of bearings in a wind turbine gearbox. *Energies* **2017**, *10*, 32. [CrossRef]
36. Lee, Y.; Cheatham, T.; Wiesner, J. Application of Correlation Analysis to the Detection of Periodic Signals in Noise. *IEEE Trans. Ind Electron. Proc. IRE* **1950**, *38*, 1165–1171. [CrossRef]
37. Dalpiaz, G.; Rivola, A.; Rubini, R. Effectiveness and sensitivity of vibration processing techniques for local fault detection in gears. *Mech. Syst. Signal Process.* **2000**, *14*, 387–412. [CrossRef]
38. Jardine, A.; Lin, D.; Banjevic, D. A review on machinery diagnostics and prognostics implementing condition-based maintenance. *Mech. Syst. Signal Process.* **2006**, *20*, 1483–1510. [CrossRef]
39. Poyhonen, S.; Jover, P.; Hyotyniemi, H. Signal processing of vibrations for condition monitoring of an induction motor. In Proceedings of the First International Symposium on Control, Communications and Signal Processing, New York, NY, USA, 21–24 March 2004. [CrossRef]
40. Baillie, D.; Mathew, J. A comparison of autoregressive modeling techniques for fault diagnosis of rolling element bearings. *Mech. Syst. Signal Process.* **1996**, *10*, 1–17. [CrossRef]
41. Zhan, Y.; Makis, V.; Jardine, A. Adaptive model for vibration monitoring of rotating machinery subject to random deterioration. *J. Qual. Maint. Eng.* **2003**, *9*, 351–375. [CrossRef]
42. Attoui, I.; Omeiri, A. Fault Diagnosis of an Induction Generator in a Wind Energy Conversion System Using Signal Processing Techniques. *Electr. Power Compon. Syst.* **2015**, *43*, 2262–2275. [CrossRef]
43. Pinto, M. Procesamiento de Señales Utilizando el Análisis Tiempo-Frecuencia. Master's Thesis, Centro de Investigación y Tecnología Digital, Instituto Politécnico Nacional, Tijuana, Mexico, 2005.
44. Serie Discreta de Fourier. Available online: http://www.ehu.eus/Procesadodesenales/tema3/71.html (accessed on 27 December 2019).
45. Artigao, E.; Honrubia-Escribano, A.; Gomez-Lazaro, E. Current signature analysis to monitor DFIG wind turbine generators: A case study. *Renew. Energy* **2018**, *116*, 5–14. [CrossRef]
46. Pons-Llinares, J.; Antonino-Daviu, J.; Riera-Guasp, M.; Lee, S.; Kang, T.; Yang, C. Advanced Induction Motor Rotor Fault Diagnosis Via Continuous and Discrete Time–Frequency Tools. *IEEE Trans. Ind. Electron.* **2015**, *63*, 1791–1801. [CrossRef]
47. Scarpazza, D. A Brief Introduction to the Wigner Distribution. 2003. Available online: http://www.scarpaz. com/Attic/Documents/TheWignerDistribution.pdf (accessed on 24 April 2020).
48. Misiti, M.; Misiti, Y.; Oppenheim, G.; Poggi, J. *Wavelet Toolbox for Use with Matlab*; The Math Works Inc.: Natick, MA, USA, 2004. Available online: https://www.ltu.se/cms_fs/1.51590!/wavelet%20toolbox%204% 20user\T1\textquoterights%20guide%20(larger%20selection).pdf (accessed on 29 January 2019).
49. Cooley, J.; Tukey, J. An algorithm for the machine calculation of complex Fourier series. *Math. Comput.* **1965**, *19*, 297–301. [CrossRef]
50. Rabiner, L.; Schafer, R. Introduction to Digital Speech Processing. *Found. Trends Signal Process.* **2007**, *1*, 1–194. [CrossRef]
51. Cohen, L. Time-Frequency Distributions: A Review. *Proc. IEEE* **1989**, *77*, 941–981. [CrossRef]
52. Amirat, Y.; Choqueuse, V.; Benbouzid, M. Condition monitoring of wind turbines based on amplitude demodulation. In Proceedings of the IEEE International Energy Conference, Manama, Bahrain, 18–22 December 2008. [CrossRef]
53. Harris, C.; Piersol, A. *Shock and Vibration Handbook*, 5th ed.; McGraw Hill: New York, NY, USA, 2002; ISBN 0-07-137081-1.
54. Shi, L.; Qiu, J.; Xu, G.; Yang, J.; Wang, J. Online detection for blade imbalance of doubly fed induction generator wind turbines based on stator current. In Proceedings of the 2nd International Conference on Power and Renewable Energy (ICPRE), Chengdu, China, 20–23 September 2017. [CrossRef]
55. Jin, X.; Cheng, F.; Peng, Y.; Qiao, W.; Qu, L. Drivetrain Gearbox Fault Diagnosis: Vibration and Current Based Approaches. *IEEE Ind. Appl. Mag.* **2018**, *24*, 56–66. [CrossRef]
56. Cheng, F.; Qu, L.; Qiao, W.; Wei, C.; Hao, L. Fault Diagnosis of Wind Turbine Gearboxes Based on DFIG Stator Current Envelope Analysis. *IEEE Trans. Sustain. Energy* **2010**, *10*, 1044–1053. [CrossRef]
57. Ibrahim, R.; Watson, S.; Djurović, S.; Crabtree, C. An Effective Approach for Rotor ElectricalAsymmetry Detection in Wind Turbine DFIGs. *IEEE Trans. Ind. Electron.* **2018**, *65*, 8872–8881. [CrossRef]
58. Mallat, S. A theory for multiresolution signal decomposition: The wavelet representation. *IEEE Trans. Pattern Anal. Mach. Intell.* **1989**, *11*, 674–693. [CrossRef]

59. Gueye-Lo, N.; Soualhi1, N.; Frini1, M.; Razik, H. Gear and Bearings Fault Detection Using Motor Current Signature Analysis. In Proceedings of the IEEE Conference on Industrial Electronics and Applications, Wuhan, China, 31 May–2 June 2018. [CrossRef]

60. Kar, C.; Mohanty, A.R. Monitoring gear vibrations through motor current signature analysis and wavelet transform. *Mech. Syst. Signal Process.* **2006**, *20*, 158–187. [CrossRef]

61. Gong, X.; Qiao, W.; Zhou, W. Incipient bearing fault detection via wind generator stator current and wavelet filter. In Proceedings of the 36th Annual Conference on IEEE Industrial Electronics Society IECON, Glendale, AZ, USA, 7–10 November 2010; pp. 2615–2620. [CrossRef]

62. Ordaz-Moreno, A.; Romero-Troncoso, R.; Vite-Frías, J.; Rivera-Gillen, J.; Garcia-Perez, A. Automatic Online Diagnosis Algorithm for Broken-Bar Detection on Induction Motors Based on Discrete Wavelet Transform for FPGA Implementation. *IEEE Trans. Ind. Electron.* **2008**, *55*, 2193–2202. [CrossRef]

63. Antonino-Daviu, J.; Riera-Guasp, M.; Roger-Folch, J.; Molina, M. Validation of a new method for the diagnosis of rotor bar failures via wavelet transformation in industrial induction machines. *IEEE Trans. Ind. Appl.* **2006**, *42*, 990–996. [CrossRef]

64. Antonino-Daviu, J.; Jover, P.; Riera-Guasp, M.; Roger-Folch, J.; Arkkio, A. DWT analysis of numerical and experimental data for the diagnosis of dynamic eccentricities in induction motors. *Mechan. Syst. Signal Process.* **2007**, *21*, 2575–2589. [CrossRef]

65. Kia, S.H.; Henao, H.; Capolino, G.A. Torsional vibration effects on induction machine current and torque signatures in gearbox-based electromechanical system. *IEEE Trans. Ind. Electron.* **2009**, *56*, 4689–4699. [CrossRef]

66. Shinde, S.; Gadre, V. An uncertainty principle for real signals in the fractional Fourier transform domain. *IEEE Xplore* **2001**, *49*, 2545–2548. [CrossRef]

67. Swami, A. HOSA-Higher Order Spectral Analysis Toolbox. Available online: https://www.mathworks.com/matlabcentral/fileexchange/3013-hosa-higher-order-spectral-analysis-toolbox (accessed on 24 March 2020).

68. Goshvarpour, A.; Goshvarpour, A.; Rahati, S.; Saadatian, V. Bispectrum estimation of electroencephalogram signals during meditation. *Iran. J. Psychiatry Behav. Sci.* **2012**, *6*, 48–54.

69. Chow, T.; Fei, G. Three phase induction machines asymmetrical faults identification using bispectrum. *IEEE Trans. Energy Convers.* **1995**, *10*, 688–693. [CrossRef]

70. Ning, T.; Bronzino, J.D. Autoregressive and Bispectral Analysis Techniques: EEG Applications. *IEEE Eng. Med. Biol. Mag.* **1990**, *9*, 47–50. [CrossRef] [PubMed]

71. Nikias, C.L.; Raghuveer, M.R. Bispectrum Estimation: A Digital Signal Processing Framework. *Proc. IEEE* **1987**, *75*, 869–891. [CrossRef]

72. Gu, F.; Shao, Y.; Hu, N. Electrical motor current signal analysis using a modified bispectrum for fault diagnosis of downstream mechanical equipment. *Mech. Syst. Signal Process.* **2011**, *25*, 360–372. [CrossRef]

73. Saidi, L.; Fnaiech, F.; Henao, H. Diagnosis of broken-bars fault in induction machines using higher order spectral analysis. *ISA Trans.* **2013**, *52*, 140–148. [CrossRef]

74. Gu, F.; Wang, T.; Alwodai, A. A new method of accurate broken rotor bar diagnosis based on modulation signal bispectrum analysis of motor current signals. *Mech. Syst. Signal Process.* **2015**, *50–51*, 400–413. [CrossRef]

75. Li, D.Z.; Wang, W.; Ismail, F. An Enhanced Bispectrum Technique with Auxiliary Frequency Injection for Induction Motor Health Condition Monitoring. *IEEE Trans. Instrum. Meas.* **2015**, *64*, 2679–2687. [CrossRef]

76. Garcia-Perez, A.; Romero-Troncoso, R.D.J.; Cabal-Yepez, E.; Osornio-Rios, R.A. The application of high-resolution spectral analysis for identifying multiple combined faults in induction motors. *IEEE Trans. Ind. Electron.* **2012**, *58*, 2002–2010. [CrossRef]

77. Cupertino, F.; de Vanna, E.; Salvatore, L.; Stasi, S. Analysis techniques for detection of IM broken rotor bars after supply disconnection. *IEEE Trans. Ind. Appl.* **2004**, *40*, 526–533. [CrossRef]

78. Kia, S.H.; Henao, H.; Capolino, G.A. A high-resolution frequency estimation method for three-phase induction machine fault detection. *IEEE Trans. Ind. Electron.* **2007**, *54*, 2305–2314. [CrossRef]

79. Chakkor, S.; Baghouri, M.; Hajraoui, A. ESPRIT method enhancement for real-time wind turbine fault recognition. *Int. J. Power Electron. Drive Syst. Porto* **2015**, *5*, 4. [CrossRef]

80. Lobos, T.; Rezmer, J.; Koglin, H. Analysis of Power System Transients using Wavelets and Prony Method. In Proceedings of the IEEE Power Tech Conference, Porto, Portugal, 10–13 September 2001. [CrossRef]

81. Janik, P.; Leonowicz, Z.; Rezmer, J. Advanced Signal Processing Methods for Evaluation of Harmonic Distortion Caused by DFIG Wind Generator. *IEEE Trans. Ind. Appl.* **2006**, *42*, 1454–1463.

82. Zidani, F.; Benbouzid, M.; Diallo, D.; Naït-Saïd, M. Induction Motor Stator Faults Diagnosis by a Current Concordia Pattern-Based Fuzzy Decision System. *IEEE Trans. Energy Convers.* **2003**, *18*, 225–2238. [CrossRef]

83. Grainger, J.; Stevenson, W. *Power System Analysis*, 1st ed.; McGraw Hill: New York, NY, USA, 2012; ISBN 007-061293-5.

84. Bacha, K.; Salem, S.; Chaari, A. An improved combination of Hilbert and Park transforms for fault detection and identification in three-phase induction motors. *Int. J. Electr. Power Energy Syst.* **2012**, *43*, 1006–1016. [CrossRef]

85. Sahraoui, M.; Ghoggal, A.; Guedidi, S.; Zouzou, S. Detection of inter-turn short-circuit in induction motors using Park-Hilbert method. *Int. J. Syst. Assur. Eng. Manag.* **2014**, *5*, 337–351. [CrossRef]

86. Cardoso, A.; Saraiva, V. Computer-aided detection of airgap eccentricity in operating three-phase induction motors by Park's vector approach. *IEEE Trans. Ind. Appl.* **1993**, *29*, 897–901. [CrossRef]

87. Cardoso, A.; Cruz, S.; Carvalho, J.; Saraiva, E. Rotor cage fault diagnosis in three phase induction motors, by Park's vector approach. *Electric Mach. Power Syst.* **1995**, *28*, 289–299. [CrossRef]

88. Cherif, B.; Bendiabdellah, A.; Khelif, M. Detection of open-circuit fault in a three-phase voltage inverter fed induction motor. *Int. Rev. Autom. Control* **2016**, *96*, 374–382. [CrossRef]

89. Vilhekar, T.; Ballal, M.; Suryawanshi, H. Application of multiple parks vector approach for detection of multiple faults in induction motors. *J. Power Electron.* **2017**, *17*, 972–982. [CrossRef]

90. Nandi, S.; Toliyat, H.A.; Li, X. Condition monitoring and fault diagnosis of electrical motors—A review. *IEEE Trans. Energy Convers.* **2001**, *20*, 719–729. [CrossRef]

91. Al Ahmar, E.; Choqueuse, V.; Benbouzid, M.E.H.; Amirat, Y.; El Assad, J.; Karam, R.; Farah, S. Advanced signal processing techniques for fault detection and diagnosis in a wind turbine induction generator drive train: A comparative study. In Proceedings of the IEEE Energy Conversion Congress and Exposition ECCE 2010, Atlanta, GA, USA, 12–16 September 2010; pp. 3576–3581. [CrossRef]

92. Taylor, J. *The Vibration Analysis Handbook*, 2nd ed.; Vibration Consultants: Cambridge, MA, USA, 2003; ISBN 9780964051720.

93. Gong, X.; Qiao, W. Imbalance Fault Detection of Direct-Drive Wind Turbines Using Generator Current Signals. *IEEE Trans. Energy Convers.* **2012**, *27*, 468–476. [CrossRef]

94. Zhivomirov, H. On the Development of STFT-analysis and ISTFT-synthesis Routines and their Practical Implementation. *TEM J.* **2019**, *8*, 56–64. [CrossRef]

95. Hafeez, S.; Zaide, C.; Siddiqui, A. Broken rotor bar detection of single phase induction motor using Wigner-Ville Distributions. In Proceedings of the IEEE 18th Conference on Emerging Technologies & Factory Automation, Cagliari, Italy, 10–13 September 2013. [CrossRef]

96. Randall, R.; Antoni, J. Rolling element bearing diagnostics—A tutorial. *Mech. Syst. Signal Process.* **2011**, *25*, 485–520. [CrossRef]

97. Da Silva, A.; Povinelli, R.; Demerdash, N. Induction machine broken bar and stator short-circuit fault diagnostics based on three-phase stator current envelopes. *IEEE Trans. Ind. Electron.* **2008**, *55*, 1310–1318. [CrossRef]

98. Jaksch, I. Faults diagnosis of three-phase induction motors using envelope analysis. In Proceedings of the 4th IEEE International Symposium on Diagnostics for Electric Machines, Power Electronics and Drives, SDEMPED 2003, Atlanta, GA, USA, 24–26 August 2003; pp. 289–293. [CrossRef]

99. Antonino-Daviu, J.A.; Riera-Guasp, M.; Roger-Folch, J.; Perez, R.B. An Analytical Comparison between DWT and Hilbert-Huang-Based Methods for the Diagnosis of Rotor Asymmetries in Induction Machines. In Proceedings of the IEEE Industry Applications Annual Meeting, New Orleans, LA, USA, 23–27 September 2007; pp. 1932–1939. [CrossRef]

100. Blödt, M.; Chabert, M.; Regnier, J.; Faucher, J. Mechanical load fault detection in induction motors by stator current time-frequency analysis. *IEEE Trans. Ind. Appl.* **2006**, *42*, 1454–1463. [CrossRef]

101. Riera-Guasp, M.; Antonino-Daviu, J.A.; Pineda-Sanchez, M.; Puche-Panadero, R.; Perez-Cruz, J. A general approach for the transient detection of slip-dependent fault components based on the discrete wavelet transform. *IEEE Trans. Ind. Electron.* **2008**, *55*, 4167–4180. [CrossRef]

102. Benbouzid, M.E.H.; Vieira, M.; Theys, C. Induction motors faults detection and localization using stator current advanced signal processing techniques. *IEEE Trans. Power Electron.* **1999**, *14*, 14–22. [CrossRef]

103. Szabo, L.; Erno, K. An Overview on Induction Machine's Diagnosis Methods. 2008. Available online: https://www.researchgate.net/publication/38112249_An_Overview_on_Induction_Machine\T1\textquoterights_Diagnosis_Methods (accessed on 24 April 2020).

Appl. Sci. **2020**, *10*, 7389

MDPI

St. Alban-Anlage 66

4052 Basel

Switzerland

Tel. +41 61 683 77 34

Fax +41 61 302 89 18

www.mdpi.com

Applied Sciences Editorial Office

E-mail: applsci@mdpi.com

www.mdpi.com/journal/applsci